INFLUENCE FUNCTIONS AND MATRICES

MECHANICAL ENGINEERING

A Series of Textbooks and Reference Books

Editor

L. L. Faulkner

*Columbus Division, Battelle Memorial Institute
and Department of Mechanical Engineering
The Ohio State University
Columbus, Ohio*

1. *Spring Designer's Handbook*, Harold Carlson
2. *Computer-Aided Graphics and Design*, Daniel L. Ryan
3. *Lubrication Fundamentals*, J. George Wills
4. *Solar Engineering for Domestic Buildings*, William A. Himmelman
5. *Applied Engineering Mechanics: Statics and Dynamics*, G. Boothroyd and C. Poli
6. *Centrifugal Pump Clinic*, Igor J. Karassik
7. *Computer-Aided Kinetics for Machine Design*, Daniel L. Ryan
8. *Plastics Products Design Handbook, Part A: Materials and Components; Part B: Processes and Design for Processes*, edited by Edward Miller
9. *Turbomachinery: Basic Theory and Applications*, Earl Logan, Jr.
10. *Vibrations of Shells and Plates*, Werner Soedel
11. *Flat and Corrugated Diaphragm Design Handbook*, Mario Di Giovanni
12. *Practical Stress Analysis in Engineering Design*, Alexander Blake
13. *An Introduction to the Design and Behavior of Bolted Joints*, John H. Bickford
14. *Optimal Engineering Design: Principles and Applications*, James N. Siddall
15. *Spring Manufacturing Handbook*, Harold Carlson
16. *Industrial Noise Control: Fundamentals and Applications*, edited by Lewis H. Bell
17. *Gears and Their Vibration: A Basic Approach to Understanding Gear Noise*, J. Derek Smith
18. *Chains for Power Transmission and Material Handling: Design and Applications Handbook*, American Chain Association
19. *Corrosion and Corrosion Protection Handbook*, edited by Philip A. Schweitzer
20. *Gear Drive Systems: Design and Application*, Peter Lynwander
21. *Controlling In-Plant Airborne Contaminants: Systems Design and Calculations*, John D. Constance
22. *CAD/CAM Systems Planning and Implementation*, Charles S. Knox
23. *Probabilistic Engineering Design: Principles and Applications*, James N. Siddall
24. *Traction Drives: Selection and Application*, Frederick W. Heilich III and Eugene E. Shube
25. *Finite Element Methods: An Introduction*, Ronald L. Huston and Chris E. Passerello
26. *Mechanical Fastening of Plastics: An Engineering Handbook*, Brayton Lincoln, Kenneth J. Gomes, and James F. Braden
27. *Lubrication in Practice: Second Edition*, edited by W. S. Robertson
28. *Principles of Automated Drafting*, Daniel L. Ryan
29. *Practical Seal Design*, edited by Leonard J. Martini
30. *Engineering Documentation for CAD/CAM Applications*, Charles S. Knox
31. *Design Dimensioning with Computer Graphics Applications*, Jerome C. Lange
32. *Mechanism Analysis: Simplified Graphical and Analytical Techniques*, Lyndon O. Barton

33. *CAD/CAM Systems: Justification, Implementation, Productivity Measurement,* Edward J. Preston, George W. Crawford, and Mark E. Coticchia
34. *Steam Plant Calculations Manual,* V. Ganapathy
35. *Design Assurance for Engineers and Managers,* John A. Burgess
36. *Heat Transfer Fluids and Systems for Process and Energy Applications,* Jasbir Singh
37. *Potential Flows: Computer Graphic Solutions,* Robert H. Kirchhoff
38. *Computer-Aided Graphics and Design: Second Edition,* Daniel L. Ryan
39. *Electronically Controlled Proportional Valves: Selection and Application,* Michael J. Tonyan, edited by Tobi Goldoftas
40. *Pressure Gauge Handbook,* AMETEK, U.S. Gauge Division, edited by Philip W. Harland
41. *Fabric Filtration for Combustion Sources: Fundamentals and Basic Technology,* R. P. Donovan
42. *Design of Mechanical Joints,* Alexander Blake
43. *CAD/CAM Dictionary,* Edward J. Preston, George W. Crawford, and Mark E. Coticchia
44. *Machinery Adhesives for Locking, Retaining, and Sealing,* Girard S. Haviland
45. *Couplings and Joints: Design, Selection, and Application,* Jon R. Mancuso
46. *Shaft Alignment Handbook,* John Piotrowski
47. *BASIC Programs for Steam Plant Engineers: Boilers, Combustion, Fluid Flow, and Heat Transfer,* V. Ganapathy
48. *Solving Mechanical Design Problems with Computer Graphics,* Jerome C. Lange
49. *Plastics Gearing: Selection and Application,* Clifford E. Adams
50. *Clutches and Brakes: Design and Selection,* William C. Orthwein
51. *Transducers in Mechanical and Electronic Design,* Harry L. Trietley
52. *Metallurgical Applications of Shock-Wave and High-Strain-Rate Phenomena,* edited by Lawrence E. Murr, Karl P. Staudhammer, and Marc A. Meyers
53. *Magnesium Products Design,* Robert S. Busk
54. *How to Integrate CAD/CAM Systems: Management and Technology,* William D. Engelke
55. *Cam Design and Manufacture: Second Edition; with cam design software for the IBM PC and compatibles, disk included,* Preben W. Jensen
56. *Solid-State AC Motor Controls: Selection and Application,* Sylvester Campbell
57. *Fundamentals of Robotics,* David D. Ardayfio
58. *Belt Selection and Application for Engineers,* edited by Wallace D. Erickson
59. *Developing Three-Dimensional CAD Software with the IBM PC,* C. Stan Wei
60. *Organizing Data for CIM Applications,* Charles S. Knox, with contributions by Thomas C. Boos, Ross S. Culverhouse, and Paul F. Muchnicki
61. *Computer-Aided Simulation in Railway Dynamics,* by Rao V. Dukkipati and Joseph R. Amyot
62. *Fiber-Reinforced Composites: Materials, Manufacturing, and Design,* P. K. Mallick
63. *Photoelectric Sensors and Controls: Selection and Application,* Scott M. Juds
64. *Finite Element Analysis with Personal Computers,* Edward R. Champion, Jr., and J. Michael Ensminger
65. *Ultrasonics: Fundamentals, Technology, Applications: Second Edition, Revised and Expanded,* Dale Ensminger
66. *Applied Finite Element Modeling: Practical Problem Solving for Engineers,* Jeffrey M. Steele
67. *Measurement and Instrumentation in Engineering: Principles and Basic Laboratory Experiments,* Francis S. Tse and Ivan E. Morse
68. *Centrifugal Pump Clinic: Second Edition, Revised and Expanded,* Igor J. Karassik
69. *Practical Stress Analysis in Engineering Design: Second Edition, Revised and Expanded,* Alexander Blake
70. *An Introduction to the Design and Behavior of Bolted Joints: Second Edition, Revised and Expanded,* John H. Bickford
71. *High Vacuum Technology: A Practical Guide,* Marsbed H. Hablanian
72. *Pressure Sensors: Selection and Application,* Duane Tandeske

73. *Zinc Handbook: Properties, Processing, and Use in Design*, Frank Porter
74. *Thermal Fatigue of Metals*, Andrzej Weronski and Tadeusz Hejwowski
75. *Classical and Modern Mechanisms for Engineers and Inventors*, Preben W. Jensen
76. *Handbook of Electronic Package Design*, edited by Michael Pecht
77. *Shock-Wave and High-Strain-Rate Phenomena in Materials*, edited by Marc A. Meyers, Lawrence E. Murr, and Karl P. Staudhammer
78. *Industrial Refrigeration: Principles, Design and Applications*, P. C. Koelet
79. *Applied Combustion*, Eugene L. Keating
80. *Engine Oils and Automotive Lubrication*, edited by Wilfried J. Bartz
81. *Mechanism Analysis: Simplified and Graphical Techniques, Second Edition, Revised and Expanded*, Lyndon O. Barton
82. *Fundamental Fluid Mechanics for the Practicing Engineer*, James W. Murdock
83. *Fiber-Reinforced Composites: Materials, Manufacturing, and Design, Second Edition, Revised and Expanded*, P. K. Mallick
84. *Numerical Methods for Engineering Applications*, Edward R. Champion, Jr.
85. *Turbomachinery: Basic Theory and Applications, Second Edition, Revised and Expanded*, Earl Logan, Jr.
86. *Vibrations of Shells and Plates: Second Edition, Revised and Expanded*, Werner Soedel
87. *Steam Plant Calculations Manual: Second Edition, Revised and Expanded*, V. Ganapathy
88. *Industrial Noise Control: Fundamentals and Applications, Second Edition, Revised and Expanded*, Lewis H. Bell and Douglas H. Bell
89. *Finite Elements: Their Design and Performance*, Richard H. MacNeal
90. *Mechanical Properties of Polymers and Composites: Second Edition, Revised and Expanded*, Lawrence E. Nielsen and Robert F. Landel
91. *Mechanical Wear Prediction and Prevention*, Raymond G. Bayer
92. *Mechanical Power Transmission Components*, edited by David W. South and Jon R. Mancuso
93. *Handbook of Turbomachinery*, edited by Earl Logan, Jr.
94. *Engineering Documentation Control Practices and Procedures*, Ray E. Monahan
95. *Refractory Linings: Thermomechanical Design and Applications*, Charles A. Schacht
96. *Geometric Dimensioning and Tolerancing: Applications and Techniques for Use in Design, Manufacturing, and Inspection*, James D. Meadows
97. *An Introduction to the Design and Behavior of Bolted Joints: Third Edition, Revised and Expanded*, John H. Bickford
98. *Shaft Alignment Handbook: Second Edition, Revised and Expanded*, John Piotrowski
99. *Computer-Aided Design of Polymer-Matrix Composite Structures*, edited by S. V. Hoa
100. *Friction Science and Technology*, Peter J. Blau
101. *Introduction to Plastics and Composites: Mechanical Properties and Engineering Applications*, Edward Miller
102. *Practical Fracture Mechanics in Design*, Alexander Blake
103. *Pump Characteristics and Applications*, Michael W. Volk
104. *Optical Principles and Technology for Engineers*, James E. Stewart
105. *Optimizing the Shape of Mechanical Elements and Structures*, A. A. Seireg and Jorge Rodriguez
106. *Kinematics and Dynamics of Machinery*, Vladimír Stejskal and Michael Valášek
107. *Shaft Seals for Dynamic Applications*, Les Horve
108. *Reliability-Based Mechanical Design*, edited by Thomas A. Cruse
109. *Mechanical Fastening, Joining, and Assembly*, James A. Speck
110. *Turbomachinery Fluid Dynamics and Heat Transfer*, edited by Chunill Hah
111. *High-Vacuum Technology: A Practical Guide, Second Edition, Revised and Expanded*, Marsbed H. Hablanian

112. *Geometric Dimensioning and Tolerancing: Workbook and Answerbook*, James D. Meadows
113. *Handbook of Materials Selection for Engineering Applications*, edited by G. T. Murray
114. *Handbook of Thermoplastic Piping System Design*, Thomas Sixsmith and Reinhard Hanselka
115. *Practical Guide to Finite Elements: A Solid Mechanics Approach*, Steven M. Lepi
116. *Applied Computational Fluid Dynamics*, edited by Vijay K. Garg
117. *Fluid Sealing Technology*, Heinz K. Muller and Bernard S. Nau
118. *Friction and Lubrication in Mechanical Design*, A. A. Seireg
119. *Influence Functions and Matrices*, Yuri A. Melnikov
120. *Machining of Ceramics and Composites*, edited by Said Jahanmir, M. Ramulu, and Philip Koshy

Additional Volumes in Preparation

Heat Exchange Design Handbook, T. Kuppan

Couplings and Joints: Second Edition, Revised and Expanded, Jon R. Mancuso

Mechanical Engineering Software

Spring Design with an IBM PC, Al Dietrich

Mechanical Design Failure Analysis: With Failure Analysis System Software for the IBM PC, David G. Ullman

INFLUENCE FUNCTIONS AND MATRICES

YURI A. MELNIKOV
Middle Tennessee State University
Murfreesboro, Tennessee

MARCEL DEKKER, INC.   NEW YORK • BASEL

ISBN: 0-8247-1941-7

This book is printed on acid-free paper.

Headquarters
Marcel Dekker, Inc.
270 Madison Avenue, New York, NY 10016
tel: 212-696-9000; fax: 212-685-4540

Eastern Hemisphere Distribution
Marcel Dekker AG
Hutgasse 4, Postfach 812, CH-4001 Basel, Switzerland
tel: 44-61-261-8482; fax: 44-61-261-8896

World Wide Web
http://www.dekker.com

The publisher offers discounts on this book when ordered in bulk quantities. For more information, write to Special Sales/Professional Marketing at the headquarters address above.

Copyright © 1999 by Marcel Dekker, Inc. All Rights Reserved.

Neither this book nor any part may be reproduced or transmitted in any form or by any means, electronic or mechanical, including photocopying, microfilming, and recording, or by any information storage and retrieval system, without permission in writing from the publisher.

Current printing (last digit):
10 9 8 7 6 5 4 3 2 1

PRINTED IN THE UNITED STATES OF AMERICA

To the memory of my parents,
Afanasy and Antonina,
my wisest teachers in this life

Preface

Notwithstanding their established significance in the qualitative theory of differential equations, Green's functions are rarely used in the realm of computational mechanics. As a working tool, the influence (Green's) function-based methods are virtually absent in texts related to numerical methods in mechanics. The cause of this discrepancy is likely to be found in the scarcity of implementable representations for Green's functions in the existing literature. This deficiency occurs because, even for the simplest boundary value problems, the construction of Green's functions is not a routine procedure. In addition, under-representation of methods based on Green's functions results from the widespread availability of algorithms based on standard approaches, such as the boundary element method.

Motivation for writing this book arose from the author's over twenty-five years experience of teaching the influence function-related courses to a variety of students at various institutions (Dnepropetrovsk State University, Brooklyn Polytechnic Institute, Northwestern University, Vanderbilt University, and Middle Tennessee State University). With this text, the author wished to popularize the influence function method among young researchers in the applied sciences and engineering.

Some classes of problems originating in applied mechanics form the basis of this book. Although the problems themselves are classical, their treatment in this context differs from that of more traditional texts in the field; the major difference being in the emphasis on influence (Green's) functions in the analysis of these problems.

A broad spectrum of problems in continuum mechanics, related to such fields as fluid flow, acoustics, electromagnetism, heat transfer, and elasticity, can be formulated as boundary value problems involving (ordinary or partial) differential equations. The Green's function is a key concept in the area of differential equations, resembling, in some sense, the inverse matrix in linear algebra. The Green's function formalism has been utilized as a powerful means of development and presentation in classical as well as contemporary treatments in the analysis of differential equations.

About two decades ago, in an attempt to make a contribution to this intriguing and challenging area, a technique was proposed [24, 42] for the construction of compact expressions of Green's functions and matrices for some elliptical PDEs that occur in mechanics. The technique is based on the classical method of eigenfunction expansion (see, for example, [63]). This productive approach is included in [45], a collection of elaborations on influence (Green's) functions methods, aimed at introducing advances in this field to the computational mechanics community.

Also included in [45] are computational implementations of influence functions with emphasis on shape complexity, on various types of nonlinearities, and on some inverse formulations, including optimal shape design. As such, [45] can serve as a source for those who are already familiar with the influence function concept and its strengths and weaknesses.

This volume is intended as a graduate text in influence (Green's) functions and is designed for students in mechanical engineering or in related fields in applied mathematics. The primary goal has been to create a supplement for existing texts in these fields, a book providing an alternative approach to boundary value problems of applied mechanics. In addition, the book is recommended to users of the boundary element method, where the use of influence functions can significantly increase the effectiveness of existing numerical procedures.

The level of presentation has been aimed at readers with standard background in calculus, linear algebra, ODEs, PDEs, and numerical methods. An elementary knowledge of integral equations is also assumed. We hope the reader will find use for the large collection of Green's functions and matrices for (systems of) ordinary and partial differential equations contained in this single volume. Many of these representations for Green's functions and matrices appears in book form for the first time.

It is with gratitude that the author acknowledges Drs. V.V. Loboda and N.V. Polyakov, Dnepropetrovsk State University and Dr. J.O. Powell, Middle Tennessee State University, for presently sharing his research interest in influence (Green's) functions and in the use of such functions for developing computational algorithms of the boundary integral equation method in engineering and science.

While writing this book, the author has received assistance from many of his colleagues in the Department of Mathematical Sciences at MTSU. And heartfelt thanks goes out to them all. Especially, the author wishes to acknowledge gratefully the many helpful suggestions about the exposition and organization, offered by Dr. Jan Zijlstra. Invaluable comments made by Drs. Leslie Aspinwall and Diane Miller during the work have substantially improved the quality of the text. The author also expresses his appreciation to Scott McDaniel, a junior faculty member, who worked through drafts of some sections of the manuscript and made valuable comments from the point of view of the student.

The author is appreciative of the outstanding job done by the following students: Yugeny Bobylov (Dnepropetrovsk State University), Lynn Roubides, Miller Hall, and Ed Slowey (Middle Tennessee State University). Their knowledge in applied mathematics and mechanics, along with highly developed skills in computer techniques, helped with the preparation of some of the test examples and illustrative materials.

The author wishes to express his personal appreciation to each of the following reviewers, whose comments, constructive criticisms, and compliments have helped to make a much improved book: Dr. Richard Dippery (Kettering University, Flint, Michigan), Dr. J. Tinsley Oden (University of Texas at Austin), and Dr. Eduard Ventsel (The Pennsylvania State University, College Park).

It is also a great pleasure to thank B. J. Clark (acquisitions editor) and Brian Black (production editor) of Marcel Dekker, Inc., for their thoughtful guidance and useful prodding at critical junctures during the development and production of the book.

During the three years period of working on this project, the author has been supported by the Foundation of Middle Tennessee State University in the form of the Faculty Research Grant in 1995, and Summer Research Grants in 1996, 1997, and 1998. This support has significantly hastened the work and is very much appreciated.

Yuri A. Melnikov
Murfreesboro, Tennessee

Contents

PREFACE		v
INTRODUCTION		1
CHAPTER 1:	**Green's Function for ODE**	7
1.1.	Defining properties and construction	7
1.2.	Symmetry of Green's functions	23
1.3.	Method of variation of parameters	30
1.4.	*Review Exercises*	45
CHAPTER 2:	**Influence Functions for Beams**	49
2.1.	Single span beams of uniform rigidity	50
2.2.	Bending under transverse loads	64
2.3.	Beams on an elastic foundation	81
2.4.	Beams of variable rigidity	92
2.5.	*Review Exercises*	104
CHAPTER 3:	**Some Beam Problems**	111
3.1.	Transverse natural vibrations	112
3.2.	Buckling (Euler formulation)	128
3.3.	Some contact problems	139
CHAPTER 4:	**Assemblies of Elements**	149
4.1.	Multi-point posed problems	149
4.2.	Bending of multi-spanned beams	160

4.3.	Problems posed on graphs	175
4.4.	*Review Exercises*	187

CHAPTER 5: Potential and Related Fields — 193

5.1.	Formulations in Cartesian coordinates	195
5.2.	Formulations in polar coordinates	219
5.3.	Potential fields on surfaces	230
5.4.	Klein-Gordon equation	238
5.5.	*Review Exercises*	251

CHAPTER 6: Problems of Solid Mechanics — 255

6.1.	Poisson-Kirchhoff's plates	257
6.2.	Reissner's plates	276
6.3.	Elastic isotropic media	289
6.4.	Orthotropic media	301
6.5.	Elastic equilibrium of thin shells	307
6.6.	*Review Exercises*	314

CHAPTER 7: Compound Media — 317

7.1.	Potential fields in compound regions	319
7.2.	Potential fields on plates and shells	328
7.3.	Plane problem for elastic media	355
7.4.	*Review Exercises*	369

CHAPTER 8: Heat Equation — 371

8.1.	Laplace transform	372
8.2.	Influence functions	376
8.3.	Influence matrices	393
8.4.	*Review Exercises*	399

APPENDIX A: Catalogue of Green's Functions — 401

APPENDIX B: Answers and Comments — 415

REFERENCES — 449

INDEX — 455

Introduction

There exists a remarkable analogy between two concepts from different areas of science, the *influence function* in mechanics and the *Green's function* in mathematics. This correspondence is of primary importance in understanding the basic idea which underlies this book. Essentially, the Green's function for a certain boundary value problem in mathematical physics (for ordinary or partial differential equation) can be identified with the influence function of that phenomenon in mechanics, for which the boundary value problem serves as a mathematical model.

To illustrate the Green's function/influence function analogy, we consider, as a first example, the differential equation

$$\frac{d}{dx}\left(m(x)\frac{dy(x)}{dx}\right) = f(x), \quad x \in (0, a) \qquad (0.1)$$

subject to the boundary conditions

$$y(0) = 0, \quad y(a) = 0 \qquad (0.2)$$

This boundary value problem can, for instance, be regarded as a model for the lateral displacement $y(x)$ of a string of length a, caused by a transverse distributed load proportional to $f(x)$ in eqn (0.1). The string has variable mass density $m(x)$, and according to the boundary conditions imposed by eqn (0.2), the string's edges $x=0$ and $x=a$ are fixed.

The Green's function $g(x, s)$ of the boundary value problem posed by the relations in eqn (0.2) for the homogeneous equation

$$\frac{d}{dx}\left(m(x)\frac{dy(x)}{dx}\right) = 0, \qquad (0.3)$$

corresponding to that of eqn (0.1), can be identified as the displacement (or response) $y_s(x)$ of the string of length a, caused by a transverse unit force concentrated at s.

This is by no means the only possible influence function interpretation of the Green's function for the boundary value problem posed by eqns (0.2) and (0.3). The problem can for instance also be associated with steady-state heat conduction in a one dimensional rod of finite length a. In this case, $y(x)$ denotes the temperature at position x in a rod made of a nonhomogeneous material with heat conductivity $m(x)$. The boundary conditions in eqn (0.2) indicate that the temperature $y(x)$ is zero at all times at the end-points $x = 0$ and $x = a$ of the rod. In this setting, the Green's function $g(x, s)$ can be identified as the influence function of a unit heat source acting permanently at position s on the rod.

These two examples illustrate the important fact that the Green's function – influence function correspondence is not necessarily a one-to-one relationship. In many instances in mathematical physics, a particular differential equation may be associated with different physical phenomena.

Another example on the Green's function – influence function correspondence stems from classical Kirchhoff's beam theory (e.g., [10, 18, 27, 30, 64, 67]) where the differential equation

$$\frac{d^2}{dx^2}\left(EI(x)\frac{d^2w(x)}{dx^2}\right) = q(x), \quad x \in (0, a) \tag{0.4}$$

subject to the boundary conditions

$$w(0) = \frac{dw(0)}{dx} = 0, \quad w(a) = \frac{d^2w(a)}{dx^2} = 0 \tag{0.5}$$

models the lateral deflection $w(x)$ of a beam of length a, caused by the transverse distributed load $q(x)$. The left end-point of the beam at $x = 0$ is assumed to be clamped, while the beam is simply supported at its right edge, $x = a$. Other physically feasible boundary conditions can be imposed at the end-points of the beam. The quantity $EI(x)$ represents the so-called flexural rigidity of the beam.

The Green's function $g(x, s)$ of the boundary value problem which consists of the homogeneous equation

$$\frac{d^2}{dx^2}\left(EI(x)\frac{d^2w(x)}{dx^2}\right) = 0, \quad x \in (0, a)$$

subjected to the boundary conditions imposed by eqn (0.5), can be identified with the influence function of a unit force applied transversely to the beam at position s. Thus, $g(x, s)$ represents the deflection of the beam at point x, in response to this transverse unit force.

A last illustration of the analogy between Green's functions and influence functions comes from classical Poisson-Kirchhoff plate theory (e.g., [72]),

INTRODUCTION

according to which the biharmonic equation

$$D\Delta\Delta w(x,y) = F(x,y), \quad (x,y) \in \Omega \qquad (0.6)$$

posed on a simply connected region Ω bounded with a smooth contour Γ and subject to the following boundary conditions

$$w(x,y) = \frac{\partial w(x,y)}{\partial n} = 0, \quad (x,y) \in \Gamma \qquad (0.7)$$

models the bending of a thin elastic clamped plate occupying the region Ω with boundary Γ. The load on the plate, $F(x,y)$, is transversely distributed and the constant D represents the flexural rigidity of the plate.

The Green's function $G(x,y;s,t)$ of the homogeneous problem corresponding to that stated with eqns (0.6) and (0.7) can be interpreted as the influence function of a transverse unit force concentrated at position (s,t) on the plate. Thus, $G(x,y;s,t)$ represents the deflection of the plate at the point (x,y), caused by this transverse unit force.

In these examples, dealing with boundary value problems from applied mechanics, one can adopt either Green's function or influence function terminology. Green's function terminology is the more suitable of the two for the discussion in Chapter 1. However, since most of the material in Chapters 2 through 8 is drawn from mechanics, influence function terminology is mainly used there.

The reader is probably aware of the important role that Green's functions play in the qualitative theory of differential equations. From the current literature on Green's (influence) functions, it becomes increasingly clear that these functions are also effective in obtaining numerical solutions, once they can be expressed explicitly in a compact manner.

There are two basic stages of the influence function method (IFM). The first stage, the focus of much of the early chapters of this book, deals with the construction of the influence function. In the second stage, a computational algorithm is developed using this influence function.

Chapter 1 is devoted to the basic concepts of the Green's function theory for linear boundary value problems for ordinary differential equations. In Section 1.1, the first of two traditional methods for the construction of the Green's function is discussed, a method based on its defining properties. Conditions for symmetry of the Green's function, in the sense that its variables are interchangeable, are considered in Section 1.2. The second traditional method for the construction of Green's functions, the method of variation of parameters, is described in Section 1.3.

Influence functions for single span Kirchhoff beams are constructed and utilized in Chapter 2. In Section 2.1 influence functions for beams of uniform flexural rigidity are presented, taking into account physically natural sets of edge conditions. In Section 2.2 it is shown how the response of a beam to transverse loads, forces or moments can be expressed in terms of the influence function of a transversely concentrated unit force. Combinations of loads are treated by means of this approach. Beams resting on elastic foundations are discussed in Section 2.3. Analytic expressions for influence functions of such beams are given and the practical use of such expressions is discussed. Section 2.4 illustrates the practical solvability of problems for beams of variable flexural rigidity by means of the influence function method.

Chapter 3 illustrates some indirect applications of beam influence functions. It is shown how one can profit from the knowledge of the influence function of a transverse concentrated unit force for a beam in solving other beam problems. Frequencies and mode shapes of natural vibrations of a beam are determined in Section 3.1. In Section 3.2, influence functions of a transverse point force are used to find such 'critical' values of axial forces applied to a beam, that cause the loss of stability of the initial equilibrium state. The classical Euler formulation of buckling problems is considered. One simple nonlinear contact problem for a beam is discussed in Section 3.3 where a Kirchhoff's beam is spaced a small distance above a Winkler foundation.

Chapter 4 deals with some specific problems from applied mechanics, which reduce to the so-called multi-point posed boundary value problems for systems of linear ordinary differential equations. These are not, however, boundary value problems for systems of equations in the conventional sense, where several unknown functions are supposed to have a common domain for an independent variable, and at least one of the equations in the system involves more than one unknown function. Instead, each equation in the systems that are considered in Chapter 4 governs a single unknown function and is formulated over an individual domain. The system is actually formed by letting those single domains interact with each other at their end-points. End-points for the individual equations become contact points in the system at which appropriate conditions are formulated.

Section 4.1 introduces the notion of a *matrix of Green's type* appropriate for a particular type of multi-point posed boundary value problems stated for a sandwich type media. Based on that, in Section 4.2 we apply the notion of a matrix of Green's type to multi-spanned Kirchhoff beams (in this case, we call it just the *influence matrix*). Several particular examples are considered where we do not only construct influence matrices but also show how they can be used in computing components of stress-strain states for particular multi-spanned beams.

The applicability of the influence matrix formalism developed in Section 4.1 is limited to a sandwich type assembly in which the material is piecewise homogeneous. To broaden its application range, the formalism is extended to a more general type of multi-point posed boundary value problems in Section 4.3. These problems, of a more complex type, represent a wider variety of situations in applied mechanics. The framework of graph theory is used for that purpose. Sets of linear ordinary differential equations are considered, formulated on finite weighted graphs in such a way that every equation in the set governs a single unknown function and is stated on a single edge of the graph. The individual equations in the set are put into system form by imposing contact and boundary conditions at the vertices and end-points of the graph, respectively. Based on this a setup, a new definition of the matrix of Green's type is introduced. Existence and uniqueness of such matrices are discussed and two methods for their construction are proposed and some particular examples are considered.

In Chapters 5 through 8, we turn to problems described by partial differential equations where the list of available Green's (influence) functions is very limited. In Chapter 5, a technique is described which was originally developed (see [24, 42, 44, 45, 53]) for the construction of Green's functions and matrices for elliptic equations in two dimensions. The technique is based on the so-called method of eigenfunction expansion [63] and has proven to be especially effective for a variety of problems in computational continuum mechanics (based on Laplace's equation, Klein-Gordon equation, biharmonic equation, as well as on Lame's system for the displacement formulation of the plane problem in the theory of elasticity).

In this technique, influence functions are first represented in terms of their Fourier expansions with respect to one of the independent variables. This consequently results in the construction of Green's functions for ordinary differential equations in the coefficients of the Fourier expansions (the first stage of the technique). This construction can be done by using either the method based on the defining properties of Green's functions (cf. Section 1.1), or by the method of variation of parameters, as described in Section 1.3. The influence function of interest is then constructed by complete or partial summation of the Fourier series (the second stage of the technique).

Potential fields and related fields are the subject matter of Chapter 5. A number of compact representations of influence functions for Laplace's and Klein-Gordon equations expressed in various coordinate systems are presented. A variety of region configurations and types of boundary conditions are considered. Some mathematical issues are addressed as they are closely related to our discussion. Among these are: (i) convergence of the series representing influence functions, (ii) splitting off singular components of influence functions, and (iii) expressing regular components in terms of uniformly convergent series.

In Chapter 6, the technique for the construction of influence functions and matrices is applied to a variety of boundary value problems occurring in solid mechanics. The material in this chapter is conceptually similar to that of Chapter 5, which may serve as a reference. Section 6.1 deals with traditional formulations from classical Poisson-Kirchhoff plate theory, in which the bending of thin plates is reduced to a boundary value problem involving the biharmonic equation. A number of influence functions are obtained for plates of various shapes and edge conditions. Section 6.2 extends the influence function formalism to thin plate problems formulated within the scope of Reissner theory (see [65]). Some of the displacement formulations for isotropic and orthotropic media from the plane problem in the theory of elasticity are discussed in Sections 6.3 and 6.4. In Section 6.5, it is shown that the elastic equilibrium of thin shells of revolution can also be treated with the suggested technique.

Chapter 7 is devoted to influence matrices for problems of continuum mechanics formulated in media whose properties are piecewise continuous functions of spatial variables. In Section 7.1, the notion of an influence matrix is introduced as that appropriate for problem classes in continuum mechanics of compound media in two dimensions. Section 7.2 considers specific problems arising in the theory of potential on thin-walled structures consisting of different plates and shells. Several closed form influence matrices are obtained for structures composed of circular and annular plates, cylindrical, spherical, and toroidal shells. Finally, in Section 7.3, it is shown how influence matrices can be constructed for plane problems which arise in the theory of elasticity for compound media.

The discussion in the final Chapter 8 focuses on the classical heat equation, a PDE of parabolic type. Explicit, easily computable expressions for influence functions of some initial-boundary value problems involving the heat equation can also be obtained by utilizing the ideas developed and widely used here. The potential of the present approach in the area of the heat equation is briefly discussed. Section 8.1 recalls some basics of the Laplace transform which this method requires. In Sections 8.2 and 8.3, the present approach is used to obtain some classical and new representations of influence functions in this area.

Each section of the text is supplied with examples illustrating how the material can be used in practice. A set of review exercises accompanies every chapter except for Chapter 3. Appendix A presents a catalogue of explicit representations of Green's functions and matrices available in the book. Appendix B contains answers and comments to most of the review exercises.

Chapter 1

Green's Function for ODE

Being a mathematical prototype of the influence function, the notion of a Green's function provides a theoretical background for comprehending the material of the present text. The development in this chapter touches upon ordinary differential equations, utilizing two classical approaches, which are commonly used in the current literature (see, for example, [7, 17, 30, 31, 68, 70]), for the construction of Green's functions for linear boundary value problems for ODEs. One of these approaches is associated with the proof of existence and uniqueness theorem for the Green's function (see Section 1.1). It appears that in the existing literature, this approach is more popular compared to the other one that is based on the Lagrange's method of variation of parameters.

In this book, the approach based on the variation of parameters method (see Section 1.3) is, however, more frequently employed because it is more universal in the sense that it does not deal with defining properties of Green's functions, which are individual for each type of an equation.

In addition, in Section 1.2 we will focus on the symmetry of Green's functions. Their symmetry is directly related to the so-called self-adjointness of the differential operator involved. This feature of a Green's function is of great theoretical and practical importance. It will be frequently discussed in the further discussion in the text.

1.1 Defining properties and construction

In this section, we introduce the definition of the Green's function for a linear boundary value problem for an ordinary differential equation of the n-th order with variable coefficients. We then give a detailed description of the traditional method for the construction of Green's functions, based on their defining properties. Many of the examples presented here are related to various problems of continuum mechanics.

The discussion that follows concerns a linear homogeneous boundary value problem for the ordinary differential equation

$$L[y(x)] \equiv p_0(x)\, y^{(n)}(x) + p_1(x)\, y^{(n-1)}(x) + p_2(x)\, y^{(n-2)}(x)$$
$$+ \ldots + p_{n-1}(x)\, y'(x) + p_n(x)\, y(x) = 0, \quad x \in (a,b) \qquad (1.1)$$

subject to the boundary conditions written as

$$M_k(y;a,b) \equiv \sum_{j=0}^{n-1}[\alpha_j^k y^{(j)}(a) + \beta_j^k y^{(j)}(b)] = 0, \quad (k=\overline{1,n}) \qquad (1.2)$$

In this formulation, the coefficients $p_j(x), (j = 0, 1, \ldots, n)$ of the governing equation are continuous functions on (a, b), where $p_0(x) \neq 0$; $M_k, (k = 1, \ldots, n)$ represent linearly independent forms with constant coefficients α_j^k and β_j^k.

The boundary conditions in eqn (1.2) are written in a general form, which implies that a certain physically natural formulation of the boundary conditions can be obtained from this form as a particular case. If, for example, the displacement formulation of a problem from the theory of elasticity is considered (beam, plate, shell problems, etc.), then the condition in eqn (1.2) may model either clamped, or simply supported, or free, or even elastically supported edge. Periodic boundary conditions can also follow from this form.

In the discussion that follows, we will be using conventional and customary notations $(a, b), [a, b]$, and $(a, b]$ or $[a, b)$ to specify open, closed, and half-open intervals, respectively.

To avoid possible confusion that may occur in regard with the term *homogeneous*, the reader must discern different meanings of this term in mathematics and mechanics. Indeed, in mathematics we usually say *homogeneous* boundary value problem, *homogeneous* equation, or *homogeneous* boundary condition, when the right-hand side in the corresponding equality is zero. In mechanics, on the other hand, when specifying properties of materials, we usually use the term *homogeneous* to indicate that an object under consideration is composed of a material whose properties do not vary with space coordinates. Within the present book, this term will be frequently used in both senses.

We now turn the reader's attention to one of the most important definitions in this study. Let us define the Green's function for the homogeneous boundary value problem that occurred in eqns (1.1) and (1.2).

Definition: The function $g(x,s)$ is said to be the *Green's function* for the boundary value problem in eqns (1.1) and (1.2), if as a function of its first variable x, it meets the following properties, for any $s \in (a,b)$:

1.1. DEFINING PROPERTIES AND CONSTRUCTION

1. On both of the intervals $[a, s)$ and $(s, b]$, $g(x, s)$ is a continuous function having continuous derivatives up to the n-th order included, and it satisfies the governing equation (1.1) on (a, s) and (s, b), i. e. :

$$L[g(x,s)] = 0, \quad x \in (a,s); \quad L[g(x,s)] = 0, \quad x \in (s,b)$$

2. For $x = s$, $g(x, s)$ is continuous along with all its derivatives up to the $(n-2)$-nd order included

$$\frac{\partial^m g(s+0,s)}{\partial x^m} - \frac{\partial^m g(s-0,s)}{\partial x^m} = 0, \quad (m = \overline{0, n-2})$$

3. The $(n-1)$-st derivative of $g(x, s)$ is discontinuous when $x = s$, providing

$$\frac{\partial^{n-1} g(s+0,s)}{\partial x^{n-1}} - \frac{\partial^{n-1} g(s-0,s)}{\partial x^{n-1}} = -\frac{1}{p_0(s)}$$

where $p_0(s)$ represents the leading coefficient of eqn (1.1);

4. It satisfies the boundary conditions in eqn (1.2), i. e. :

$$M_k(g) = 0, \quad (k = \overline{1, n})$$

The following theorem is valid specifying the conditions of existence and uniqueness for the Green's function.

Theorem 1.1 (of existence and uniqueness): If the homogeneous boundary value problem in eqns (1.1) and (1.2) has only the trivial (zero) solution, then there exists its unique Green's function $g(x, s)$.

The reader is insistently suggested to carefully go through this proof because it actually provides a straightforward algorithm for the practical construction of Green's functions. Throughout the present text, we will be frequently using this algorithm.

Proof. Let functions $y_i(x), (i = 1, \ldots, n)$ represent the fundamental solution set for eqn (1.1). That is, $y_i(x)$ are linearly independent on (a, b) particular solutions of eqn (1.1).

In numerous practical situations, one can find an analytic form for $y_i(x)$. This can, in particular, be easily done for equations with constant coefficients. If, however, the governing differential equation does not allow an analytical solution, then appropriate numerical procedures may be employed for obtaining approximate ones. Later in this book we will discuss this point in more detail.

In compliance with property 1 of the definition, for any arbitrarily fixed value of $s \in (a, b)$, the Green's function $g(x, s)$ must be a solution of eqn

(1.1) in (a, s) (on the left of s), as well as in (s, b) (on the right of s). Since any solution of eqn (1.1) can be expressed as a linear combination of the components $y_i(x)$ of the fundamental solution set, one may write $g(x, s)$ in the following form

$$g(x, s) = \sum_{i=1}^{n} \begin{cases} y_i(x) A_i(s), & \text{for } a \leq x \leq s \\ y_i(x) B_i(s), & \text{for } s \leq x \leq b \end{cases} \quad (1.3)$$

where $A_i(s)$ and $B_i(s)$ represent the functions to be determined. Clearly, the number of these functions is $2n$ and the number of the relations for them, which can be derived from properties 2, 3, and 4 of the definition, is also $2n$. Thus, the situation is promising so far. Indeed, we are going to derive a system of $2n$ equations in $2n$ unknowns ($(n-1)$ equations can be obtained from property 2, one equation comes from property 3, and n equations follow from property 4). Hence, the key issues to be highlighted in the remaining part of this proof are whether that system is going to be consistent and whether it has a unique solution.

By virtue of property 2, which stipulates the continuity of $g(x, s)$ itself and its partial derivatives with respect to x up to the $(n-2)$-nd order, as $x = s$, one derives the following system of $(n-1)$ linear algebraic equations

$$\sum_{i=1}^{n} C_i(s) y_i^{(j)}(s) = 0, \quad (j = \overline{0, n-2}) \quad (1.4)$$

in n unknown functions

$$C_i(s) = B_i(s) - A_i(s), \quad (i = \overline{1, n}) \quad (1.5)$$

The superscript j on $y_i(s)$ in eqn (1.4) specifies the differentiation order.

The system in eqn (1.4) is underdetermined, because the number of equations in it $(n-1)$ is fewer than the number of unknowns (n) involved. This drawback can be eluded, however, by applying property 3 to the expression in eqn (1.3). This yields one more linear algebraic equation

$$\sum_{i=1}^{n} C_i(s) y_i^{(n-1)}(s) = -\frac{1}{p_0(s)} \quad (1.6)$$

in the same set $C_i(s)$ of unknowns. Hence, the relations in eqn (1.4) together with those in eqn (1.6) form a system of n simultaneous linear algebraic equations in n unknowns. The determinant of the coefficient matrix of this system is not zero, because it represents the Wronskian for the fundamental solution set $y_i(s)$. Thus, this system has a unique solution. So, one can readily obtain the explicit expressions for $C_i(s)$ from eqns (1.4) and (1.6).

1.1. DEFINING PROPERTIES AND CONSTRUCTION

In order to obtain the values of $A_i(s)$ and $B_i(s)$, we take advantage of property 4. In doing so, let us first break down the forms $M_k(y)$ in the boundary conditions in eqn (1.2) into two parts as

$$M_k(y) = P_k(y) + Q_k(y), \quad (k = \overline{1,n}) \tag{1.7}$$

with $P_k(y)$ and $Q_k(y)$ being defined as

$$P_k(y) = \sum_{j=0}^{n-1} \alpha_j^k y^{(j)}(a), \qquad Q_k(y) = \sum_{j=0}^{n-1} \beta_j^k y^{(j)}(b)$$

In compliance with property 4, we now substitute the expression for $g(x,s)$ from eqn (1.3) into eqn (1.2)

$$M_k(g) \equiv P_k(g) + Q_k(g) = 0, \quad (k = \overline{1,n})$$

Since P_k in eqn (1.7) governs the values of $g(x,s)$ at the left-end point $x = a$ of the interval $[a,b]$, while Q_k governs the values of $g(x,s)$ at the right-end point $x = b$, the upper branch

$$\sum_{i=1}^{n} y_i(x) A_i(s)$$

of $g(x,s)$ from eqn (1.3) goes to $P_k(g)$, while the lower branch

$$\sum_{i=1}^{n} y_i(x) B_i(s)$$

must be substituted into $Q_k(g)$, resulting in

$$M_k(g) \equiv \sum_{i=1}^{n} [P_k(y_i) A_i(s) + Q_k(y_i) B_i(s)] = 0, \quad (k = \overline{1,n})$$

Replacing the values of $A_i(s)$ in the above equation with $B_i(s) - C_i(s)$ in accordance with eqn (1.5), one rewrites it in the form

$$\sum_{i=1}^{n} [P_k(y_i)(B_i(s) - C_i(s)) + Q_k(y_i) B_i(s)] = 0, \quad (k = \overline{1,n})$$

Combining then the terms with $B_i(s)$ and taking the term with $C_i(s)$ to the right-hand side, one obtains

$$\sum_{i=1}^{n} [P_k(y_i) + Q_k(y_i)] B_i(s) = \sum_{i=1}^{n} P_k(y_i) C_i(s), \quad (k = \overline{1,n})$$

Upon recalling the partitioning from eqn (1.7), the above relations can finally be rewritten in the form

$$\sum_{i=1}^{n} M_k(y_i) B_i(s) = \sum_{i=1}^{n} P_k(y_i) C_i(s), \quad (k = \overline{1,n}) \qquad (1.8)$$

These relations constitute a system of n linear algebraic equations in the n unknowns $B_i(s)$. The coefficient matrix of this system is not singular, since the forms M_k are linearly independent. The right-hand side vector in eqn (1.8) is defined in terms of the known values of $C_i(s)$. This system has, consequently, a unique solution for $B_i(s)$. Based on this, the values of $A_i(s)$ can readily be obtained from eqn (1.5). Hence, this final step completes the proof of Theorem 1.1, because upon substituting the values of $A_i(s)$ and $B_i(s)$ into eqn (1.3), we finally obtain an explicit expression for $g(x,s)$.

As we have already mentioned, the proof just completed suggests a consistent way to practically construct the Green's function. This point is illustrated below with a series of particular examples.

In each of the following examples, we present and analyze different peculiarities in statements of boundary value problems, which may occur while considering practical situations in computational mechanics.

EXAMPLE 1   Consider the following differential equation

$$\frac{d^2 y(x)}{dx^2} = 0, \quad x \in (0, a) \qquad (1.9)$$

subject to the boundary conditions written as

$$y(0) = y(a) = 0 \qquad (1.10)$$

This boundary value problem can be associated with many phenomena in continuum mechanics (it represents a particular case of the problem in eqns (0.2) and (0.3) from the introduction to this text, as $m(x)$ is set to a constant).

The most elementary set of functions constituting a fundamental solution set for eqn (1.9) is represented by

$$y_1(x) \equiv 1, \quad y_2(x) \equiv x$$

Therefore, the general solution $y_g(x)$ for this equation can be written as

$$y_g(x) = D_1 + D_2 x$$

where D_1 and D_2 represent arbitrary constants.

A substitution of this function into the boundary conditions in eqn (1.10) yields the homogeneous system of linear algebraic equations in D_1 and D_2,

1.1. DEFINING PROPERTIES AND CONSTRUCTION

with a well-posed coefficient matrix. Hence, the problem in eqns (1.9) and (1.10) has only the trivial solution.

Thus, there exists a unique Green's function for this problem. According to the procedure described earlier, it can be sought in the form

$$g(x,s) = \begin{cases} A_1(s) + xA_2(s), & \text{for } 0 \leq x \leq s \\ B_1(s) + xB_2(s), & \text{for } s \leq x \leq a \end{cases} \quad (1.11)$$

Introducing then, as it is suggested in eqn (1.5), $C_1(s) = B_1(s) - A_1(s)$ and $C_2(s) = B_2(s) - A_2(s)$, we form a system of linear algebraic equations in these unknowns (see the system in eqns (1.4) and (1.6)) written as

$$\begin{cases} C_1(s) + sC_2(s) = 0 \\ C_2(s) = -1 \end{cases} \quad (1.12)$$

Its obvious solution is $C_1(s) = s$ and $C_2(s) = -1$.

The first boundary condition $y(0) = 0$ in eqn (1.10), being satisfied with the upper branch of $g(x,s)$, results in $A_1(s) = 0$. The upper branch is chosen because $x = 0$ belongs to its domain $0 \leq x \leq s$. Since $B_1(s) = C_1(s) + A_1(s)$, it follows that $B_1(s) = s$.

The second condition $y(a) = 0$ in eqn (1.10), being treated with the lower branch of $g(x,s)$, yields $B_1(s) + aB_2(s) = 0$. Hence, $B_2(s) = -s/a$, and finally, since $A_2(s) = B_2(s) - C_2(s)$, it follows that $A_2(s) = 1 - s/a$. Substituting these into eqn (1.11), we ultimately obtain the Green's function that we are looking for in the form

$$g(x,s) = \begin{cases} a^{-1}x(a-s), & \text{for } 0 \leq x \leq s \\ a^{-1}s(a-x), & \text{for } s \leq x \leq a \end{cases} \quad (1.13)$$

EXAMPLE 2 We now formulate another boundary value problem

$$\frac{dy(0)}{dx} = 0, \quad \frac{dy(a)}{dx} = 0$$

for eqn (1.9) over the interval $(0, a)$.

This problem is not uniquely solvable. Indeed, any constant function represents the solution to it. Hence, the condition of existence and uniqueness for Green's function does not hold for the above statement. Therefore, a Green's function cannot be constructed in this case, because it does not exist.

EXAMPLE 3 Consider one more boundary value problem

$$\frac{dy(0)}{dx} = 0, \quad \frac{dy(a)}{dx} + my(a) = 0 \quad (1.14)$$

for equation (1.9) over $(0, a)$, where m is thought to be a nonzero constant.

It can easily be shown (see exercise 1.1(a) of this chapter) that the problem in eqns (1.9) and (1.14) has only the trivial solution. Consequently, there exists a unique Green's function for this problem.

The first part of the construction procedure precisely resembles that from the problem stated with eqns (1.9) and (1.10). The Green's function is again expressed by eqn (1.11), the coefficients $C_1(s)$ and $C_2(s)$ again satisfy the system in eqn (1.12), resulting in $C_1(s) = s$ and $C_2(s) = -1$.

The first boundary condition in eqn (1.14), being treated by the upper branch in eqn (1.11), yields $A_2(s)=0$. This immediately results in $B_2(s)= -1$. The second condition in (1.14), being treated by the lower branch in eqn (1.11), yields the following equation

$$B_2(s) + m[B_1(s) + aB_2(s)] = 0$$

in $B_1(s)$ and $B_2(s)$. Based on the known value of $B_2(s)$, one obtains $B_1(s) = (1+ma)/m$. This in turn yields $A_1(s) = [1 + m(a-s)]/m$.

Substituting the values of $A_i(s)$ and $B_i(s)$ just found, into eqn (1.11), we finally obtain the Green's function to the boundary value problem posed by eqns (1.9) and (1.14) in the form

$$g(x,s) = \begin{cases} (a-s) + m^{-1}, & \text{for } 0 \leq x \leq s \\ (a-x) + m^{-1}, & \text{for } s \leq x \leq a \end{cases} \quad (1.15)$$

Notice that as m is taken to infinity, the second term m^{-1} in (1.15) vanishes yielding the Green's function

$$g(x,s) = \begin{cases} a-s, & \text{for } 0 \leq x \leq s \\ a-x, & \text{for } s \leq x \leq a \end{cases}$$

for equation (1.9) subject to the following boundary conditions

$$\frac{dy(0)}{dx} = 0, \quad y(a) = 0$$

In applied mechanics, one frequently is required to work out research projects for phenomena occurring in infinite media. The influence (Green's) function formalism can successfully be applied to the associated boundary value problems formulated over infinite intervals. As our next example, we construct the Green's function for such a problem.

EXAMPLE 4   Consider the following differential equation

$$\frac{d^2y(x)}{dx^2} - k^2 y(x) = 0, \quad x \in (0, \infty) \quad (1.16)$$

1.1. DEFINING PROPERTIES AND CONSTRUCTION

subject to the boundary conditions imposed as

$$y(0) = 0, \quad |y(\infty)| < \infty \qquad (1.17)$$

It can be shown (see exercise 1.1(b)) that the conditions of existence and uniqueness for the Green's function are met in this case assuring a unique Green's function of the above formulation.

Since the following two functions

$$y_1(x) \equiv \exp(kx), \quad y_2(x) \equiv \exp(-kx)$$

represent the fundamental solution set for eqn (1.16), one can express the Green's function for the boundary value problem in eqns (1.16) and (1.17) in the following form

$$g(x,s) = \begin{cases} A_1(s)\exp(kx) + A_2(s)\exp(-kx), & \text{for } x \leq s \\ B_1(s)\exp(kx) + B_2(s)\exp(-kx), & \text{for } s \leq x \end{cases} \qquad (1.18)$$

Denoting $C_i(s) = B_i(s) - A_i(s), (i = 1,2)$, one obtains the following system of linear algebraic equations

$$\begin{cases} \exp(ks)C_1(s) + \exp(-ks)C_2(s) = 0 \\ k\exp(ks)C_1(s) - k\exp(-ks)C_2(s) = -1 \end{cases}$$

in $C_1(s)$ and $C_2(s)$. Its solution is expressed as

$$C_1(s) = -\frac{1}{2k}\exp(-ks), \quad C_2(s) = \frac{1}{2k}\exp(ks) \qquad (1.19)$$

The first condition in eqn (1.17) implies

$$A_1(s) + A_2(s) = 0 \qquad (1.20)$$

while the second condition results in $B_1(s) = 0$, because the exponential function $\exp(kx)$ is unbounded as x approaches infinity. And the only way to satisfy the second condition in eqn (1.17) is to set $B_1(s)$ to zero. This immediately yields

$$A_1(s) = \frac{1}{2k}\exp(-ks)$$

and the relation in eqn (1.20) consequently provides

$$A_2(s) = -\frac{1}{2k}\exp(-ks)$$

Hence, based on the known values of $C_2(s)$ and $A_2(s)$, one obtains

$$B_2(s) = \frac{1}{2k}[\exp(ks) - \exp(-ks)]$$

Upon substituting the values of the coefficients $A_i(s)$ and $B_i(s)$ just found into eqn (1.18), one finally obtains the Green's function to the problem posed by eqns (1.16) and (1.17) in the form

$$g(x,s) = \frac{1}{2k} \begin{cases} \exp(k(x-s)) - \exp(-k(x+s)), & \text{for } x \leq s \\ \exp(k(s-x)) - \exp(-k(s+x)), & \text{for } s \leq x \end{cases} \quad (1.21)$$

EXAMPLE 5 Consider a boundary value problem for the same equation as in the previous example but formulated over a different domain. Let

$$\frac{d^2y(x)}{dx^2} - k^2 y(x) = 0, \quad x \in (0, a) \quad (1.22)$$

be subjected to the boundary conditions written as

$$y(0) = y(a), \quad \frac{dy(0)}{dx} = \frac{dy(a)}{dx} \quad (1.23)$$

This boundary value problem represents one more important type of formulations in applied mechanics. The relations in eqn (1.23) specify conditions of the a-periodicity of the solution.

It can be shown (see exercise 1.1(c) of this chapter) that this boundary value problem has only the trivial solution, providing existence of the unique Green's function for it.

Since the formulation in eqns (1.22) and (1.23) again entails the same differential equation which was considered in *EXAMPLE 4*, the beginning of the construction procedure for the Green's function resembles that from the previous problem. We again express the Green's function by eqn (1.18), and the coefficients $C_1(s)$ and $C_2(s)$ are again given with eqn (1.19).

Satisfying the first condition in eqn (1.23), we utilize the upper branch in eqn (1.18) in order to compute the value of $y(0)$, while its lower branch is used for computing the value of $y(a)$. This results in

$$A_1(s) + A_2(s) = B_1(s)\exp(ka) + B_2(s)\exp(-ka) \quad (1.24)$$

Satisfying the second condition in eqn (1.23), we compute the derivative of $y(x)$ at $x = 0$ by using the upper branch in eqn (1.18), while the value of the derivative of $y(x)$ at $x = a$ is computed by using the lower branch of eqn (1.18). This yields

$$A_1(s) - A_2(s) = B_1(s)\exp(ka) - B_2(s)\exp(-ka) \quad (1.25)$$

1.1. DEFINING PROPERTIES AND CONSTRUCTION

So the relations in eqns (1.24) and (1.25) along with those in eqn (1.19) form a system of four linear algebraic equations in $A_1(s)$, $A_2(s)$, $B_1(s)$, and $B_2(s)$. To find the values of $A_1(s)$ and $B_1(s)$, we add eqns (1.24) and (1.25) to each other. This provides

$$A_1(s) - B_1(s)\exp(ka) = 0 \tag{1.26}$$

The first relation in eqn (1.19) can be rewritten in the form

$$-A_1(s) + B_1(s) = -\frac{1}{2k}\exp(-ks) \tag{1.27}$$

Solving eqns (1.26) and (1.27) simultaneously, one obtains

$$A_1(s) = \frac{\exp(k(a-s))}{2k[\exp(ka)-1]}, \quad B_1(s) = \frac{\exp(-ks)}{2k[\exp(ka)-1]}$$

To find the values of $A_2(s)$ and $B_2(s)$, we subtract eqn (1.25) from eqn (1.24) resulting in

$$A_2(s) - B_2(s)\exp(-ka) = 0 \tag{1.28}$$

Rewriting then the second relation from eqn (1.19) in the form

$$-A_2(s) + B_2(s) = \frac{1}{2k}\exp(ks) \tag{1.29}$$

we solve eqns (1.28) and (1.29) simultaneously. This yields

$$A_2(s) = \frac{\exp(ks)}{2k[\exp(ka)-1]}, \quad B_2(s) = \frac{\exp(k(s+a))}{2k[\exp(ka)-1]}$$

Substituting the values of $A_1(s)$, $A_2(s)$, $B_1(s)$, and $B_2(s)$ just found into eqn (1.18), we finally obtain

$$g(x,s) = K_0 \begin{cases} \exp(k(x-s+a)) + \exp(k(s-x)), & \text{for } x \leq s \\ \exp(k(s-x+a)) + \exp(k(x-s)), & \text{for } s \leq x \end{cases} \tag{1.30}$$

where $K_0 = \{2k[\exp(ka)-1]\}^{-1}$.

In all of the examples considered so far, we have dealt with ordinary differential equations having constant coefficients. Clearly, variable coefficients do not bring any limitations to the algorithm described, if the fundamental solution set of the equation under consideration is obtainable in terms of elementary functions. In other words, if the governing differential equation allows exact solution, one can readily construct a Green's function by means of this algorithm.

EXAMPLE 6 To address the last issue, consider the equation

$$\frac{d}{dx}\left((mx+p)\frac{dy}{dx}\right) = 0, \quad x \in (0, a) \tag{1.31}$$

with the boundary conditions imposed as

$$\frac{dy(0)}{dx} = 0, \quad y(a) = 0 \tag{1.32}$$

where we assume $m > 0$ and $p > 0$, which means $mx + p \neq 0$ on $x \in [0, a]$. The fundamental solution set

$$y_1(x) \equiv 1, \quad y_2(x) \equiv \ln(mx + p)$$

required for the construction of the Green's function for the problem in eqns (1.31) and (1.32) can be obtained by two successive integrations of eqn (1.31).

In view of exercise 1.1(d), the problem in eqns (1.31) and (1.32) has only the trivial solution. Hence, there exists a unique Green's function which can be presented in the form

$$g(x, s) = \begin{cases} A_1(s) + \ln(mx+p)A_2(s), & \text{for } 0 \le x \le s \\ B_1(s) + \ln(mx+p)B_2(s), & \text{for } s \le x \le a \end{cases} \tag{1.33}$$

Following then our customary procedure, one obtains the system of linear algebraic equations

$$\begin{cases} C_1(s) + \ln(ms + p)C_2(s) = 0 \\ m(ms + p)^{-1}C_2(s) = -(ms + p)^{-1} \end{cases}$$

in $C_i(s) = B_i(s) - A_i(s), (i = 1, 2)$. Its solution is

$$C_1(s) = \frac{1}{m}\ln(ms + p), \quad C_2(s) = -\frac{1}{m} \tag{1.34}$$

The first boundary condition in eqn (1.32) yields $A_2(s) = 0$. Consequently, $B_2(s) = -1/m$. The second condition in eqn (1.32) gives

$$B_1(s) + \ln(ma + p)B_2(s) = 0$$

resulting in $B_1(s) = [\ln(ma + p)]/m$, which provides

$$A_1(s) = \frac{1}{m}\ln\frac{ma + p}{ms + p}$$

1.1. DEFINING PROPERTIES AND CONSTRUCTION

Substituting the values of $A_i(s)$ and $B_i(s)$ just found into eqn (1.33), one obtains the Green's function that we are looking for in the form

$$g(x,s) = \frac{1}{m} \begin{cases} \ln[(ma+p)(ms+p)^{-1}], & \text{for } 0 \leq x \leq s \\ \ln[(ma+p)(mx+p)^{-1}], & \text{for } s \leq x \leq a \end{cases} \qquad (1.35)$$

Sometimes in applied mechanics, we consider boundary value problems formulated over finite intervals, where one of the end-points is a singular point for the governing differential equation. The algorithm described in this section can also be used to construct Green's functions for such problems.

EXAMPLE 7   As an illustrative example on this issue, we consider a boundary value problem for the following differential equation

$$\frac{d}{dx}\left(x\frac{dy(x)}{dx}\right) = 0, \quad x \in (0,a) \qquad (1.36)$$

subject to the boundary conditions written as

$$|y(0)| < \infty, \quad \frac{dy(a)}{dx} + hy(a) = 0 \qquad (1.37)$$

Clearly, the left end-point $x = 0$ of the domain is a point of singularity for eqn (1.36). Therefore, instead of formulating a traditional boundary condition at this point, we require in eqn (1.37) for $y(0)$ to be bounded.

Integrating eqn (1.36) successively two times, one obtains its fundamental solution set that can be written as

$$y_1(x) \equiv 1, \quad y_2(x) \equiv \ln x \qquad (1.38)$$

The problem in eqns (1.36) and (1.37) has only the trivial solution (see exercise 1.1(e)), allowing a unique Green's function in the form

$$g(x,s) = \begin{cases} A_1(s) + \ln x\, A_2(s), & \text{for } 0 \leq x \leq s \\ B_1(s) + \ln x\, B_2(s), & \text{for } s \leq x \leq a \end{cases} \qquad (1.39)$$

In compliance with our customary procedure, we form a system of linear algebraic equations

$$\begin{cases} C_1(s) + \ln s\, C_2(s) = 0 \\ s^{-1} C_2(s) = -s^{-1} \end{cases}$$

whose solution is $C_1(s) = \ln s$ and $C_2(s) = -1$.

The boundedness of the Green's function at $x = 0$ implies $A_2(s) = 0$. Consequently, $B_2(s) = -1$. The second condition in eqn (1.37) yields

$$B_2(s)/a + h[B_1(s) + \ln a B_2(s)] = 0$$

Hence, $B_1(s) = 1/ah + \ln a$, and ultimately, $A_1(s) = 1/ah - \ln s/a$. Thus, we finally obtain

$$g(x,s) = \begin{cases} (ah)^{-1} - \ln[(a)^{-1}s], & \text{for } 0 \leq x \leq s \\ (ah)^{-1} - \ln[(a)^{-1}x], & \text{for } s \leq x \leq a \end{cases} \quad (1.40)$$

Notice that as the value of h is taken to infinity, the first term $(ah)^{-1}$ in eqn (1.40) vanishes, yielding the Green's function

$$g(x,s) = \begin{cases} -\ln[(a)^{-1}s], & \text{for } 0 \leq x \leq s \\ -\ln[(a)^{-1}x], & \text{for } s \leq x \leq a \end{cases}$$

for eqn (1.36) subject to the boundary conditions $|y(0)| < \infty$ and $y(a) = 0$.

EXAMPLE 8   For the next example, we formulate a boundary value problem for the equation of the fourth order

$$\frac{d^4 y(x)}{dx^4} = 0, \quad x \in (0,1) \quad (1.41)$$

with boundary conditions written as

$$y(0) = \frac{dy(0)}{dx} = 0, \quad y(1) = \frac{d^2 y(1)}{dx^2} = 0 \quad (1.42)$$

As is known, this formulation relates to the bending phenomenon of a beam of unit length, if its left edge is clamped while the right edge is simply supported. In Chapters 2 and 3, we consider a number of other problems from the beam theory.

The following set of functions

$$y_1(x) \equiv 1, \ y_2(x) \equiv x, \ y_3(x) \equiv x^2, \ y_4(x) \equiv x^3 \quad (1.43)$$

constitutes the simplest fundamental solution set for eqn (1.41). Hence, its general solution is

$$y_g(x) = D_1 + D_2 x + D_3 x^2 + D_4 x^3$$

Applying the boundary conditions from eqn (1.42), one derives a homogeneous system of linear algebraic equations in D_i. The coefficient matrix

1.1. DEFINING PROPERTIES AND CONSTRUCTION

of that system is not singular, providing only the trivial solution for the system. Consequently, there exists a unique Green's function for the problem posed by eqns (1.41) and (1.42).

Based on the fundamental solution set presented in eqn (1.43), the Green's function can be written in the form

$$g(x,s) = \begin{cases} A_1(s) + A_2(s)x + A_3(s)x^2 + A_4(s)x^3, & \text{for } x \leq s \\ B_1(s) + B_2(s)x + B_3(s)x^2 + B_4(s)x^3, & \text{for } s \leq x \end{cases} \quad (1.44)$$

From properties 2 and 3 of the definition of the Green's function, one derives the following system of linear equations in $C_i(s) = B_i(s) - A_i(s)$, written in a matrix form

$$\begin{pmatrix} 1 & s & s^2 & s^3 \\ 0 & 1 & 2s & 3s^2 \\ 0 & 0 & 2 & 6s \\ 0 & 0 & 0 & 6 \end{pmatrix} \times \begin{pmatrix} C_1(s) \\ C_2(s) \\ C_3(s) \\ C_4(s) \end{pmatrix} = \begin{pmatrix} 0 \\ 0 \\ 0 \\ -1 \end{pmatrix}$$

whose solution

$$C_1 = \frac{1}{6}s^3, \quad C_2 = -\frac{1}{2}s^2, \quad C_3 = \frac{1}{2}s, \quad C_4 = -\frac{1}{6} \quad (1.45)$$

is easily obtained because of the triangular form of its coefficient matrix.

By virtue of property 4 in the definition, the boundary conditions in eqn (1.42) provide

$$A_1 = 0, \quad A_2 = 0, \quad B_1 + B_2 + B_3 + B_4 = 0, \quad 2B_3 + 6B_4 = 0$$

while the rest of the coefficients for $g(x,s)$

$$A_3 = -\frac{1}{4}s^3 + \frac{3}{4}s^2 - \frac{1}{2}s, \quad A_4 = \frac{1}{12}s^3 - \frac{1}{4}s^2 + \frac{1}{6}$$

$$B_1 = \frac{1}{6}s^3, \quad B_2 = -\frac{1}{2}s^2, \quad B_3 = -\frac{1}{4}s^3 + \frac{3}{4}s^2, \quad B_4 = \frac{1}{12}s^3 - \frac{1}{4}s^2$$

are computed through the values of $C_i(s)$ presented in eqn (1.45).

Substituting all the coefficients $A_i(s)$ and $B_i(s)$ just obtained into eqn (1.44), we obtain the Green's function $g(x,s)$ for the boundary value problem posed by eqns (1.41) and (1.42). For $x \leq s$, it is found in the form

$$g(x,s) = -\left(\frac{1}{4}s^3 - \frac{3}{4}s^2 + \frac{1}{2}s\right)x^2 + \left(\frac{1}{12}s^3 - \frac{1}{4}s^2 + \frac{1}{6}\right)x^3 \quad (1.46)$$

while for $x \geq s$, its expression is

$$g(x,s) = -\left(\frac{1}{4}x^3 - \frac{3}{4}x^2 + \frac{1}{2}x\right)s^2 + \left(\frac{1}{12}x^3 - \frac{1}{4}x^2 + \frac{1}{6}\right)s^3$$

This example shows that even for equations of higher order, the procedure for the construction of Green's functions utilized here results in a reasonable amount of computation.

Analyzing the form of all of the Green's functions constructed so far in this section, one may notice their common property (they are symmetric). That is, $g(x,s) = g(s,x)$. Indeed, the interchange of x with s in the expression valid for $x \leq s$ yields that valid for $x \geq s$ and vice versa. In the next section, we will discuss this issue in more detail. The conditions will be found under which the symmetry takes place.

EXAMPLE 9   We close the discussion in this section with a problem whose Green's function, contrary to all previous ones, appears to be in a nonsymmetrical form. Namely, consider the following equation

$$\frac{d^2y(x)}{dx^2} + \frac{dy(x)}{dx} - 2y(x) = 0, \quad x \in (0, \infty) \tag{1.47}$$

subject to the boundary conditions

$$y(0) = 0, \quad |y(\infty)| < \infty \tag{1.48}$$

Clearly, this problem has only the trivial solution (see exercise 1.1(f)), allowing, subsequently, a unique Green's function. Since $y_1(x) = \exp(x)$ and $y_2(x) = \exp(-2x)$ represent the fundamental solution set to eqn (1.47), one can express the Green's function to this problem in the form

$$g(x,s) = \begin{cases} A_1(s)\exp(x) + A_2(s)\exp(-2x), & \text{for } x \leq s \\ B_1(s)\exp(x) + B_2(s)\exp(-2x), & \text{for } s \leq x \end{cases} \tag{1.49}$$

This results in the system of linear equations in $C_i(s) = B_i(s) - A_i(s)$

$$\begin{pmatrix} \exp(s) & \exp(-2s) \\ \exp(s) & -2\exp(-2s) \end{pmatrix} \times \begin{pmatrix} C_1(s) \\ C_2(s) \end{pmatrix} = \begin{pmatrix} 0 \\ -1 \end{pmatrix}$$

whose solution is found as

$$C_1(s) = -\frac{1}{3}\exp(-s), \quad C_2(s) = \frac{1}{3}\exp(2s)$$

The first condition in eqn (1.48) provides $A_1(s) + A_2(s) = 0$, while the second condition implies $B_1(s) = 0$. Therefore $A_1(s) = [\exp(-s)]/3$, resulting in $A_2(s) = -[\exp(-s)]/3$, and, finally, $B_2(s) = [\exp(2s) - \exp(-s)]/3$. Substituting these values into eqn (1.49), one obtains

$$g(x,s) = \frac{1}{3} \begin{cases} \exp(-s)\,[\exp(x) - \exp(-2x)], & \text{for } x \leq s \\ \exp(-2x)\,[\exp(2s) - \exp(-s)], & \text{for } s \leq x \end{cases} \quad (1.50)$$

It is clearly seen that this Green's function fails to be symmetric. Why? What makes the statement of this last problem different from all the ones considered earlier in this section? The reader will find the reasoning for this occurrence in the next section.

1.2 Symmetry of Green's functions

In order to address the basic issue of this section, a certain preparatory work has to be carried out. Let us write down the linear homogeneous differential equation of the n-th order

$$\begin{aligned} L[y(x)] &\equiv p_0(x)\,y^{(n)}(x) + p_1(x)\,y^{(n-1)}(x) + p_2(x)\,y^{(n-2)}(x) \\ &\quad + \ldots + p_{n-1}(x)\,y'(x) + p_n(x)\,y(x) = 0 \end{aligned}$$

From the general theory of linear ODEs (see, for example, [31, 68]), it is known that the equation

$$\begin{aligned} L_a[y(x)] &\equiv (-1)^n [p_0(x)\,y(x)]^{(n)} + (-1)^{n-1}[p_1(x)\,y(x)]^{(n-1)} \\ &\quad + \ldots - [p_{n-1}(x)\,y(x)]' + p_n(x)\,y(x) = 0 \end{aligned}$$

is said to be *adjoint* to $L[y(x)] = 0$. The operator L_a is called *adjoint* to L, and if $L \equiv L_a$, then L is said to be a *self-adjoint* operator and the equation $L[y(x)] = 0$ is said to be a *self-adjoint* equation.

For the sake of simplicity, the discussion herein is limited to equations of the second order

$$L[y(x)] \equiv p_0(x)\frac{d^2y(x)}{dx^2} + p_1(x)\frac{dy(x)}{dx} + p_2(x)y(x) = 0 \quad (1.51)$$

The leading coefficient $p_0(x)$ is not supposed to equal zero at any single point in (a,b) except, maybe, for one of its end-points. In addition, in the discussion that follows, we require the coefficient $p_0(x)$ to be two times differentiable and $p_1(x)$ to be just differentiable on (a,b).

According to what we recently recalled, the following equation

$$L_a[y(x)] \equiv \frac{d^2}{dx^2}[p_0(x)y(x)] - \frac{d}{dx}[p_1(x)y(x)] + p_2(x)y(x) = 0 \quad (1.52)$$

is adjoint to that in eqn (1.51).

We will review here a brief discussion on the self-adjointness of differential equations and relevant issues, which are required just to analyze the symmetry of Green's functions. The more detailed discussion on this subject can be found, for example, in [17, 30, 31, 35, 63, 66, 68].

Using the product rule for the differentiation in eqn (1.52), the operator L_a can be rewritten in the form

$$L_a[y(x)] \equiv \frac{d}{dx}\left(y\frac{dp_0}{dx} + p_0\frac{dy}{dx}\right) - \left(y\frac{dp_1}{dx} + p_1\frac{dy}{dx}\right) + p_2 y$$

Differentiating and combining the like terms, one obtains

$$L_a[y(x)] \equiv p_0\frac{d^2 y}{dx^2} + \left(2\frac{dp_0}{dx} - p_1\right)\frac{dy}{dx} + \left(\frac{d^2 p_0}{dx^2} - \frac{dp_1}{dx} + p_2\right) y \quad (1.53)$$

Suppose eqn (1.51) is self-adjoint, that is $L[y(x)] \equiv L_a[y(x)]$. If so, then upon comparing the coefficients of y' in $L[y(x)]$ and $L_a[y(x)]$, one obtains the following relation for the coefficients $p_0(x)$ and $p_1(x)$

$$2\frac{dp_0(x)}{dx} - p_1(x) = p_1(x)$$

which must hold for the self-adjointness of eqn (1.51). This implies

$$p_1(x) = \frac{dp_0(x)}{dx} \quad (1.54)$$

Recall then the coefficient $p_0''(x) - p_1'(x) + p_2(x)$ of $y(x)$ in eqn (1.53). Since the sum of the first two terms equals zero (to realize this, one needs to differentiate the relation in eqn (1.54)), it follows that the self-adjointness puts no additional constraints on the coefficient $p_2(x)$ in eqn (1.51). Hence, if eqn (1.51) is self-adjoint, it can be written in the form

$$p_0(x)\frac{d^2 y(x)}{dx^2} + \frac{dp_0(x)}{dx}\frac{dy(x)}{dx} + p_2(x)y(x) = 0$$

The first two terms in this equation can be combined, providing the following compact form

$$\frac{d}{dx}\left(p_0(x)\frac{dy(x)}{dx}\right) + p_2(x)y(x) = 0 \quad (1.55)$$

This form is usually referred to as the standard form of a self-adjoint equation of the second order.

1.2. SYMMETRY OF GREEN'S FUNCTIONS

Hence, if the coefficients $p_0(x)$ and $p_1(x)$ satisfy the relation in eqn (1.54), then eqn (1.51) is in a self-adjoint form, regardless of the form of the coefficient $p_2(x)$. This prompts an idea of how eqn (1.51) can be reduced to a self-adjoint form. Indeed, multiplying eqn (1.51) through by a certain nonzero function (the integrating factor) and applying then the relation in eqn (1.54) to the coefficients of $y''(x)$ and $y'(x)$ of the resultant equation, one can readily formulate a simple relation from which the integrating factor can afterwards be found. We leave the completion of this procedure for one of the exercises of this section.

Assume now that L represents a self-adjoint operator of the second order

$$L \equiv \frac{d}{dx}\left(p_0(x)\frac{d}{dx}\right) + p_2(x)$$

Consider two functions $u(x)$ and $v(x)$ both being two times continuously differentiable on (a, b), and form the following bilinear combination of them

$$u(x)\,L[v(x)] - v(x)\,L[u(x)]$$

which can be rewritten explicitly

$$u\left(\frac{d}{dx}\left(p_0(x)\frac{dv}{dx}\right) + p_2(x)\,v\right) - v\left(\frac{d}{dx}\left(p_0(x)\frac{du}{dx}\right) + p_2(x)\,u\right)$$

Removing the outer parentheses in both the components above and cancelling the terms $p_0(x)uv$, we have

$$uL(v) - vL(u) = u\frac{d}{dx}\left(p_0(x)\frac{dv}{dx}\right) - v\frac{d}{dx}\left(p_0(x)\frac{du}{dx}\right)$$

Applying the product rule to the exterior differentiation for both the terms above, one obtains

$$u\frac{d}{dx}\left(p_0(x)\frac{dv}{dx}\right) - v\frac{d}{dx}\left(p_0(x)\frac{du}{dx}\right) = \frac{d}{dx}\left(p_0(x)\left(u\frac{dv}{dx} - v\frac{du}{dx}\right)\right)$$

Hence, this bilinear combination is finally reduced to

$$u\,L(v) - v\,L(u) = \frac{d}{dx}\left(p_0(x)\left(u\frac{dv}{dx} - v\frac{du}{dx}\right)\right) \tag{1.56}$$

Integrating both sides of eqn (1.56) throughout the interval $[a, b]$, one obtains the following relation

$$\int_a^b [u\,L(v) - v\,L(u)]\,dx = p_0(x)\left(u\frac{dv}{dx} - v\frac{du}{dx}\right)\Big|_a^b \tag{1.57}$$

which is usually referred to as *Green's formula* for a self-adjoint operator.

If in addition to being two times continuously differentiable on (a, b), $u(x)$ and $v(x)$ are functions for which the right-hand side in eqn (1.57) vanishes. That is, when

$$p_0(x)\left(u\frac{dv}{dx} - v\frac{du}{dx}\right)\bigg|_a^b = 0 \qquad (1.58)$$

then Green's formula reduces to the very compact form

$$\int_a^b [u\,L(v) - v\,L(u)]\,dx = 0 \qquad (1.59)$$

So, the Green's formula in eqn (1.59) is valid for a self-adjoint operator L, with $u(x)$ and $v(x)$ being two times continuously differentiable on (a, b) satisfying the relation in eqn (1.58). This relation is, however, implicit in nature, which makes it too cumbersome to deal with over and over again in actual computation. Therefore, it is important to find some of its explicit equivalents which are more convenient to use in practice.

In doing so, we rewrite the relation in eqn (1.58) in the extended form

$$p_0(b)\,[u(b)v'(b) - v(b)u'(b)] - p_0(a)\,[u(a)v'(a) - v(a)u'(a)] = 0 \qquad (1.60)$$

Since this relation contains the values of $u(x)$, $v(x)$, and their derivatives $u'(x)$ and $v'(x)$ at the end-points of the interval $[a, b]$, it should be directly seen that the equality in eqn (1.60) holds, if $u(x)$ and $v(x)$ both satisfy either of the following sets of boundary conditions at $x = a$ and $x = b$:

- $y(a) = 0, \quad y(b) = 0$
- $y(a) = 0, \quad y'(b) = 0$
- $y'(a) = 0, \quad y'(b) = 0$

It should also be directly seen that the condition in eqn (1.60) is valid in the so-called *singular* case, when the leading coefficient $p_0(x)$ in eqn (1.55) equals zero at one of the end-points of $[a, b]$. In such a case we usually require $y(x)$ to be bounded at that end-point, with a value of either $y(x)$ or $y'(x)$ being prescribed at the other end-point, that is:

- $|y(a)| < \infty, \quad y(b) = 0$
- $|y(a)| < \infty, \quad y'(b) = 0$

In addition, from exercises 1.5(a)–1.5(e), it follows that the condition in eqn (1.60) holds also for both $u(x)$ and $v(x)$ satisfying one of the following sets of boundary conditions:

1.2. SYMMETRY OF GREEN'S FUNCTIONS

- $y(a) = 0, \quad y'(b) + hy(b) = 0$
- $y'(a) = 0, \quad y'(b) + hy(b) = 0$
- $y'(a) + h_1 y(a) = 0, \quad y'(b) + h_2 y(b) = 0$
- $y(a) = y(b), \quad p_0(a) y'(a) = p_0(b) y'(b)$
- $|y(a)| < \infty, \quad y'(b) + hy(b) = 0$

The last conditions set presumes that the leading coefficient $p_0(x)$ of eqn (1.55) equals zero at $x = a$.

Boundary value problems formulated for eqn (1.55) subject to either one of the sets of boundary conditions listed above, belong to the important class of the so-called *self-adjoint boundary value problems*.

To provide the definition of the self-adjointness for a boundary value problem, we consider an equation in the self-adjoint form

$$\frac{d}{dx}\left(p_0(x)\frac{dy(x)}{dx}\right) + p_2(x)y(x) = 0, \quad x \in (a,b) \tag{1.61}$$

subject to the boundary conditions

$$B_1(y;a,b) = 0, \quad B_2(y;a,b) = 0 \tag{1.62}$$

The boundary conditions in eqn (1.62) are supposed to match the relation in eqn (1.58) in the sense that for any allowable functions $u(x)$ and $v(x)$, each satisfying the conditions in eqn (1.62), the relation in eqn (1.58) is also satisfied.

With these assumptions in mind, we say that the formulation in eqns (1.61) and (1.62) represents a *self-adjoint boundary value problem*, if Green's formula in eqn (1.59) is valid for any two times continuously differentiable functions $u(x)$ and $v(x)$ satisfying the boundary conditions in eqn (1.62).

We now turn the reader's attention to the basic question in this section, that is, the symmetry of a Green's function. Based on the self-adjointness of a boundary value problem, we formulate the condition for a Green's function to be symmetric by the following theorem.

Theorem 1.2: If the boundary value problem in eqns (1.61) and (1.62) is self-adjoint and has only the trivial solution, then its Green's function $g(x, s)$ is symmetric, provided that $g(x, s) = g(s, x)$.

Proof. This proof is based on a slight modification of that procedure which has been used in the proof of Theorem 1.1 in Section 1.1. Here we also consider two linearly independent particular solutions $y_1(x)$ and $y_2(x)$

of the governing equation in eqn (1.61). But contrary to Theorem 1.1, we put some restrictions on them choosing these solutions in a special way.

First, let $y_1(x)$ satisfy the first boundary condition in eqn (1.62), and let $y_2(x)$ satisfy the second condition in eqn (1.62). Clearly, $y_1(x)$ cannot satisfy the second boundary condition in eqn (1.62) and $y_2(x)$ cannot satisfy the first of those conditions, because according to what we have stated earlier, the trivial solution is the only solution of eqn (1.61) satisfying both of the boundary conditions in eqn (1.62).

Second, based on $y_1(x)$ and $y_2(x)$, let us form the bilinear combination

$$y_1(x)\, L[y_2(x)] - y_2(x)\, L[y_1(x)]$$

which identically equals zero on (a, b), since $L[y_1(x)] \equiv 0$ and $L[y_2(x)] \equiv 0$ for $x \in (a, b)$.

Recalling the relation in eqn (1.56), derived earlier in this section, and rewriting it in terms of $y_1(x)$ and $y_2(x)$ yields

$$y_1\, L(y_2) - y_2\, L(y_1) = \frac{d}{dx}\left(p_0(x)\left(y_1\frac{dy_2}{dx} - y_2\frac{dy_1}{dx}\right)\right)$$

Hence, in the case of a self-adjoint boundary value problem, $y_1(x)$ and $y_2(x)$ must satisfy the following relation

$$\frac{d}{dx}\left(p_0(x)\left(y_1\frac{dy_2}{dx} - y_2\frac{dy_1}{dx}\right)\right) = 0$$

which implies

$$p_0(x)\left(y_1\frac{dy_2}{dx} - y_2\frac{dy_1}{dx}\right) = C \tag{1.63}$$

where C is a constant.

Notice that $y_1(x)$ and $y_2(x)$ are determined up to a constant multiple. Indeed, if $y_1(x)$, for example, satisfies both the governing equation in eqn (1.61) and the first boundary condition in eqn (1.62), then, for any nonzero constant α, the product $\alpha\, y_1(x)$ also satisfies both of these relations. This is equally true for $y_2(x)$. Hence, we can rewrite the relation in eqn (1.63) in the form

$$p_0(x)\left(y_1\frac{dy_2}{dx} - y_2\frac{dy_1}{dx}\right) = 1 \tag{1.64}$$

Thus, without losing generality, we can assume that $y_1(x)$ and $y_2(x)$ meet the condition in eqn (1.64) throughout (a, b). We will return to this point later in this section.

1.2. SYMMETRY OF GREEN'S FUNCTIONS

Fix now an arbitrary point $s \in (a,b)$ and express the Green's function $g(x,s)$ to the problem in eqns (1.61) and (1.62) in the form

$$g(x,s) = \begin{cases} c_1(s)\, y_1(x), & \text{for } a \leq x \leq s \\ c_2(s)\, y_2(x), & \text{for } s \leq x \leq b \end{cases} \qquad (1.65)$$

This function satisfies the boundary conditions in eqn (1.62) regardless of the values of $c_1(s)$ and $c_2(s)$. This occurs because $y_1(x)$ and $y_2(x)$ satisfy the first and second of those boundary conditions, respectively. Hence, $g(x,s)$ in the form of eqn (1.65) already meets properties 1 and 4 of the definition of Green's function.

By virtue of properties 2 and 3 of the definition, we obtain the following system of linear algebraic equations

$$\begin{pmatrix} y_2(s) & -y_1(s) \\ y_2'(s) & -y_1'(s) \end{pmatrix} \times \begin{pmatrix} c_2(s) \\ c_1(s) \end{pmatrix} = \begin{pmatrix} 0 \\ -p_0^{-1}(s) \end{pmatrix}$$

in $c_1(s)$ and $c_2(s)$. The coefficient matrix of this system is not singular, because its determinant $y_1(s)y_2'(s) - y_2(s)y_1'(s)$ is the Wronskian of the two linearly independent functions $y_1(s)$ and $y_2(s)$. Hence, this system has a unique solution which can be written in the form

$$c_1(s) = -\frac{y_2(s)}{p_0(s)\, W(s)}, \quad c_2(s) = -\frac{y_1(s)}{p_0(s)\, W(s)}$$

Upon substituting these values of $c_1(s)$ and $c_2(s)$ into eqn (1.65), one obtains, for the upper branch of the Green's function

$$g(x,s) = -\frac{y_1(x)\, y_2(s)}{p_0(s)\, W(s)}, \quad x \leq s \qquad (1.66)$$

while for the lower branch, we have

$$g(x,s) = -\frac{y_2(x)\, y_1(s)}{p_0(s)\, W(s)}, \quad s \leq x \qquad (1.67)$$

According to the relation in eqn (1.64), the denominator in eqns (1.66) and (1.67) meets the condition

$$p_0(s)\, W(s) \equiv p_0(s)\left(y_1(s)\frac{dy_2(s)}{dx} - y_2(s)\frac{dy_1(s)}{dx}\right) \equiv 1$$

In view of this fact, we can finally write the Green's function $g(x,s)$ for the self-adjoint boundary value problem posed by eqns (1.61) and (1.62) in the following symmetric compact form

$$g(x,s) = \begin{cases} -y_2(s)\,y_1(x), & \text{for } a \leq x \leq s \\ -y_1(s)\,y_2(x), & \text{for } s \leq x \leq b \end{cases} \quad (1.68)$$

From this representation, it follows that $g(x,s)$ is invariant to the interchange of x with s. In other words, the Green's function is symmetric in the case of a self-adjoint boundary value problem.

In the next section, we will return to the basic issue of this chapter, which is the construction of Green's functions. One more procedure available for this construction in the current literature will be discussed in detail.

1.3 Method of variation of parameters

So far in this text, we have discussed boundary value problems stated for homogeneous equations with homogeneous boundary conditions imposed.

In this section, we will recall [17, 30, 31, 63, 66, 68, 70] an important theorem that establishes a theoretical background for the utilization of Green's functions in solving boundary value problems for nonhomogeneous equations and boundary conditions. Then we will present the procedure for construction of Green's functions, which is based on that theorem and Lagrange's method of variation of parameters. As it is known, Lagrange's method is traditionally used to analytically solve nonhomogeneous linear differential equations if the fundamental solution set is available for the corresponding homogeneous equation.

Consider a boundary value problem for the nonhomogeneous equation

$$L(y) \equiv p_0(x)y^{(n)} + p_1(x)y^{(n-1)} + \ldots + p_{n-1}(x)y' + p_n(x)y = -f(x) \quad (1.69)$$

on (a,b), subject to the homogeneous boundary conditions written as

$$M_k(y; a, b) \equiv \sum_{j=0}^{n-1} \left(\alpha_j^k y^{(j)}(a) + \beta_j^k y^{(j)}(b) \right) = 0, \quad (k = \overline{1,n}) \quad (1.70)$$

where the coefficients $p_i(x)$ are continuous functions on (a,b), with $p_0(x) \neq 0$ and $M_k(y; a, b)$ represent linearly independent forms with constant coefficients.

Suppose the problem posed by eqns (1.69) and (1.70) has a unique solution. This consequently implies that the corresponding homogeneous boundary value problem has only the trivial solution.

1.3. METHOD OF VARIATION OF PARAMETERS

The following theorem supports a direct method for solving boundary value problems that are formulated for nonhomogeneous equations subject to homogeneous boundary conditions.

Theorem 1.3: If $g(x,s)$ represents the Green's function of the homogeneous boundary value problem corresponding to that posed by eqns (1.69) and (1.70), then the unique solution of the problem in eqns (1.69) and (1.70) itself can be expressed by the integral

$$y(x) = \int_a^b g(x,s) f(s) \, ds \qquad (1.71)$$

Proof. It is clear that two independent facts need to be proven. First, that the integral in eqn (1.71) satisfies the equation (1.69), and second, that it satisfies the boundary conditions in eqn (1.70).

Since the Green's function $g(x,s)$ is defined in pieces, we break down the integral in eqn (1.71) into two integrals as shown

$$y(x) = \int_a^x g(x,s) f(s) \, ds + \int_x^b g(x,s) f(s) \, ds \qquad (1.72)$$

In order to differentiate $y(x)$, we should take into account its occurrence. Indeed, it is defined in terms of integrals containing a parameter and having variable limits. Therefore, one has to recall from calculus [68] that if

$$I(x) = \int_{a(x)}^{b(x)} F(x,s) ds$$

then the derivative of this function is written as

$$\frac{dI(x)}{dx} = F(x,b(x))b'(x) - F(x,a(x))a'(x) + \int_{a(x)}^{b(x)} F'_x(x,s) ds \qquad (1.73)$$

Hence, since both of the integrals in eqn (1.72) contain x as a parameter and their limits depend on x, we obtain

$$y'(x) = \int_a^x g'_x(x,s) f(s) ds + g(x, x-0) f(x)$$

$$+ \int_x^b g'_x(x,s) f(s) ds - g(x, x+0) f(x)$$

The above integrals can be combined and nonintegral terms are eliminated due to the continuity of the Green's function as $x = s$. This yields

$$y'(x) = \int_a^b g'_x(x,s) f(s) ds$$

Recalling the continuity of the derivatives of the Green's function up to the $(n-2)$-nd order included as $x=s$ (see property 2 of the definition), the higher order derivatives of the integral in eqn (1.72) up to the $(n-1)$-st order included can be computed analogously to the first derivative

$$y^{(j)}(x) = \int_a^b g_x^{(j)}(x,s)f(s)ds, \quad (j=\overline{2,n-1})$$

Thus, the boundary conditions in eqn (1.70) are satisfied with $y(x)$ expressed by eqn (1.71), since all the differentiations in $M_k(y;a,b)$ can be carried out under the integration sign.

Additionally, in order to substitute $y(x)$ into eqn (1.69), we compute its n-th order derivative

$$y^{(n)}(x) = \int_a^b g_x^{(n)}(x,s)f(s)ds + [g_x^{(n-1)}(x,x-0) - g_x^{(n-1)}(x,x+0)]\,f(x)$$

which, in accordance with property 3 of the definition of Green's function, yields

$$y^{(n)}(x) = \int_a^b g_x^{(n)}(x,s)f(s)ds - f(x)p_0^{-1}(x)$$

Upon substituting $y(x)$ and its derivatives found above into eqn (1.69) and grouping all the integral terms together, one finally obtains

$$\int_a^b L[g(x,s)]\,f(s)\,ds - f(x) = -f(x)$$

The above equality is an identity, since $L[g(x,s)]=0$ on (a,b). Thus, the theorem is proved.

Based on this theorem, we describe below one more approach which can be used for the construction of Green's functions. The idea behind this approach is to employ Lagrange's method of variation of parameters which is traditionally used to solve nonhomogeneous linear differential equations. For the sake of simplicity, we again consider a boundary value problem for the equation of the second order

$$L[y(x)] \equiv p_0(x)\frac{d^2y(x)}{dx^2} + p_1(x)\frac{dy(x)}{dx} + p_2(x)y(x) = -f(x) \quad (1.74)$$

subject to the simplest set of boundary conditions

$$y(a) = 0, \quad y(b) = 0 \quad (1.75)$$

Assume the above boundary value problem has a unique solution or, in other words, the corresponding homogeneous problem has only the trivial solution. Let $y_1(x)$ and $y_2(x)$ represent two linearly independent particular

1.3. METHOD OF VARIATION OF PARAMETERS

solutions of the homogeneous equation associated with that in eqn (1.74). Express then the general solution of eqn (1.74) in the form

$$y(x) = C_1(x)y_1(x) + C_2(x)y_2(x) \qquad (1.76)$$

where $C_1(x)$ and $C_2(x)$ are differentiable functions to be found in what follows.

The relation in eqn (1.74) represents the only relation on (a, b) available at this point for determining $C_1(x)$ and $C_2(x)$. This presumes a certain degree of freedom in choosing the second relation. Lagrange's method provides the most effective and elegant choice of such a relation.

The direct substitution of $y(x)$ from eqn (1.76) into eqn (1.74) would result in a cumbersome single differential equation of the second order in two unknown functions $C_1(x)$ and $C_2(x)$. In order to avoid such an unfortunate complication, the procedure in Lagrange's method suggests as follows. First differentiate eqn (1.76) using the product rule

$$y'(x) = C_1'(x)y_1(x) + C_1(x)y_1'(x) + C_2'(x)y_2(x) + C_2(x)y_2'(x)$$

and, keeping in mind the degree of freedom mentioned above, we make then a simplifying assumption

$$C_1'(x)y_1(x) + C_2'(x)y_2(x) = 0 \qquad (1.77)$$

resulting in

$$y'(x) = C_1(x)y_1'(x) + C_2(x)y_2'(x) \qquad (1.78)$$

Hence, the second derivative of $y(x)$ is now expressed as follows

$$y''(x) = C_1'(x)y_1'(x) + C_1(x)y_1''(x) + C_2'(x)y_2'(x) + C_2(x)y_2''(x) \qquad (1.79)$$

Substitute $y(x)$, $y'(x)$, and $y''(x)$ from eqns (1.76), (1.78), and (1.79), respectively, into eqn (1.74). This yields

$$p_0(C_1'y_1' + C_1y_1'' + C_2'y_2' + C_2y_2'') + p_1(C_1y_1' + C_2y_2')$$

$$+ p_2(C_1y_1 + C_2y_2) = -f(x)$$

Rearranging the order of terms, it follows that

$$C_1(p_0y_1'' + p_1y_1' + p_2y_1) + C_2(p_0y_2'' + p_1y_2' + p_2y_2)$$

$$+ p_0(C_1'y_1' + C_2'y_2') = -f(x)$$

Since $y_1(x)$ and $y_2(x)$ represent particular solutions of the homogeneous equation associated with eqn (1.74), the first and second terms in the left-hand side above are zero. This yields

$$C_1'(x)y_1'(x) + C_2'(x)y_2'(x) = -f(x)p_0^{-1}(x) \qquad (1.80)$$

Solving eqns (1.77) and (1.80) simultaneously, we obtain

$$C_1'(x) = -\frac{y_2(x)f(x)}{p_0(x)W(x)}, \quad C_2'(x) = \frac{y_1(x)f(x)}{p_0(x)W(x)}$$

where $W(x) = y_1(x)y_2'(x) - y_2(x)y_1'(x)$ is the Wronskian of the fundamental solution set $y_1(x)$ and $y_2(x)$. Straightforward integration of the derivatives $C_1'(x)$ and $C_2'(x)$ yields

$$C_1(x) = -\int_a^x \frac{y_2(s)f(s)}{p_0(s)W(s)}ds + H_1, \quad C_2(x) = \int_a^x \frac{y_1(s)f(x)}{p_0(s)W(s)}ds + H_2$$

Substituting these values of $C_1(x)$ and $C_2(x)$ into eqn (1.76), we take $y_1(x)$ and $y_2(x)$ inside of the integrals and combine then the two integrals together to obtain

$$y(x) = \int_a^x \frac{y_1(s)y_2(x) - y_1(x)y_2(s)}{p_0(s)W(s)} + H_1 y_1(x) + H_2 y_2(x) \qquad (1.81)$$

Let us satisfy the boundary conditions in eqn (1.75) with $y(x)$ as expressed above. This yields the following system of linear equations

$$\begin{pmatrix} y_1(a) & y_2(a) \\ y_1(b) & y_2(b) \end{pmatrix} \times \begin{pmatrix} H_1 \\ H_2 \end{pmatrix} = \begin{pmatrix} 0 \\ P(a,b) \end{pmatrix} \qquad (1.82)$$

in the coefficients H_1 and H_2, where $P(a,b)$ is defined as

$$P(a,b) = \int_a^b \frac{R(b,s)}{p_0(s)W(s)} f(s) ds$$

where $R(b,s) = y_1(b)y_2(s) - y_1(s)y_2(b)$. This provides the solution to the system in eqn (1.82) in the form

$$H_1 = -\int_a^b \frac{y_2(a)R(b,s)f(s)}{p_0(s)R(a,b)W(s)}ds, \quad H_2 = \int_a^b \frac{y_1(a)R(b,s)f(s)}{p_0(s)R(a,b)W(s)}ds$$

Upon substituting these quantities into eqn (1.81), we readily obtain the solution of the boundary value problem in eqns (1.74) and (1.75) as

$$y(x) = -\int_a^x \frac{R(x,s)f(s)}{p_0(s)W(s)}ds + \int_a^b \frac{R(a,x)R(b,s)f(s)}{p_0(s)R(a,b)W(s)}ds$$

1.3. METHOD OF VARIATION OF PARAMETERS

Combining then the above integrals in the form of a single integral, we finally obtain

$$y(x) = \int_a^b K(x,s)\,f(s)\,ds \qquad (1.83)$$

Clearly, the kernel function $K(x,s)$ in this integral representation can be written, for the range $x \leq s$, in the form

$$K(x,s) = \frac{R(a,x)R(b,s)}{p_0(s)R(a,b)W(s)}$$

while for the range $x \geq s$, one easily obtains

$$K(x,s) = \frac{R(a,x)R(b,s) - R(x,s)R(a,b)}{p_0(s)R(a,b)W(s)}$$

After a trivial but quite cumbersome transformation, the above expression can be simplified as follows

$$K(x,s) = \frac{R(a,s)R(b,x)}{p_0(s)R(a,b)W(s)}$$

Thus, by virtue of Theorem 1.3, the kernel function $K(x,s)$ in eqn (1.83) does in fact represent the Green's function to the problem formulated in eqns (1.74) and (1.75). So, the suggested approach can be used for practical construction of Green's functions as an alternative method to that described in Section 1.1. We present below a number of examples illustrating some peculiar points of this approach in practical situations.

EXAMPLE 1 Consider the following integral

$$I(x) = \int_{x^2}^{x} (x+s)^2 ds \qquad (1.84)$$

containing a parameter and having variable limits. To find its derivative $I'(x)$, let us apply the formula from eqn (1.73). This yields

$$\frac{dI(x)}{dx} = (x+x)^2 - (x+x^2)^2 2x + \int_{x^2}^{x} 2(x+s)\,ds$$

Integrating then and carrying out a trivial transformation, one finally obtains

$$\frac{dI(x)}{dx} = 7x^2 - 4x^3 - 5x^4 - 2x^5$$

The same expression for $I'(x)$ can be obtained by directly integrating eqn (1.84) and differentiating the result afterwards.

Recall one more integral of the considered feature

$$I(x) = \int_0^x (x-s)^2 ds$$

In an attempt to compute the derivative $I'(x)$ of this integral, one realizes that the value of the integrand $(x-s)^2$ equals zero if $x=s$. The derivative $I'(x)$ in this case can consequently be obtained by formal differentiation under the integral sign

$$I'(x) = \int_0^x 2(x-s) ds = x^2$$

Such a situation is usually applicable when we use Lagrange's method for the construction of Green's functions.

EXAMPLE 2 To demonstrate in detail the procedure of the method of variation of parameters being applied to the construction of Green's functions, we formulate a boundary value problem for the nonhomogeneous equation

$$\frac{d^2 y(x)}{dx^2} + k^2 y(x) = -f(x), \quad x \in (0, a) \tag{1.85}$$

with the homogeneous boundary conditions imposed as

$$y'(0) = 0, \quad y'(a) = 0 \tag{1.86}$$

We assume that the right-hand side function $f(x)$ in eqn (1.85) is continuous throughout (a, b).

It can easily be shown (see exercise 1.8(a)) that the homogeneous problem ($f(x)=0$) associated with that in eqns (1.85) and (1.86) has only the trivial solution. This implies that the conditions of existence and uniqueness of the Green's function are met for this problem.

Since the functions $y_1(x) \equiv \sin(kx)$ and $y_2(x) \equiv \cos(kx)$ represent the fundamental solution set for the homogeneous equation corresponding to that in eqn (1.85), it follows that the general solution for eqn (1.85) can be expressed as

$$y(x) = C_1(x) \sin(kx) + C_2(x) \cos(kx) \tag{1.87}$$

The system of linear algebraic equations in $C_1'(x)$ and $C_2'(x)$, which has been derived in eqns (1.77) and (1.80), in this case is

$$\begin{pmatrix} \sin(kx) & \cos(kx) \\ k\cos(kx) & -k\sin(kx) \end{pmatrix} \times \begin{pmatrix} C_1'(x) \\ C_2'(x) \end{pmatrix} = \begin{pmatrix} 0 \\ -f(x) \end{pmatrix}$$

1.3. METHOD OF VARIATION OF PARAMETERS

providing us with the following solution

$$C_1'(x) = \frac{1}{k}\cos(kx)f(x), \quad C_2'(x) = -\frac{1}{k}\sin(kx)f(x)$$

Integrating these expressions, one obtains

$$C_1(x) = \int_0^x \frac{1}{k}\cos(ks)f(s)ds + H_1, \quad C_2(x) = -\int_0^x \frac{1}{k}\sin(ks)f(s)ds + H_2$$

Upon substituting these values into eqn (1.87) and carrying out an obvious transformation, we obtain

$$y(x) = \int_0^x \frac{1}{k}\sin k(x-s)f(s)ds + H_1\sin(kx) + H_2\cos(kx) \qquad (1.88)$$

To determine the values of H_1 and H_2, let us first differentiate $y(x)$ above

$$y'(x) = \int_0^x \cos k(x-s)f(s)ds + H_1 k\cos(kx) - H_2 k\sin(kx)$$

In performing this differentiation, it happens that the nonintegral terms, which can be seen in eqn (1.73), do not appear in this case because of the specific form of the integrand $\sin k(x-s)$ which vanishes as $x = s$ (see the comment provided in *EXAMPLE 1*).

It follows then that from the first condition $y'(0) = 0$ in eqn (1.86), we have $H_1 = 0$, while the second condition $y'(a) = 0$ yields

$$\int_0^a \cos k(a-s)f(s)ds - H_2 k\sin(ka) = 0$$

from which it immediately follows that

$$H_2 = \int_0^a \frac{\cos k(a-s)}{k\sin(ka)} f(s)ds$$

Upon substituting the values of H_1 and H_2 just found into eqn (1.88) and correspondingly grouping the integrals in it, one obtains

$$y(x) = \int_0^x \frac{\sin k(x-s)}{k}f(s)ds + \int_0^a \cos(kx)\frac{\cos k(a-s)}{k\sin(ka)}f(s)ds \qquad (1.89)$$

Both of the above integrals can be combined and written in the form of a compact single integral. In doing so, we add formally the term

$$\int_x^a 0 \cdot f(s)ds$$

to the first of the two integrals in eqn (1.89) and break down the second of the integrals as
$$\int_0^a \cos(kx)\frac{\cos k(a-s)}{k\sin(ka)}f(s)ds$$
$$= \int_0^x \cos(kx)\frac{\cos k(a-s)}{k\sin(ka)}f(s)ds + \int_x^a \cos(kx)\frac{\cos k(a-s)}{k\sin(ka)}f(s)ds.$$

This finally yields
$$y(x) = \int_0^a K(x,s)f(s)ds \qquad (1.90)$$

whose kernel $K(x,s)$ is defined in pieces as
$$K(x,s) = \frac{1}{k\sin(ka)}\begin{cases} \cos(kx)\cos k(a-s), & \text{for } x \le s \\ \cos(ks)\cos k(a-x), & \text{for } s \le x \end{cases} \qquad (1.91)$$

Hence, in light of Theorem 1.3, it follows that, since the solution of the boundary value problem stated with eqns (1.85) and (1.86) is expressed as the integral representation in eqn (1.90), its kernel $K(x,s)$ in eqn (1.91) represents the Green's function to the homogeneous problem associated with that occurring in eqns (1.85) and (1.86).

EXAMPLE 3 Consider again the boundary value problem for the nonhomogeneous equation
$$\frac{d^2y(x)}{dx^2} - k^2 y(x) = -f(x) \qquad (1.92)$$

subject to the homogeneous boundary conditions
$$y'(0) = 0, \quad y(a) = 0 \qquad (1.93)$$

The homogeneous ($f(x)=0$) boundary value problem corresponding to that of eqns (1.92) and (1.93) has only the trivial solution (see exercise 1.8(b)). This justifies the existence and uniqueness of the Green's function.

As we have already mentioned in Section 1.1, the following set of functions
$$y_1(x) \equiv \exp(kx), \quad y_2(x) \equiv \exp(-kx)$$

represents the fundamental solution set for the homogeneous equation corresponding to that in eqn (1.92). Hence, the general solution of eqn (1.92) can be represented by
$$y(x) = C_1(x)\exp(kx) + C_2(x)\exp(-kx) \qquad (1.94)$$

1.3. METHOD OF VARIATION OF PARAMETERS

In compliance with the procedure of Lagrange's method, one obtains the values of $C_1(x)$ and $C_2(x)$ in the form

$$C_1(x) = -\int_0^x \frac{1}{2k} \exp(-ks)f(s)ds + H_1, \quad C_2(x) = \int_0^x \frac{1}{2k} \exp(ks)f(s)ds + H_2$$

The substitution of these into eqn (1.94) yields

$$y(x) = -\int_0^x \frac{1}{k} \sinh k(x-s)f(s)ds + H_1 \exp(kx) + H_2 \exp(-kx) \quad (1.95)$$

The first boundary condition $y'(0)=0$ in eqn (1.93) implies that $H_1 = H_2$, while the second condition $y(a)=0$ yields

$$H_1 = H_2 = \int_0^a \frac{\sinh k(a-s)}{2k \cosh(ka)} f(s)ds$$

Substituting these into eqn (1.95), one obtains

$$y(x) = \int_0^a \frac{\cosh(kx)\sinh k(a-s)}{k\cosh(ka)} f(s)ds - \int_0^x \frac{1}{k}\sinh k(x-s)f(s)ds$$

Hence, following again the same transformation of the above integrals as in the previous example, the Green's function $g(x,s)$ to the homogeneous boundary value problem associated with that of eqns (1.92) and (1.93) is expressed as

$$g(x,s) = \frac{1}{k\cosh(ka)} \begin{cases} \cosh(kx)\sinh k(a-s), & \text{for } x \leq s \\ \cosh(ks)\sinh k(a-x), & \text{for } s \leq x \end{cases} \quad (1.96)$$

EXAMPLE 4 Let us again consider eqn (1.92), but subject to a different set of boundary conditions. Namely, we consider the case

$$y'(0) - hy(0) = 0, \quad |y(\infty)| < \infty \quad (1.97)$$

This example is designed to show how Lagrange's method manages to treat the boundedness conditions of the type that occurred in eqn (1.97).

It can be easily shown (see exercise 1.8(c)) that there exists a unique Green's function for the homogeneous boundary value problem corresponding to that posed by eqns (1.92) and (1.97).

The general solution of eqn (1.92) is derived in eqn (1.95). In this case, however, we prefer its pure exponential form

$$y(x) = \int_0^x \frac{1}{2k}[e^{k(s-x)} - e^{k(x-s)}]f(s)ds + H_1 e^{kx} + H_2 e^{-kx} \quad (1.98)$$

rather than the mixed hyperbolic-exponential form used in eqn (1.95). The above form is more practical in view of the necessity to treat the boundedness conditions in the discussion that follows. Indeed, splitting off both of the exponential terms under the integral sign and grouping then together both of the terms containing $\exp(kx)$ and both of the terms containing $\exp(-kx)$, we rewrite eqn (1.98) in the form

$$y(x) = \left(H_1 - \int_0^x \frac{e^{-ks}}{2k} f(s)ds \right) e^{kx} + \left(H_2 + \int_0^x \frac{e^{ks}}{2k} f(s)ds \right) e^{-kx} \quad (1.99)$$

It is clearly seen that the condition of boundedness $|y(\infty)|<\infty$ implies that the coefficient of the positive exponential term $\exp(kx)$ in eqn (1.99) must equal zero. This yields

$$H_1 = \int_0^\infty \frac{1}{2k} \exp(-ks) f(s) ds$$

while the first condition in eqn (1.97) subsequently yields

$$H_2 = \frac{k-h}{k+h} H_1 = \int_0^\infty \frac{k-h}{2k(k+h)} \exp(-ks) f(s) ds$$

Upon substituting the values of H_1 and H_2 just found into eqn (1.98) and rewriting then its first term again in a more compact hyperbolic form, we obtain

$$y(x) = -\int_0^x \frac{1}{k} \sinh k(x-s) f(s) ds$$

$$+ \int_0^\infty \frac{1}{2k} \exp(-ks) \left(\exp(kx) + h^* \exp(-kx) \right) f(s) ds$$

where $h^* = (k-h)/(k+h)$. From this representation it follows that the Green's function $g(x,s)$ to the problem in eqns (1.92) and (1.97) is written as

$$g(x,s) = \frac{1}{2k} \begin{cases} \exp(-ks) \left(\exp(kx) + h^* \exp(-kx) \right), & \text{for } x \leq s \\ \exp(-kx) \left(\exp(ks) + h^* \exp(-ks) \right), & \text{for } s \leq x \end{cases} \quad (1.100)$$

EXAMPLE 5 Here we construct the Green's function for the homogeneous boundary value problem associated with the following equation

$$\frac{d^4 y(x)}{dx^4} - 2k^2 \frac{d^2 y(x)}{dx^2} + k^4 y(x) = -f(x) \quad (1.101)$$

subject to the following boundary conditions

$$y(0) = 0, \quad y'(0) = 0, \quad |y(\infty)|<\infty, \quad |y'(\infty)|<\infty \quad (1.102)$$

1.3. METHOD OF VARIATION OF PARAMETERS

The existence and uniqueness of the Green's function for the problem above is justified by exercise 1.8(d).

Since the characteristic equation

$$m^4 - 2k^2 m^2 + k^4 = 0$$

associated with eqn (1.101) has two pairs of repeated roots: $m_{1,2} = k$ and $m_{3,4} = -k$, the general solution for eqn (1.101) can be expressed as

$$y(x) = C_1(x)e^{kx} + C_2(x)e^{-kx} + C_3(x)xe^{kx} + C_4(x)xe^{-kx} \qquad (1.103)$$

The coefficient matrix for the system of linear equations in $C_i'(x), (i = 1, 2, 3, 4)$ is obtained in this case as

$$\begin{pmatrix} e^{kx} & e^{-kx} & xe^{kx} & xe^{-kx} \\ ke^{kx} & -ke^{-kx} & (1+kx)e^{kx} & (1-kx)e^{-kx} \\ k^2 e^{kx} & k^2 e^{-kx} & k(2+kx)e^{kx} & -k(2-kx)e^{-kx} \\ k^3 e^{kx} & -k^3 e^{-kx} & k^2(3+kx)e^{kx} & k^2(3-kx)e^{-kx} \end{pmatrix}$$

while the right-hand side vector is $(0, 0, 0, -f(x))^T$. Clearly, the above matrix is not singular, because its determinant represents the Wronskian of the fundamental solution set used in eqn (1.103). Therefore, one can readily obtain

$$C_1'(x) = \frac{1+kx}{4k^3} e^{-kx} f(x), \quad C_2'(x) = -\frac{1-kx}{4k^3} e^{kx} f(x)$$

$$C_3'(x) = -\frac{1}{4k^2} e^{-kx} f(x), \quad C_4'(x) = -\frac{1}{4k^2} e^{kx} f(x)$$

Upon integrating these, the functions $C_i(x)$ are found in the form

$$C_1(x) = \int_0^x \frac{1+ks}{4k^3} e^{-ks} f(s) ds + H_1, \quad C_2(x) = -\int_0^x \frac{1-ks}{4k^3} e^{ks} f(s) ds + H_2$$

$$C_3(x) = -\int_0^x \frac{1}{4k^2} e^{-ks} f(s) ds + H_3, \quad C_4(x) = -\int_0^x \frac{1}{4k^2} e^{ks} f(s) ds + H_4$$

Hence, $y(x)$ in eqn (1.103), with $C_i(x)$ just obtained, provides the general solution to eqn (1.101). The constants of integration H_i are to be computed when satisfying the boundary conditions in eqn (1.102). Before going any further with this, for the better clarity in the development that follows, we first differentiate $y(x)$ in eqn (1.103) by using the product rule

$$y'(x) = ke^{kx} C_1(x) + e^{kx} \underbrace{\left(\frac{1+kx}{4k^3} e^{-kx} f(x) \right)}$$

$$- ke^{-kx}C_2(x) + \underbrace{e^{-kx}\left(-\frac{1-kx}{4k^3}e^{kx}f(x)\right)}$$

$$+ (1+kx)e^{kx}C_3(x) + \underbrace{xe^{kx}\left(-\frac{1}{4k^2}e^{-kx}f(x)\right)}$$

$$+ (1-kx)e^{-kx}C_4(x) + \underbrace{xe^{-kx}\left(-\frac{1}{4k^2}e^{kx}f(x)\right)}$$

Clearly, the sum of the underbraced terms happens to vanish. Substituting then the values of $C_i(x)$ into the remaining part of $y'(x)$, one obtains

$$y'(x) = ke^{kx}\left(\int_0^x \frac{1+ks}{4k^3}e^{-ks}f(s)ds + H_1\right)$$

$$+ ke^{-kx}\left(-\int_0^x \frac{1-ks}{4k^3}e^{ks}f(s)ds + H_2\right)$$

$$+ (1+kx)e^{kx}\left(-\int_0^x \frac{1}{4k^2}e^{-ks}f(s)ds + H_3\right)$$

$$+ (1-kx)e^{-kx}\left(-\int_0^x \frac{1}{4k^2}e^{ks}f(s)ds + H_4\right).$$

Let us now return to the boundary conditions imposed by eqn (1.102). The first of them $y(0)=0$ yields

$$H_1 + H_2 = 0 \tag{1.104}$$

while the second $y'(0)=0$ results in

$$kH_1 + H_3 - kH_2 + H_4 = 0 \tag{1.105}$$

The boundedness conditions $|y(\infty)|<\infty$ and $|y'(\infty)|<\infty$ formulated in eqn (1.102) provide

$$H_1 = -\int_0^\infty \frac{1+ks}{4k^3}e^{-ks}f(s)ds, \quad H_3 = \int_0^\infty \frac{1}{4k^2}e^{-ks}f(s)ds$$

Substituting these into eqns (1.104) and (1.105), one obtains

$$H_2 = \int_0^\infty \frac{1+ks}{4k^3}e^{-ks}f(s)ds, \quad H_4 = \int_0^\infty \frac{1+2ks}{4k^2}e^{-ks}f(s)ds$$

Upon substituting the values of H_1, H_2, H_3, and H_4 into eqn (1.103) and combining then the like integrals, one finally obtains

$$y(x) = \int_0^x \left(\frac{1-k(x-s)}{4k^3}e^{k(x-s)} - \frac{1+k(x-s)}{4k^3}e^{-k(x-s)}\right)f(s)ds$$

1.3. METHOD OF VARIATION OF PARAMETERS

$$+ \int_0^\infty \left(\frac{1+k(x+s)+2k^2xs}{4k^3} e^{-k(x+s)} - \frac{1-k(x-s)}{4k^3} e^{k(x-s)} \right) f(s) ds.$$

This representation for $y(x)$ can be rewritten as a single integral

$$y(x) = \int_0^\infty K(x,s) f(s) ds$$

whose kernel-function $K(x,s)$ is found in the form

$$K(x,s) = \frac{1}{4k^3} \begin{cases} (1+k(x+s)+2k^2xs)e^{-k(x+s)} - (1-k(x-s))e^{k(x-s)} \\ (1+k(x+s)+2k^2xs)e^{-k(x+s)} - (1-k(s-x))e^{k(s-x)} \end{cases} \quad (1.106)$$

where the upper branch is defined for $x \leq s$, while the lower branch is defined for $s \leq x$. Hence, in view of Theorem 1.3, the kernel-function $K(x,s)$ represents the Green's function for the boundary value problem posed by eqns (1.101) and (1.102).

EXAMPLE 6 Based on a corresponding Green's function and on Theorem 1.3, let us solve a boundary value problem for the nonhomogeneous equation

$$\frac{d^4 y(x)}{dx^4} = -\sqrt{x} \quad (1.107)$$

subjected to the homogeneous boundary conditions

$$y(0) = \frac{dy(0)}{dx} = 0, \quad y(1) = \frac{d^2 y(1)}{dx^2} = 0 \quad (1.108)$$

The Green's function for the homogeneous problem associated with that in eqns (1.107) and (1.108) was derived in Section 1.1 (see eqn (1.46)). Its branch valid for $x \leq s$ was found as

$$g(x,s) = -\left(\frac{1}{4}s^3 - \frac{3}{4}s^2 + \frac{1}{2}s\right)x^2 + \left(\frac{1}{12}s^3 - \frac{1}{4}s^2 + \frac{1}{6}\right)x^3$$

Due to the self-adjointness of the original statement, the second branch of $g(x,s)$ that is valid for $x \geq s$, can be obtained from that above by interchanging x with s.

Hence, in compliance with Theorem 1.3, the solution of the problem in eqns (1.107) and (1.108) can be found by the straightforward integration

$$y(x) = -\int_0^1 g(x,s)(-\sqrt{s}) ds$$

In order to evaluate this integral, we recall the piecewise format of the Green's function $g(x,s)$ and break the integral down as

$$y(x) = \int_0^x \left((\frac{x^3}{12} - \frac{x^2}{4} + \frac{1}{6})s^3 - (\frac{x^3}{4} - \frac{3x^2}{4} + \frac{x}{2})s^2 \right) \sqrt{s}\,ds$$

$$+ \int_x^1 \left((\frac{s^3}{12} - \frac{s^2}{4} + \frac{1}{6})x^3 - (\frac{s^3}{4} - \frac{3s^2}{4} + \frac{s}{2})x^2 \right) \sqrt{s}\,ds \qquad (1.109)$$

Clearly, since the lower branch in $g(x,s)$ is defined for the range of variables $s \leq x$, it must serve as the kernel of the first integral in eqn (1.109), while the upper branch serves as the kernel of the second integral in eqn (1.109).

Computation of the integrals in eqn (1.109) is a routine procedure, upon accomplishing which we finally obtain

$$y(x) = \frac{11x^3}{189} - \frac{13x^2}{315} - \frac{16}{945}x^4\sqrt{x}$$

The straightforward integration of the type shown above is especially efficient if we are required to compute a number of solutions of the same boundary value problem with a variety of different right-hand side functions. In such a situation, the direct use of Theorem 1.3 provides a notable computational convenience.

As the reader can easily realize, the material in this chapter touches upon classical boundary value problems for ODEs, for which the Green's function formalism is well developed and its use is a quite straightforward procedure. Later in this text, we will present a new sphere of possible applications where this method can be extended. That is, in Chapter 4, we will extend the Green's function formalism to a specific class of problems, which is frequently applicable in various areas of applied mechanics. The so-called *multi-point posed boundary value problems* for some systems of ordinary differential equations will be treated by means of the Green's function method.

Chapters 2, 3, and 4 describe direct applications of the material of this chapter to some problems of applied mechanics modeled by ODEs. In Chapters 5 through 8, boundary value problems for PDEs of applied mechanics are considered. Influence functions and matrices for such problems are constructed there by reducing to corresponding boundary value problems for ODEs. So, the material of this chapter is essential for the entire text.

1.4 Review Exercises

1.1 Determine whether or not the following boundary value problems have only the trivial solution:

(a) $y''(x) = 0$, with $y'(0) = 0$ and $y'(a) + my(a) = 0$;

(b) $y''(x) - k^2 y(x) = 0$, with $y(0) = 0$ and $|y(\infty)| < \infty$;

(c) $y''(x) - k^2 y(x) = 0$, with $y(0) = y(a)$ and $y'(0) = y'(a)$;

(d) $((mx + p)y'(x))' = 0$, with $y'(0) = 0$ and $y(a) = 0$;

(e) $(xy'(x))' = 0$, with $|y(0)| < \infty$ and $y'(a) + hy(a) = 0$;

(f) $y''(x) + y'(x) - 2y(x) = 0$, with $y(0) = 0$ and $|y(\infty)| < \infty$;

(g) $y''(x) + y'(x) = 0$, with $y'(0) = 0$ and $y'(a) = 0$;

(h) $y''''(x) = 0$, with $y''(0) = y'''(0) = 0$ and $y''(a) = y'''(a) = 0$.

1.2 Construct Green's functions for the following boundary value problems on the indicated interval:

(a) $y''(x) = 0$, with $y(0) = 0$ and $y'(a) = 0$;

(b) $y''(x) = 0$, with $y(0) = 0$ and $y'(a) + hy(a) = 0$, $(h \geq 0)$. Show that if $h = 0$, the Green's function for this problem reduces to that in exercise 1.2(a);

(c) $y''(x) = 0$, with $y'(0) - h_1 y(0) = 0$ and $y'(a) + h_2 y(a) = 0$, when h_1 and h_2 both are not zero. Show that if $h_1 = 0$, the Green's function for this problem reduces to that of *EXAMPLE 3* in Section 1.1;

(d) $((mx + p)y'(x))' = 0$, with $y(0) = 0$ and $y(a) = 0$, when $m > 0$ and $p > 0$;

(e) $(\exp(\beta x) y'(x))' = 0$, with $y(0) = 0$ and $y(a) = 0$;

(f) $(\exp(\beta x) y'(x))' = 0$, with $y(0) = 0$ and $y'(a) = 0$;

(g) $y''(x) + k^2 y(x) = 0$, with $y(0) = 0$ and $y(a) = 0$;

(h) $y^{IV}(x) = 0$ for $x \in (0,1)$, with $y(0) = y'(0) = 0$ and $y''(1) = y'''(1) = 0$.

1.3 Determine whether the following equations are in a self-adjoint form:

(a) $y''(x) + k^2 y(x) = 0$;

(b) $x^2 y''(x) + 2xy'(x) - (x^2 - 1)y(x) = 0$;

(c) $x^2 y''(x) - 2xy'(x) + y(x) = 0$;

(d) $y''(x) + 3y'(x) + 9y(x) = 0$;

(e) $\sin^2(x) y''(x) + \sin(2x) y'(x) - y(x) = 0$.

1.4 By introducing integrating factors, reduce the following differential equations to a self-adjoint form:

(a) $y''(x) - 2y'(x) + 4y(x) = 0$;

(b) $y''(x) + x y'(x) - x^2 y(x) = 0$;

(c) $x^2 y''(x) - x y'(x) + y(x) = 0$;

(d) $x^2 y''(x) + x y'(x) - y(x) = 0$.

1.5 Determine whether the following boundary value problems are self-adjoint:

(a) $y''(x) + y(x) = 0$, with $y(a) = 0$ and $y'(b) + hy(b) = 0$;

(b) $y''(x) - y(x) = 0$, with $y'(a) = 0$ and $y'(b) + hy(b) = 0$;

(c) $x y''(x) + y'(x) - y(x) = 0$, $y'(a) + h_1 y(a) = 0$, $y'(b) + h_2 y(b) = 0$;

(d) $x y''(x) + y'(x) = 0$, with $y(a) = y(b)$ and $a y'(a) = b y'(b)$;

(e) $(x-a) y''(x) + y'(x) - y(x) = 0$, with $|y(a)| < \infty$ and $y'(b) + hy(b) = 0$;

(f) $y''(x) + y(x) = 0$, with $y'(0) + y(0) + y(a) = 0$ and $y'(0) - y(0) + y'(a) = 0$.

1.6 Construct the Green's function for the following problem

$$y''(x) + 3y'(x) - 10y(x) = 0, \quad y(0) = 0, \quad |y(\infty)| < \infty$$

Reduce this problem to a self-adjoint form and construct the Green's function again. Observe how this reducing affects the symmetry of the Green's function.

1.7 Construct the Green's function for the problem

$$y''(x) + k^2 y(x) = 0, \quad y(0) = 0, \quad y'(1) = 0$$

by the approach discussed in the proofs of Theorem 1.1. Compare this to the method used in Theorem 1.2.

1.8 Determine whether the following boundary value problems have only the trivial solution:

(a) $y''(x) + k^2 y(x) = 0$, with $y'(0) = y'(a) = 0$;

(b) $y''(x) - k^2 y(x) = 0$, with $y'(0) = y(a) = 0$;

1.4. REVIEW EXERCISES

(c) $y''(x) - k^2 y(x) = 0$, with $y'(0) - hy(0) = 0$, $|y(\infty)| < \infty$;

(d) $y^{IV}(x) - 2k^2 y''(x) + k^4 y(x) = 0$, with $y(0) = y'(0) = 0$, $|y(\infty)| < \infty$, $|y'(\infty)| < \infty$.

1.9 Use Lagrange's method to construct Green's functions for the following boundary value problems:

(a) $(xy'(x))' = 0$, with $|y(0)| < \infty$, $y(a) = 0$;

(b) $y''(x) - k^2 y(x) = 0$, with $y'(0) - hy(0) = 0$, $y(a) = 0$;

(c) $y^{IV}(x) = 0$, with $y(0) = y'(0) = 0$, $y(1) = y'(1) = 0$;

(d) $y^{IV}(x) = 0$, with $y(0) = y''(0) = 0$, $y(1) = y''(1) = 0$.

1.10 Based on Theorem 1.3, compute solutions for the following boundary value problems by utilizing corresponding Green's functions:

(a) $y''(x) + y(x) = \exp(2x)$, with $y(a) = 0$, $y'(b) + hy(b) = 0$;

(b) $y''(x) - y(x) = x$, with $y'(a) = 0$, $y'(b) + hy(b) = 0$;

(c) $y''(x) + 2y'(x) + y(x) = \sin x$, $y'(a) = 0$, $y'(b) = 0$;

(d) $y^{IV}(x) = x^2$, with $y(0) = y'(0) = 0$, $y(1) = y'(1) = 0$;

(e) $y^{IV}(x) - 2y''(x) + y(x) = 1$, with $y(0) = y'(0) = 0$, $|y(\infty)| < \infty$, $|y'(\infty)| < \infty$.

Chapter 2

Influence Functions for Beams

In this book, the reader will find and enjoy a variety of applications of the influence (Green's) function method to some basic classical problems in applied mechanics. We will be involved here with problems that are traditionally discussed in *mechanics of materials, strength of materials, structural analysis,* and relevant courses. Some nontraditional implementations of this method are also discussed.

The idea of utilizing influence functions for solving beam problems within the scope of Kirchhoff's theory can be associated with the so-called *singularity method* and with the concept of *influence lines* (see, for example, [10, 18, 27, 64]). The methodology of the singularity method was developed many years ago. However, it has never been suggested as a universal approach to all of the problem classes from beam theory discussed in this book.

This chapter opens a systematic debate on possible ways of utilizing the influence function method in computational mechanics. We will bring to the reader's attention one of the most traditional applications of this method. Namely, the linear bending of elastic Kirchhoff beams will be considered, as they undergo various combinations of transverse concentrated and transverse distributed forces as well as bending moments.

It is shown here that in the computational procedure based on this method, the solution of any problem of a considered class can be as standard as either analytical or numerical computation of a definite integral.

We precede the forthcoming discussion in this chapter with a description of the construction procedures for influence functions for a beam of uniform flexural rigidity, with various types of edge conditions imposed. We then show how influence functions may be utilized for computation of the most important components of a stress-strain state of a beam subjected to various types of loading. The influence function's formalism is further extended to a beam resting on a simple elastic foundation. And finally, we show how this approach can be adjusted to a beam of variable flexural rigidity.

2.1 Single span beams of uniform rigidity

A number of influence functions will be constructed here for a simple, single span beam whose cross-section has a uniform flexural rigidity. We assume that the beam is composed of a homogeneous elastic isotropic material and is subjected to various combinations of edge conditions.

To construct the Green's function, which is identified with the influence function to be found, we will use the slightly modified version of the classical method utilized in the proof of Theorem 1.2 in Section 1.2 of Chapter 1. That proof was applied to equations of the second order. We will here extend it to equations of higher order. As before, the method of variation of parameters will also be used to construct influence functions.

The bending of beams is considered here within the scope of the classical Kirchhoff theory. For a beam of length a, this theory [18, 67], yields a boundary value problem for the so-called Euler–Bernoulli equation

$$\frac{d^2}{dx^2}\left(EI(x)\frac{d^2w(x)}{dx^2}\right) = q(x), \quad x \in (0, a) \tag{2.1}$$

where $w(x)$ represents the beam's deflection, $E(x)$ is the elastic modulus of the material of which the beam is composed, $I(x)$ is the moment of inertia of the cross-section x (the product $EI(x)$ is usually referred to as the flexural rigidity of the beam), and $q(x)$ is the transverse load applied to the beam.

To ensure a unique solution, eqn (2.1) must be subjected to an appropriate set of boundary conditions at the end-points of the interval $[0, a]$. These conditions are usually formulated in terms of the deflection function $w(x)$. We present them here in general form

$$B_{0,i}[w(0)] = 0, \quad B_{a,i}[w(a)] = 0, \quad i = 1, 2 \tag{2.2}$$

from which, for a beam in statics, any natural set of edge conditions (clamped – clamped, clamped – free, clamped – simply supported, clamped – sliding, clamped – elastically supported, simply supported – sliding, simply supported – elastically supported, simply supported – simply supported, elastically supported – elastically supported, and elastically supported – sliding) can be obtained as a particular case.

The operators $B_{0,i}$ and $B_{a,i}$ can specify either clamped edge condition preventing the beam from rotation and displacement

$$w = 0, \quad \frac{dw}{dx} = 0,$$

or simply supported edge condition preventing displacement but allowing rotation (the bending moment is zero at the simple support)

$$w = 0, \quad M = EI\frac{d^2w}{dx^2} = 0,$$

2.1. SINGLE SPAN BEAMS OF UNIFORM RIGIDITY

or free edge condition corresponding to no geometric restraint or load (the bending moment and shear force are zero at the free edge)

$$M = 0, \quad Q = EI\frac{d^3w}{dx^3} = 0,$$

or sliding edge condition preventing the beam from rotation but allowing displacement (the shear force is zero at the sliding edge)

$$\frac{dw}{dx} = 0, \quad Q = EI\frac{d^3w}{dx^3} = 0,$$

or elastically supported edge condition allowing rotation but restraining an arbitrary displacement by posing the linear relation between the shear force and displacement (the bending moment is also zero at the edge)

$$M = 0, \quad EI\frac{d^3w}{dx^3} + k_0 w = 0.$$

The coefficient k_0 above represents the constant of elastic support.

Let $g(x,s)$ represent the Green's function of the homogeneous equation

$$\frac{d^2}{dx^2}\left(EI(x)\frac{d^2w(x)}{dx^2}\right) = 0, \quad x \in (0, a)$$

corresponding to that of eqn (2.1), subjected to the boundary conditions in eqn (2.2). In accordance with Theorem 1.3 from Section 1.3, the solution $w(x)$ of the problem stated by eqns (2.1) and (2.2) can be expressed in terms of $g(x,s)$ as

$$w(x) = -\int_0^a g(x,s) q(s) ds, \quad x \in [0, a] \tag{2.3}$$

To comprehend the physical interpretation of the Green's function $g(x,s)$, let the beam be subjected to a single transverse force of magnitude P_0 concentrated at an arbitrary point $x = s_0$. The right-hand term in eqn (2.1) can subsequently be written in this case as a scalar multiple of the generalized Dirac delta function

$$q(x) = P_0 \delta(x - s_0)$$

yielding as follows

$$w(x) = -\int_0^a g(x,s) P_0 \delta(s - s_0) ds = -P_0 g(x, s_0) \tag{2.4}$$

The essential properties of the Dirac delta function, which are required to follow the material of this book, can be found, for example, in [30, 31,

66]. The rigorous mathematical approach to the theory of this function can be found in specialized sources on generalized functions.

The transformation in eqn (2.4) has been carried out by virtue of the property of the Dirac delta function, which states that if a function $f(x)$ is bounded on the interval $[a, b]$ and has a finite number of local maxima and minima on it (such a function is usually referred to as the *function of a limited variation* [31]), then the following relation holds

$$\int_0^b f(s)\, \delta(s - s_0)\, ds = f(s_0)$$

In view of the relation in eqn (2.4), the Green's function $g(x, s)$ is usually called *the influence function of a transverse concentrated unit force*, because it actually determines the deflection of a beam at the point x, caused by a transverse unit force applied to the point s.

If the flexural rigidity $EI(x)$ of the beam does not vary with x, that is, EI is constant, then we can take it out of the differentiation sign in eqn (2.1). This substantially simplifies equation (2.1) providing

$$\frac{d^4 w(x)}{dx^4} = \frac{q(x)}{EI}, \quad x \in (0, a) \tag{2.5}$$

At this point, we recall the modification of the classical method for the construction of Green's functions, which was used in Section 1.2 for analysis of the symmetry of Green's functions. We will now extend this modification to the class of boundary value problems posed by eqn (2.2) and the homogeneous equation of the fourth order

$$\frac{d^4 w(x)}{dx^4} = 0, \quad x \in (0, a) \tag{2.6}$$

corresponding to that in eqn (2.5).

We assume that the boundary value problem posed by eqns (2.6) and (2.2) has only the trivial solution. This implies that there exists its unique Green's function. The latter is identified with the influence function for the associated elastic beam.

Let two functions $w_1(x)$ and $w_2(x)$ represent particular solutions of eqn (2.6). We assume, in addition, that both $w_1(x)$ and $w_2(x)$ satisfy the edge conditions imposed at $x=0$ by the first relation in eqn (2.2). Let $w_3(x)$ and $w_4(x)$ represent another two particular solutions of eqn (2.6), and assume that both $w_3(x)$ and $w_4(x)$ satisfy the edge conditions imposed at $x=a$ by the second relation in eqn (2.2).

Assume also that the set $w_i(x)$, $(i = \overline{1,4})$ is linearly independent on $[0, a]$. This implies that they form the fundamental solution set for eqn

2.1. SINGLE SPAN BEAMS OF UNIFORM RIGIDITY

(2.6). Based on this set, we seek the Green's function to the problem in eqns (2.2) and (2.6) in the form

$$g(x,s) = \begin{cases} a_1(s)w_1(x) + a_2(s)w_2(x), & \text{for } x \leq s \\ b_1(s)w_3(x) + b_2(s)w_4(x), & \text{for } s \leq x \end{cases} \quad (2.7)$$

From this representation, it follows that the entire set of boundary conditions in eqn (2.2) is satisfied by $g(x,s)$ in this form, regardless of the values of the coefficients $a_i(s)$ and $b_i(s)$, $(i=1,2)$. This occurs because the upper branch in $g(x,s)$ is a linear combination of $w_1(x)$ and $w_2(x)$, each of which satisfies the boundary conditions at $x=0$, while the lower branch is a linear combination of $w_3(x)$ and $w_4(x)$, satisfying respectively the boundary conditions at $x=a$. Hence, $g(x,s)$ in eqn (2.7) meets properties 1 and 4 in the definition of the Green's function.

To compute the coefficients $a_i(s)$ and $b_i(s)$ in eqn (2.7), we take advantage of the remaining defining properties of the Green's function. In compliance with property 2, one recalls that $g(x,s)$ is continuous as $x=s$, that is

$$g(s+0,s) - g(s-0,s) = 0$$

This consequently yields

$$w_3(s)b_1(s) + w_4(s)b_2(s) - w_1(s)a_1(s) - w_2(s)a_2(s) = 0 \quad (2.8)$$

According to the same property 2, the first derivative of $g(x,s)$ with respect to x is also continuous as $x=s$

$$g'_x(s+0,s) - g'_x(s-0,s) = 0$$

resulting in

$$w'_3(s)b_1(s) + w'_4(s)b_2(s) - w'_1(s)a_1(s) - w'_2(s)a_2(s) = 0 \quad (2.9)$$

The second derivative $g''_x(x,s)$ is also continuous as $x=s$

$$g''_x(s+0,s) - g''_x(s-0,s) = 0$$

Hence,

$$w''_3(s)b_1(s) + w''_4(s)b_2(s) - w''_1(s)a_1(s) - w''_2(s)a_2(s) = 0 \quad (2.10)$$

And finally, in compliance with property 3, the third derivative $g'''_x(x,s)$ is discontinuous as $x=s$, providing

$$g'''_x(s+0,s) - g'''_x(s-0,s) = -1$$

This yields as follows

$$w_3'''(s)b_1(s) + w_4'''(s)b_2(s) - w_1'''(s)a_1(s) - w_2'''(s)a_2(s) = -1 \quad (2.11)$$

Clearly, the relations in eqns (2.8)–(2.11) constitute a system of linear algebraic equations in $a_i(s)$ and $b_i(s)$. It is well-posed, because the determinant of its coefficient matrix represents the Wronskian for a set of linearly independent functions. Thus, upon solving this system and substituting the values of $a_i(s)$ and $b_i(s)$ into eqn (2.7), we complete the construction procedure for the influence function $g(x,s)$ of the beam whose equilibrium is modeled by the boundary value problem in eqns (2.2) and (2.6).

In the series of instructive examples below we discuss some key points of the algorithm for the construction of influence functions for beams with various types of edge conditions imposed.

EXAMPLE 1 Consider a beam of length a with both edges clamped. To construct the influence function for such a beam, let us formulate the boundary value problem written as

$$w(0) = \frac{dw(0)}{dx} = 0, \quad w(a) = \frac{dw(a)}{dx} = 0 \quad (2.12)$$

for the homogeneous equation (2.6).

It can easily be seen that the homogeneous boundary value problem posed by eqns (2.6) and (2.12) has only the trivial solution. Hence, there exists the unique Green's function of this problem, which actually represents the influence function of a transverse concentrated unit force for the clamped beam shown in Figure 2.1.

Notice that for problems allowing natural physical interpretation, one can draw on intuition to decide whether the corresponding boundary value problem has a unique Green's function. Speaking of the present problem, for example, it is intuitively clear that the clamped beam uniquely responds to a concentrated transverse force regardless of the point of its application. This observation indirectly justifies the existence and uniqueness of the Green's function of the problem under consideration.

As we recall from *EXAMPLE 8* of Section 1.1 in Chapter 1, the fundamental solution set for eqn (2.6) can be exhibited in the simplest form

$$w_1(x) \equiv 1, \quad w_2(x) \equiv x, \quad w_3(x) \equiv x^2, \quad w_4(x) \equiv x^3 \quad (2.13)$$

However, in view of the foregoing discussion, we would rather take advantage of a different system of functions

$$\begin{array}{ll} w_1(x) \equiv x^2, & w_2(x) \equiv x^3 \\ w_3(x) \equiv (x-a)^2, & w_4(x) \equiv (x-a)^3 \end{array} \quad (2.14)$$

2.1. SINGLE SPAN BEAMS OF UNIFORM RIGIDITY

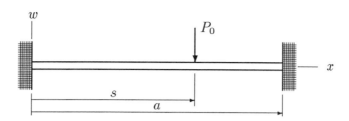

Figure 2.1 A beam subjected to a concentrated force

which can also be utilized as the fundamental solution set for eqn (2.6). Indeed, each of the functions in eqn (2.14), being a polynomial of degree less than or equal to three, represents a particular solution of eqn (2.6). Moreover, the Wronskian of this system

$$\begin{vmatrix} x^2 & x^3 & (x-a)^2 & (x-a)^3 \\ 2x & 3x^2 & 2(x-a) & 3(x-a)^2 \\ 2 & 6x & 2 & 6(x-a) \\ 0 & 6 & 0 & 6 \end{vmatrix} \equiv 12a^4$$

does not identically equal zero. This consequently implies that the functions in eqn (2.14) are indeed linearly independent.

To easily follow the development in this section, it is important to comprehend the principal distinction between the two fundamental solution sets presented in eqns (2.13) and (2.14). What makes the choice of the second of them more promissory in view of the construction of the influence function? The answer to this question can be given in conjunction with the boundary conditions imposed in eqn (2.12).

The point is that, whereas the components of the system in eqn (2.13) are not at all associated with the boundary conditions in eqn (2.12), both $w_1(x)$ and $w_2(x)$ in eqn (2.14) satisfy the conditions imposed at $x=0$ in eqn (2.12), while both $w_3(x)$ and $w_4(x)$ satisfy the boundary conditions at $x=a$. As a result, the actual computing of the influence function promises to be more compact, because there will be no need to directly treat the boundary conditions when going through the construction procedure.

We now seek the Green's function $g(x,s)$ of the boundary value problem in eqns (2.6) and (2.12), which is the influence function for the clamped beam, in the following form

$$g(x,s) = \begin{cases} a_1(s)x^2 + a_2(s)x^3, & \text{for } x \leq s \\ b_1(s)(x-a)^2 + b_2(s)(x-a)^3, & \text{for } s \leq x \end{cases} \quad (2.15)$$

From the preceding representation, it follows that the boundary conditions in eqn (2.12) are satisfied by $g(x,s)$ in this form, because the upper branch in $g(x,s)$ is a linear combination of $w_1(x)$ and $w_2(x)$, each satisfying the boundary conditions at $x=0$, while the lower branch is a linear combination of $w_3(x)$ and $w_4(x)$, satisfying the boundary conditions at $x=a$.

Hence, $g(x,s)$ in eqn (2.15) meets the defining properties 1 and 4 in the definition of the Green's function.

To compute the values of the coefficients $a_i(s)$ and $b_i(s)$, we take advantage of the remaining defining properties of the Green's function. In compliance with property 2 in the definition, it follows that

$$(s-a)^2 b_1(s) + (s-a)^3 b_2(s) - s^2 a_1(s) - s^3 a_2(s) = 0 \qquad (2.16)$$

According to the same property 2, the first derivative of $g(x,s)$ with respect to x is also continuous as $x=s$, providing

$$2(s-a)b_1(s) + 3(s-a)^2 b_2(s) - 2s a_1(s) - 3s^2 a_2(s) = 0 \qquad (2.17)$$

The second derivative $g''_x(x,s)$ is also continuous as $x=s$. This consequently implies that

$$2b_1(s) + 6(s-a)b_2(s) - 2a_1(s) - 6s a_2(s) = 0 \qquad (2.18)$$

And finally, in compliance with property 3, the third derivative $g'''_x(x,s)$ is discontinuous as $x=s$, yielding

$$6b_2(s) - 6a_2(s) = -1 \qquad (2.19)$$

Grouping then equations (2.16)–(2.19), we obtain the following well-posed system of linear algebraic equations

$$\begin{pmatrix} (s-a)^2 & (s-a)^3 & -s^2 & -s^3 \\ 2(s-a) & 3(s-a)^2 & -2s & -3s^2 \\ 2 & 6(s-a) & -2 & -6s \\ 0 & 6 & 0 & -6 \end{pmatrix} \times \begin{pmatrix} b_1(s) \\ b_2(s) \\ a_1(s) \\ a_2(s) \end{pmatrix} = \begin{pmatrix} 0 \\ 0 \\ 0 \\ -1 \end{pmatrix}$$

in the coefficients $a_i(s)$ and $b_i(s)$ of $g(x,s)$ in eqn (2.15). The solution of this system is found to be

$$a_1(s) = -\frac{s(s-a)^2}{2a^2}, \quad a_2(s) = \frac{(s-a)^2(2s+a)}{6a^3},$$

2.1. SINGLE SPAN BEAMS OF UNIFORM RIGIDITY

$$b_1(s) = \frac{s^2(s-a)}{2a^2}, \quad b_2(s) = \frac{s^2(2s-3a)}{6a^3}.$$

Upon substituting these values into eqn (2.15) and performing some routine algebra, one finally obtains the influence function for the clamped beam in the following form

$$g(x,s) = \frac{1}{6a^3} \begin{cases} x^2(s-a)^2[2s(x-a)+a(x-s)], & \text{for } x \leq s \\ s^2(x-a)^2[2x(s-a)+a(s-x)], & \text{for } s \leq x \end{cases} \quad (2.20)$$

Thus, $g(x,s)$ represents the deflection of the clamped beam at the point x, associated with the concentrated unit force applied to the point s. This implies that $g(x,s)$ determines the response of the clamped beam, at the point x, to the unit force concentrated at the point s. In specialized literature dealing with influence (Green's) functions, x is often referred to as the *observation* or *field* point, while s is referred to as the *source* point.

The representation appeared in eqn (2.20) is symmetric, that is $g(x,s) = g(s,x)$. As we know from Chapter 1, this reflects the self-adjointness of the boundary value problem in eqns (2.6) and (2.12) for which $g(x,s)$ is the Green's function. In some sources (see, for example, [31]), this property is called Maxwell's reciprocity. From the mechanics standpoint, it can be read as: *the response of the beam at x due to a transverse concentrated force at s, is the same as the response at s due to a force at x*.

In order to run the procedure for the construction of Green's function, a set of n linearly independent particular solutions is required for the governing differential equation of the n-th order. Recalling that modification of the classical method, which has been used in the foregoing discussion, let us highlight one peculiarity in it.

The reader has perhaps already grasped the significance of finding the pairs of linearly independent particular solutions for the governing equation, which *a priori* satisfy appropriate boundary conditions. Of course, it was not hard to find such solutions for the example recently completed, because the boundary conditions of a clamped edge can be readily visualized.

This aspect, however, is not trivial in more complicated cases. The question remains, how the appropriate choice of such particular solutions can be carried out in more complex situations when boundary conditions are more complicated. Addressing this issue, we consider the next example where the universal approach is proposed for creating appropriate fundamental solution sets.

EXAMPLE 2 Let us construct the influence function for the beam of length a, whose left edge is elastically supported, while the right edge is clamped (Figure 2.2).

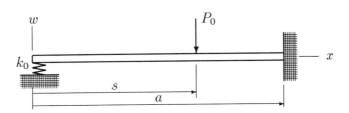

Figure 2.2 An elastically supported – clamped beam

It is intuitively obvious that there exists a unique influence (Green's) function in this case. In other words, the beam should uniquely respond to the transverse concentrated force P_0 regardless of the point of its application. The reader, of course, can easily check it out by directly solving the corresponding homogeneous boundary value problem for eqn (2.6).

From the beam theory [18, 67], it follows that the edge conditions of elastic support for the beam can be formulated in the form

$$\frac{d^2w(0)}{dx^2} = 0, \quad EI\frac{d^3w(0)}{dx^3} + k_0 w(0) = 0 \qquad (2.21)$$

where k_0 represents the elastic constant of the support.

Clearly, as k_0 approaches zero, the conditions in eqn (2.21) reduce to the classical free edge conditions, whereas with k_0 going to infinity, we face the simply supported edge.

Conditions at the clamped edge $x = a$, in turn, are obviously

$$w(a) = 0, \quad \frac{dw(a)}{dx} = 0 \qquad (2.22)$$

Thus, the influence function that we are looking for represents the Green's function of the boundary value problem posed by eqns (2.6), (2.21), and (2.22).

Clearly, the pair of linearly independent particular solutions $w_3(x)$ and $w_4(x)$ for equation (2.6), satisfying the edge conditions at $x = a$ in eqn (2.22), as we have suggested within the preceding example, can again be taken in the form

$$w_3(x) \equiv (x-a)^2, \quad w_4(x) \equiv (x-a)^3 \qquad (2.23)$$

It turns out, however, that the choice of the two other particular solutions $w_1(x)$ and $w_2(x)$ for eqn (2.6), which satisfy the edge conditions that

2.1. SINGLE SPAN BEAMS OF UNIFORM RIGIDITY

appeared in eqn (2.21), is not that obvious. Therefore, we show below how such a choice can be made on the formal basis.

Namely, to compute one of these solutions, say $w_1(x)$, we formulate the following *initial* value problem

$$\frac{d^4 w_1(x)}{dx^4} = 0, \quad x > 0 \tag{2.24}$$

$$\frac{d^2 w_1(0)}{dx^2} = 0, \quad EI\frac{d^3 w_1(0)}{dx^3} + k_0 w_1(0) = 0 \tag{2.25}$$

$$\frac{dw_1(0)}{dx} = 0, \quad w_1(0) = 1 \tag{2.26}$$

whose solution is easily within reach by the standard approach. Let us write the general solution of eqn (2.24) in the form

$$w_1(x) = C_1 + C_2 x + C_3 x^2 + C_4 x^3$$

The coefficients C_i above can be obtained upon satisfying the initial conditions in eqn (2.25) and (2.26). Thus, finally we have

$$w_1(x) = 1 - kx^3, \quad k = \frac{k_0}{6EI} \tag{2.27}$$

Clearly, $w_1(x)$, being a solution of the problem in eqns (2.24)–(2.26), represents a particular solution of eqn (2.6) and satisfies the edge conditions of elastic support at the left edge of the beam (see the relations in eqn (2.25)). The relations in eqn (2.26) may vary. These are imposed just to provide the unique solvability for $w_1(x)$.

To obtain the second particular solution, $w_2(x)$, of eqn (2.6), which satisfies the boundary conditions in eqn (2.21), we formulate one more initial value problem as follows

$$\frac{d^4 w_2(x)}{dx^4} = 0, \quad x > 0$$

$$\frac{d^2 w_2(0)}{dx^2} = 0, \quad EI\frac{d^3 w_2(0)}{dx^3} + k_0 w_2(0) = 0$$

$$\frac{dw_2(0)}{dx} = 1, \quad w_2(0) = 0$$

Upon utilizing the standard procedure for the above problem, one obtains its solution as

$$w_2(x) = x \tag{2.28}$$

Notice that the set of particular solutions of eqn (2.6) presented in (2.23), (2.27), and (2.28) is linearly independent on $[0, a]$. This allows the Green's function to the boundary value problem in eqns (2.6), (2.21), and (2.22) (that is the influence function of the beam whose left edge is elastically supported and the right is clamped) to be sought in the form

$$g(x,s) = \begin{cases} a_1(s)x + a_2(s)(1-kx^3), & \text{for } x \leq s \\ b_1(s)(x-a)^2 + b_2(s)(x-a)^3, & \text{for } s \leq x \end{cases} \quad (2.29)$$

Since the above representation holds for properties 1 and 4 of the definition of the Green's function, we satisfy the continuity and discontinuity conditions at $x = s$ (see properties 2 and 3 of the definition). This yields the following well-posed system of linear algebraic equations

$$\begin{pmatrix} (s-a)^2 & (s-a)^3 & -s & ks^3-1 \\ 2(s-a) & 3(s-a)^2 & -1 & 3ks^2 \\ 2 & 6(s-a) & 0 & 6ks \\ 0 & 6 & 0 & 6k \end{pmatrix} \times \begin{pmatrix} b_1(s) \\ b_2(s) \\ a_1(s) \\ a_2(s) \end{pmatrix} = \begin{pmatrix} 0 \\ 0 \\ 0 \\ -1 \end{pmatrix}$$

in the coefficients $a_1(s)$, $a_2(s)$, $b_1(s)$ and $b_2(s)$ of the representation occurring in eqn (2.29). The solution of this system is found as

$$a_1(s) = -\frac{(s-a)^2(ka^2s-1)}{2(1+2ka^3)}, \quad b_1(s) = \frac{(s-a)[kas(s+a)+1]}{2(1+2ka^3)}$$

$$a_2(s) = \frac{(s-a)^2(s+2a)}{6(1+2ka^3)}, \quad b_2(s) = \frac{1+2ks^3-3ks(s^2-a^2)}{6(1+2ka^3)}$$

Upon substituting these values into eqn (2.29) and performing some routine algebraic transformations, one finally obtains the influence function of the transverse unit force for the beam under consideration in the following form

$$g(x,s) = \frac{1}{6(1+2ka^3)} \begin{cases} (s-a)^2[3x(1-ka^2s)+(kx^3-1)(s+2a)], & x \leq s \\ (x-a)^2[3s(1-ka^2x)+(ks^3-1)(x+2a)], & s \leq x \end{cases} \quad (2.30)$$

The reader should notice that this representation is symmetric. This fact reflects the self-adjointness of the boundary value problem in eqns (2.6), (2.21), and (2.22), for which the above is the Green's function.

2.1. SINGLE SPAN BEAMS OF UNIFORM RIGIDITY

Two particular cases follow from eqn (2.30). If the coefficient k_0 of elastic support in the second of the boundary conditions in eqn (2.21) approaches zero, then they reduce to the free edge conditions

$$\frac{d^2w(0)}{dx^2}=0, \quad \frac{d^3w(0)}{dx^3}=0$$

and, consequently, the influence function in eqn (2.30), as $k_0=0$, is transformed into the influence function

$$g(x,s)=\frac{1}{6}\begin{cases}(s-a)^2[2(x-a)+(x-s)], & x\leq s\\ (x-a)^2[2(s-a)+(s-x)], & s\leq x\end{cases}$$

for a cantilever beam whose right edge is fixed.

One more particular case of the influence function in eqn (2.30) can readily be derived. That is, if the value of k_0 approaches infinity, then one obtains

$$g(x,s)=\frac{1}{12a^3}\begin{cases}x(s-a)^2[s(a^2-x^2)-2a(x^2-as)], & x\leq s\\ s(x-a)^2[x(a^2-s^2)-2a(s^2-ax)], & s\leq x\end{cases}$$

the influence function for the beam whose left edge is simply supported while the right edge is clamped.

In our last example in this section, we will use a different approach for the construction of influence functions. Instead of the procedure based on the defining properties, the method of variation of parameters will be employed for the cantilever beam.

EXAMPLE 3 Consider the following boundary value problem

$$\frac{d^4w(x)}{dx^4}=-q(x), \quad x\in(0,a) \tag{2.31}$$

$$w(0)=\frac{dw(0)}{dx}=0, \quad \frac{d^2w(a)}{dx^2}=\frac{d^3w(a)}{dx^3}=0 \tag{2.32}$$

that models the bending of the cantilever beam of length a, subjected to the transverse load which is directly proportional to $q(x)$.

And again, it is intuitively clear that there exists a unique influence function for this beam. That is, this beam uniquely responds to a transverse concentrated force (see Figure 2.3).

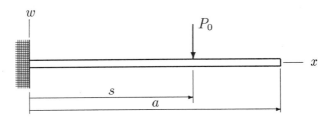

Figure 2.3 A cantilever beam subjected to a point force

The influence function of a unit transverse concentrated force for the beam shown above represents the Green's function for the homogeneous boundary value problem corresponding to that posed by eqns (2.31) and (2.32). To construct this function in compliance with the procedure of Lagrange's method of variation of parameters described in Section 1.3 of Chapter 1, we seek the general solution of eqn (2.31) in the form

$$w(x) = C_1(x) + C_2(x)x + C_3(x)x^2 + C_4(x)x^3 \qquad (2.33)$$

The system of linear algebraic equations in $C_i'(x)$, $(i=1,2,3,4)$, which follows from the standard procedure of Lagrange's method, in this case appears in the form

$$\begin{pmatrix} 1 & x & x^2 & x^3 \\ 0 & 1 & 2x & 3x^2 \\ 0 & 0 & 2 & 6x \\ 0 & 0 & 0 & 6 \end{pmatrix} \times \begin{pmatrix} C_1'(x) \\ C_2'(x) \\ C_3'(x) \\ C_4'(x) \end{pmatrix} = \begin{pmatrix} 0 \\ 0 \\ 0 \\ -q(x) \end{pmatrix}$$

A triangular structure of the coefficient matrix makes the solution of this system as simple as a backward substitution. Thus, we have

$$C_1'(x) = \frac{x^3}{6}q(x), \quad C_2'(x) = -\frac{x^2}{2}q(x)$$

$$C_3'(x) = \frac{x}{2}q(x), \quad C_4'(x) = -\frac{1}{6}q(x)$$

Hence, values of the coefficients $C_i(x)$ themselves can be obtained by integrating $C_i'(x)$. This yields

$$C_1(x) = \int_0^x \frac{s^3}{6}q(s)ds + H_1, \quad C_2(x) = -\int_0^x \frac{s^2}{2}q(s)ds + H_2$$

2.1. SINGLE SPAN BEAMS OF UNIFORM RIGIDITY

$$C_3(x) = \int_0^x \frac{s}{2} q(s) ds + H_3, \quad C_4(x) = -\int_0^x \frac{1}{6} q(s) ds + H_4$$

Upon substituting these values into eqn (2.33) and then grouping all of the integral terms, one obtains

$$w(x) = \int_0^x \frac{(s-x)^3}{6} q(s) ds + H_1 + H_2 x + H_3 x^2 + H_4 x^3 \qquad (2.34)$$

To compute the values of the coefficients H_i, we take advantage of the boundary conditions in eqn (2.32). In doing so, the condition $w(0) = 0$ yields $H_1 = 0$, while $w'(0) = 0$ results in $H_2 = 0$. Satisfying the boundary conditions at the right edge $x = a$, one obtains

$$H_3 = -\int_0^a \frac{s}{2} q(s) ds, \quad H_4 = \int_0^a \frac{1}{6} q(s) ds$$

Thus, the final expression for the solution to the problem in eqns (2.31) and (2.32) is found in the form

$$w(x) = \int_0^x \frac{(s-x)^3}{6} q(s) ds + \int_0^a \frac{x^2(x-3s)}{6} q(s) ds \qquad (2.35)$$

This representation of the deflection function can now be rewritten in the form of a single integral

$$w(x) = \int_0^a K(x,s) q(s) ds, \quad x \in [0, a] \qquad (2.36)$$

where the kernel function $K(x,s)$ is defined in pieces as follows

$$K(x,s) = -\frac{1}{6} \begin{cases} x^2(3s - x), & \text{for } x \leq s \\ s^2(3x - s), & \text{for } s \leq x \end{cases} \qquad (2.37)$$

By virtue of Theorem 1.3 in Chapter 1, it follows that $K(x,s)$ does in fact represent the Green's function of the homogeneous boundary value problem corresponding to that in eqns (2.31) and (2.32). That is, $K(x,s)$ represents the influence function of a transverse concentrated unit force for the cantilever beam clamped at the left edge (see Figure 2.3).

In this section one can find everything required for developing a proper understanding of the influence function concept. The material herein is also helpful in the actual constructing of such functions for single span beams. The only thing the reader still misses is the experience which can be gained by going through a set of good exercises. Therefore, the reader is advised to work through each of the exercises in this chapter, related to the material of the present section. They represent a carefully chosen set of problems developed to highlight all the peculiarities in the constructing procedure. This will develop the basis for comprehending the material of the remaining part of this book.

In the next section, we propose solutions to a number of problems for beams with the standard edge conditions, undergoing various combinations of transverse and bending loads. The discussion will be based on Theorem 1.3 and will utilize the influence functions available in the present section.

2.2 Bending under transverse loads

We begin this discussion on possible applications of influence functions to the solution of various problems in beam theory. A number of classical problems will be discussed here on the bending of Kirchhoff beams in statics. Since those problems can be simulated with a linear ordinary differential equation allowing analytic integration, their exact solutions are well-known. We include such problems in the present text just to show how their classical solutions can be obtained by means of the influence function method.

Consider, for example, a beam of a uniform flexural rigidity EI, undergoing a transverse continuously distributed load $q(x)$. Recall again the displacement formulation of this problem

$$\frac{d^4 w(x)}{dx^4} = \frac{q(x)}{EI}, \quad x \in (0, a) \tag{2.38}$$

$$B_{0,i}[w(0)] = 0, \quad B_{a,i}[w(a)] = 0, \quad (i = 1, 2) \tag{2.39}$$

where the relations in eqn (2.39) present any combination of edge conditions that is feasible for statics.

Let $g(x, s)$ represent the influence function of the beam under consideration. That is, $g(x, s)$ represents the beam's deflection at the point x, caused by a transverse unit force concentrated at the point s. In other words, we assume that $g(x, s)$ is the Green's function of the homogeneous boundary value problem associated with that in eqns (2.38) and (2.39).

Hence, in compliance with Theorem 1.3 in Section 1.3, the deflection caused by the transverse load $q(x)$, distributed over the entire beam, can be expressed as follows

$$w_q(x) = -\frac{1}{EI} \int_0^a g(x, s) q(s) ds, \quad x \in [0, a] \tag{2.40}$$

To practically compute this integral, we recall that the influence function is defined in pieces. Therefore, one is required to break down the above integral into two integrals as shown

$$w_q(x) = -\frac{1}{EI} \left\{ \int_0^x g^-(x,s) q(s) ds + \int_x^a g^+(x,s) q(s) ds \right\}, \quad x \in [0, a] \tag{2.41}$$

2.2. BENDING UNDER TRANSVERSE LOADS

where $g^-(x,s)$ and $g^+(x,s)$ represent the branches of the influence function defined for $s \leq x$ and $x \leq s$, respectively.

Let us now discuss particular cases of loads that may occur in practice (such as concentrated and distributed forces and moments) and find how the integral representation in eqn (2.41) can be utilized to handle each of them as well as any reasonable combination of them.

Clearly, if the load $q(x)$ is applied to only a certain part $[\alpha, \beta]$ of $(0, a)$, then, breaking down the interval $[0, a]$ into three parts, we obtain

$$w_q(x) = -\frac{1}{EI} \left\{ \int_0^\alpha g(x,s) \cdot 0 ds + \int_\alpha^\beta g(x,s) q(s) ds + \int_\beta^a g(x,s) \cdot 0 ds \right\}$$

$$= -\frac{1}{EI} \int_\alpha^\beta g(x,s) q(s) ds, \quad x \in [0, a] \tag{2.42}$$

Thus, the deflection at any point of the beam is obtained in this case by the integration over the loaded interval $[\alpha, \beta]$. But in computing this integral, one must distinguish three different options for the location of the observation point x. Namely, for $x \leq \alpha$ (on the left of the loaded interval), eqn (2.42) is transformed into

$$w_q(x) = -\frac{1}{EI} \int_\alpha^\beta g^+(x,s) q(s) ds, \quad x \in [0, \alpha]$$

because the variable of integration s, ranging from α to β, remains in this case greater than or equal to x, and $g^+(x,s)$ is defined just for $x \leq s$.

For the values of x determined by $\alpha \leq x \leq \beta$, in turn, the integral in eqn (2.42) simply has to be split off as shown

$$w_q(x) = -\frac{1}{EI} \left\{ \int_\alpha^x g^-(x,s) q(s) ds + \int_x^\beta g^+(x,s) q(s) ds \right\}, \quad x \in [\alpha, \beta]$$

For $x \geq \beta$ (on the right of the loaded interval), we finally obtain

$$w_q(x) = -\frac{1}{EI} \int_\alpha^\beta g^-(x,s) q(s) ds, \quad x \in [\beta, a]$$

because in this case the variable of integration s, ranging from α to β, remains less than or equal to x, and $g^-(x,s)$ is defined just for $s \leq x$.

Let us consider some other possible types of loading. If, for example, the beam of length a is subjected to only a single transverse force of magnitude P_0, concentrated at the point $x = s_0 \in (0, a)$, then (as we have mentioned earlier) the right-hand side of eqn (2.38) can be thought of as a scalar multiple of the Dirac delta function

$$q(x) = P_0 \delta(x - s_0)$$

yielding as follows

$$w_{p_0}(x) = -\frac{1}{EI}\int_0^a g(x,s)\,P_0\,\delta(s-s_0)ds = -\frac{P_0}{EI}g(x,s_0) \qquad (2.43)$$

Notice that the above transformation has been carried out by virtue of the property of the Dirac function, which states that, if $f(x)$ is a function of a limited variation [31], then

$$\int_0^a f(s)\,\delta(s-s_0)\,ds = f(s_0)$$

The reader should recall this property from the development in Section 2.1, where we have introduced the Dirac delta function.

Let us now return to the relation in eqn (2.40), which provides the response of a beam to a load $q(x)$ continuously distributed over the entire beam. The reader should recall that this relation was derived in Section 1.3 (see eqn (1.71)) in proving Theorem 1.3. At this point, we can derive the relation in eqn (2.40) in a different manner. Namely, we will derive it by taking advantage of eqn (2.43) for the deflection caused by a transverse concentrated force.

In doing so, we consider a beam of length a subjected to a continuously distributed load $q(x)$. Let $g(x,s)$ represent the influence function for the beam. We partition the interval $[0,a]$ into n subintervals by choosing a set of distinct points $0 = x_0 < x_1 < x_2 < \cdots < x_n = a$. Then we pick up an arbitrary point s_k inside of every subinterval $[x_{k-1}, x_k]$ of the partitioning, and replace $q(s)$ within each of these subintervals with a concentrated force whose magnitude P_k is obtained as follows

$$P_k = q(s_k)\Delta s_k, \quad (k=\overline{1,n})$$

Assume that P_k is applied to the point s_k. In compliance with eqn (2.43), the beam's response to each of the forces P_k (the deflection $w_{p_k}(x)$ caused by P_k) can be written in the form

$$w_{p_k}(x) = -\frac{1}{EI}g(x,s_k)q(s_k)\Delta s_k, \quad (k=\overline{1,n})$$

Summing up the responses to each force P_k throughout the entire partitioning, we obtain the resultant response $w_n(x)$ in the form

$$w_n(x) = -\frac{1}{EI}\sum_{k=1}^{n}g(x,s_k)q(s_k)\Delta s_k$$

Clearly, the limit of $w_n(x)$ as n goes to infinity

$$-\frac{1}{EI}\lim_{n\to\infty}\sum_{k=1}^{n}g(x,s_k)q(s_k)\Delta s_k \qquad (2.44)$$

2.2. BENDING UNDER TRANSVERSE LOADS

represents the deflection $w_q(x)$ caused by the distributed load $q(x)$. But, the limit in eqn (2.44) represents a definite integral of $g(x,s)q(s)$ taken from 0 to a. In other words, we have shown that $w_q(x)$ can indeed be written as the integral in eqn (2.40).

We now consider other types of loading which can be treated by means of the influence function method. If, for instance, the beam is subjected not to a single concentrated transverse force but to a finite number of such forces with magnitudes P_i, $(i = \overline{1,k})$ and respective locations at the points s_i, then the deflection of the beam, caused by such a loading, can be obtained in the form

$$w(x) = -\frac{1}{EI} \int_0^a g(x,s) \sum_{i=1}^k P_i \, \delta(s - s_i) ds =$$

$$-\frac{1}{EI} \sum_{i=1}^k P_i \int_0^a g(x,s) \delta(s - s_i) ds = -\frac{1}{EI} \sum_{i=1}^k P_i \, g(x, s_i) \qquad (2.45)$$

Thus, within the scope of the geometrically linear statement of the problem, the response of the beam to a finite number of concentrated forces can be obtained as a sum of the responses to each of those forces.

Let us now consider the case when the beam is subjected to a bending couple of magnitude M_0, which is applied to the point s_0. It can readily be shown that this case is also solvable in terms of the influence function of a single transverse concentrated force.

To show how it can be realized, we consider two equal but opposing transverse forces of magnitude P_0, which are concentrated at two closely spaced points s_0 and $s_0 + h$. In addition, we assume that the following relation $M_0 = P_0 h$ holds for the quantities M_0, P_0, and h.

Thus, in compliance with eqn (2.43), the deflection $w_h(x)$ of the beam, caused by the two forces recently introduced can consequently be written as

$$w_h(x) = -\frac{P_0}{EI} [g(x, s_0 + h) - g(x, s_0)]$$

As we replace P_0 with the quotient M_0/h that follows from the assumption recently made, the above equation becomes

$$w_h(x) = -\frac{M_0}{EIh} [g(x, s_0 + h) - g(x, s_0)] \qquad (2.46)$$

From what we assumed earlier, it follows that the limit of w_h, as h approaches zero, equals the value of the deflection $w_{m_0}(x)$ of the beam, caused by the concentrated bending couple M_0, that is

$$w_{m_0}(x) = \lim_{h \to 0} w_h(x)$$

Hence, taking into account the expression for w_h in eqn (2.46), one obtains

$$w_{m_0}(x) = -\frac{M_0}{EI} \lim_{h \to 0} \frac{g(x, s_0+h) - g(x, s_0)}{h}$$

It turns out that the above limit represents the value of the partial derivative of $g(x,s)$ with respect to s at $s=s_0$. Hence, for the deflection of the beam, caused by the concentrated bending moment M_0, we finally have

$$w_{m_0}(x) = -\frac{M_0}{EI} \frac{\partial g(x, s_0)}{\partial s}, \quad x \in [0, a] \tag{2.47}$$

Thus, to obtain the value of the deflection function at any point x of the beam, caused by the bending moment concentrated at a certain point s_0, we need to know the value of the partial derivative

$$\frac{\partial g(x, s_0)}{\partial s}$$

of the influence function at $s=s_0$. In other words, the derivative

$$\frac{\partial g(x, s)}{\partial s}$$

being a function of x and s, represents the response (output) of the beam at the point x on the unit concentrated bending moment (input) applied to the point s. In the discussion that follows, it will be referred to as the *influence function of the second order*.

In view of the foregoing discussion, one can readily derive the following expression

$$w(x) = -\frac{1}{EI} \sum_{i=1}^{k} M_i \frac{\partial g(x, s_i)}{\partial s}, \quad x \in [0, a] \tag{2.48}$$

for the deflection of the beam subjected to a finite number of the concentrated bending moments $Mi, (i=\overline{1,k})$ applied to the points s_i, respectively.

Based on the present discussion, it follows that the influence function $g(x,s)$ of the transverse concentrated unit force can be successfully employed for computing analytic expressions for the deflection of a beam, caused by a simultaneous action of several concentrated and distributed transverse loads and bending moments. The resultant deflection of the beam in such a case is to be obtained by a superposition of the deflections associated with each individual input.

In other words, the resultant deflection is a proper combination of those in eqns (2.40)–(2.43), (2.45), (2.47), and (2.48). This is true, of course, if the combined result of the individual loads does not cause either physical non-linearity (violation of the Hooke's law for the material of which the beam is

2.2. BENDING UNDER TRANSVERSE LOADS

composed, resulting in the nonlinear stress-strain relationship) or geometric nonlinearity (a high level of deflection). In either case, the aforementioned boundary value problem in eqns (2.2) and (2.5) is no longer adequately applicable to the physical problem.

Thus, the influence function method provides a universal pattern for computing the deflection function in an analytic form, for all of the physically feasible combinations of transverse and bending loads. This makes it possible to obtain both of the most required outputs in engineering practice for the problem: the bending moment $M(x)$ and the shear force $Q(x)$

$$M(x) = EI\frac{d^2w(x)}{dx^2}, \quad Q(x) = EI\frac{d^3w(x)}{dx^3}$$

at any cross-section of the beam, also in analytic form. This occurs because in doing so one can analytically differentiate the corresponding deflection function. Indeed, for a beam undergoing a transverse load $q(x)$ distributed over the interval (α, β), for example, upon differentiating $w(x)$ in equation (2.42), one obtains

$$M(x) = -\int_\alpha^\beta \frac{\partial^2 g(x,s)}{\partial x^2} q(s)ds, \quad x \in [0, a] \qquad (2.49)$$

$$Q(x) = -\int_\alpha^\beta \frac{\partial^3 g(x,s)}{\partial x^3} q(s)ds, \quad x \in [0, a] \qquad (2.50)$$

In view of the foregoing discussion, which touched upon the integral in eqn (2.42), it follows that to practically compute $M(x)$ and $Q(x)$, one again must account for three different possible locations of the field point x with respect to the interval of integration in eqns (2.49) and (2.50).

Notice that in obtaining the derivatives in eqns (2.49) and (2.50), we have differentiated a definite integral containing a parameter x, with respect to this parameter. Clearly, such an operation affects only the influence function, because that is the only function of x under the integral sign.

Notice also that the relations in eqns (2.49) and (2.50) are valid for computing bending moments and shear forces for a beam subjected to a continuously distributed transverse load, regardless of the edge conditions prescribed. This highlights the nature of the influence function method in which the edge conditions are usually treated when the influence function is constructed. And when that function is used in integrals like that in eqn (2.40), for example, the entire integral satisfies the prescribed edge conditions.

The differentiation in eqns (2.49) and (2.50) was accomplished analytically. This aspect becomes crucial when the function $q(x)$ is complex in form and the boundary value problem in eqns (2.38) and (2.39) cannot, consequently, be solved exactly. This implies that numerical methods are required and the approximate differentiation cannot be avoided. Clearly, the influence function method, for such cases, is significantly superior compared to pure numerical techniques like the finite difference or finite element methods. Later in this section, we present some data illustrating the last point.

We will now turn to other types of elementary loads that can possibly be applied to the beam. Upon the corresponding differentiation of $w(x)$ in eqn (2.43), one readily obtains the bending moment

$$M(x) = -P_0 \frac{\partial^2 g(x, s_0)}{\partial x^2}, \quad x \in [0, a] \tag{2.51}$$

and the shear force

$$Q(x) = -P_0 \frac{\partial^3 g(x, s_0)}{\partial x^3}, \quad x \in [0, a] \tag{2.52}$$

at any cross-section x of the beam subjected to a single transverse force of magnitude P_0, concentrated at the point s_0.

For a finite number of concentrated transverse forces P_i, $(i=\overline{1,k})$ acting simultaneously, the bending moment

$$M(x) = -\sum_{i=1}^{k} P_i \frac{\partial^2 g(x, s_i)}{\partial x^2}, \quad x \in [0, a] \tag{2.53}$$

and the shear force

$$Q(x) = -\sum_{i=1}^{k} P_i \frac{\partial^3 g(x, s_i)}{\partial x^3}, \quad x \in [0, a] \tag{2.54}$$

are to be computed at any cross-section of the beam by the corresponding differentiation of $w(x)$ in eqn (2.45).

For the input in the form of a single concentrated bending moment of magnitude M_0, the bending moment and the shear force at any cross-section can be obtained by the appropriate differentiation of the influence function of the second order. This yields

$$M(x) = -M_0 \frac{\partial^3 g(x, s_0)}{\partial s \, \partial x^2}, \quad x \in [0, a] \tag{2.55}$$

and

$$Q(x) = -M_0 \frac{\partial^4 g(x, s_0)}{\partial s \, \partial x^3}, \quad x \in [0, a] \tag{2.56}$$

2.2. BENDING UNDER TRANSVERSE LOADS

by properly differentiating $w(x)$ in eqn (2.47).

Given the input in the form of a finite sum of concentrated bending moments M_i, $(i=\overline{1,k})$, the values of the bending moment $M(x)$

$$M(x) = -\sum_{i=1}^{k} M_i \frac{\partial^3 g(x, s_i)}{\partial s\, \partial x^2}, \quad x \in [0, a] \tag{2.57}$$

and the shear force $Q(x)$

$$Q(x) = -\sum_{i=1}^{k} M_i \frac{\partial^4 g(x, s_i)}{\partial s\, \partial x^3}, \quad x \in [0, a] \tag{2.58}$$

at any cross-section are derived from eqn (2.48).

Notice that formulas in eqns (2.49)–(2.58) are derived for corresponding loads without any regard to the edge conditions imposed in a particular problem. The edge conditions in turn are satisfied by an appropriate choice of the influence function.

We below exhibit a number of instructive examples that show how the classical solutions from Kirchhoff's beam theory can be obtained by means of influence functions.

EXAMPLE 1 Consider a cantilever beam of length a, undergoing a transverse load $q(x)$ written as

$$q(x) = q_0 x(a-x), \quad q_0 = \text{const} \tag{2.59}$$

distributed throughout the entire beam as shown in Figure 2.4.

From what we have shown earlier in this section, it follows that the deflection function $w(x)$ of a beam undergoing a continuously distributed load $q(x)$ and subject to standard edge conditions can be expressed in terms of the corresponding influence function $g(x, s)$ by eqn (2.40)

$$w(x) = -\frac{1}{EI} \int_0^a g(x, s) q(s) ds, \quad x \in [0, a]$$

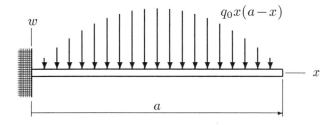

Figure 2.4 A cantilever beam subjected to a distributed load

Since the influence function $g(x,s)$ for any beam problem is defined in a piecewise fashion by different expressions for $x \leq s$ and $s \leq x$, we break down the above integral into two integrals as shown

$$w(x) = -\frac{1}{EI}\left\{\int_0^x g(x,s)q(s)ds + \int_x^a g(x,s)q(s)ds\right\}, \quad x \in [0,a] \quad (2.60)$$

Recalling then the influence function $g(x,s)$ for the cantilever beam from eqn (2.37), we match its lower branch ($s \leq x$) with the first of the integrals in eqn (2.60), because s represents its variable of integration, while x represents its upper limit and bounds, therefore, the variable of integration s from above. The upper branch ($x \leq s$) of the influence function in turn goes into the second integral in eqn (2.60). Recalling also the load function $q(x)$ from eqn (2.59) and substituting $g(x,s)$ and $q(x)$ into eqn (2.60), one obtains

$$w(x) = -\frac{q_0}{6EI}\left\{\int_0^x s^2(s-3x)s(a-s)ds + \int_x^a x^2(x-3s)s(a-s)ds\right\} \quad (2.61)$$

Computing the elementary integrals in eqn (2.61), we obtain the deflection function of the cantilever beam of length a, undergoing the transverse load specified in eqn (2.59), in the form

$$w(x) = -\frac{q_0}{360EI}x^2(x^4 - 3ax^3 + 10a^3x - 15a^4), \quad x \in [0,a] \quad (2.62)$$

Since the integration in this case has been accomplished analytically, one can compute the output bending moment $M(x)$ and shear force $Q(x)$ by properly differentiating the deflection function just obtained in eqn (2.62). This finally yields for the bending moment

$$M(x) = -\frac{q_0}{12}(x^4 - 2ax^3 + 2a^3x - a^4)$$

and for the shear force

$$Q(x) = -\frac{q_0}{6}(2x^3 - 3ax^2 + a^3)$$

EXAMPLE 2   Consider a beam of length a, whose left edge is elastically supported (with k being the constant of elastic support), while the right edge is clamped. Assume that the load is in the form of a single bending moment M_0 spaced at the point $x = s_0$ (see Figure 2.5).

2.2. BENDING UNDER TRANSVERSE LOADS

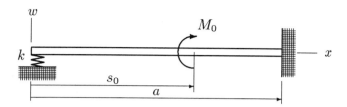

Figure 2.5 A beam subjected to a concentrated moment

From what we have shown earlier in this section, it follows that for a beam subjected to any standard set of edge conditions and undergoing a single bending moment of magnitude M_0, concentrated at point s_0, the deflection function $w_{m_0}(x)$ can be found in terms of the value of the corresponding influence function of the second order (see eqn (2.47))

$$w_{m_0}(x) = -\frac{M_0}{EI}\frac{\partial g(x, s_0)}{\partial s}, \quad x \in [0, a] \qquad (2.63)$$

The influence function $g(x, s)$ (of a unit transverse concentrated force for the beam subject to the edge conditions imposed in the statement above) can be found in turn in eqn (2.30) of Section 2.1

$$g(x, s) = \frac{1}{6(1+2ka^3)}\begin{cases}(s-a)^2[3x(1-ka^2s)+(kx^3-1)(s+2a)], & x \leq s \\ (x-a)^2[3s(1-ka^2x)+(ks^3-1)(x+2a)], & s \leq x\end{cases}$$

Upon differentiating $g(x, s)$ with respect to s and substituting the resultant expression for the derivative into eqn (2.63), one obtains the deflection of the beam under consideration in the form

$$w_{m_0}(x) = -\frac{M_0 EI^{-1}}{2(1+2ka^3)}\begin{cases}(s_0-a)[(kx^3-1)(s_0+a)+(ka^3-3ka^2s_0+2)x], & x \leq s_0 \\ (x-a)^2[k(x+2a)s_0^2+(1-ka^2x)], & s_0 \leq x\end{cases}$$

The output bending moment $M_{m_0}(x)$

$$M_{m_0}(x) = -\frac{M_0}{1+2ka^3}\begin{cases}3k(s_0^2-a^2)x, & x \leq s_0 \\ 3k(s_0^2-a^2)x+(1+2ka^3), & s_0 \leq x\end{cases}$$

in this case is computed by using eqn (2.55). It should be clearly seen that $M_{m_0}(x)$ makes a jump of discontinuity of magnitude M_0 at the point s_0.

Equation (2.56) provides the following expression

$$Q_{m_0}(x) = -\frac{3M_0 k(s_0^2 - a^2)}{1 + 2ka^3}$$

for the shear force output $Q_{m_0}(x)$ caused by M_0, which is uniform throughout the intervals $(0, s_0)$ and (s_0, a). It should be stated, however, that it is undefined at s_0, because $M_{m_0}(x)$ is discontinuous at that point.

For the next example, we examine a beam undergoing a combination of several elementary loads.

EXAMPLE 3   Consider a beam of length a with both edges simply supported, undergoing a combination of loads (a uniform load q_0 distributed over the interval $[s_1, s_2]$, a bending moment M_0 concentrated at the point s_3, and a transverse force of magnitude P_0 concentrated at the point s_4) applied in the fashion depicted in Figure 2.6.

The influence function

$$g(x,s) = \frac{1}{6a} \begin{cases} x(a-s)[(a-s)^2 + (x^2-a^2)], & x \leq s \\ s(a-x)[(a-x)^2 + (s^2-a^2)], & s \leq x \end{cases}$$

of a unit force for the simply supported beam is to be obtained by the reader in Section 2.1 (see exercise 2.4).

The outputs (deflection, bending moment, and shear force) caused by each of the inputs specified in the above statement can readily be found in terms of the influence function $g(x,s)$. Indeed, in accordance with the relation in eqn (2.41), the deflection $w_{q_0}(x)$ caused by the load q_0 uniformly distributed over the interval $[s_1, s_2]$, can be obtained as

$$w_{q_0}(x) = -\frac{1}{EI} \int_{s_1}^{s_2} g(x,s) q_0 ds$$

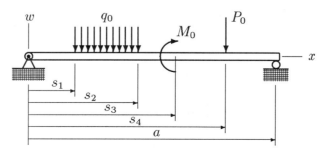

Figure 2.6   A beam subjected to a combination of loads

2.2. BENDING UNDER TRANSVERSE LOADS

In compliance with the piecewise format of the influence function, three separate formulas ought to be derived for the above integral. For $x \leq s_1$, we have

$$w_{q_0}(x) = -\frac{q_0}{6EIa} \int_{s_1}^{s_2} x(a-s)[(a-s)^2 + (x^2-a^2)]ds, \quad x \in [0, s_1]$$

If the observation point x is located within the loaded interval (that is $s_1 < x < s_2$), then one obtains the deflection in the form

$$w_{q_0}(x) = -\frac{q_0}{6EIa} \left\{ \int_{s_1}^{x} s(a-x)[(a-x)^2 + (s^2-a^2)]ds \right.$$
$$\left. + \int_{x}^{s_2} x(a-s)[(a-s)^2 + (x^2-a^2)]ds \right\}, \quad x \in [s_1, s_2]$$

And for $x \geq s_2$, one obtains

$$w_{q_0}(x) = -\frac{q_0}{6EIa} \int_{s_1}^{s_2} s(a-x)[(a-x)^2 + (s^2-a^2)]ds, \quad x \in [s_2, a]$$

In accordance with eqn (2.47), one obtains the deflection caused by the bending moment M_0 concentrated at the point s_3 by means of the influence function of the second order in the form

$$w_{m_0}(x) = \frac{M_0}{6EIa} \begin{cases} x[3(a-s_3)^2 - (a^2-x^2)], & x \leq s_3 \\ (a-x)[x(2a-a) - 3s_3], & x \geq s_3 \end{cases}$$

And, finally, from eqn (2.43), it follows that the deflection, caused by the transverse force P_0 concentrated at the point s_4, can be obtained as

$$w_{p_0}(x) = -\frac{P_0}{6EIa} \begin{cases} x(a-s_4)[(a-s_4)^2 + (x^2-a^2)], & x \leq s_4 \\ s_4(a-x)[(a-x)^2 + (s_4^2-a^2)], & s_4 \leq x \end{cases}$$

The resultant deflection function $w(x)$ caused by a collection of the loads specified in the statement of the problem, is now expressed as a sum of the components $w_{q_0}(x), w_{m_0}(x)$, and $w_{p_0}(x)$.

The bending moment and shear force caused by the given collection of loads can also be expressed analytically. In compliance with the relations in eqns (2.49), (2.51), and (2.55), the resultant bending moment $M(x)$, for example, is computed in this case as

$$M(x) = -\frac{\partial^2}{\partial x^2} \left(P_0 g(x, s_4) + M_0 \frac{\partial g(x, s_3)}{\partial s} + \int_{s_1}^{s_2} g(x,s) q_0 ds \right)$$

From the material discussed in this section it follows that, if a boundary value problem of the type in eqns (2.38) and (2.39) allows an exact solution, then the influence function method can be readily used as one possible way to practically obtain such a solution. Moreover, this method suggests a standard approach for the treatment of distributed as well as concentrated loads in a similar manner. The reader could definitely benefit from this feature of the influence function method.

Beneficiary features of the influence function method become even more significant in more complicated situations, especially when the exact solution of the problem formulated in eqns (2.38) and (2.39) cannot be attained. If, for instance, the loading function $q(x)$ is complex in the sense that only a numerical solution is possible, then it is hard to find a viable alternative to the influence function method in terms of the computational efficiency. To illustrate this point, we present the following example.

EXAMPLE 4 Compute the deflection function $w(x)$, bending moment $M(x)$, and shear force $Q(x)$ for the clamped beam of length a, subject to the transverse load $q(x) = q_0 \sin(\pi x^2/a^2)$ distributed over the entire beam, with q_0 being a constant (see Figure 2.7).

Based on this statement, we formulate the boundary value problem

$$\frac{d^4 w(x)}{dx^4} = -\frac{q_0}{EI} \sin(\frac{\pi x^2}{a^2}), \quad x \in (0, a) \tag{2.64}$$

$$w(0) = \frac{dw(0)}{dx} = 0, \quad w(a) = \frac{dw(a)}{dx} = 0 \tag{2.65}$$

By virtue of Theorem 1.3 of Chapter 1, one can express the solution of the problem posed by eqns (2.64) and (2.65) in the form

$$w(x) = \frac{q_0}{EI} \int_0^a g(x, s) \sin(\frac{\pi s^2}{a^2}) ds \tag{2.66}$$

where $g(x, s)$ represents the influence function of a clamped beam, which can be found in eqn (2.20) of Section 2.1.

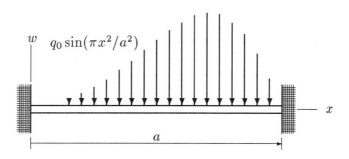

Figure 2.7 A beam subjected to a distributed load

2.2. BENDING UNDER TRANSVERSE LOADS

Unfortunately, the integral that occurrs in eqn (2.66) cannot be taken in closed form although it can of course be computed numerically. And in the discussion that follows, we will show that such a computation is substantially more efficient compared to other classical numerical alternatives when solving the boundary value problem in eqns (2.64) and (2.65).

The point is that, because of the specific form of the right-hand side of eqn (2.64), the exact solution of the problem posed by eqns (2.64) and (2.65) cannot at all be expressed in terms of elementary functions. Indeed, in an attempt to integrate eqn (2.64), we subsequently face the so-called Fresnal integrals

$$\int \sin(x^2)dx, \quad \text{or} \quad \int \cos(x^2)dx$$

that cannot [68] be taken in closed form.

Thus, numerical methods are required to approximately solve the problem in eqns (2.64) and (2.65).

Before going any further with the statement in eqns (2.64) and (2.65), we pose a test problem allowing an exact solution. That is the boundary value problem in eqn (2.65) for the following equation

$$\frac{d^4w(x)}{dx^4} = -\frac{q_0}{EI}\sin(\frac{\pi x}{a}), \quad x \in (0,a) \tag{2.67}$$

The exact solution to this problem

$$w(x) = \frac{a^2 q_0}{\pi^4 EI}\left[\pi x^2 - a\pi x + a^2 \sin(\frac{\pi x}{a})\right] \tag{2.68}$$

is easily within reach by either the method of undetermined coefficients or Lagrange's method of variation of parameters, or one can simply successively integrate equation (2.67) four times.

On the other hand, analogously to the integral representation shown in eqn (2.66), one can express the exact solution of the problem in eqns (2.65) and (2.67) in terms of the influence function taken from eqn (2.20) by the following integral

$$w(x) = \frac{q_0}{EI}\int_0^a g(x,s)\sin(\frac{\pi s}{a})ds \tag{2.69}$$

Contrary to what we experience with the integral in eqn (2.66), the above integral can be taken in closed form, yielding the representation in eqn (2.68). However, in view of the foregoing discussion we will not take advantage of such an opportunity. Instead we will rather compute the integral in eqn (2.69) numerically. Any of the classical quadrature formulae are appropriate for that computation.

For the sake of simplicity, we utilize herein the trapezoid rule based on a uniform partitioning of the interval $[0, a]$ into n subintervals with partitioning points x_k defined as

$$x_k = \frac{ak}{n}, \quad (k=\overline{0,n}) \tag{2.70}$$

In doing so, one obtains approximate values of the deflection function $w(x)$ at the partitioning points x_i in the form

$$w(x_i) \approx \frac{q_0 a}{2nEI} \sum_{k=1}^{n} \left[g(x_i, x_k) \sin(\frac{\pi x_k}{a}) + g(x_i, x_{k-1}) \sin(\frac{\pi x_{k-1}}{a}) \right], \tag{2.71}$$

where $i = \overline{1, n}$.

Since the exact solution of the problem posed by eqns (2.65) and (2.67) is available, it is clear that upon carrying out the actual computation with various partitioning parameters n, one can test the practical convergence of the influence function-based numerical procedure and control the level of accuracy that can be attained.

To control the relative efficiency of the numerical influence function approach, we have conducted a computational experiment on the comparison of its numerical results against those computed by the finite difference method [4, 16, 61]. We have used the simplest version of this method, one that reduces the boundary value problem in eqns (2.65) and (2.67) to the following system of linear algebraic equations

$$w_{k+2} - 4w_{k+1} + 6w_k - 4w_{k-1} + w_{k-2} = -\frac{q_0 a^2}{n^2 EI} \sin(\frac{ax_k}{n}), \tag{2.72}$$

where $k = \overline{2, n-2}$.

To carry out the actual computation, we utilize the same uniform partitioning as defined in eqn (2.70).

Such a primitive finite difference scheme has been chosen on purpose, as it is equivalent to the trapezoid rule in terms of the order of accuracy $O(h^2)$ provided. Hence, from the error estimation viewpoint, it follows that the outputs of both of the finite difference and influence function methods must be equivalently accurate. Practice, however, shows differently.

The results exhibited in Tables 2.1 and 2.2, for example, show the level of accuracy attained by the finite difference and influence function methods applied to the boundary value problem in eqns (2.65) and (2.67). For both the methods, the same partitioning from eqn (2.70) has been used. Because of the symmetry of the formulation about the midpoint of the interval $[0, a]$, we have observed only a half of it, that is $(0, a/2]$.

2.2. BENDING UNDER TRANSVERSE LOADS

Table 2.1 Approximate values w_k of the deflection function for the problem in eqns (2.65) and (2.67), computed by the FDM

Observation point, x/a	Partitioning number, n				Exact solution
	10	20	50	100	
0.1	.000000	.000123	.000182	.000239	.000270
0.2	.000357	.000600	.000729	.000817	.000874
0.3	.000836	.001170	.001325	.001459	.001533
0.4	.001234	.001619	.001782	.001942	.002023
0.5	.001406	.001799	.001947	.002122	.002203

Table 2.2 Approximate values w_k of the deflection function for the problem in eqns (2.65) and (2.67), computed by the IFM

Observation point, x/a	Partitioning number, n				Exact solution
	5	10	20	100	
0.1	.000256	.000265	.000269	.000270	.000270
0.2	.000832	.000859	.000870	.000874	.000874
0.3	.001501	.001515	.001528	.001533	.001533
0.4	.002031	.002017	.002021	.002023	.002023
0.5	.002215	.002204	.002203	.002203	.002203

From the above data, it follows, in particular, that when computing the deflection of the beam, the IFM with the partitioning number $n=5$ happens to be notably more accurate than the FDM with $n=100$. This phenomenon is not accidental. Moreover, it is predictable. Indeed, the IFM must be more superior computationally because the numerical integration, generally speaking, is more 'accurate' in practice than numerical differentiation.

This observation in no means contradicts the theoretical prediction that claims an equivalent level of accuracy for both the finite difference scheme in eqn (2.72) and the trapezoid rule-based influence function procedure which follows from eqn (2.71). The presented data just show that quite often *a priori* estimations of errors (which are necessarily required when the qualitative analysis is accomplished) are not directly applicable in practice.

The superiority of the influence function method compared to the finite difference approach becomes even more evident when computing values of the bending moment and shear force. As is known, these are expressed in terms of the corresponding derivatives of the deflection function. In order to address this issue in more detail, we present the output bending moment $M(x)$ (see Table 2.3) and shear force $Q(x)$ (see Table 2.4) again for the clamped beam whose stress–strain state is simulated by the boundary value problem posed by eqns (2.65) and (2.67).

CHAPTER 2. INFLUENCE FUNCTIONS FOR BEAMS

Table 2.3 Approximate values of the bending moment $M(x)$ computed by the IFM and FDM

Observation point, x/a	$n=10$ IFM	$n=10$ FDM	$n=100$ IFM	$n=100$ FDM	Exact values
0.0	.064460	—	.064503	—	.064503
0.1	.032723	.035682	.033189	.033740	.033193
0.2	.004515	.012233	.004944	.005832	.004948
0.3	−.017984	−.008127	−.017473	−.016500	−.017467
0.4	−.032588	−.022608	−.031866	−.030995	−.031859
0.5	−.037658	−.029001	−.036826	−.036157	−.036818

Table 2.4 Approximate values of the shear force $Q(x)$ computed by the IFM and FDM

Observation point, x/a	$n=10$ IFM	$n=10$ FDM	$n=100$ IFM	$n=100$ FDM	Exact values
0.0	−.307768	—	−.318205	—	−.318309
0.1	−.295323	−.234494	−.302657	−.296469	−.302731
0.2	−.253782	−.203592	−.257480	−.252654	−.257518
0.3	−.186009	−.144814	−.187086	−.183346	−.187098
0.4	−.098362	−.063912	−.098363	−.095329	−.098363
0.5	−.000000	.031193	.000000	−.002781	.000000

From the data presented above, it follows that for the bending moment and shear force, the numerical influence function method is practically as accurate as for the deflection function. This feature of the influence function method is not surprising, because following the IFM procedure, we compute values of $M(x)$ and $Q(x)$ again by numerical integration (see eqns (2.49) and (2.50)) completely avoiding the numerical differentiation.

Table 2.5 Approximate values of the deflection function and the bending moment obtained for the problem in eqns (2.64) and (2.65) by IFM.

Field point, x/a	Deflection, $w(x)$			Bending moment, $M(x)$		
	$n=10$	$n=20$	$n=100$	$n=10$	$n=20$	$n=100$
0.0	.00000	.00000	.00000	.04025	.04012	.04011
0.2	.00057	.00058	.00058	.00710	.00710	.00700
0.4	.00145	.00146	.00146	−.02034	−.02016	−.02010
0.6	.00161	.00160	.00160	−.02864	−.02797	−.02776
0.8	.00073	.00075	.00075	−.00146	−.00079	−.00059
1.0	.00000	.00000	.00000	.05927	.05951	.05952

2.3. BEAMS ON AN ELASTIC FOUNDATION

On the contrary, the finite difference scheme that was discussed earlier (see eqn (2.72)) is not recommended for computing either bending moments or shear forces, unless some adjustments are made. For example, being within the scope of the finite difference method, the simplest way to improve the solution accuracy would be to radically increase the partitioning parameter n. Because of the five-diagonal structure of the coefficient matrix of the system in eqn (2.72), such an increase can easily be afforded in practice.

Let us now return to the original formulation of the problem presented in eqns (2.64) and (2.65). Table 2.5 shows some results of solving this problem by the influence function procedure. In the computation that follows, we assume $q_0/EI = 1$. The deflection and bending moment are computed throughout the entire segment $[0, 1]$ for three different values of the partitioning parameter (number of partitionings). In this case, we cannot compare approximate results against the exact solution since the latter is not available. Therefore, in view of the foregoing computation, practical convergence of the influence function procedure could be used as a guarantee for the accuracy of the approximate solution.

From what we observe in this table, it follows that for the problem under consideration, the numerical procedure of the influence function method is rapidly converging for all of the solution components required in applications. The author's experience on the use of this method for a broad variety of problems has always shown its high computational potential. Based on this experience, we recommend it to users without reservations, with the partitioning parameter n being well below one hundred range.

The next section is devoted to one more problem class that can successfully be treated in terms of the influence function method. We will show that beams resting on elastic foundations also allow the productive influence function analysis.

2.3 Beams on an elastic foundation

Within the scope of Kirchhoff theory, we consider an infinite beam having a uniform flexural rigidity and resting on an elastic foundation. Such a model (infinite beam) is frequently appropriate as a simplifying assumption for a really long beam when the boundary conditions imposed at its edges do not practically affect a local stress-strain state of areas remote from the edges, unless we are interested in areas that are situated just near the edges. In such cases, however, a beam of finite length can be considered.

Let EI and k_0 represent the flexural rigidity of the beam and the elastic coefficient of the foundation, respectively. Let the beam undergo a transverse distributed load $q(x)$.

A differential equation modeling the equilibrium state of this beam can be presented (see [18, 27, 64, 67]) in the form

$$\frac{d^4w(x)}{dx^4} + 4k^4 w(x) = -\frac{q(x)}{EI}, \quad x \in (-\infty, \infty) \tag{2.73}$$

where $k = (k_0/4EI)^{1/4}$. This specific form of the coefficient of the above equation allows the fundamental solution set for the corresponding homogeneous equation to be expressed in the most compact form.

The right-hand side function $q(x)$ in the above equation is defined over a finite interval $[a, b]$. Beyond this interval, $q(x)$ identically equals zero. Clearly, this limitation on $q(x)$ is realistic from the application standpoint, because a loading function is usually applied to a finite part of a beam.

The edge conditions for the infinite beam can be thought of as the boundedness conditions for all the solution components as x approaches both positive and negative infinity.

Let $g(x, s)$ represent the influence function of a transverse unit force concentrated at point s of the beam. In other words, $g(x, s)$ is the Green's function to the homogeneous equation associated with that in eqn (2.73), subjected to the boundedness conditions at positive and negative infinity. Based on Theorem 1.3 of Section 1.3, such a solution of eqn (2.73), which vanishes at infinities, can be represented as the integral

$$w(x) = \frac{1}{EI} \int_a^b g(x, s) q(s) ds, \quad x \in (-\infty, \infty) \tag{2.74}$$

Hence, by using this representation, one can compute any component of the stress-strain state of the beam in terms of the relations provided in Section 2.2. Therefore, in the discussion that follows, we first focus on the construction of the influence function $g(x, s)$. As we already mentioned, $g(x, s)$ represents the Green's function for the homogeneous equation

$$\frac{d^4w(x)}{dx^4} + 4k^4 w(x) = 0, \quad x \in (-\infty, \infty) \tag{2.75}$$

corresponding to that in eqn (2.73). The boundary conditions in the definition of the Green's function (see Section 1.1) must be replaced, in this case, with the boundedness conditions.

Upon utilizing the standard analysis for linear homogeneous differential equations with constant coefficients, one can obtain the fundamental solution set for eqn (2.75) in the following form

$$\begin{aligned} w_1(x) = e^{kx} \cos(kx), \quad w_2(x) = e^{kx} \sin(kx) \\ w_3(x) = e^{-kx} \cos(kx), \quad w_4(x) = e^{-kx} \sin(kx) \end{aligned} \tag{2.76}$$

2.3. BEAMS ON AN ELASTIC FOUNDATION

Clearly, the first two components, $w_1(x)$ and $w_2(x)$ in this set approach zero as x approaches negative infinity, whereas $w_3(x)$ and $w_4(x)$ vanish as x goes to positive infinity. These properties of the components of the fundamental solution set are very convenient for the discussion that follows. Tracing out the technique suggested in Section 2.1, we find $g(x,s)$ to be in the form

$$g(x,s) = \begin{cases} a_1(s)e^{kx}\cos(kx) + a_2(s)e^{kx}\sin(kx), & x \leq s \\ b_1(s)e^{-kx}\cos(kx) + b_2(s)e^{-kx}\sin(kx), & s \leq x \end{cases} \quad (2.77)$$

From this representation, it clearly follows (due to the vanishing of $w_1(x)$ and $w_2(x)$ at negative infinity as well as the vanishing of $w_3(x)$ and $w_4(x)$ at infinity) that the defining properties 1 and 4 of a Green's function are met by $g(x,s)$. To compute the coefficients $a_i(s)$ and $b_i(s)$, we take advantage of the two remaining defining properties. By virtue of property 2, which claims a continuity of $g(x,s)$, as $x=s$, one obtains

$$b_1(s)e^{-ks}\cos(ks) + b_2(s)e^{-ks}\sin(ks)$$

$$-a_1(s)e^{ks}\cos(ks) - a_2(s)e^{ks}\sin(ks) = 0 \quad (2.78)$$

The continuity of the first order partial derivative of $g(x,s)$ with respect to x, as $x=s$, required in property 2, yields

$$-b_1(s)e^{-ks}(\cos(ks)+\sin(ks)) + b_2(s)e^{-ks}(\cos(ks)-\sin(ks))$$

$$-a_1(s)e^{ks}(\cos(ks)-\sin(ks)) - a_2(s)e^{ks}(\cos(ks)+\sin(ks)) = 0 \quad (2.79)$$

Claiming a continuity of the second order partial derivative of $g(x,s)$ with respect to x, as $x=s$, one derives the following equation

$$b_1(s)e^{-ks}\sin(ks) - b_2(s)e^{-ks}\cos(ks)$$

$$+ a_1(s)e^{ks}\sin(ks) - a_2(s)e^{ks}\cos(ks) = 0 \quad (2.80)$$

And lastly, by virtue of defining property 3, which claims a discontinuity of the third order partial derivative of $g(x,s)$ with respect to x, as $x=s$, we obtain one more relation for the coefficients of $g(x,s)$ in eqn (2.77)

$$b_1(s)e^{-ks}(\cos(ks)-\sin(ks)) + b_2(s)e^{-ks}(\cos(ks)+\sin(ks)) + a_1(s)e^{ks}$$

$$\times(\cos(ks)+\sin(ks)) - a_2(s)e^{ks}(\cos(ks)-\sin(ks)) = -1/2k^3 \quad (2.81)$$

The relations in eqns (2.78)–(2.81) form a well-posed system of linear algebraic equations in $a_1(s)$, $a_2(s)$, $b_1(s)$, and $b_2(s)$. This follows from the fact that the set of functions in eqn (2.76) form a fundamental solution set for eqn (2.75). Upon solving the system, one obtains

$$a_1(s) = -\frac{e^{-ks}}{8k^3}(\cos(ks)+\sin(ks)), \quad a_2(s) = \frac{e^{-ks}}{8k^3}(\cos(ks)-\sin(ks))$$

$$b_1(s) = -\frac{e^{ks}}{8k^3}(\cos(ks)-\sin(ks)), \quad b_2(s) = -\frac{e^{ks}}{8k^3}(\cos(ks)+\sin(ks))$$

Substituting these values into eqn (2.77) and accomplishing a rather straightforward algorithm, we ultimately obtain the influence function of a transverse unit force concentrated at point s for the infinite beam resting on an elastic foundation (see Figure 2.8) in the form

$$g(x,s) = -\frac{1}{8k^3} \begin{cases} e^{k(x-s)}\left[\cos(k(x-s))-\sin(k(x-s))\right], & x \leq s \\ e^{k(s-x)}\left[\cos(k(x-s))+\sin(k(x-s))\right], & s \leq x \end{cases} \quad (2.82)$$

And again analogously to the development carried out in Section 2.2 for a simple beam, one can apply the influence function formalism to analytically compute the required outputs (deflections, bending moments, and shear forces) for the beam resting on an elastic foundation.

Due to the compactness of the representation of the influence function occurring in eqn (2.82), one can readily account for any load subjected in the form of either concentrated transverse forces and bending moments or continuously distributed loads as well as in the form of a reasonable collection of those (given that the linearity of the problem is not violated). Thus, the response of an infinite beam resting on an elastic foundation to any conventional load can be computed in compliance with the discussion that we had in Section 2.2.

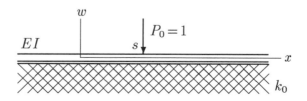

Figure 2.8 An infinite beam on an elastic foundation

2.3. BEAMS ON AN ELASTIC FOUNDATION

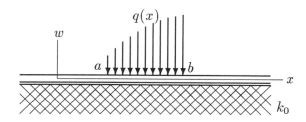

Figure 2.9 A load applied to a finite interval

If, for example, a transverse distributed load $q(x)$ is applied to a finite interval $[a, b]$ as depicted in Figure 2.9, then analogously to what has been done to eqn (2.42) in Section 2.2, we obtain the deflection at any point located on the left of a by the integral

$$w(x) = \frac{1}{8k^3 EI} \int_a^b e^{k(x-s)}[\cos(k(x-s)) - \sin(k(x-s))]q(s)ds \qquad (2.83)$$

For $a \leq x \leq b$, in turn, the deflection can be computed by

$$w(x) = \frac{1}{8k^3 EI} \left\{ \int_a^x e^{k(s-x)}[\cos(k(x-s)) + \sin(k(x-s))]q(s)ds \right.$$

$$\left. + \int_x^b e^{k(x-s)}[\cos(k(x-s)) - \sin(k(x-s))]q(s)ds \right\} \qquad (2.84)$$

When the deflection is to be computed on the right of the loaded interval (as $x \geq b$), then we finally obtain

$$w(x) = \frac{1}{8k^3 EI} \int_a^b e^{k(s-x)}[\cos(k(x-s)) + \sin(k(x-s))]q(s)ds \qquad (2.85)$$

Clearly, if the form of the loading function $q(x)$ is simple enough (polynomial, exponential, trigonometric function, or their reasonable combinations), then the integrals in eqns (2.83)–(2.85) can be computed analytically. If, however, the analytic integration is either impossible or results in a too time-consuming computation, then quite accurate numerical results can be obtained by using numerical integration instead. And in the latter case, as we have shown in Section 2.2, the accuracy attained by this approach is anticipated to be of a much higher level than that of the finite difference method.

The rest of components of the stress-strain state of an infinite beam resting on an elastic foundation can also be obtained by the recommendations given in Section 2.2. Thus, in order to compute values of the bending moment $M(x)$ and shear force $Q(x)$ caused by the load $q(x)$, we recall [18, 27] their expressions in terms of the deflection function $w(x)$

$$M(x) = EI\frac{d^2w(x)}{dx^2}, \quad Q(x) = EI\frac{d^3w(x)}{dx^3} \tag{2.86}$$

and apply these to the integrals in eqns (2.83)–(2.85). In doing so, we appropriately differentiate the kernel functions of those integrals (the corresponding branches of the influence function $g(x,s)$) with respect to the observation variable x. Hence, on the left of the loaded interval, one obtains for the bending moment and shear force caused by $q(x)$

$$M(x) = \frac{1}{4k}\int_a^b e^{k(x-s)}[\cos(k(x-s)) + \sin(k(x-s))]q(s)ds$$

and

$$Q(x) = \frac{1}{2}\int_a^b e^{k(x-s)}\cos(k(x-s))q(s)ds$$

Within the loaded interval, as $a \leq x \leq b$, one obtains

$$M(x) = \frac{1}{4k}\left\{\int_a^x e^{k(s-x)}[\cos(k(x-s)) - \sin(k(x-s))]q(s)ds\right.$$

$$\left. + \int_x^b e^{k(x-s)}[\cos(k(x-s)) + \sin(k(x-s))]q(s)ds\right\}$$

and

$$Q(x) = \frac{1}{2}\left\{\int_a^x e^{k(x-s)}\cos(k(x-s))q(s)ds - \int_x^b e^{k(s-x)}\cos(k(x-s))q(s)ds\right\}.$$

For $x \geq b$, we finally have

$$M(x) = \frac{1}{4k}\int_a^b e^{k(s-x)}[\cos(k(x-s)) - \sin(k(x-s))]q(s)ds$$

and

$$Q(x) = -\frac{1}{2}\int_a^b e^{k(s-x)}\cos(k(x-s))q(s)ds$$

The above integrals can properly be computed either analytically or numerically. As it is shown in *EXAMPLE 4* of Section 2.2, in the latter case, the level of accuracy attained must be very high.

From the material of Section 2.2, it also follows that one can use the influence function method to compute a response of an infinite beam resting on an elastic foundation to any load other than continuously distributed. This material is left for the exercises.

2.3. BEAMS ON AN ELASTIC FOUNDATION

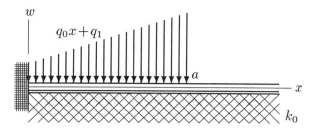

Figure 2.10 A beam subjected to a linear load

In what follows, we will formulate and solve two problems on the bending of a semi–infinite ($x \in [0, \infty)$) beam resting on an elastic foundation, with different edge conditions at $x = 0$.

EXAMPLE 1 Let a semi–infinite beam $0 \leq x < \infty$ of a uniform flexural rigidity EI rest on an elastic foundation whose elastic constant equals k_0. Let the edge $x = 0$ be clamped and let the beam be subjected to a transverse load $q(x) = q_0 x + q_1$ applied to the interval $[0, a]$ as shown in Figure 2.10.

As an introductory step for solving this problem, we will construct the influence function for the beam under consideration. Clearly, such an influence function can be identified with the Green's function $g(x, s)$ of the boundary value problem posed by eqn (2.75) subjected to the following set of boundary conditions

$$w(0) = 0, \quad \frac{dw(0)}{dx} = 0, \quad |w(\infty)| < \infty, \quad \left|\frac{dw(\infty)}{dx}\right| < \infty \qquad (2.87)$$

To precede the construction itself, we first obtain two linearly independent particular solutions of eqn (2.75), both satisfying the above written boundary conditions at $x = 0$. In order to obtain the first of such solutions, $w_1(x)$, let us formulate the following initial value problem

$$\frac{d^4 w(x)}{dx^4} + 4k^4 w(x) = 0, \quad x > 0 \qquad (2.88)$$

$$w(0) = 0, \quad \frac{dw(0)}{dx} = 0, \quad \frac{d^2 w(0)}{dx^2} = 0, \quad \frac{d^3 w(0)}{dx^3} = 1 \qquad (2.89)$$

It can easily be seen that the solution of this problem as well as any of its scalar multiples could be taken as $w_1(x)$.

Clearly, the general solution of eqn (2.88) can be written as the following linear combination

$$C_1 e^{kx} \cos(kx) + C_2 e^{kx} \sin(kx) + C_3 e^{-kx} \cos(kx) + C_4 e^{-kx} \sin(kx) \qquad (2.90)$$

of its four linearly independent particular solutions available in eqn (2.76)

Satisfying the initial conditions in eqn (2.89), one obtains the following well-posed system of linear algebraic equations

$$\begin{pmatrix} 1 & 0 & 1 & 0 \\ 1 & 1 & -1 & 1 \\ 0 & 1 & 0 & -1 \\ -1 & 1 & 1 & 1 \end{pmatrix} \times \begin{pmatrix} C_1 \\ C_2 \\ C_3 \\ C_4 \end{pmatrix} = \begin{pmatrix} 0 \\ 0 \\ 0 \\ 1/2k^3 \end{pmatrix}$$

in C_1, C_2, C_3, and C_4. Its solution is

$$C_1 = -\frac{1}{8k^3}, \quad C_2 = \frac{1}{8k^3}, \quad C_3 = \frac{1}{8k^3}, \quad C_4 = \frac{1}{8k^3}$$

Upon substituting these coefficients into eqn (2.90), we obtain the following expression for the solution

$$w(x) = \frac{1}{8k^3}[-e^{kx}\cos(kx) + e^{kx}\sin(kx) + e^{-kx}\cos(kx) + e^{-kx}\sin(kx)]$$

of the initial value problem stated in eqns (2.88) and (2.89).

Recalling then the recently made comment about a scalar multiple of $w_1(x)$, we present the component $w_1(x)$ needed in the construction of the Green's function in the form

$$w_1(x) = -e^{kx}\cos(kx) + e^{kx}\sin(kx) + e^{-kx}\cos(kx) + e^{-kx}\sin(kx) \qquad (2.91)$$

The same approach can be used for obtaining the second particular solution $w_2(x)$ of eqn (2.88), which satisfy the boundary conditions at $x=0$ (see eqn (2.87)). Namely, let us formulate the initial value problem

$$w(0) = 0, \quad \frac{dw(0)}{dx} = 0, \quad \frac{d^2w(0)}{dx^2} = 1, \quad \frac{d^3w(0)}{dx^3} = 0$$

for eqn (2.88). And again, the solution of this problem as well as any of its scalar multiples could be taken as $w_2(x)$.

Proceeding analogously to the case with $w_1(x)$, one finally obtains $w_2(x)$ in the form

$$w_2(x) = -e^{kx}\sin(kx) + e^{-kx}\sin(kx)$$

The set of functions consisting of $w_1(x)$ from eqn (2.91), $w_2(x)$ just found, $w_3(x) = e^{-kx}\cos(ks)$, and $w_4(x) = e^{-kx}\sin(kx)$ taken from eqn (2.76),

2.3. BEAMS ON AN ELASTIC FOUNDATION

is linearly independent on any finite interval. This can easily be tested with their Wronskian. Thus, the Green's function for the boundary value problem in eqns (2.75) and (2.87) or, in other words, the influence function for a semi-infinite beam whose edge is clamped, can be written as

$$g(x,s) = \begin{cases} a_1(s)w_1(x) + a_2(s)w_2(x), & x \leq s \\ b_1(s)e^{-kx}\cos(kx) + b_2(s)e^{-kx}\sin(kx), & s \leq x \end{cases} \quad (2.92)$$

Clearly, the upper branch of $g(x,s)$ satisfies the boundary conditions at $x=0$ as those shown in eqn (2.87), since $w_1(x)$ and $w_2(x)$ do so. The lower branch of $g(x,s)$, in turn, satisfies the boundedness conditions at infinity, since both $w_3(x)$ and $w_4(x)$ approach zero as x goes to infinity. Hence, $g(x,s)$ in the above form meets defining properties 1 and 4 for the Green's function.

Satisfying then properties 2 and 3, one derives a system of linear algebraic equations in $a_1(s)$, $a_2(s)$, $b_1(s)$, and $b_2(s)$. That system is well-posed because the determinant of its coefficient matrix, representing the Wronskian for $w_1(x)$, $w_2(x)$, $w_3(x)$, and $w_4(x)$, is not zero. This, subsequently, yields a unique solution for the system. Substituting the values of $a_1(s)$, $a_2(s)$, $b_1(s)$, and $b_2(s)$ into eqn (2.92), one completes the construction process.

Ultimately, the expression of the influence function of a semi-infinite beam with clamped edge, which is valid for $x \leq s$, is obtained in the form

$$g^+(x,s) = \frac{1}{8k^3} \left\{ e^{k(x-s)}[\sin(k(x-s)) - \cos(k(x-s))] \right.$$

$$\left. + e^{-k(s+x)}[\sin(k(x+s)) + 2\sin(kx)\sin(ks) + \cos(k(x-s))] \right\} \quad (2.93)$$

while for $x \geq s$, we have

$$g^-(x,s) = \frac{1}{8k^3} \left\{ e^{k(s-x)}[\sin(k(s-x)) - \cos(k(x-s))] \right.$$

$$\left. + e^{-k(s+x)}[\sin(k(x+s)) + 2\sin(kx)\sin(ks) + \cos(k(x-s))] \right\} \quad (2.94)$$

The reader should readily realize that the influence function just obtained enables the computation of output characteristics of the beam under consideration if it is subjected to any conventional loading. Therefore, let us now return to the original problem. In order to compute the response of the clamped semi-infinite beam to the transverse load $q(x)$ continuously distributed over the finite interval $[0, a]$, we recall the formula in eqn (2.42) of Section 2.2, which in this case is written as

$$w(x) = -\frac{1}{EI}\left\{\int_0^x g^-(x,s)q(s)ds + \int_x^a g^+(x,s)q(s)ds\right\}, \quad x \leq a \quad (2.95)$$

for the observation point x located within the loaded interval. If, however, the deflection function is to be computed aside from the loaded interval (if $x \geq a$), then the deflection function must be computed as

$$w(x) = -\frac{1}{EI}\int_0^a g^-(x,s)q(s)ds, \quad x \geq a$$

Recalling from the statement of this problem that $q(x) = q_0 x + q_1$ and utilizing the expressions for $g^+(x,s)$ and $g^-(x,s)$ from eqns (2.93) and (2.94), we substitute all of these into eqn (2.95). For $x \leq a$, this yields

$$w(x) = -\frac{1}{8k^3 EI}\int_0^x \left\{e^{k(s-x)}[\sin(k(s-x)) - \cos(k(x-s))]\right.$$
$$\left. + e^{-k(s+x)}[\sin(k(x+s)) + 2\sin(kx)\sin(ks) + \cos(k(x-s))]\right\}$$
$$\times [q_0(s) + q_1]ds - \frac{1}{8k^3 EI}\int_x^a \left\{e^{k(x-s)}[\sin(k(x-s)) - \cos(k(x-s))]\right.$$
$$\left. + e^{-k(x+s)}[\sin(k(x+s)) + 2\sin(kx)\sin(ks) + \cos(k(x-s))]\right\}$$
$$\times [q_0(s) + q_1]ds$$

If, however, $x \geq a$, then we obtain

$$w(x) = -\frac{1}{8k^3 EI}\int_0^a \left\{e^{k(s-x)}[\sin(k(s-x)) - \cos(k(x-s))]\right.$$
$$\left. + e^{-k(s+x)}[\sin(k(x+s)) + 2\sin(kx)\sin(ks) + \cos(k(x-s))]\right\}[q_0(s) + q_1]ds$$

To compute the bending moment $M(x)$ and the shear force $Q(x)$ caused by $q(x)$, one simply recalls the relations in eqn (2.86) and carries out the appropriate analytical differentiation of the above representations.

We below present one more example on the application of this method to the bending of a semi-infinite beam resting on an elastic foundation. Different edge conditions at $x = 0$ and different loading will be considered.

EXAMPLE 2 Construct the influence function for a semi–infinite ($x \in [0, \infty)$) beam resting on an elastic foundation, if its edge at $x = 0$ is free. Then utilize it to obtain the response to two concentrated transverse forces of magnitudes P_1 and P_2 applied at the points $x = a_1$ and $x = a_2$, $(a_1 < a_2)$, respectively (see Figure 2.11).

2.3. BEAMS ON AN ELASTIC FOUNDATION

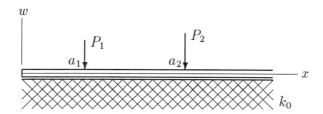

Figure 2.11 A beam freely resting on a foundation

Clearly, the influence function of a transverse unit concentrated force for this beam represents the Green's function of the following boundary value problem

$$\frac{d^2w(0)}{dx^2}=0, \quad \frac{d^3w(0)}{dx^3}=0, \quad |w(\infty)|<\infty, \quad \left|\frac{dw(\infty)}{dx}\right|<\infty$$

for eqn (2.88) formulated over the interval $(0,\infty)$.

Tracing out the procedure used in the previous example, one ultimately obtains the expression for the influence function of the semi-infinite beam with free edge, which is valid for $x \leq s$, in the form

$$g^+(x,s) = \frac{1}{8k^3}\left\{e^{k(x-s)}[\sin(k(x-s))-\cos(k(x-s))]\right.$$

$$\left. + e^{-k(s+x)}[\sin(k(x+s))-2\cos(kx)\cos(ks)-\cos(k(x-s))]\right\}, \quad (2.96)$$

while for $x \geq s$, we have

$$g^-(x,s) = \frac{1}{8k^3}\left\{e^{k(s-x)}[\sin(k(s-x))-\cos(k(x-s))]\right.$$

$$\left. + e^{-k(s+x)}[\sin(k(x+s))-2\cos(kx)\cos(ks)-\cos(k(x-s))]\right\} \quad (2.97)$$

From eqn (2.43) of Section 2.2, it follows that based on the expressions for the influence function in eqns (2.96) and (2.97), one can compute the beam's response to the forces P_1 and P_2 placed at the points a_1 and a_2, respectively, in the form

$$w(x) = -\frac{1}{EI}[P_1 g(x,a_1) + P_2 g(x,a_2)]$$

The actual computation of $w(x)$, however, must be accomplished on the basis of the piecewise format of the influence function $g(x,s)$. For $x < a_1$

(on the left of the point of application of the force P_1), for example, the deflection function caused by these forces is expressed as

$$w(x) = -\frac{1}{EI}\left[P_1 g^+(x, a_1) + P_2 g^+(x, a_2)\right]$$

For $a_1 \leq x \leq a_2$, the computation should be carried out by

$$w(x) = -\frac{1}{EI}\left[P_1 g^-(x, a_1) + P_2 g^+(x, a_2)\right]$$

Whereas, on the right of the point of application of the force P_2 (for $x > a_2$), we have

$$w(x) = -\frac{1}{EI}\left[P_1 g^-(x, a_1) + P_2 g^-(x, a_2)\right]$$

Clearly, by using these expressions for the deflection function, the bending moment $M(x)$ and the shear force $Q(x)$ in any cross-section of the beam can be immediately computed by the corresponding differentiation (see eqn (2.86))

$$M(x) = EI\frac{d^2 w(x)}{dx^2}, \quad Q(x) = EI\frac{d^3 w(x)}{dx^3}$$

Thus, all required components of the stress-strain state of the beam in this case can be analytically computed, as can be those caused by different conventional loads.

2.4 Beams of variable rigidity

In this section, we reveal one more problem class that can be successfully tackled by the influence function approach. The extension of this approach is here proposed to problems modeling static equilibrium of beams with variable flexural rigidity, undergoing transverse and bending loads.

It is shown herein that once the influence function for the beam is available, the computational procedure of this method, as applied to a beam of variable flexural rigidity, is analogous to that described earlier in this chapter for a beam of uniform flexural rigidity. Influence functions are constructed for beams with linear and exponential variation of flexural rigidity and utilized in practical computations.

The bending of beams is again treated within the scope of Kirchhoff theory that implies a boundary value problem for the Euler–Bernoulli equation

$$\frac{d^2}{dx^2}\left(EI(x)\frac{d^2 w(x)}{dx^2}\right) = -q(x), \quad (0, a) \qquad (2.98)$$

2.4. BEAMS OF VARIABLE RIGIDITY

subjected to boundary conditions written as

$$B_{0,i}[w(0)] = 0, \quad B_{a,i}[w(a)] = 0, \quad i = 1, 2 \qquad (2.99)$$

One can find below a number of particular statements of such problems. The technique for finding the solution is fairly standard for all of them. We first focus on the construction of the influence function related to the statement and then actually solve the problem by the influence function method.

Before going any further with actual solutions of particular problems, let us reveal the relations for components of the stress-strain state in terms of the influence function of a unit transverse concentrated force for a beam having a variable flexural rigidity $EI(x)$. Those relations are going to be slightly different from the corresponding ones derived in Section 2.2 for a beam with a uniform flexural rigidity.

Let $g(x, s)$ be the influence function for the beam under consideration. That is, $g(x, s)$ represents the Green's function for the homogeneous boundary value problem associated with that in eqns (2.98) and (2.99).

In compliance with Theorem 1.3 of Chapter 1, for a beam having a variable flexural rigidity, the deflection caused by the transverse load $q(x)$ continuously distributed over the interval $[\alpha, \beta]$ can then be computed, for $x \leq \alpha$ (on the left of the loaded interval), as

$$w(x) = \int_\alpha^\beta g^+(x,s) q(s) ds, \quad x \in [0, \alpha] \qquad (2.100)$$

For $\alpha \leq x \leq \beta$, in turn, it can be obtained as

$$w(x) = \int_\alpha^x g^-(x,s) q(s) ds + \int_x^\beta g^+(x,s) q(s) ds, \quad x \in [\alpha, \beta] \qquad (2.101)$$

For $x \geq \beta$ (on the right of the loaded interval) we have

$$w(x) = \int_\alpha^\beta g^-(x,s) q(s) ds, \quad x \in [\beta, a] \qquad (2.102)$$

where $g^-(x, s)$ and $g^+(x, s)$ represent the branches of the influence function $g(x, s)$ defined for $s \leq x$ and $x \leq s$, respectively.

Following the standard technique described in Section 2.2, one can readily derive the formula

$$w(x) = \sum_{i=1}^k P_i g(x, s_i), \quad x \in [0, a] \qquad (2.103)$$

for the deflection (of a beam having a variable flexural rigidity) caused by a set of transverse concentrated forces of magnitudes P_i, $(i = \overline{1,k})$, located at the points s_i respectively.

For a beam having a variable flexural rigidity, the deflection caused by a set of concentrated bending moments of magnitudes M_i, $(i = \overline{1,k})$, located at the points s_i respectively, can be computed by means of the influence function of the second order as

$$w(x) = \sum_{i=1}^{k} M_i \frac{\partial g(x, s_i)}{\partial s}, \quad x \in [0, a] \tag{2.104}$$

And analogously to the situation with a beam of uniform flexural rigidity, the response to a reasonable combination of elementary loads can be computed as a linear combination of the corresponding elementary responses. Saying *reasonable* we again mean that the simultaneous action of individual loads must not cause either geometrical or physical nonlinearity.

Notice that for a beam having a variable flexural rigidity, the output bending moment $M(x)$ and shear force $Q(x)$ are expressed in terms of the deflection function $w(x)$ written as

$$M(x) = EI(x) \frac{d^2 w(x)}{dx^2}, \quad Q(x) = \frac{d}{dx}\left(EI(x) \frac{d^2 w(x)}{dx^2}\right) \tag{2.105}$$

Hence, the expressions for $M(x)$ and $Q(x)$ in any particular problem for a beam of a variable flexural rigidity can be obtained by a proper differentiation of the deflection function. For example, the output bending moment $M(x)$ caused by a set of concentrated bending moments of magnitudes M_i, $(i = \overline{1,k})$, located at the points s_i respectively, can be computed by formally taking the second derivative of $w(x)$ in eqn (2.104) and substituting it into the first equation in (2.105). This yields

$$M(x) = EI(x) \sum_{i=1}^{k} M_i \frac{\partial^3 g(x, s_i)}{\partial s \partial x^2}, \quad x \in [0, a]$$

Below one finds some particular examples where we demonstrate the practical solvability of problems for beams of variable flexural rigidity by means of the influence function method.

EXAMPLE 1 Consider a cantilever beam of length a whose edge at $x = 0$ is clamped, if its flexural rigidity $EI(x)$ is represented by a linear function $EI(x) = mx + b$ of the observation variable x.

It should be intuitively clear that a transverse concentrated force applied to an internal point of the beam under consideration causes its unique response. In other words, there is no doubt that there exists a unique influence function for this beam, which must be identified with the Green's function of the homogeneous equation

$$\frac{d^2}{dx^2}\left((mx + b) \frac{d^2 w(x)}{dx^2}\right) = 0 \tag{2.106}$$

2.4. BEAMS OF VARIABLE RIGIDITY

subjected to the following set of boundary conditions

$$w(0)=0, \quad \frac{dw(0)}{dx}=0, \quad \frac{d^2w(a)}{dx^2}=0, \quad \frac{d^3w(a)}{dx^3}=0, \qquad (2.107)$$

The method of variation of parameters will be used here for the construction of the Green's function. To obtain the fundamental solution set for eqn (2.106), we first rewrite this equation in the form

$$(mx+b)\frac{d^4w(x)}{dx^4} + 2m\frac{d^3w(x)}{dx^3} = 0 \qquad (2.108)$$

by simply removing the parenthesis in eqn (2.106).

Because of the specific form of this equation (it contains only the third and the fourth derivatives of $w(x)$), one can easily obtain the first three components of its fundamental solution set

$$w_1(x) \equiv 1, \quad w_2(x) \equiv x, \quad w_3(x) \equiv x^2$$

To determine the fourth component $w_4(x)$ of the fundamental solution set, it is convenient to introduce a new function $u(x)$ written as

$$u(x) = \frac{d^3w(x)}{dx^3} \qquad (2.109)$$

and to reduce eqn (2.108) to the separable type of the first order

$$(mx+b)\frac{du(x)}{dx} + 2mu(x) = 0$$

A particular solution of this equation can readily be obtained by straightforward integration. This yields

$$u(x) = \frac{1}{(mx+b)^2} \qquad (2.110)$$

And in compliance with the relation in eqn (2.109), by integrating $u(x)$ in eqn (2.110) three times successively, we eventually obtain $w_4(x)$ in the form

$$w_4(x) \equiv (mx+b)\ln(mx+b)$$

Based on the fundamental solution set consisting of $w_1(x)$, $w_2(x)$, $w_3(x)$, and $w_4(x)$ just obtained, we can start the actual construction of the influence function. In doing so by the method of variation of parameters, we seek the general solution of eqn (2.98) in the form

$$w(x) = C_1(x) + C_2(x)x + C_3(x)x^2 + C_4(x)(mx+b)\ln(mx+b) \quad (2.111)$$

According to the conventional procedure of this method, one obtains the following system

$$\begin{pmatrix} 1 & x & x^2 & (mx+b)\ln(mx+b) \\ 0 & 1 & 2x & m\ln(mx+b)+m \\ 0 & 0 & 2 & m^2/(mx+b) \\ 0 & 0 & 0 & -m^3/(mx+b)^2 \end{pmatrix} \times \begin{pmatrix} C_1'(x) \\ C_2'(x) \\ C_3'(x) \\ C_4'(x) \end{pmatrix} = \begin{pmatrix} 0 \\ 0 \\ 0 \\ q^*(x) \end{pmatrix}$$

of linear algebraic equations in $C_i'(x)$, $(i=\overline{1,4})$. Here $q^*(x)=-q(x)/(mx+b)$. The well-posedness of this system follows from the linear independence of the components $w_i(x)$, $(i=\overline{1,4})$ in the fundamental solution set.

From the above system, it follows that

$$C_1'(x) = \frac{1}{2m^3}[mx(mx+2b) - b(mx+b)\ln(mx+b)]q(x),$$

$$C_2'(x) = -\frac{1}{m^2}[b + (mx+b)\ln(mx+b)]q(x),$$

and

$$C_3'(x) = -\frac{1}{2m}q(x), \quad C_4'(x) = \frac{mx+b}{m^3}q(x).$$

Integrating these relations, one obtains

$$C_1(x) = \int_0^x \frac{1}{2m^3}[ms(ms+2b) - b(ms+b)\ln(ms+b)]q(s)ds + H_1,$$

$$C_2(x) = -\int_0^x \frac{1}{m^2}[b + (ms+b)\ln(ms+b)]q(s)ds + H_2,$$

and

$$C_3(x) = -\int_0^x \frac{1}{2m}q(s)ds + H_3, \quad C_4(x) = \int_0^x \frac{(ms+b)}{m^3}q(s)ds + H_4.$$

Upon substituting these coefficients into eqn (2.111) and accomplishing some routine algebra, one finally obtains the general solution of eqn (2.98)

$$w(x) = H_1 + H_2 x + H_3 x^2 + H_4(mx+b)\ln(mx+b) +$$

$$\int_0^x \frac{1}{2m^3}\left\{2(mx+b)(ms+b)\ln\frac{mx+b}{ms+b} - m(x-s)[m(x+s)+2b]\right\}q(s)ds$$

Completing the construction procedure for the influence function, we should compute the coefficients H_i, upon satisfying the boundary conditions

2.4. BEAMS OF VARIABLE RIGIDITY

imposed with eqn (2.107). Substituting then these coefficients into the above equation, the deflection $w(x)$ should be expressed in the form of a single integral over the interval $[0, a]$. And then, recalling Theorem 1.3 from Chapter 1, one finally obtains the explicit expression for the influence function in question. In doing so, that branch of the influence function, which is valid for $x \leq s$, is finally presented in the form

$$g^+(x,s) = \frac{1}{2m^3} \left\{ mx[mx + 2(ms+b)] + 2(mx+b)(ms+b) \ln \frac{b}{mx+b} \right\}$$

while, for $x \geq s$, we have

$$g^-(x,s) = \frac{1}{2m^3} \left\{ ms[ms + 2(mx+b)] + 2(mx+b)(ms+b) \ln \frac{b}{ms+b} \right\}$$

Based on the influence function just obtained, one can compute any component of a stress–strain state of the beam under consideration, caused by a combination of transverse and bending loads.

Let the beam under consideration be subjected to a combination of two loads. Suppose a concentrated transverse force of magnitude P_0 is applied at $x = a_1$ and a concentrated bending moment of magnitude M_0 is applied at $x = a_2$, with $a_1 < a_2$ (see Figure 2.12).

In accordance with the relations in eqns (2.103) and (2.104), the resultant deflection $w(x)$ of this beam can be expressed as

$$w(x) = P_0 g(x, a_1) + M_0 \frac{\partial g(x, a_2)}{\partial s}$$

To compute this function, one should account for a piecewise format of $g(x, s)$. That is, on the left of a_1, the resultant deflection is expressed by

$$w(x) = \frac{P_0}{2m^3} \left\{ mx[mx + 2(ma_1+b)] + 2(mx+b)(ma_1+b) \ln \frac{b}{mx+b} \right\}$$
$$+ \frac{M_0}{m^2} \left[mx + (mx+b) \ln \frac{b}{mx+b} \right], \quad x \leq a_1$$

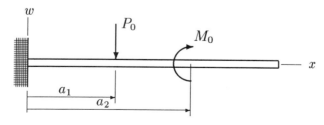

Figure 2.12 A cantilever beam of a variable rigidity

Between a_1 and a_2, $w(x)$ can be found as

$$w(x) = \frac{P_0}{2m^3}\left\{ma_1[ma_1+2(mx+b)]+2(mx+b)(ma_1+b)\ln\frac{b}{ma_1+b}\right\}$$

$$+\frac{M_0}{m^2}\left[mx+(mx+b)\ln\frac{b}{mx+b}\right], \quad a_1 < x < a_2$$

Whereas, for $x \geq a_2$, we have

$$w(x) = \frac{P_0}{2m^3}\left\{ma_1[ma_1+2(mx+b)]+2(mx+b)(ma_1+b)\ln\frac{b}{ma_1+b}\right\}$$

$$+\frac{M_0}{m^2}\left[ma_2+(mx+b)\ln\frac{b}{ma_2+b}\right], \quad x \geq a_2$$

To compute either the bending moment $M(x)$ or the shear force $Q(x)$ caused by P_0 and M_0, one is required to properly differentiate the above expressions for the deflection function in accordance with the relations in eqn (2.105).

EXAMPLE 2   Determine the deflection $w(x)$, the bending moment $M(x)$, and the shear force $Q(x)$ of a simply-supported beam of length a, with the flexural rigidity $EI(x)$ being an exponential function me^{bx}, subjected to a transverse load $q^*(x)$ continuously distributed over the interval $[\alpha, \beta]$, as shown in Figure 2.13.

Clearly, the boundary value problem modeling the equilibrium state of the beam in this statement can be formulated as follows

$$\frac{d^2}{dx^2}\left(me^{bx}\frac{d^2w(x)}{dx^2}\right) = -q(x), \quad x \in (0,a) \tag{2.112}$$

$$w(0) = \frac{d^2w(0)}{dx^2} = 0, \quad w(a) = \frac{d^2w(a)}{dx^2} = 0 \tag{2.113}$$

where

$$q(x) = \begin{cases} 0, & x < \alpha \\ q^*(x) & \alpha \leq x \leq \beta \\ 0, & x > \beta \end{cases}$$

Since the construction procedure for an influence function (which represents in this case the Green's function for the homogeneous boundary value problem corresponding to that posed by eqns (2.112) and (2.113)) can be developed on the standard basis, we will not provide its detailed description here.

2.4. BEAMS OF VARIABLE RIGIDITY

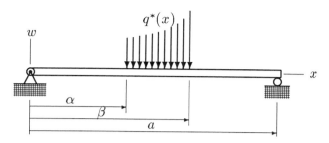

Figure 2.13   A beam of a variable rigidity

One issue in this procedure, however, is worth focusing on: how to obtain the fundamental solution set for the homogeneous equation associated with that in eqn (2.112). To address this issue, we accomplish the differentiation in eqn (2.112) by using the product rule. This yields

$$\frac{d^4w(x)}{dx^4} + 2b\frac{d^3w(x)}{dx^3} + b^2\frac{d^2w(x)}{dx^2} = -\frac{1}{m}q(x)e^{-bx}$$

Hence, eqn (2.112) is reduced to one with constant coefficients. Since its characteristic equation has two roots ($k=0$ and $k=-b$, each of multiplicity two), it follows that the fundamental solution set for the homogeneous equation corresponding to that in eqn (2.112) can be represented by the following set of functions

$$w_1(x) \equiv 1, \quad w_2(x) \equiv x, \quad w_3(x) \equiv e^{-bx}, \quad w_4(x) \equiv xe^{-bx}$$

Upon using this set, one readily obtains the influence function $g(x,s)$ for the simply supported beam having an exponential flexural rigidity. In doing so, we follow the standard procedure. The branch of this function which is defined for $x \le s$ is finally expressed as

$$g^+(x,s) = \frac{1}{mb^3a^2}\{2xse^{-ba} - 2(a-x)(a-s)$$

$$+ xa[b(a-s)-2]e^{-bs} + a(a-s)(2+bx)e^{-bx}\}$$

while for $x \ge s$, we obtain

$$g^-(x,s) = \frac{1}{mb^3a^2}\{2xse^{-ba} - 2(a-s)(a-x)$$

$$+ sa[b(a-x)-2]e^{-bx} + a(a-x)(2+bs)e^{-bs}\}$$

Since the analytic expression for the influence function is already available, we now can return to the original statement of the problem posed by eqns (2.112) and (2.113). Upon utilizing this influence function, one determines the solution of this problem (that is the deflection, caused by the transverse load $q(x)$ applied to the interval $[\alpha, \beta]$) in accordance with the relations given by eqns (2.100)–(2.102). It follows that in this case, for $x \leq \alpha$ (on the left of the loaded interval), we have

$$w(x) = \frac{1}{mb^3 a^2} \int_\alpha^\beta \{2xse^{-ba} - 2(a-x)(a-s)$$

$$+ xa[b(a-s) - 2]e^{-bs} + a(a-s)(2+bx)e^{-bx}\} q(s) ds, \quad x \in [0, \alpha]$$

For $\alpha \leq x \leq \beta$, in turn, $w(x)$ can be obtained as

$$w(x) = \frac{1}{mb^3 a^2} \int_\alpha^x \{2xse^{-ba} - 2(a-s)(a-x)$$

$$+ sa[b(a-x) - 2]e^{-bx} + a(a-x)(2+bs)e^{-bs}\} q(s) ds$$

$$+ \frac{1}{mb^3 a^2} \int_x^\beta \{2xse^{-ba} - 2(a-x)(a-s)$$

$$+ xa[b(a-s) - 2]e^{-bs} + a(a-s)(2+bx)e^{-bx}\} q(s) ds, \quad x \in [\alpha, \beta]$$

For $x \geq \beta$ (on the right of the loaded interval), we finally obtain

$$w(x) = \frac{1}{mb^3 a^2} \int_\alpha^\beta \{2xse^{-ba} - 2(a-s)(a-x)$$

$$+ sa[b(a-x) - 2]e^{-bx} + a(a-x)(2+bs)e^{-bs}\} q(s) ds, \quad x \in [\beta, a]$$

If the loading function $q(x)$ is simple enough, the integrals above can be computed analytically. When $q(x)$ is too complicated to practically obtain an analytic solution, one should numerically integrate by choosing an appropriate quadrature formula. The choice is determined by the level of accuracy that is desired. For our numerical computations, the trapezoid rule provides sufficiently accurate results.

To compute either bending moments $M(x)$ or shear forces $Q(x)$, caused by $q(x)$, one is required to differentiate the expression for $w(x)$ in compliance with the relations in eqn (2.105).

So far in this section, we have considered equations (see eqns (2.106) or (2.112)) whose fundamental solution sets are expressed in terms of elementary functions. Let us now pose the beam-related homogeneous boundary value problem

$$\frac{d^2}{dx^2}\left(EI(x)\frac{d^2 w(x)}{dx^2}\right) = 0 \tag{2.114}$$

$$B_{0,i}[w(0)] = 0, \quad B_{a,i}[w(a)] = 0, \quad i = 1, 2 \tag{2.115}$$

2.4. BEAMS OF VARIABLE RIGIDITY

corresponding to that in eqns (2.98) and (2.99). Suppose eqn (2.114) does not allow an analytical solution. That is, the fundamental solution set of this equation is not available in the form consisting of elementary functions. Clearly, the Green's function for the above problem represents the associated influence function.

The construction procedure that has been developed in Section 2.1 for the influence function of a beam having uniform flexural rigidity and has been used earlier in this section for some particular cases of eqn (2.114), can be directly applied to the problem in eqns (2.114)–(2.115). The point is that the fundamental solution set $\{w_i(x)\}, (i = 1, 2, 3, 4)$ for eqn (2.114), which is to be utilized in the construction procedure, must not be necessarily available in analytical form. Accurate numerical solutions can be used instead. In order to illustrate the latter point, we present the next example.

EXAMPLE 3 Let us formulate the following boundary value problem for the nonhomogeneous equation

$$\frac{d^2}{dx^2}\left((mx+b)\frac{d^2w(x)}{dx^2}\right) = -q(x), \quad x \in (0, a) \qquad (2.116)$$

subjected to the boundary conditions

$$w(0) = \frac{d^2w(0)}{dx^2} = 0, \quad w(a) = \frac{d^2w(a)}{dx^2} = 0 \qquad (2.117)$$

This problem is associated with a continuously loaded beam of length a. The edges of the beam are simply supported and its flexural rigidity is a linear function of x.

As it is shown in *EXAMPLE 1* of this section, the homogeneous equation corresponding to that in eqn (2.116) allows an analytical solution. We will take advantage of this fact when comparing the two Green's function solutions of the above boundary value problem. One of these solutions will be computed analytically based on the fundamental solution set that has been used in eqn (2.111) when we constructed the influence function for a cantilever beam whose flexural rigidity linearly depends on x. The other Green's function solution of the problem in eqns (2.116) and (2.117) will be obtained based on the numerically computed fundamental solution set.

To numerically obtain the four linearly independent particular solutions of the homogeneous equation

$$\frac{d^2}{dx^2}\left((mx+b)\frac{d^2w(x)}{dx^2}\right) = 0 \qquad (2.118)$$

corresponding to that in eqn (2.116), we pose the following four separate sets of initial conditions

$$w(0) = 1, \quad \frac{dw(0)}{dx} = \frac{d^2w(0)}{dx^2} = \frac{d^3w(0)}{dx^3} = 0 \qquad (2.119)$$

$$\frac{dw(0)}{dx}=1, \quad w(0)=\frac{d^2w(0)}{dx^2}=\frac{d^3w(0)}{dx^3}=0 \qquad (2.120)$$

$$\frac{d^2w(0)}{dx^2}=1, \quad \frac{dw(0)}{dx}=w(0)=\frac{d^3w(0)}{dx^3}=0 \qquad (2.121)$$

$$\frac{d^3w(0)}{dx^3}=1, \quad \frac{dw(0)}{dx}=\frac{d^2w(0)}{dx^2}=w(0)=0 \qquad (2.122)$$

It can be easily shown that the solutions of the four initial value problems posed by eqns (2.118) and (2.119), (2.118) and (2.120), (2.118) and (2.121), and (2.118) and (2.122) are linearly independent on [0,a]. Indeed, let us denote by $w_1(x)$ the solution of the problem posed by eqns (2.118) and (2.119). Let $w_2(x)$ represent the solution of the problem in eqns (2.118) and (2.120). Let $w_3(x)$ be the solution of the problem posed by eqns (2.118) and (2.121). And finally, let $w_4(x)$ represent the solution of the problem posed by eqns (2.118) and (2.122). If so, then the linear combination of these functions

$$W(x)=C_1w_1(x)+C_2w_2(x)+C_3w_3(x)+C_4w_4(x)$$

with arbitrary coefficients $C_i, (i=1,2,3,4)$, represents a solution of the initial value problem written as

$$\frac{d^2}{dx^2}\left((mx+b)\frac{d^2W(x)}{dx^2}\right)=0, \quad x\geq 0$$

$$W(0)=C_1, \quad \frac{dW(0)}{dx}=C_2, \quad \frac{d^2W(0)}{dx^2}=C_3, \quad \frac{d^3W(0)}{dx^3}=C_4$$

However, this initial value problem has a nontrivial solution if at least one of the four constants C_i is nonzero. And the only case for which $W(x)\equiv 0$ on $[0,a]$ is that with all C_i equal to zero. Hence, the functions $w_i(x), (i=1,2,3,4)$, which represent solutions of the initial value problems posed by eqns (2.118) and (2.119); (2.118) and (2.120); (2.118) and (2.121); and (2.118) and (2.122), respectively, are linearly independent on $[0,a]$. They could therefore constitute the fundamental solution set for eqn (2.118).

Practical computing of such components $w_i(x)$ of the fundamental solution set for eqn (2.118) can be carried out numerically by utilizing either one of the existing standard procedures [4, 16, 17, 61] for solving initial value problems. And then, once the Green's function $g(x,s)$ for the boundary value problem in eqns (2.117) and (2.118) is obtained, the solution of the problem posed by eqns (2.116) and (2.117) can be found in terms of $g(x,s)$

2.4. BEAMS OF VARIABLE RIGIDITY

and the right-hand side term of eqn (2.116), in compliance with Theorem 1.3 of Chapter 1.

In Table 2.6 some results are presented on the comparison of the analytical and numerical Green's function treatments of the problem in eqns (2.116) and (2.117), where we assumed: $a = 1, m = 3, b = 2$, and the loading function is defined as $q(x) = -\csc(2 + x/2)$. Values of the deflection function $w(x)$ are exhibited.

Notably, the analytical version yields the exact solution of the problem, whereas in the numerical version, the entire procedure is almost identical to that in the analytical version, except for the manner in which to get the components $w_i(x)$ of the fundamental solution set of eqn (2.118). We numerically obtained $w_i(x)$ by applying the standard fourth-order Runge–Kutta procedure [4, 16, 61] to each of the initial value problems associated with eqns (2.119), (2.120), (2.121), and (2.122) (with n being the partitioning parameter of the interval (0,1)).

Both the analytical and numerical Green's function solutions have been computed in the Maple V computer algebra system environment. Clearly, when computing the analytical solutions, Maple V is utilized only for analytic transformation, whereas when being applied to the numerical version, it also runs the Runge–Kutta procedure for each of the components $w_i(x)$ of the fundamental solution set.

The relatively rapid convergence of the numerical version of the Green's function method should be evident from the data of Table 2.6. It brings a confidence in high efficiency of the influence function method applied to problems that are related to beams having variable flexural rigidity, if the governing differential equation cannot be solved analytically.

Table 2.6 The computational effectiveness of the numerical version of the Green's function method applied to the problem in eqns (2.116) and (2.117)

Observation point, x	Analytical solution	Numerical solution		
		$n = 10$	$n = 20$	$n = 100$
0.1	.0016815	.0016785	.0016798	.0016815
0.2	.0031287	.0031209	.0031251	.0031288
0.3	.0041962	.0041824	.0041895	.0041962
0.4	.0048088	.0047872	.0047967	.0048088
0.5	.0049432	.0049014	.0049301	.0049432
0.6	.0046155	.0046899	.0046632	.0046155
0.7	.0038732	.0038553	.0038682	.0038732
0.8	.0027897	.0027794	.0027869	.0027897
0.9	.0014604	.0014566	.0014593	.0014604
Percent of error		1.0	.20	.01

Notice that the numerical version does not look computationally expensive unless the partitioning parameter n exceeds the level of 30–40. CPU time to run the numerical version for $n=20$, for instance, is about the same as that to run the analytical version. For $n=100$, however, the numerical version becomes five to eight times as computationally expensive as the analytical version. But the accuracy level attained with $n=20$ is high enough to satisfy most needs of practice.

2.5 Review Exercises

2.1 For the equation $w''(x) + k^2 w(x) = 0$, on $[a, b]$, analytically determine the fundamental solution set $\{w_1(x), w_2(x)\}$ such that:

(a) $w_1(a) = 0$ and $w_2(b) = 0$.

(b) $w_1(a) = 0$ and $w_2'(b) = 0$.

(c) $w_1'(a) = 0$ and $w_2'(b) - hw_2(b) = 0$.

(d) $w_1'(a) + hw_1(a) = 0$ and $w_2(b) = 0$.

2.2 For the equation $w''(x) - k^2 w(x) = 0$, on $[a, b]$, analytically determine the fundamental solution set $\{w_1(x), w_2(x)\}$ such that:

(a) $w_1(a) = 0$ and $w_2(b) = 0$.

(b) $w_1(a) = 0$ and $w_2'(b) = 0$.

(c) $w_1'(a) = 0$ and $w_2'(b) + hw_2(b) = 0$.

(d) $w_1'(a) + hw_1(a) = 0$ and $w_2(b) = 0$.

2.3 For the equation $w^{IV}(x) + 4k^4 w(x) = 0$, on $[0, a]$, analytically determine the fundamental solution set $w_i(x)$, $(i = 1, 2, 3, 4)$ such that:

(a) $w_{1,2}(0) = w_{1,2}'(0) = 0$ and $w_{3,4}(a) = w_{3,4}''(a) = 0$.

(b) $w_{1,2}(0) = w_{1,2}''(0) = 0$ and $w_{3,4}(a) = w_{3,4}''(a) = 0$.

(c) $w_{1,2}''(0) = w_{1,2}'''(0) = 0$ and $w_{3,4}(a) = w_{3,4}'(a) = 0$.

2.4 For a simply supported beam of length a, construct the influence function of a transverse unit force (Figure 2.14) by the classical method described in Section 1.1 of Chapter 1.

2.5. REVIEW EXERCISES

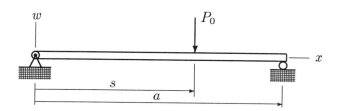

Figure 2.14   A beam subjected to a point force

2.5 For a simply supported beam, construct the influence function by the modification of the classical method proposed in Section 2.1. Compare your routine against that of the classical method.

2.6 Construct the influence function for the simply supported – clamped beam depicted in Figure 2.15. Use the classical approach as well as its modification discussed earlier in Section 2.1. Compare your routine against that of the classical method.

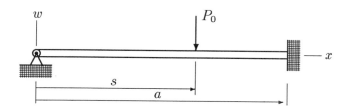

Figure 2.15   A simply supported – clamped beam

2.7 For a beam of length a with both edges elastically supported (see Figure 2.16), construct the influence function by the modification of the classical method proposed in Section 2.1.

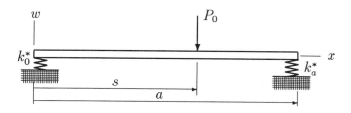

Figure 2.16   An elastically supported beam

2.8 Consider the following particular cases of the influence function constructed in exercise 2.7, and explain what physical models stand behind each of these statements:

(a) either k_0^* or k_a^* approach zero;

(b) both k_0^* and k_a^* approach infinity;

(c) k_0^* approaches infinity, while k_a^* remains finite.

2.9 For a beam of length a, with the left edge being simply supported, while the right edge is elastically supported (k_a^*) (see Figure 2.17), construct the influence function by the method of variation of parameters. What happens to this influence function as k_a^* approaches zero (as it approaches infinity)?

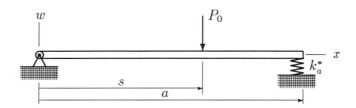

Figure 2.17 Simply and elastically supported beam

2.10 Construct the influence function for a beam of length a, whose left edge is simply supported, while the right edge is sliding (see Figure 2.18).

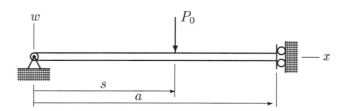

Figure 2.18 A simply supported – sliding beam

2.11 Construct the influence function for a beam of length a, with the left edge being clamped, while the right edge is sliding (see Figure 2.19).

2.5. REVIEW EXERCISES

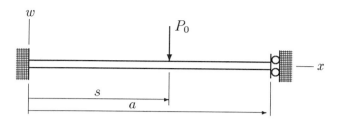

Figure 2.19 A clamped – sliding beam

2.12 For a beam of length a with the left and right edges being simply supported and clamped, respectively, determine the deflection, the bending moment, and the shear force caused by:

(a) a transverse load $q(x) = q_0(x - a/4)(x - 3a/4)$ distributed over the interval $[a/4, 3a/4]$;

(b) two concentrated forces P_1 and P_2, spaced at $x = a/3$ and $x = 2a/3$, respectively;

(c) two concentrated bending moments M_1 and M_2, spaced at $x = a/3$ and $x = 2a/3$, respectively;

2.13 For a beam of length a with the left and right edges being simply and elastically supported (k_a^*), respectively, determine the deflection, the bending moment, and the shear force caused by:

(a) a transverse load $q(x) = q_0 x(x - a)$ distributed over the entire beam;

(b) two concentrated bending moments M_1 and M_2, spaced at $x = a/4$ and $x = a/2$, respectively;

(c) two concentrated forces P_1 and P_2, spaced at $x = a/3$ and $x = 2a/3$, respectively.

2.14 For a beam of length a with both the edges being elastically supported (k_0^* and k_a^*), determine the deflection function caused by:

(a) a transverse load $q(x) = q_0 x(x - a)$ distributed over the entire beam;

(b) a combination of the bending moment M_0 spaced at $x = a/2$ and a uniform load $q(x) \equiv q_0$ distributed over the entire beam;

(c) a combination of the concentrated force P_0 and the bending moment M_0 spaced at $x = a/4$ and $x = a/2$, respectively;

(d) a transverse load $q(x) = q_0 \exp(-x^2)$ distributed over the entire beam.

2.15 For the infinite beam resting on an elastic foundation, with EI and k_0 being the flexural rigidity of the beam and the elastic constant of the foundation, respectively, determine its deflection function caused by the loads shown below:

(a) a transverse load $q(x) = q_0(1+x^2)$ continuously distributed over the interval $[a, b]$;

(b) two transverse concentrated forces P_1 and P_2, spaced at $x = a$ and $x = b$, respectively, with $b > a$;

(c) a combination of the transverse load q_0 uniformly distributed over the interval $[a_1, a_2]$ and the concentrated force P_0 spaced at $x = b$, where $b > a_2$;

(d) two concentrated bending moments M_1 and M_2, spaced at $x = a$ and $x = b$, respectively, with $b > a$.

2.16 For a semi-infinite beam resting on an elastic foundation and having a free edge, determine its response to the loads shown below:

(a) two concentrated bending moments, M_1 and M_2, spaced at $x = a$ and $x = b$, respectively, with $b > a$;

(b) a combination of the concentrated force P_0 and the bending moment M_0 spaced at $x = a$ and $x = b$, respectively, with $b > a$.

2.17 Construct the influence function for a semi-infinite beam (EI) resting on an elastic foundation (k_0), if its edge is simply supported.

2.18 For the beam in exercise 2.17, determine its deflection function caused by the loads shown below:

(a) a combination of the concentrated moment M_0 spaced at $x = a$ and the concentrated force P_0 spaced at $x = b$, with $b > a$;

(b) two concentrated bending moments M_1 and M_2, spaced at $x = a$ and $x = b$, respectively, with $b > a$.

2.19 For a cantilever beam of length a, whose left edge is clamped, with $EI(x) \equiv mx + b$ representing the flexural rigidity, determine its response to the loads shown below:

(a) a uniform transverse load q_0 distributed over the entire beam;

2.5. REVIEW EXERCISES

(b) two concentrated forces P_1 and P_2, spaced at $x=a_1$ and $x=a_2$, respectively, with $a_2 > a_1$;

(c) two concentrated bending moments M_1 and M_2, spaced at $x=a_1$ and $x=a_2$, respectively, with $a_2 > a_1$.

2.20 Construct the influence function for a cantilever beam of length a, whose right edge is clamped, with $EI(x) \equiv mx$ representing the flexural rigidity.

2.21 For the beam in exercise 2.20, determine its response to a combination of the transverse concentrated force P_0 spaced at $x = a_1$, and the concentrated bending moment M_0 spaced at $x = a_2$, with $a_1 < a_2$.

2.22 Construct the influence function for a cantilever beam of length a, whose left edge is clamped, with $EI(x) \equiv me^{bx}$ representing the flexural rigidity.

2.23 For the beam in exercise 2.22, determine its response to a combination of the concentrated force P_0 spaced at $x = a/4$ and the uniform transverse load q_0 distributed over the interval $[a/2, a]$.

Chapter 3

Some Beam Problems

The first two chapters of this text have revealed the fundamental aspects of the influence function method as applied to Kirchhoff's beam problems. Chapter 1 introduced the notion of a Green's function, while Chapter 2 highlighted the ways of possible utilization of influence (Green's) functions in solving some elementary beam problems for which the practicality of this method can be easily shown.

In this chapter, the reader will experience the extension of this method to some other problem classes that occur in Kirchhoff's beam theory. Some of those problems have traditionally been formulated in solid mechanics and their solutions are usually described in detail in texts covering this subject. We will consider, for example, the determination of natural frequencies of transverse vibrations of beams as well as some classical buckling beam problems.

The reader can readily recall that each of the problem classes mentioned above is usually modeled with a certain boundary value problem for an ordinary differential equation. Each of those classes is usually tackled by a specific individual approach which is appropriate only for that problem class for which it is developed. In other words, there is no universal method available that could be equally effective for all of those problem classes.

With this in mind, we demonstrate in this chapter that the influence function method could actually serve as a universal approach for a variety of beam problem classes. By including among those the traditional natural vibrations and buckling problems, we show that their practical solvability can also be achieved by means of the influence function method that has never been used before in tackling these problem classes.

In addition, one nontraditional (for this type of text) problem will be posed and analyzed. We will discuss one specific nonlinear formulation of a contact beam problem for which the influence function formalism also appears to be fairly productive.

3.1 Transverse natural vibrations

This and the two following sections are designed as special units in the present text. They will illustrate a possibility of offering some nontraditional undergraduate research projects within the existing curricula in mechanical engineering or in other related fields. The author's intention here is to initiate the reader's interest in implementing the influence function method for some linear and nonlinear formulations which could be even more complicated compared to those actually treated in this section.

So far in this text, we have implemented influence functions to problems for which these functions have been constructed. Indeed, we have computed components of a stress-strain state that is caused by a variety of loads applied to a beam, in terms of its response to a transverse concentrated unit force applied to an arbitrary point in the beam. In other words, direct applications of influence functions have been considered so far.

In this section we are going to advance a little further. The possibility of some indirect applications of influence functions will be explored. We will utilize influence functions, constructed for problems of a linear bending of beams, for the solution of some more complicated problems.

Expanding the sphere of productive application of the influence function method, we will recall in this section one more problem class from Kirchhoff beam theory. This method happens to be fairly efficient when computing frequencies and mode shapes of free transverse vibrations of a beam subjected to any physically feasible combination of edge conditions.

As is known (see, for example, [73]), this problem class for beams of variable flexural rigidity is associated with the so-called *eigenvalue problem* written as

$$\frac{d^2}{dx^2}\left(EI(x)\frac{d^2w(x)}{dx^2}\right) - p^2 m(x)A(x)w(x) = 0, \quad x \in [0, a] \tag{3.1}$$

$$B_{0,k}[w(0)] = 0, \quad B_{a,k}[w(a)] = 0, \quad k = 1, 2 \tag{3.2}$$

where, in addition to our customary notations, $m(x)$ and $A(x)$ represent the mass density of the material and the cross-sectional area of the beam, respectively, and p is a parameter to be determined. The boundary conditions in (3.2) are again presented in general form, since we do not need them to be specified at the moment. However, as we have mentioned in Chapter 2, any certain physically feasible set of edge conditions can be derived from eqn (3.2) as a particular case.

In mathematics, those values of the parameter p that yield nontrivial solutions for the above problem, are referred to as the *eigenvalues* of this problem, while the corresponding nontrivial solutions themselves are called

3.1. TRANSVERSE NATURAL VIBRATIONS

the *eigenfunctions* of this problem. Physical interpretation of eigenvalues and eigenfunctions directly leads to the natural frequencies and modes of transverse vibration for the beam under consideration.

From the homogeneity of eqn (3.1), it follows that if a certain function represents its solution, then any scalar multiple of this function is also a solution. In other words, an eigenfunction is defined up to a scalar multiple.

In showing how the influence function method could be applied to the problem posed by eqns (3.1) and (3.2), let us reduce this problem to a regular integral equation whose approximate solution would be easily within reach by standard numerical methods.

Let $g(x,s)$ represent the Green's function for the boundary value problem posed by the homogeneous equation

$$\frac{d^2}{dx^2}\left(EI(x)\frac{d^2w(x)}{dx^2}\right) = 0, \quad x \in [0,a] \tag{3.3}$$

(associated with that in eqn (3.1)) subject to the boundary conditions imposed by eqn (3.2). By this we assume that $g(x,s)$ represents the influence function of a transverse unit concentrated force for the beam under consideration.

If we now rewrite eqn (3.1) in the equivalent form

$$\frac{d^2}{dx^2}\left(EI(x)\frac{d^2w(x)}{dx^2}\right) = p^2 m(x) A(x) w(x), \quad x \in [0,a] \tag{3.4}$$

and formally treat its term $p^2 m(x) A(x) w(x)$ as the right-hand side term, then in compliance with Theorem 1.3 of Chapter 1, one can readily express the solution of the problem in eqns (3.4) and (3.2) (which is actually an equivalent of the original problem posed by eqns (3.1) and (3.2)) in the form

$$w(x) = -\int_0^a g(x,s)[p^2 m(s) A(s) w(s)] ds$$

from which by factoring out the parameter p^2, we obtain

$$w(x) = -p^2 \int_0^a g(x,s) m(s) A(s) w(s) ds \tag{3.5}$$

The relation in eqn (3.5) represents the so-called homogeneous Fredholm integral equation of the second kind. In [17, 61, 68] the reader can find introductions to the basic concepts of integral equations. The homogeneous integral equation (3.5) poses an eigenvalue problem, analogously to homogeneous boundary value problems for either ordinary or partial differential equations involving a parameter, or to homogeneous systems of linear algebraic equations with a parameter in the coefficient matrix.

Thus, as a result of the development just completed, the eigenvalue problem formulated in eqns (3.1) and (3.2) eventually reduces to the eigenvalue problem for the integral equation (3.5). So, we are again looking for those values of p that yield nontrivial solutions to eqn (3.5).

In general, eqn (3.1), as an equation with variable coefficients, does not allow an analytical solution; neither does, of course, eqn (3.5). Therefore, in the developing a computational procedures for this equation, one ought to rely only on approximate methods.

An integral equation of the type in eqn (3.5) can successfully be solved by a variety of traditional numerical methods. Let, for example, the quadrature formulae method [4, 61] be utilized, where x_k and B_k, $(k=\overline{1,n})$ represent the partitioning points and the quadrature coefficients, respectively. Let in addition w_k represent the approximate values of $w(x)$ at $x = x_k$. In accordance with the standard scheme of the quadrature formulae method, at every partitioning point x_j, eqn (3.5) is approximated with the following homogeneous linear algebraic equation

$$w_j = -p^2 \sum_{k=1}^{n} B_k g(x_j, s_k) m(s_k) A(s_k) w_k, \quad (j=\overline{1,n}) \qquad (3.6)$$

in n unknowns w_k, so that the entire set of the above relations (as $j=\overline{1,n}$) constitutes a standard eigenvalue problem of linear algebra and can therefore be solved in a standard way. This ultimately yields approximate values of the n lowest components of the eigenvalue spectrum of eqn (3.5) along with approximate values of the corresponding eigenfunctions whose values are computed at the partitioning points.

In the discussion that follows, the reader will find some data indicating a high level of accuracy attained when the approach described here is used in practice. The results appeared to be very accurate even if one uses such a primitive quadrature technique as the trapezoid rule with an equally spaced set of a limited number n of partitioning points.

Later in this section, we will also conduct a computational experiment on the comparison of the results obtained by the finite difference method directly applied to the problem posed by eqns (3.1) and (3.2) against those obtained by the direct tackling of eqn (3.5) by the quadrature formulae method.

EXAMPLE 1 To continue the discussion, we first pose a validation example. This represents a classical formulation. Let us seek natural frequencies and mode shapes of transverse vibration of a single-span simply supported beam of length a, having a uniform flexural rigidity EI.

This problem results in the standard eigenvalue formulation

$$\frac{d^4w(x)}{dx^4} - \lambda^4 w(x) = 0, \quad x \in [0, a] \qquad (3.7)$$

3.1. TRANSVERSE NATURAL VIBRATIONS

$$w(0) = \frac{d^2w(0)}{dx^2} = 0, \quad w(a) = \frac{d^2w(a)}{dx^2} = 0 \tag{3.8}$$

where a specific notation λ^4 for the parameter in equation (3.7) is used just for the notational convenience.

Physical interpretation of the eigenvalues and eigenfunctions of this problem directly leads to the natural frequencies and modes of transverse vibration of the beam under consideration. Namely, the circular natural frequency f (in hertz) can be found for this beam in terms of λ as

$$f = \frac{\lambda^2}{2\pi}\sqrt{\frac{EI}{mA}}$$

where the constants m and A represent the mass density of the material and the cross-sectional area of the beam, respectively. The eigenfunctions of the problem in eqns (3.7) and (3.8) represent, in turn, the mode shapes.

By observation, it follows that each of the following functions

$$w_l(x) = C\sin\left(\frac{l\pi x}{a}\right), \quad l = 1, 2, 3, \ldots \tag{3.9}$$

where C is an arbitrary constant, represents a solution of eqn (3.7) if the parameter λ takes on the values

$$\lambda_l = \frac{l\pi}{a} \tag{3.10}$$

Moreover, it can easily be seen that each of the functions in eqn (3.9) satisfies the boundary conditions imposed with eqn (3.8). Hence, there exist infinitely many eigenvalues λ_l of the problem posed by eqns (3.7) and (3.8). They are given by eqn (3.10) and each of them, in turn, is associated with an infinite set of eigenfunctions defined by eqn (3.9).

Thus, equations (3.9) and (3.10) present the exact solution to the problem posed by eqns (3.7) and (3.8). In what follows, while computing the approximate solution to this problem, we will take advantage of the fact that its exact solution (eigenvalues and eigenfunctions) is available. This makes it convenient to test computational algorithms to be developed and to estimate the accuracy level to be attained.

For developing the numerical procedure based on the influence function method, let us reduce the eigenvalue problem posed by eqns (3.7) and (3.8) to the corresponding integral equation. This reduction can be accomplished upon implementing the approach introduced earlier in this section. In doing so, let $g(x,s)$ represent the Green's function for the boundary value problem posed by the following homogeneous equation

$$\frac{d^4w(x)}{dx^4} = 0, \quad x \in [0, a]$$

subject to the boundary conditions formulated in eqn (3.8).

That is, we assume that $g(x,s)$ represents the influence function of a unit transverse force concentrated at point s for a simply supported beam of length a, with a uniform flexural rigidity. In Chapter 2 (see exercise 2.4), the reader was assigned to construct this influence function which appears to be in the form

$$g(x,s) = \frac{1}{6a} \begin{cases} x(a-s)[(a-s)^2+(x^2-a^2)], & x \leq s \\ s(a-x)[(a-x)^2+(s^2-a^2)], & s \leq x \end{cases}$$

Based on this compact representation of the influence function, the integral equation that brings an alternative formulation to the eigenvalue problem posed by eqns (3.7) and (3.8), in the sense introduced earlier, is written in this case as

$$w(x) = -\lambda^4 \int_0^a g(x,s)w(s)ds \qquad (3.11)$$

In Table 3.1 we exhibit the partial eigenvalue spectrum (the first six eigenvalues) for the integral equation (3.11) with $a=1$. The trapezoid rule has been used with a uniform partitioning of the interval $[0,a]$ into $n=10$ subintervals with partitioning points defined by

$$x_k = \frac{ak}{n}, \quad (k=\overline{0,n}) \qquad (3.12)$$

From the data in Table 3.1, it is clearly seen that although a very rough partitioning ($n=10$) has been used, the IFM solution happens nevertheless to be fairly accurate. Indeed, the accuracy attained for the lowest eigenvalue λ_1 (fundamental natural frequency) is at the level of 99.999%. The accuracy gradually drops down for the upper components of the spectrum, though remaining at a relatively high level of 99% for λ_6. The eigenfunctions are not shown here, but they were computed in fact with the same high accuracy level as the eigenvalues.

Table 3.1 Eigenvalues of the problem in eqns (3.7) and (3.8), computed by the influence function (IFM) and finite difference (FDM) methods

Method used	Eigenvalue, λ_m					
	$m=1$	$m=2$	$m=3$	$m=4$	$m=5$	$m=6$
IFM	3.14158	6.28283	9.42164	12.5508	15.6508	18.6738
FDM	4.63605	7.52141	10.2228	12.6720	14.8237	16.6413
exact	3.14159	6.28319	9.42478	12.5664	15.7080	18.8496

3.1. TRANSVERSE NATURAL VIBRATIONS

To determine the accuracy level potentially attainable by our version of the IFM, a computational experiment was conducted on the comparison of its results against those computed by the finite difference method (FDM).

To run the actual computation for the eigenvalue problem in eqns (3.7) and (3.8) by the FDM, we utilized the extended version

$$x_k = \frac{ak}{n}, \quad (k = \overline{-1, n+1})$$

of the uniform partitioning of the interval [0,a]. The two extra grid-points x_{-1} and x_{n+1} are added to the partitioning (3.12) for the sake of methodological convenience. This helps to obtain a consistent system of linear algebraic equations. In the development that follows, the partitioning step a/n is denoted with h, and the approximate value of the deflection function $w(x)$ at $x = x_k$ is denoted with w_k.

Upon recalling the simplest finite difference scheme of the order of accuracy $O(h^2)$, which has been described earlier in Section 2.2, we reduce the boundary value problem in eqns (3.7) and (3.8) to the well-posed eigenvalue problem of linear algebra. In doing so, eqn (3.7) is approximated at the interior $(k = \overline{1, n-1})$ grid-points with the system

$$w_{k+2} - 4w_{k+1} + 6w_k - 4w_{k-1} + w_{k-2} - (\lambda h)^4 w_k = 0, \quad k = \overline{1, n-1} \quad (3.13)$$

This system is not well-posed because the number of unknowns in it is four units greater then the number of equations. This inconsistency, however, can easily be rectified. Four additional equations are derived upon approximating the boundary conditions in eqn (3.8). This yields

$$w_0 = w_n = 0, \quad w_{-1} - 2w_0 + w_1 = 0, \quad w_{n-1} - 2w_n + w_{n+1} = 0$$

Incorporating these into eqn (3.13), we obtain a standard formulation of the eigenvalue problem of linear algebra.

Recall again that this primitive finite difference scheme has been chosen on purpose, as it is equivalent to the trapezoid rule in terms of the order of accuracy $O(h^2)$ provided. Hence, from the error estimation viewpoint, it follows that computed outputs of both the finite difference (FDM) and influence function (IFM) methods in this experiment ought to be equivalently accurate. However, the data in Table 3.1 show a different result. The low accuracy level of the FDM solution is not tolerable at all with $n = 10$. Hence, the level of partitioning required for more accurate results should be essentially increased for this method. Whereas the IFM, as we have seen earlier, works fine.

EXAMPLE 2 In this example, we will present some more numerical results obtained by the influence function method in solving eigenvalue problems for single-span beams with different edge conditions.

We again consider a beam of length a, having a uniform flexural rigidity EI. If, for example, the left edge $x=0$ of the beam is simply supported while the right edge $x=a$ is clamped, then we obtain the following standard eigenvalue formulation

$$\frac{d^4 w(x)}{dx^4} - \lambda^4 w(x) = 0, \quad x \in [0, a]$$

$$w(0) = \frac{d^2 w(0)}{dx^2} = 0, \quad w(a) = \frac{dw(a)}{dx} = 0$$

In the discussion that follows, this problem will be referred to as the S–C problem. Contrary to the previous case of a simply supported beam, the exact solution of the S–C problem is not available, but its approximate solution is well tabulated (see, for example, [11, 73]) and can be used for testing purposes.

Following the IFM procedure, we reduce the S–C problem again to the integral equation (3.11). The influence function

$$g(x, s) = \frac{1}{12a^3} \begin{cases} x(a-s)^2[s(a^2-x^2) - 2a(x^2-as)], & x \leq s \\ s(a-x)^2[x(a^2-s^2) - 2a(s^2-ax)], & s \leq x \end{cases}$$

of a transverse unit concentrated force, required in this case, is supposed to be constructed by the reader in Section 2.1 (see exercise 2.6).

Table 3.2 exhibits approximate eigenvalues λ_m of the integral equation (3.11) for beams of a unit length. We again used the trapezoid rule with $n=10$. The upper line of the table presents, in particular, the values of λ_1 through λ_5 for the simply supported – clamped beam (S–C problem).

Table 3.2 Approximate eigenvalues λ_m of eqn (3.11), computed by the IFM for various combinations of the edge conditions

Problem solved	Method used	Eigenvalue, λ_m				
		$m=1$	$m=2$	$m=3$	$m=4$	$m=5$
S–C	IFM	3.92652	7.06841	10.2065	13.3182	16.4021
	exact	3.92660	7.06858	10.2102	13.3518	16.4934
C–Sd	IFM	2.35708	5.53224	8.82739	12.1274	15.3958
	exact	2.36502	5.49781	8.63937	11.7810	14.9226
S–Sd	IFM	1.56642	4.77013	8.06684	11.3506	14.6214
	exact	1.57080	4.71239	7.85398	10.9956	14.1372
C–C	IFM	4.72990	7.85172	10.9862	14.0972	17.1455
	exact	4.73004	7.85319	10.9956	14.1372	17.2788

3.1. TRANSVERSE NATURAL VIBRATIONS

In addition to the S–C problem, one can also find in this table data for: (i) C–Sd problem (one edge is clamped while the other is subjected to the sliding conditions), (ii) S–Sd problem (one edge is simply supported while the other is sliding), and (iii) C–C problem (beam clamped at both edges). Each of these problems reduces to the integral equation (3.11) with the corresponding influence functions involved. Eqn (3.11) reduces, in turn, to an eigenvalue problem of linear algebra (by the trapezoid rule with $n=10$) and was solved then numerically in the standard way.

The accuracy of the presented eigenvalues varies slightly from case to case, but remains on the relatively high level agreeing with the validation problem discussed in *EXAMPLE 1*. The lowest eigenvalues, for example, for all of the problems have been computed with an accuracy level that is well above 99.5%. It is worth noting again that one of the most primitive quadrature formulas has been used. This indicates the great computational potential of the IFM in the eigenvalue analysis.

All the influence functions that have been utilized for the data presented in Table 3.2 are available in this text. Indeed, we assume that the reader has already constructed the influence function

$$g(x,s) = \frac{1}{12a} \begin{cases} x^2[3s(s-2a)+2ax], & x \leq s \\ s^2[3x(x-2a)+2as], & s \leq x \end{cases}$$

required to run the computational procedure for the approximate solution of the C–Sd problem (see exercise 2.11). Whereas the influence function

$$g(x,s) = \frac{1}{6a} \begin{cases} x[3s(s-2a)+x^2], & x \leq s \\ s[3x(x-2a)+s^2], & s \leq x \end{cases}$$

required for the S–Sd problem, follows from exercise 2.10. The influence function for the C–C problem is available in Section 2.1 (see eqn (2.20)).

We now return to the original formulation in eqns (3.1) and (3.2) for a beam of variable flexural rigidity. The influence function formalism has already been applied to reduce that problem to the integral equation (3.5). Section 2.4 describes the procedure for the numerical construction of required influence functions, which has been used in computing results for the next example.

EXAMPLE 3 Let us compute natural frequencies of transverse vibrations for a simply supported beam of length a, with a rectangular cross-section $b \times h(x)$, whose height is a linear function of x. The material of which the beam is built is isotropic and homogeneous. That is, $E = const$ and $m = const$ (see eqn (3.1)). The configuration of the beam in reference to the operative coordinate system is shown in Figure 3.1.

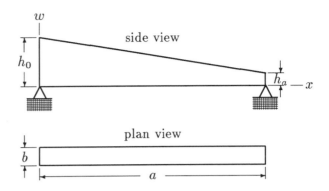

Figure 3.1 A beam of a variable cross-section

Clearly, the cross-sectional properties of the beam depicted in Figure 3.1 vary with x. The flexural rigidity represents a cubic function of the longitudinal coordinate x. That is

$$EI(x) = \frac{Eb}{12}\left(h_0 - \frac{h_0-h_a}{a}x\right)^3$$

while its cross-sectional area is a linear function of x

$$A(x) = b\left(h_0 - \frac{h_0-h_a}{a}x\right)$$

The actual computation, the results of which are discussed in this example, has been accomplished for a beam of a unit length $a=1$ with unit flexural rigidity at the left edge, that is $EI(0)=1$. The influence function of a transverse unit concentrated force for this beam (that is the Green's function of the boundary value problem posed by eqns (3.3) and (3.8)) has been obtained by numerically computing the four linearly independent solutions of eqn (3.3).

Some results computed for this beam are shown in Table 3.3. They are obtained by implementing the trapezoid rule for eqn (3.5) where the number n of uniform partitioning equals 10.

Table 3.3 Approximate eigenvalues p_k of the simply supported beam whose configuration is shown in Figure 3.1

h_0/h_a	1	2	3	4	5	6
p_1	9.8687	7.0226	5.6711	5.3015	4.9147	4.7118
p_2	39.4805	27.9486	23.7583	21.6953	20.7092	20.5619
p_3	88.8269	64.1278	54.8396	49.9238	46.9872	45.3128

3.1. TRANSVERSE NATURAL VIBRATIONS

Table 3.4 Approximate eigenvalues p_k of the clamped – simply supported beam whose configuration is shown in Figure 3.1

h_0/h_a	1	2	3	4	5	6
p_1	15.4216	12.7739	10.8024	9.9846	9.9729	9.9678
p_2	49.9682	37.4928	31.4519	28.3651	27.2130	27.1544
p_3	104.2542	77.5147	65.9788	59.8972	55.3747	54.4665

The case $h_0/h_a = 1$ relates to a beam of a uniform flexural rigidity $EI \equiv 1$. The exact eigenvalues of a simply supported beam in this case are well-known $p_k = (k\pi)^2$. Hence, the exact values corresponding to those of the first column in Table 3.3 are $p_1 = 9.8696$, $p_2 = 39.4784$, and $p_3 = 88.8264$. Thus, the relative accuracy attained is about 99.9%. The rest of the results in this table also compare fairly well with data available in the literature (see, for example, [11]).

In Table 3.4 we present the IFM numerical results for one more eigenvalue problem. We again consider the beam whose configuration is depicted in Figure 3.1. In this case, however, different boundary conditions are subjected. That is, a clamped – simply supported beam is considered.

These results are also in a good agreement with corresponding data available in the relevant sources (see, for example, [11]). The same computational procedure as before has been utilized. Other problems of this type for a single-span beam with different edge conditions and different configurations can also be reduced to the integral equation (3.5).

Other classical natural vibration problems in the beam theory could be tackled by means of the influence function method. The example that follows is designed to illustrate this issue.

EXAMPLE 4 Consider the natural vibration of a beam of uniform flexural rigidity, subjected to either tensile or compressive axial forces. To be specific, the tensile forces are shown in Figure 3.2. The simply supported beam is depicted here. However, in what follows, we also consider some other types of the edge conditions.

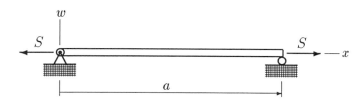

Figure 3.2 A beam subjected to axial forces

Probably, in the case when the compressive force is applied, the buckling phenomenon could come to the reader's mind. This issue, however, is not going to be a subject of the present consideration, though some buckling problems are perfectly suitable for the influence function treatment. We will focus on buckling problems in our next section.

While considering a compressive force, we here assume its magnitude to be well below the value of the so-called Euler elastic buckling force. Consequently, the question to discuss here is how axial forces, acting on a beam, affect the spectrum of its natural frequencies. This classical statement is well known in mechanics. What we are going to do in this study is to answer this question within the scope of the influence function method.

As it is known [73], the following eigenvalue formulation

$$\frac{d^4w(x)}{dx^4} - \frac{S}{EI}\frac{d^2w(x)}{dx^2} = p^2\frac{mA}{EI}w(x), \quad x \in (0, a) \tag{3.14}$$

$$w(0) = \frac{d^2w(0)}{dx^2} = 0, \quad w(a) = \frac{d^2w(a)}{dx^2} = 0 \tag{3.15}$$

models the natural vibrations of the beam, with S being the magnitude of a tensile axial force. If a compressive force is applied instead, then the sign of the second term in eqn (3.14) has to be changed to a plus.

The exact solution (eigenvalues and corresponding eigenfunctions) of this classical problem can be easily obtained by inspection. Indeed, it follows that each of the following functions

$$w_l(x) = C \sin\left(\frac{l\pi x}{a}\right), \quad l = 1, 2, 3, \ldots \tag{3.16}$$

where C is an arbitrary constant, satisfies all of the boundary conditions in eqn (3.15). Upon substituting these functions into eqn (3.14), one obtains the following algebraic equation

$$\left(\frac{l\pi}{a}\right)^4 + \frac{S}{EI}\left(\frac{l\pi}{a}\right)^2 = p_l^2\frac{mA}{EI}$$

in p_l. Solving this equation for p_l, one obtains

$$p_l = \left(\frac{l\pi}{a}\right)^2\sqrt{R\left(1 + \left(\frac{qa}{l\pi}\right)^2\right)} \tag{3.17}$$

where R and q are introduced as $R = EI/mA$ and $q = \sqrt{S/EI}$.

3.1. TRANSVERSE NATURAL VIBRATIONS

Thus, eqn (3.16) presents the eigenfunctions (natural modes of the vibration of the beam), while the eigenvalues p_l (angular frequencies of the vibration) for a tensile force S are computed by eqn (3.17).

Recall that for a compressive force S, the sign of the second term in eqn (3.14) must be changed to a plus. Thus, the sign in the radicand in eqn (3.17) has to be consequently changed to a minus. That is

$$p_l = \left(\frac{l\pi}{a}\right)^2 \sqrt{R\left(1-\left(\frac{qa}{l\pi}\right)^2\right)}$$

It is appropriate to make an important comment concerning the magnitude of the compressive force acting on the beam. From the above expression it clearly follows that, since the radicand in it must not be negative

$$1-\left(\frac{qa}{l\pi}\right)^2 = 1 - \frac{S}{EI}\left(\frac{a}{l\pi}\right)^2 > 0$$

the magnitude of the compressive force is bounded from above. That is,

$$S < EI\left(\frac{l\pi}{a}\right)^2$$

The physics behind this condition is obvious. It follows that if the value S of the compressive force exceeds the magnitude of the so-called Euler elastic buckling force (the right-hand side of the above inequality actually represents this force for the beam under consideration), then the statement of the corresponding eigenvalue problem becomes physically meaningless.

Before proceeding any further with the development of the influence function method for the eigenvalue problem posed by eqns (3.14) and (3.15), we present the Green's function

$$g(x,s) = \frac{1}{aq^3\sinh(qa)}\begin{cases} qx(s-a)\sinh(qa) - a\sinh(qx)\sinh(q(s-a)) \\ qs(x-a)\sinh(qa) - a\sinh(qs)\sinh(q(x-a)), \end{cases} \quad (3.18)$$

(where, as usual, the upper branch is derived for $x \leq s$) of the following boundary value problem

$$\frac{d^4w(x)}{dx^4} - q^2\frac{d^2w(x)}{dx^2} = 0, \quad x \in (0,a)$$

$$w(0) = \frac{d^2w(0)}{dx^2} = 0, \quad w(a) = \frac{d^2w(a)}{dx^2} = 0$$

The Green's function in eqn (3.18) represents the influence function of a simply supported beam undergoing a tensile axial force S.

In compliance with Theorem 1.3, the problem in eqns (3.14) and (3.15) reduces to the following homogeneous integral equation

$$w(x) = -\lambda^4 \int_0^a g(x,s)w(s)ds \tag{3.19}$$

where

$$\lambda^4 = \frac{p^2 mA}{EI} = \frac{p^2}{R}$$

This notation enables us to match results of the current development with those of *EXAMPLE 1* in this section.

For the compressive force S, the influence function method again yields the integral equation (3.19). However, its kernel $g(x,s)$ represents in this case the Green's function

$$g(x,s) = \frac{1}{aq^3\sin(qa)} \begin{cases} qx(a-s)\sin(qa) + a\sin(qx)\sin(q(s-a)), & x \leq s \\ qs(a-x)\sin(qa) + a\sin(qs)\sin(q(x-a)), & s \leq x \end{cases} \tag{3.20}$$

of the boundary value problem written as

$$\frac{d^4w(x)}{dx^4} + q^2\frac{d^2w(x)}{dx^2} = 0, \quad x \in (0,a)$$

$$w(0) = \frac{d^2w(0)}{dx^2} = 0, \quad w(a) = \frac{d^2w(a)}{dx^2} = 0$$

Notice that the Green's function in eqn (3.20) is identified with the influence function of a transverse unit concentrated force for a simply supported beam undergoing the compressive axial force S.

The influence functions presented by eqns (3.18) and (3.20) have been constructed based on different fundamental solution sets. Namely, to construct the first of them, we utilized the fundamental solution set

$$1, \quad x, \quad \sinh(qx), \quad \cosh(qx)$$

whereas for the influence function in eqn (3.20), we utilized the fundamental solution set written as

$$1, \quad x, \quad \sin(qx), \quad \cos(qx)$$

The numerical solution of the integral equation (3.19) has been accomplished in this study by the quadrature formulae method. The trapezoid rule with a uniform partitioning of the interval $[0,a]$ has been implemented. We again purposely used such a primitive numerical routine (with a limited number of partitionings) in order to bring to the reader's attention the high potential of the influence function approach.

3.1. TRANSVERSE NATURAL VIBRATIONS

The reader finds below some data computed by means of the influence function method for the problem stated by eqns (3.14) and (3.15) and also for some relevant problems.

The eigenvalue formulation for the integral equation (3.19) (to which the original problem is converted) has been approximately reduced to a corresponding eigenvalue problem of linear algebra by the trapezoid rule-based quadrature formulae method, with $n = 10$. The data in Tables 3.5 through 3.8 are obtained for a beam of a unit length with $R=1$.

The accuracy level of the approximate eigenvalues in Tables 3.5 and 3.6 is well above 99.5%. Notice again that one of the most primitive quadrature formulae with a limited number of partitioning was used. These data reveal a high potential of the IFM in the numerical eigenvalue analysis.

To provide some additional validation examples to this computational procedure, we have conducted computations for a simply supported – sliding beam subjected to axial forces as shown in Figure 3.3. Exact solutions in this case are also available (see, for example, [11]).

Table 3.5 Approximate eigenvalues λ_l of the integral equation (3.19) for a simply supported beam subjected to a tensile axial force

Method used	Parameter q^2	Eigenvalue, λ_l				
		$l=1$	$l=2$	$l=3$	$l=4$	$l=5$
IFM	1	3.2183	6.3224	9.4458	12.5535	15.6659
	10	3.7421	6.6476	9.6759	12.7440	15.8050
	100	5.7382	8.6125	11.3723	14.1783	17.0188
exact	1	3.2183	6.3226	9.4512	12.5862	15.7239
	10	3.7422	6.6480	9.6795	12.7608	15.8648
	100	5.7384	8.6142	11.3802	14.2060	17.1026

Table 3.6 Approximate eigenvalues λ_l of the integral equation (3.19) for a simply supported beam subjected to a compressive axial force

Method used	Parameter q^2	Eigenvalue, λ_l				
		$l=1$	$l=2$	$l=3$	$l=4$	$l=5$
IFM	1	3.0588	6.2427	9.3950	12.5310	15.6352
	3	2.8695	6.1599	9.3411	12.4911	15.6037
	9	1.7116	5.8896	9.1742	12.3769	15.5076
exact	1	3.0588	6.2430	9.3981	12.5465	15.6920
	3	2.8695	6.1603	9.3442	12.5063	15.6600
	9	1.7116	5.8896	9.1764	12.3834	15.5627

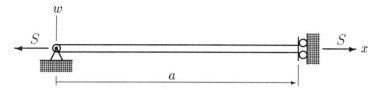

Figure 3.3 A simply supported – sliding beam

The boundary eigenvalue problems, associated with the natural vibration of a simply supported – sliding beam subjected to axial forces, are again converted to the homogeneous integral equation (3.19). The influence functions required for this conversion are shown below. For the tensile force, we have

$$g(x,s) = \frac{1}{q^3 \sinh(qa)} \begin{cases} -qx\sinh(qa) + \sinh(qx)\cosh(q(s-a)), & x \leq s \\ -qs\sinh(qa) + \sinh(qs)\cosh(q(x-a)), & s \leq x \end{cases}$$

and for the compressive force, we obtain

$$g(x,s) = \frac{1}{q^3 \sin(qa)} \begin{cases} qx\cos(qa) - \sin(qx)\cos(q(s-a)), & x \leq s \\ qs\cos(qa) - \sin(qs)\cos(q(x-a)), & s \leq x \end{cases}$$

Integral equation (3.19), in turn, has again been reduced to an eigenvalue problem of linear algebra with the aid of the quadrature formulae method (trapezoid rule, with $n=10$). Some results of this computation for the beam subjected to tensile axial forces are exhibited in Table 3.7, while Table 3.8 contains some results obtained for the beam subjected to comprehensive axial forces.

Table 3.7 Approximate eigenvalues λ_l of integral eqn (3.19) for a simply supported – sliding beam subjected to a tensile axial force

Method used	Parameter q^2	Eigenvalue, λ_l				
		$l=1$	$l=2$	$l=3$	$l=4$	$l=5$
IFM	1	1.7066	4.8116	8.0882	11.3580	14.6361
	10	2.3486	5.1849	8.3012	11.3281	14.7684
	100	3.9767	7.2213	10.0653	12.9815	15.9876
exact	1	1.7103	4.7646	7.8856	11.0182	14.1374
	10	2.3551	5.1714	8.1546	10.9994	14.3119
	100	3.9876	7.2176	9.9934	12.7838	15.6463

3.1. TRANSVERSE NATURAL VIBRATIONS

Table 3.8 Approximate eigenvalues λ_l of integral eqn (3.19) for a simply supported – sliding beam subjected to a compressive axial force

Method used	Parameter q^2	Eigenvalue, λ_l				
		$l=1$	$l=2$	$l=3$	$l=4$	$l=5$
IFM	1	1.3756	4.7334	8.0467	11.3334	14.6068
	2	1.0334	4.7089	8.0201	11.3067	14.5023
	2.3	0.7995	4.6259	8.0128	11.3119	14.5126
exact	1	1.3794	4.6584	7.8220	10.9500	14.1018
	2	1.0363	4.6025	7.7895	10.9043	14.0663
	2.3	0.8017	4.5853	7.7797	10.9956	14.0963

The accuracy of approximate eigenvalues exhibited in the above two tables dropped slightly compared to the data in Tables 3.5 and 3.6. Nevertheless it still remains at the level well above 99.5% for the lowest eigenvalues and 97% for the highest eigenvalues. This level is acceptable for the trapezoid rule with ten partitioning points used.

Many other problems of natural vibrations of beams with more complicated statements can also be tackled by direct utilization of the IFM computational procedure described above. This point can be clarified by the consideration of a beam of variable flexural rigidity $EI(x)$, subjected to a continuously distributed axial force $S(x)$. The governing boundary value problem of natural vibration in this case can be written [73] in the form

$$\frac{d^2}{dx^2}\left(EI(x)\frac{d^2w(x)}{dx^2}\right) - \frac{d}{dx}\left(S(x)\frac{dw(x)}{dx}\right) - p^2 m(x)A(x)w(x) = 0 \qquad (3.21)$$

$$B_{0,k}[w(0)] = 0, \quad B_{a,k}[w(a)] = 0, \quad k = 1,2 \qquad (3.22)$$

Hence, if $g(x,s)$ is the Green's function for the corresponding homogeneous boundary value problem written as

$$\frac{d^2}{dx^2}\left(EI(x)\frac{d^2w(x)}{dx^2}\right) - \frac{d}{dx}\left(S(x)\frac{dw(x)}{dx}\right) = 0 \qquad (3.23)$$

$$B_{0,k}[w(0)] = 0, \quad B_{a,k}[w(a)] = 0, \quad k = 1,2$$

then in compliance with Theorem 1.3, we customarily obtain the following integral representation

$$w(x) = -p^2 \int_0^a g(x,s)m(s)A(s)w(s)\,ds \qquad (3.24)$$

for the solution of the problem in eqns (3.21) and (3.22). This relation does not explicitly present the solution to the problem in eqns (3.21) and (3.22), because $w(x)$ is actually expressed in terms of itself. Hence, the relation in eqn (3.24) is an integral equation in $w(x)$. With certain constrains put on the function $g(x,s)m(s)A(s)$ (which are met in our case), such equations are called (see [4, 9, 17, 36, 61, 68]) homogeneous Fredholm integral equations of the second kind.

Eqn (3.24) contains a parameter (p^2) and it poses therefore the eigenvalue problem whose solution (eigenvalues and eigenfunctions) can be computed with the aid of standard numerical procedures. Note that the quadrature formulae method which is used in this study is not the only possible method for such problems, it represents just one of several options [61].

The Green's function $g(x,s)$ of the boundary value problem in eqns (3.23) and (3.22) represents, in fact, the influence function of a transverse unit concentrated force for a beam under consideration. One can readily construct $g(x,s)$ numerically by using the procedure introduced in Section 2.4 for equations with variable coefficients.

Thus, the influence function method is easily adaptive to eigenvalue problems posed by differential equations with variable coefficients. Such an adaptivity should make this method attractive to users of numerical methods in engineering due to potentially high accuracy level attainable.

Clearly, the described approach is directly applicable to particular situations when, for example, either the flexural rigidity of a beam is uniform while the axial distributed force is variable, or conversely.

Buckling problems in beam theory represent one more problem class for which the influence function method is found to be very effective. This issue will be discussed in the next section.

3.2 Buckling (Euler formulation)

In the preceding sections of this book, we have presented some beam problem classes in which the influence function method is found to be fairly efficient compared to other more traditional approaches. In this section, we are going to explore one more classical problem class from Kirchhoff beam theory, for which influence functions, constructed for problems of the linear bending, could be useful as well. From this section, it follows that the buckling phenomenon for elastic Kirchhoff's beams (known as the classical Euler problem) can also be successfully tackled by means of this method.

To show a possible way of doing so, we consider a clamped – simply supported elastic beam of variable flexural rigidity $EI(x)$, subjected to the axial compressive force N as shown in Figure 3.4. Any other type of supports can also be assumed within the scope of the present consideration.

3.2. BUCKLING (EULER FORMULATION)

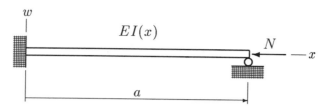

Figure 3.4 A clamped – simply supported beam

The static equilibrium of such a beam can be modeled [73] by the following boundary value problem

$$\frac{d^2}{dx^2}\left(EI(x)\frac{d^2w(x)}{dx^2}\right) + N\frac{d^2w(x)}{dx^2} = 0, \quad x \in [0,a] \tag{3.25}$$

$$w(0) = \frac{dw(0)}{dx} = 0, \quad w(a) = \frac{d^2w(a)}{dx^2} = 0 \tag{3.26}$$

Clearly, the beam depicted in Figure 3.4 was chosen just to be specific. If different supports of the end-points are assumed, then different boundary conditions have to be imposed at $x=0$ and $x=a$. And boundary conditions in a statement, as we know, are taken care of by the influence function which can be constructed for any physically feasible edge conditions.

The magnitude of the parameter N is very essential for the above statement. There exists a certain value N_{cr} of the compressive force N (that is usually referred to as the *critical value* of the compressive force or the *Euler elastic buckling force*), which yields a sudden change of the undeformed stable equilibrium state to a bent unstable one.

In looking for the value of N_{cr}, one usually reduces the problem shown in eqns (3.25) and (3.26) to a corresponding eigenvalue problem of linear algebra and then computes the lowest component of its spectrum as well as the corresponding eigenvector that represents *the buckling failure shape*. Such a reduction can be accomplished in many ways by appropriate numerical methods. In this study, the eigenvalue problem of linear algebra is usually derived through converting the problem in eqns (3.25) and (3.26) to an integral equation by means of the corresponding influence function of a transverse unit concentrated force.

For a beam of uniform flexural rigidity ($EI(x) \equiv EI_0 = const$), for instance, eqn (3.25) can be rewritten in the form

$$\frac{d^4w(x)}{dx^4} + \frac{N}{EI_0}\frac{d^2w(x)}{dx^2} = 0, \quad x \in [0,a] \tag{3.27}$$

for which it follows that, if the beam is, for example, simply supported at both edges, then the Euler elastic buckling force can be taken as

$$N_{cr} = \frac{\pi^2}{a^2} EI_0 \qquad (3.28)$$

and the corresponding equilibrium shape is given by

$$w_{cr}(x) = \sin\left(\frac{\pi x}{a}\right) \qquad (3.29)$$

In the general case of a variable flexural rigidity, however, the analytic solution of the problem stated with eqns (3.25) and (3.26) is impossible and the approximate value of N_{cr} should be computed numerically. There exists a broad variety of approximate methods available for such a computation. It is worth noting, however, that the influence function method has not even been mentioned yet as an option in this area. Filling in this unfortunate gap, we are going to show here how naturally various versions of this method can be implemented. A computational experiment will be conducted later to demonstrate the accuracy of this method.

We will present here two different algorithms of the influence function method that we developed for solving the eigenvalue problem posed by eqns (3.25) and (3.26). These algorithms are found to be equivalently accurate, although they are based on different ideas and utilize different influence functions in solving the same problem.

To introduce the first of these algorithms, let $g(x,s)$ represent the Green's function for the boundary value problem posed by the equation

$$\frac{d^2}{dx^2}\left(EI(x)\frac{d^2w(x)}{dx^2}\right) = 0, \quad x \in [0, a] \qquad (3.30)$$

subject to the corresponding set of boundary conditions. In other words, by this we again assume that $g(x,s)$ represents the influence function of a unit transverse concentrated force for the beam to be considered.

Treating formally the second term of eqn (3.25) as its right-hand side, in compliance with Theorem 1.3 in Chapter 1, one can express the solution of the problem in eqns (3.25) and (3.26) by the following integral

$$w(x) = \int_0^a g(x,s)[N\frac{d^2w(s)}{ds^2}]ds$$

Factoring out the parameter N, we rewrite this integral representation in the form

$$w(x) = N\int_0^a g(x,s)\frac{d^2w(s)}{ds^2}ds \qquad (3.31)$$

3.2. BUCKLING (EULER FORMULATION)

The relation in eqn (3.31) represents the so-called integral–differential equation in $w(x)$. Due to the presence of the second order derivative of $w(s)$ in the integrand, this equation cannot be reduced to an eigenvalue problem of linear algebra by directly applying quadrature formulae. This, however, can easily be fixed upon reducing eqn (3.31) to a regular homogeneous integral equation of the second kind. This implies that, afterwards, one can directly implement the computational procedure described in Section 3.1.

To accomplish such a reduction, we apply integration by parts to the integral in eqn (3.31) two times. For the first time, the integrand in it is partitioned in the following manner

$$g(x,s) = u(s); \quad \frac{d^2 w(s)}{ds^2} ds = dv(s);$$

then we consequently have

$$du(s) = \frac{\partial g(x,s)}{\partial s} ds, \quad v(s) = \frac{dw(s)}{ds}$$

This eventually reduces the integral in eqn (3.31) to

$$\int_0^a g(x,s) \frac{d^2 w(s)}{ds^2} ds = \left. g(x,s) \frac{dw(s)}{dx} \right|_0^a - \int_0^a \frac{\partial g(x,s)}{\partial s} \frac{dw(s)}{ds} ds$$

By virtue of the defining properties of the influence function, it follows that

$$g(x,0) = g(x,a) = 0$$

These relations reflect an obvious physical interpretation. They imply that a simply supported beam is not deflected if a transverse concentrated force is applied at either of its edge points $s=0$ and $s=a$. Thus, the first integration by parts in eqn (3.31) finally provides

$$\int_0^a g(x,s) \frac{d^2 w(s)}{ds^2} ds = - \int_0^a \frac{\partial g(x,s)}{\partial s} \frac{dw(s)}{ds} ds \tag{3.32}$$

In performing the second integration by parts, we partition the integrand of the integral in the above right-hand side as

$$\frac{\partial g(x,s)}{\partial s} = u(s); \quad \frac{dw(s)}{ds} ds = dv(s);$$

which results in

$$du(s) = \frac{\partial^2 g(x,s)}{\partial s^2} ds, \quad v(s) = w(s)$$

This subsequently yields

$$\int_0^a \frac{\partial g(x,s)}{\partial s} \frac{dw(s)}{ds} ds = \left. \frac{\partial g(x,s)}{\partial s} w(s) \right|_0^a - \int_0^a \frac{\partial^2 g(x,s)}{\partial s^2} w(s) ds$$

From this relation, it follows that the first (nonintegral) term in the above right-hand side vanishes in compliance with the boundary conditions in eqn (3.26). Thus, as a result of the development just completed, the integral–differential equation (3.31) is finally replaced with the following regular integral equation

$$w(x) = N \int_0^a \frac{\partial^2 g(x,s)}{\partial s^2} w(s) ds \qquad (3.33)$$

Hence, the eigenvalue problem posed by the integral–differential equation (3.31) is eventually reduced to the eigenvalue problem for the integral equation (3.33). Clearly, the latter is an integral equation of the same kind as that of eqn (3.11). Therefore, this makes it possible to directly apply the quadrature formulae method to reduce eqn (3.33) to the corresponding eigenvalue problem of linear algebra.

The situation here is even less demanding computationally compared to that for the natural vibration problem discussed in Section 3.1. Indeed, in this case, we do not require the entire spectrum of the coefficient matrix to be computed because the critical value of the compressive force is associated only with the lowest eigenvalue of that matrix. Due to this fact, obtaining the critical force should not be computationally problematic at all. And when computing, we are able to take advantage of this fact.

Some data are presented below to illustrate the computational potential of the influence function algorithm for solving buckling problems.

EXAMPLE 1 Consider first a validation example, one that has already been mentioned in this section. That is, the Euler problem for a simply supported (S–S) beam of a uniform flexural rigidity EI_0. The value N_{cr} of the Euler elastic buckling force and the buckling failure shape $w_{cr}(x)$ are presented in this case by eqns (3.28) and (3.29), respectively.

To formulate the integral equation posed in eqn (3.33) for this problem, we recall, from Section 3.1, the influence function

$$g(x,s) = \frac{1}{6a} \begin{cases} x(a-s)[(a-s)^2 + (x^2 - a^2)], & x \leq s \\ s(a-x)[(a-x)^2 + (s^2 - a^2)], & s \leq x \end{cases} \qquad (3.34)$$

of a transverse concentrated unit force for the simply supported beam of length a and take then its second derivative with respect to s to obtain the kernel of that integral equation.

To develop the computational procedure, we partition the interval $[0, a]$ with the set of grid-points s_j, $(j = \overline{0, n})$, shown with the solid dots in Figure 3.5, into n uniform subintervals of length $\Delta s = a/n$ as shown.

3.2. BUCKLING (EULER FORMULATION)

Figure 3.5 Partitioning and setting up the grid-points

Based on the above partitioning, we break down the integral in eqn (3.33) into n elementary integrals resulting in

$$w(x) = N \sum_{j=0}^{n-1} \int_{s_j}^{s_{j+1}} \frac{\partial^2 g(x,s)}{\partial s^2} w(s) ds$$

Applying then the second mean value theorem for integrals (see, for example, [17, 68]) to each of the elementary integrals in the above relation, we obtain

$$\frac{1}{N} w(x) = \sum_{j=0}^{n-1} w(\overline{s}_{j+1}) \int_{s_j}^{s_{j+1}} \frac{\partial^2 g(x,s)}{\partial s^2} ds \qquad (3.35)$$

where $s_j \leq \overline{s}_{j+1} \leq s_{j+1}$.

We now fix the values of $\overline{s}_{j+1}$ in eqn (3.35) at the midpoints of the elementary segments (s_j, s_{j+1}). Let then the observation point x go through the same set of values (we denote it as x_i, $(i=\overline{1,n})$ and depict both $\overline{s}_j$ and x_i with the empty dots in Figure 3.5). This yields the following homogeneous system of linear algebraic equations

$$\frac{1}{N} w_i = \sum_{j=0}^{n-1} w_{j+1} \int_{s_j}^{s_{j+1}} \frac{\partial^2 g(x_i,s)}{\partial s^2} ds, \quad (i=\overline{1,n}) \qquad (3.36)$$

in the approximate values w_i of the deflection function $w(x)$, defined at the grid-points $x_i, (i=\overline{1,n})$.

The elementary integrals in the system in eqn (3.36) represent the entries of the coefficient matrix. They can be approximated, for example, as

$$A_{i,j} = \int_{s_j}^{s_{j+1}} \frac{\partial^2 g(x_i,s)}{\partial s^2} ds \approx (G_{i,j+1} + G_{i,j}) \frac{\Delta s}{2} \qquad (3.37)$$

where, for the notational convenience, we denote

$$G_{i,j} = \frac{\partial^2 g(x_i, s_j)}{\partial s^2}$$

It follows that for a beam of uniform flexural rigidity, with any traditional type of the edge conditions imposed, the integrals A_{ij} do not require numerical treatment because they can be computed analytically. Whereas, for a beam of variable rigidity, numerical methods have to be applied, and eqn (3.37) suggests a possible way of doing so in such cases.

In any case, the homogeneous system of linear algebraic equations in eqn (3.36) is finally presented in the standard (for eigenvalue problems) form

$$(A - \mu I) W = 0 \qquad (3.38)$$

where A represents an $n \times n$ matrix whose entries $A_{i,j}$ are, in general, determined by eqn (3.37), I is an identity matrix of the same $n \times n$ dimension, W represents a vector whose components are the values w_i introduced earlier, and the parameter μ represents the reciprocal value of N.

Thus, the original homogeneous integral equation (3.33) is ultimately approximated by the eigenvalue problem of linear algebra (see eqn (3.38)), whose lowest eigenvalue μ_1 represents the reciprocal of the critical force N_{cr}. The numerical solution of such a problem is a routine procedure.

To estimate the accuracy level and the rate of convergence of the described algorithm, we have conducted a computational experiment on how the partitioning parameter n affects the accuracy attained. We have computed approximate values of the critical force N_{cr} for the simply supported beam of a unit length and unit flexural rigidity, with $n = 4, 6$, and 10. The results of this experiment are shown in Table 3.9.

From these data it clearly follows that the influence function based algorithm described here: (i) provides an extremely high accuracy level even for $n = 4$ and (ii) converges fairly fast.

Notice that this algorithm can be used for beams with different supports. But in dealing with those, different influence functions $g(x,s)$ are required. For a simply supported – clamped (S–C) beam of length a, for example, we recall its influence function

$$g(x,s) = \frac{1}{12a^3} \begin{cases} x(a-s)^2[s(a^2-x^2) - 2a(x^2-as)], & x \leq s \\ s(a-x)^2[x(a^2-s^2) - 2a(s^2-ax)], & s \leq x \end{cases}$$

from Section 3.1. For a clamped at both edges (C–C) beam of length a and cantilever (C–F) beam, we recall their influence functions from Section 2.1 (see eqns (2.20) and (2.37), respectively).

Table 3.9 Dependence of the relative error of the critical force N_{cr} on the partitioning parameter n for a simply supported beam

Partitioning parameter, n	4	6	10
Approximate values of N_{cr}	9.8696037	9.8696043	9.8696044
Relative error, %	$.7 \times 10^{-4}$	$.1 \times 10^{-5}$	$.1 \times 10^{-8}$

3.2. BUCKLING (EULER FORMULATION)

In what follows, we present the second influence function algorithm for computing the critical value of the compressive force acting on a beam. In doing so, we again formulate the corresponding Euler problem

$$\frac{d^2}{dx^2}\left(EI(x)\frac{d^2w(x)}{dx^2}\right) + N\frac{d^2w(x)}{dx^2} = 0, \quad x \in [0,a] \tag{3.39}$$

$$B_{0,k}[w(0)] = 0, \quad B_{a,k}[w(a)] = 0, \quad k = 1, 2 \tag{3.40}$$

and add formally the term $-w(x)$ to both sides of eqn (3.39). This yields

$$\frac{d^2}{dx^2}\left(EI(x)\frac{d^2w(x)}{dx^2}\right) + N\frac{d^2w(x)}{dx^2} - w(x) = -w(x), \quad x \in [0,a] \tag{3.41}$$

At this point, let $g(x, s; N)$ represent the Green's function of the following boundary value problem

$$\frac{d^2}{dx^2}\left(EI(x)\frac{d^2w(x)}{dx^2}\right) + N\frac{d^2w(x)}{dx^2} - w(x) = 0, \quad x \in [0,a]$$

$$B_{0,k}[w(0)] = 0, \quad B_{a,k}[w(a)] = 0, \quad k = 1, 2$$

We put the parameter N as an argument of $g(x, s; N)$ to highlight the fact that it is a function of N. It is important to notice that $g(x, s; N)$ is a nonlinear function of N. This will be crucial in the discussion that follows.

In compliance with Theorem 1.3, the boundary value problem in eqns (3.40) and (3.41) reduces then to a regular homogeneous integral equation

$$w(x) = \int_0^a g(x, s; N) w(s) ds \tag{3.42}$$

with respect to $w(x)$. However, since $g(x, s; N)$ represents a nonlinear function of N, the above integral equation poses a nonlinear eigenvalue problem. This yields an obvious difficulty in computing its entire spectrum. But since we are interested only in the lowest eigenvalue N_{cr}, the situation can be easily resolved. Indeed, based on the partitioning depicted in Figure 3.5 and using the approach described earlier in this section, one can approximate eqn (3.42) by the following system of linear algebraic equations

$$w_i = \sum_{j=0}^{n-1} w_{j+1} \int_{s_j}^{s_{j+1}} g(x_i, s; N) ds, \quad (i = \overline{1, n})$$

in approximate values w_i of the deflection function $w(x)$, defined on the set of grid-points x_i depicted with the empty dots in Figure 3.5. The above system can be written in a matrix form as

$$(A(N) - I)W = 0 \tag{3.43}$$

where W represents a vector whose components are w_i. Thus, the coefficient matrix of a linear system nonlinearly depends on a parameter. So the eigenvalues of the coefficient matrix $A(N)-I$ in eqn (3.43) represent the roots of the transcendental equation

$$\det(A(N)-I) = 0 \qquad (3.44)$$

The smallest root of this equation represents the approximate value of N_{cr}. This root can be obtained by directly computing the determinant of the coefficient matrix $A(N)-I$ with a fixed value of N. One can always make a quite educated guess for the initial approximation of N_{cr} and by successive iterations obtain it then more accurately.

EXAMPLE 2 Based on the algorithm just described, determine approximate values of the critical force N_{cr} for beams of uniform flexural rigidity EI_0, subjected to various edge conditions.

The Green's functions required for the computation that follows have been obtained analytically for all considered types of boundary conditions. We below present just one of them. Namely, for the boundary value problem

$$\frac{d^4w(x)}{dx^4} + \frac{N}{EI_0}\frac{d^2w(x)}{dx^2} - w(x) = 0, \quad x \in [0, a]$$

$$w(0) = \frac{d^2w(0)}{dx^2} = 0, \quad w(a) = \frac{d^2w(a)}{dx^2} = 0$$

which relates to a simply supported beam. The branch of the Green's function, which is valid for $0 \le x \le s \le a$, is obtained in this case in the form

$$g(x,s;N) = \frac{r_0}{r_2}\left[\frac{r_2 e^{r_1 a} - r_1 e^{-r_1 a}}{2r_1^2 \sinh(r_1 a)}\sinh(r_1 x) + \frac{\sin(r_2 x)}{\sin(r_2 a)}\sin(r_2(a-s))\right] \qquad (3.45)$$

where

$$r_0 = \frac{1}{\sqrt{p^2+4}}, \quad p = \frac{N}{EI_0}$$

and

$$r_1 = \frac{\sqrt{2}}{2}\sqrt{-p+\sqrt{p^2+4}}, \quad r_2 = \frac{\sqrt{2}}{2}\sqrt{p+\sqrt{p^2+4}}$$

Due to the self-adjointness of this problem, the other branch of $g(x,s;N)$ can be obtained from that in eqn (3.45) by interchanging of x with s.

In Table 3.10 we present the approximate values of N_{cr} computed for the simply supported (S–S), simply supported – clamped (S–C), and clamped (C–C) beam by reducing eqn (3.42) to the transcendental eqn (3.44). The partitioning parameter is fixed at $n = 10$. We again assume a unit length and a unit flexural rigidity of a beam.

3.2. BUCKLING (EULER FORMULATION)

Table 3.10 Approximate values of the critical force N_{cr}, computed for beams of uniform flexural rigidities by solving eqn (3.44)

Method	Edge conditions		
used	S–S	S–C	C–C
Approximate values of N_{cr}	9.8696044	20.1907290	39.4784176
Exact values	9.8696044	20.1997	39.4784176
Relative error, %	$.1 \times 10^{-8}$	—	$.5 \times 10^{-7}$

The accuracy of approximate values of N_{cr} in this table varies slightly from case to case, remaining nevertheless on a very high level for the relatively rough partitioning ($n=10$) which we used. Notice that for the simply supported – clamped (S–C) beam the exact value of the critical force is not available and the value 20.1997 in this table represents the classical value (see, for example [11]) which is obtained numerically. Subsequently, the relative error for this beam is not shown.

In what follows, we will present the influence function algorithm for one more buckling problem. A beam of variable flexural rigidity $EI(x)$, resting on a Winkler foundation (whose elastic coefficient is k_0) is subjected to an axial compressive force N as shown in Figure 3.6.

And again, like in the problem posed by eqns (3.25) and (3.26), we are looking for the critical value N_{cr} that causes the loss of stability of the system. Any other edge supports can readily be imagined and taken care of by the procedure we are going to present.

As is known [73], the static equilibrium of this beam can be modeled by the following boundary value problem

$$\frac{d^2}{dx^2}\left(EI(x)\frac{d^2w(x)}{dx^2}\right) + N\frac{d^2w(x)}{dx^2} + k_0 w(x) = 0, \quad x \in [0, a] \quad (3.46)$$

$$w(0) = \frac{dw(0)}{dx} = 0, \quad \frac{d^2w(a)}{dx^2} = \frac{d^3w(a)}{dx^3} = 0 \quad (3.47)$$

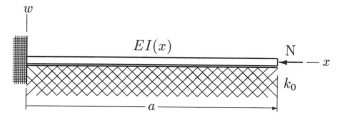

Figure 3.6 A beam on an elastic foundation

The formulation in eqns (3.46) and (3.47) represents an eigenvalue problem with the parameter N to be found. This problem can be reduced to a homogeneous integral equation whose kernel contains a parameter.

In doing so, let $g(x, s; N)$ represent the Green's function of the following homogeneous boundary value problem

$$\frac{d^2}{dx^2}\left(EI(x)\frac{d^2w(x)}{dx^2}\right) + N\frac{d^2w(x)}{dx^2} = 0, \quad x \in [0, a]$$

$$w(0) = \frac{dw(0)}{dx} = 0, \quad \frac{d^2w(a)}{dx^2} = \frac{d^3w(a)}{dx^3} = 0$$

Such a Green's function represents the influence function of a transverse unit concentrated force for the beam under consideration, subjected to the axial compressive force N. Generally, such an influence function is to be constructed numerically.

In compliance with Theorem 1.3, it follows that the solution of the problem in eqns (3.46) and (3.47) is expressed in terms of $g(x, s; N)$ as

$$w(x) = k_0 \int_0^a g(x, s; N)w(s)ds \tag{3.48}$$

Thus, the eigenvalue problem in eqns (3.46) and (3,47) is reduced to the linear homogeneous integral equation (3.48) that nevertheless constitutes a nonlinear eigenvalue formulation. By the example that follows, we show the accuracy level attainable when the latter is solved numerically.

EXAMPLE 3 As a validation example, we will compute the approximate value of N_{cr} for a simply supported beam of a uniform flexural rigidity EI_0 resting on the Winkler foundation whose elastic coefficient is k_0.

Clearly, the eigenvalue problem

$$EI_0\frac{d^4w(x)}{dx^4} + N\frac{d^2w(x)}{dx^2} + k_0 w(x) = 0, \quad x \in [0, a] \tag{3.49}$$

$$w(0) = \frac{d^2w(0)}{dx^2} = 0, \quad w(a) = \frac{d^2w(a)}{dx^2} = 0 \tag{3.50}$$

which is associated with the above statement, allows an exact solution. Indeed, the buckling failure shape in this case can be sought in the form

$$w_{cr}(x) = \sin\left(\frac{\pi x}{a}\right)$$

and the critical value of N is then found as

$$N_{cr} = \frac{EI_0\pi^4 + a^4 k_0}{(a\pi)^2}$$

3.3. SOME CONTACT PROBLEMS

Table 3.11 Approximate values of the critical force N_{cr}, computed for a simply supported beam of a uniform flexural rigidity, resting on an elastic foundation whose coefficient is k_0

Elastic coefficient, k_0	50	100	150	200
Approximate values, N_{cr}	14.9356	20.0016	25.0675	30.1333
Exact values	14.9357	20.0017	25.0678	30.1338
Relative error, %	$.3 \times 10^{-3}$	$.1 \times 10^{-2}$	$.3 \times 10^{-2}$	$.6 \times 10^{-2}$

In Table 3.11 one finds some approximate results computed for the problem in eqns (3.49) and (3.50) by reducing it to the eigenvalue problem for the integral equation (3.48). The influence function $g(x, s; N)$ that serves in this case as a kernel of eqn (3.48), is available in Section 3.1 (see eqn (3.20), where q^2 has to be replaced with N/EI_0). In the present computation, we assumed a beam of unit flexural rigidity $EI_0 = 1$. The lowest eigenvalue N_{cr} of eqn (3.48) has been computed by the algorithm described earlier in this section, with the partitioning parameter $n = 10$.

From the above data it follows that the accuracy of computing the values of N_{cr} slightly drops as k_0 increases, but even for $k_0 = 200$ it still remains at a very high level of $.6 \times 10^{-2}$%.

3.3 Some contact problems

A number of applications of the influence function method to beam problems have been presented in this text. All of them are linear. The nonlinear eigenvalue problems that appeared in eqns (3.42) and (3.48) in the last section do not count, because the original formulations that reduce to eqns (3.42) and (3.48) are linear in nature. So the nonlinearity results from the approach used.

However, the influence function of a transverse unit concentrated force being constructed for a Kirchhoff's beam could also be utilized in the development of computational algorithms for numerical solutions of some actually nonlinear beam problems. In this section, we will present just one particular nonlinear formulation which is designed as an illustrative example to show how this idea can be realized in practice. The nonlinearity that we will be involved with is of the type in the contact problem where the contact zone is not specified in advance and is to be found along with other characteristics of the problem.

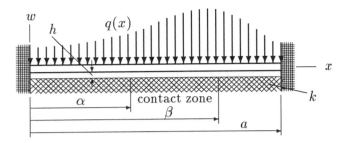

Figure 3.7 A beam spaced above a Winkler foundation

We will consider a contact interaction of a beam with an elastic foundation where there is an initial distance between them and a partial contact is caused by the transverse load subjected to the beam. Thus, the contact zone is not specified by the statement of the problem. This clearly results in a nonlinearity of the problem, even if the maximum value of the beam's deflection is assumed to be of a permissible order for Kirchhoff theory, in other words, if the geometrically linear statement is assumed.

A word of caution is appropriate in relation to this statement. Notably, Kirchhoff theory is not perfectly adequate to the physics of such contact problems. To obtain more accurate solutions, more advanced beam theories should be implemented instead. However, the model proposed here can definitely be utilized at least for initial estimation purposes.

Let an elastic single-span beam of a variable flexural rigidity $EI(x)$ be spaced a distance h just above a Winkler foundation whose elastic coefficient is k. Let the beam undergo a transverse distributed load $q(x)$ as shown in Figure 3.7.

The clamped beam is depicted just for the presentation convenience. Any other standard type of supports can readily be considered.

The distance h is assumed to be small enough to ensure the geometric linearity of the statement (the interaction is to be analyzed within the scope of Kirchhoff theory). The size and location of the contact zone (parameters α and β) are to be determined along with other components of the stress-strain state.

Clearly, within the contact zone (for $\alpha \leq x \leq \beta$), the elastic equilibrium of the beam can be described by the following governing equation

$$\frac{d^2}{dx^2}\left(EI(x)\frac{d^2w(x)}{dx^2}\right) + k[w(x) - h] = -q(x), \quad x \in (\alpha, \beta) \qquad (3.51)$$

since within the contact zone, the beam is actually resting on the Winkler foundation whose elastic coefficient is k.

3.3. SOME CONTACT PROBLEMS

On the left and on the right of the contact zone (for $0 \leq x \leq \alpha$ and $\beta \leq x \leq a$), the elastic equilibrium of the beam can be defined in a standard way by the Euler-Bernoulli equation

$$\frac{d^2}{dx^2}\left(EI(x)\frac{d^2w(x)}{dx^2}\right) = -q(x)$$

The elastic equilibrium of the entire beam can subsequently be described by a single governing differential equation

$$\frac{d^2}{dx^2}\left(EI(x)\frac{d^2w(x)}{dx^2}\right) = -q^*(x), \quad x \in (0, a) \quad (3.52)$$

where the right-hand side $q^*(x)$ represents a continuous function that is formally defined on $(0, a)$ in pieces as

$$q^*(x) = \begin{cases} q(x), & 0 \leq x \leq \alpha \\ q(x) + k(w(x) - h), & \alpha \leq x \leq \beta \\ q(x), & \beta \leq x \leq a \end{cases} \quad (3.53)$$

In accordance with the statement of the problem (see Figure 3.7), the boundary conditions of a clamped edge

$$w(0) = \frac{dw(0)}{dx} = 0, \quad w(a) = \frac{dw(a)}{dx} = 0, \quad (3.54)$$

are to be subjected in this case. However, it is worth noting again that these could be replaced with other boundary conditions if necessary.

Let $g(x, s)$ represent the influence function of a transverse unit concentrated force for the clamped beam of length a, with flexural rigidity $EI(x)$. Since the right-hand side $q^*(x)$ of eqn (3.52) is a continuous function on the interval $(0, a)$ (see eqn (3.53)), the solution of the boundary value problem in eqns (3.52) and (3.54) can be expressed, in compliance with Theorem 1.3 of Chapter 1, in the form of a single integral

$$w(x) = \int_0^a g(x, s) q^*(s) ds$$

Because the right-hand side $q^*(x)$ of eqn (3.52) is defined in pieces (see eqn (3.53)), the above integral can be broken down into three separate integrals as

$$w(x) = \int_0^\alpha g(x, s) q(s) ds$$
$$+ \int_\alpha^\beta g(x, s)\{q(s) + k[w(s) - h]\} ds + \int_\beta^a g(x, s) q(s) ds$$

For the notational convenience, we now regroup the integral terms in the above relation and write it in a more compact form

$$w(x) = \int_0^a g(x,s)q(s)ds + k\int_\alpha^\beta g(x,s)\left[w(s)-h\right]ds \qquad (3.55)$$

The first term on the right-hand side of the above relation is a definite integral depending on a parameter x. Hence, it represents a function of x that can be computed either analytically (if the loading function $q(x)$ is simple enough), or numerically. Whereas the second term represents a definite integral whose integrand contains the deflection function $w(x)$ to be determined. So, the entire relation in eqn (3.55) represents an integral equation with the deflection function being the basic unknown.

One more peculiarity of the second integral term in eqn (3.55) should be outlined. This term represents a definite integral whose limits are not fixed (the size and location of the contact zone are unknown). Thus, the relation in eqn (3.55) represents a nonlinear integral equation with respect to the deflection function $w(x)$. The nonlinearity of this equation is conditioned by the fact that the parameters α and β are not known and should be determined along with the deflection function while solving eqn (3.55).

In this study, the computational procedure for the nonlinear integral equation (3.55) is developed on the iteration basis. The idea behind this is that once the lower and the upper limits (α and β) in the second integral term of eqn (3.55) are fixed, the entire relation becomes a regular Fredholm integral equation of the second kind. Various standard numerical methods could be applied for solution of such a regular equation. Based on the author's experience, the method of quadrature formulae as well as various iteration methods are highly recommended, providing inexpensive computations with a relatively high level of accuracy.

Several issues of a practical importance arise while developing the computational procedure for the nonlinear problem posed by eqn (3.55). We below address the most significant of those, namely:

- Appropriate choice of the initial approximations α_0 and β_0 for the parameters α and β;

- Establishing of the criteria for terminating the iteration process;

- Choice of the productive strategy to meet the above criteria.

Clearly, the convergence rate of the iteration procedure for the nonlinear problem posed by eqn (3.55) could be significantly affected by the choice of the initial values α_0 and β_0. To find the appropriate way of choosing these,

3.3. SOME CONTACT PROBLEMS

we pose and solve the following subsidiary linear problem

$$\frac{d^2}{dx^2}\left(EI(x)\frac{d^2w_s(x)}{dx^2}\right) = -q(x), \quad x \in (0, a) \tag{3.56}$$

$$w_s(0) = \frac{dw_s(0)}{dx} = 0, \quad w_s(a) = \frac{dw_s(a)}{dx} = 0, \tag{3.57}$$

stated for the same beam as depicted in Figure 3.7 and subjected to the same load and boundary conditions, having, however, no elastic foundation beneath it (as shown in Figure 3.8).

Clearly, if the maximal value of the deflection function in this subsidiary problem exceeds the distance h between the beam and the foundation in the original nonlinear problem, then the two values x_1 and x_2 of x, for which $w_s(x_1) = h$ and $w_s(x_2) = h$, can be utilized as the initial approximations α_0 and β_0 of the values of α and β, respectively in the iteration procedure for the nonlinear problem posed by eqn (3.55).

Using the values of α_0 and β_0, the first approximation $w_1(x)$ of the deflection function $w(x)$ can be computed from the regular Fredholm integral equation of the second kind

$$w_1(x) = \int_0^a g(x,s)q(s)ds + k\int_{\alpha_0}^{\beta_0} g(x,s)\left[w_1(s) - h\right]ds$$

that can then be solved numerically providing the next approximation α_1 and β_1 for the true values of the contact zone end-points.

Hence, the iteration procedure is designed in which every consecutive approximation $w_{i+1}(x)$ of the deflection function can be computed based on the values α_i and β_i obtained from the preceding approximation as a solution of the following linear integral equation

$$w_{i+1}(x) = \int_0^a g(x,s)q(s)ds + k\int_{\alpha_i}^{\beta_i} g(x,s)[w_{i+1}(s) - h]ds, \quad (i = 0, 1, 2, \ldots) \tag{3.58}$$

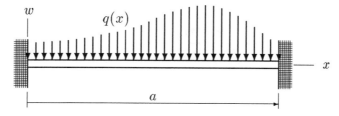

Figure 3.8 The beam modeled by eqns (3.56) and (3.57)

This yields the sequence $\{w_1(x), w_2(x), \ldots, w_n(x), \ldots\}$ of successive approximations for the deflection function $w(x)$ as well as the sequence of pairs $\{\alpha_0, \beta_0; \alpha_1, \beta_1; \ldots; \alpha_{n-1}, \beta_{n-1}; \ldots\}$ of the coordinates for the end-points α and β of the contact zone.

If the iteration process described above practically converges, then it can be truncated upon attaining the required accuracy level for the size and location of the contact zone and for the deflection function. If the convergence is too slow, then one may accomplish a manual correction at any stage of the process. Our experience shows an extremely high rate of convergence where the required accuracy level is attained after just several iterations.

We below present some numerical results illustrating the potential solvability of this problem class by means of the proposed approach.

EXAMPLE 1 Consider a clamped beam of uniform flexural rigidity EI, spaced a distance h above the elastic foundation and subjected to a uniform pressure $q(x) \equiv q_0$ as depicted in Figure 3.9.

Due to the perfect (geometrical and physical) symmetry of this statement, the resultant stress-strain state of the beam has to be symmetric about the vertical line passing through the beam's midpoint. Taking advantage of this feature of the problem, we can appropriately rewrite the problem formulation. Thus, we are looking for the solution of the following differential equation

$$EI\frac{d^4w(x)}{dx^4} = -q^*(x), \quad x \in (0, \frac{a}{2}) \tag{3.59}$$

where $q^*(x)$ is defined on $(0, a/2)$ in a piecewise manner as

$$q^*(x) = \begin{cases} q_0, & 0 \leq x \leq \alpha \\ q_0 + k[w(x) - h], & \alpha \leq x \leq a/2 \end{cases}$$

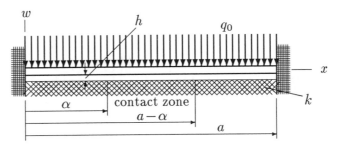

Figure 3.9 Symmetric statement of the contact problem

3.3. SOME CONTACT PROBLEMS

and $w(x)$ satisfies the boundary conditions imposed in the form

$$w(0) = \frac{dw(0)}{dx} = 0, \quad \frac{dw(a/2)}{dx} = \frac{d^3w(a/2)}{dx^3} = 0 \qquad (3.60)$$

By employing either one of the two standard techniques described in Chapter 2, one can readily construct the influence function of a transverse concentrated unit force for half of the clamped beam, with conditions of symmetry imposed at the midpoint $x = a/2$. Clearly, that influence function represents the Green's function $g(x, s)$ of the homogeneous boundary value problem corresponding to that posed by eqns (3.59) and (3.60). This Green's function is found to be in the form

$$g(x, s) = \frac{1}{6aEI} \begin{cases} x^2[3s(s-a)+ax], & \text{for } x \le s \\ s^2[3x(x-a)+as], & \text{for } s \le x \end{cases} \qquad (3.61)$$

Following the pattern used earlier in this section, one obtains the following expression for the deflection function $w(x)$

$$w(x) = q_0 \int_0^{a/2} g(x, s) ds + k \int_\alpha^{a/2} g(x, s)[w(s) - h] ds \qquad (3.62)$$

in terms of the influence function $g(x, s)$ presented in eqn (3.61).

For the sake of simplicity in the computation that follows, we take a beam of unit length $(a = 1)$, with unit flexural rigidity $(EI = 1)$, and apply the load $q_0 = 1$ to it. In this case the subsidiary problem posed by eqns (3.56) and (3.57) is formulated as follows

$$\frac{d^4 w_s(x)}{dx^4} = -1, \quad x \in (0, 1)$$

$$w_s(0) = \frac{dw_s(0)}{dx} = 0, \quad w_s(1) = \frac{dw_s(1)}{dx} = 0,$$

and its solution is given by

$$w_s(x) = -\frac{1}{24} x^2 (x-1)^2 \qquad (3.63)$$

This function takes on its maximum value, which approximately equals .0026, at the midpoint $x = .5$ of the unit segment. Thus, the maximum value of the deflection function $w_s(x)$ for the beam under consideration, having however no foundation beneath, approximately equals .0026. Based on this and assuring a reasonable contact zone for the original problem, we assume $h = .001$ and $k = 5,000$ (see the formulation in eqns (3.59) and (3.60)).

From eqn (3.63), one can easily estimate the initial approximation α_0 for the parameter α in eqn (3.62). Indeed, equating the deflection function $w_s(x)$ to the value of h, we derive as follows

$$\frac{1}{24}x^2(1-x)^2 = .001$$

Taking the square root of the both-hand sides of this equation, we obtain the following quadratic equation

$$x^2 - x + .15492 = 0$$

whose meaningful root is .19165. Clearly, this can be taken as the value of α_0 to start the iteration process for eqn (3.62).

As it is seen from the first line of Table 3.12, it took just four iterations of eqn (3.62) to accurately obtain the true size ($\alpha = .32390$) of the contact zone in this case.

EXAMPLE 2 Consider various single-span beams of unit length, having uniform flexural rigidity EI, undergoing continuously distributed transverse loads $q(x) = x$, and spaced a distance h above the elastic foundation.

Note that, contrary to the previous example, the statements of the problems considered here are not symmetric about the vertical line passing through the midpoint $x = a/2$ of a beam. Therefore, the influence functions utilized here have to satisfy the imposed boundary conditions at $x = 0$ and $x = a$. This is true for all the three cases to be discussed (notice that the geometrically symmetric case (S–S) fails nevertheless to be physically symmetric due to the nonsymmetric loading function $q(x) = x$).

The influence functions required to proceed with the computation are available in the present text (for the clamped – simply supported (C–S) and simply supported (S–S) beam, see *EXAMPLES 2* and *1*, respectively, in Section 3.2, while for the cantilever (C–F) beam, see eqn (2.37) in Section 2.1).

In Table 3.12 in addition to the results for the clamped beam, one also finds the sizes of contact zones and the number n of iterations required for the clamped – simply supported, cantilever, and simply supported beam. The values of the initial data $q(x)$, EI, h, and k for each of those problems have been appropriately chosen in order to make the statements reasonable physically as well as mathematically.

From these data, it particularly follows that a fairly high convergence rate of the iteration process occurred for all of the cases considered. This reflects the integral nature of the operators to be approximated at each step of the process.

3.3. SOME CONTACT PROBLEMS

Table 3.12 Approximate size of the contact zone and number n of iterations accomplished for beams of unit length, spaced a distance h above a Winkler foundation whose coefficient is k

Type of beam	$q(x)$	k/EI	h	α	β	Num. of iter.
C–C	1	5,000	-.001	.3239	.6761	4
C–S	1	100	-.002	.2437	.8813	4
C–F	1	10	-.030	.4804	—	5
S–S	x	100	-.002	.1583	.8668	4

Note that any required component of the stress-strain state discussed here is also computed highly accurately, because the numerical differentiation (which represents a source of large computational errors in many of the conventional methods) is avoided within the scope of our algorithm where all the differentiations can be performed analytically. This feature of the algorithm provides in fact high accuracy level.

Summarizing the material of this chapter, note that relatively primitive statements from Kirchhoff beam theory, considered here, have been chosen to just help the reader to more easily learn possible strategies of computational implementations based on influence functions.

The significance of influence functions and matrices in computational mechanics can be grasped by reading, for example, Chapters 4 through 6 of monograph [45]. We hope that using our experience, the reader will be well prepared for studying more specific texts on influence (Green's) functions and for doing really complicated projects where the possibility of influence functions is not obvious from the very beginning.

Chapter 4

Assemblies of Elements

In this chapter, we are concerned with some nontraditional implementations of the influence function method in applied mechanics. We will deal here with such statements that reduce to the so-called multi-point posed boundary value problems for systems of linear ordinary differential equations. These are not, however, boundary value problems for systems of equations in a common sense, where several unknown functions have a common domain for an independent variable, and at least one of the equations in the system involves more than one unknown function.

Each equation in the systems that are discussed in this chapter governs a single unknown function and is formulated over an individual domain. The system is formed by letting the single domains contact each other at their end-points which become contact points. Subsequently, the single differential equations are united in a system format by appropriate contact conditions imposed at the contact points.

Based on the development of Section 4.1, where the notion of a matrix of Green's type is introduced for a piecewise homogeneous media of sandwich type, in Section 4.2 we construct influence matrices for multi-spanned beams. By implementing finite weighted graphs, in Section 4.3 we extend the notion of a matrix of Green's type to problems of applied mechanics stated on complex assemblies of one-dimensional elements.

4.1 Multi-point posed problems

To present a typical formulation of a multi-point posed boundary value problem of the kind to be considered, let the interval $[a_0, a_k]$ be partitioned with a set of internal points $a_i, (i = \overline{1, k-1})$ into k arbitrary subintervals

(a_{i-1}, a_i). Consider a set of nonhomogeneous linear differential equations

$$L_i[y_i(x)] \equiv \sum_{j=0}^{n} p_{ij}(x) y_i^{(n-j)}(x) = f_i(x), \quad x \in (a_{i-1}, a_i), \quad (i=\overline{1,k}) \qquad (4.1)$$

each of which is stated over an individual subinterval. The coefficients $p_{ij}(x)$ of the operators L_i represent continuous functions on $[a_{i-1}, a_i]$, with the leading coefficients $p_{i0}(x)$ being nonzero at any single point on $[a_{i-1}, a_i]$. Each of the right-hand side functions $f_i(x)$ in eqn (4.1) is also continuous on the corresponding subinterval (a_{i-1}, a_i).

In order to put eqns (4.1) into a system format, the following set of boundary and contact conditions

$$M_q[y_1(a_0), y_1(a_1), y_2(a_1), \ldots, y_k(a_k)] = 0, \quad (q = \overline{1, n \times k}) \qquad (4.2)$$

is imposed, where M_q represents linearly independent forms modeling the contact conditions at the internal points $a_i, (i = \overline{1, (k-1)})$ as well as the endpoint conditions for $y_1(x)$ imposed at a_0 and for $y_k(x)$ imposed at a_k.

Assume that the homogeneous boundary value problem corresponding to that stated with eqns (4.1) and (4.2) has only the trivial solution.

Introduce a vector-function $\mathbf{Y}(x)$ whose components $Y_i(x)$ are defined in terms of $y_i(x)$ in the following fashion

$$Y_i(x) = \begin{cases} y_i(x), & \text{for } x \in (a_{i-1}, a_i) \\ 0, & \text{for } x \in (a_0, a_k) \backslash (a_{i-1}, a_i) \end{cases}$$

Based on the right-hand side functions $f_i(x)$ in eqn (4.1), we also introduce a vector-function $\mathbf{F}(x)$ whose components $F_i(x)$ are defined by

$$F_i(x) = \begin{cases} f_i(x), & \text{for } x \in (a_{i-1}, a_i) \\ 0, & \text{for } x \in (a_0, a_k) \backslash (a_{i-1}, a_i) \end{cases}$$

We are now in a position to formally extend the definition of a Green's function so as to make it valid for the multi-point posed boundary value problems of the type in eqns (4.1) and (4.2).

Definition: If for any allowable vector-function $\mathbf{F}(x)$, the vector-function $\mathbf{Y}(x)$ is expressible in the integral form

$$\mathbf{Y}(x) = -\int_{a_0}^{a_k} G(x, s) \mathbf{F}(s) \, ds \qquad (4.3)$$

then let us call the kernel matrix

$$G(x, s) = (g_{ij}(x, s))_{i,j=\overline{1,k}}$$

4.1. MULTI-POINT POSED PROBLEMS

in eqn (4.3) the *matrix of Green's type* for the homogeneous multi-point posed boundary value problem corresponding to that of eqns (4.1) and (4.2).

The components $g_{ij}(x,s)$ in this matrix are defined for $x \in [a_{i-1}, a_i]$ and $s \in [a_{j-1}, a_j]$, holding the following properties:

1. For $i \neq j$, the functions $g_{ij}(x,s)$ are continuous along with their derivatives with respect to x up to the n-th order included.

2. For $i = j$ (as $x \neq s$), $g_{ij}(x,s)$ are also continuous along with their derivatives with respect to x up to the n-th order included, but as $x = s$, $g_{ij}(x,s)$ are continuous along with their derivatives with respect to x up to the $(n-2)$-nd order included, whereas their $(n-1)$-st derivatives make jumps, the magnitudes of which equal $-p_{i0}^{-1}(s)$.

3. For $x \neq s$, $g_{ij}(x,s)$ as functions of x satisfy the homogeneous equations
$$L_i[g_{ij}(x,s)] = 0, \quad x \in (a_{i-1}, a_i), \quad (i = \overline{1,k})$$

4. $g_{ij}(x,s)$ satisfy all of the boundary and contact conditions posed by eqn (4.2), i. e. :
$$M_q[g_{ij}(x,s)] = 0, \quad (q = \overline{1, n \times k})$$

We will show how matrices of Green's type can practically be constructed by presenting several particular examples.

EXAMPLE 1 Consider the simplest three-point posed boundary value problem written as

$$\frac{d^2 y_1(x)}{dx^2} = f_1(x), \quad x \in (-1, 0) \tag{4.4}$$

$$\frac{d^2 y_2(x)}{dx^2} = f_2(x), \quad x \in (0, 1) \tag{4.5}$$

$$y_1(-1) = 0, \quad y_2(1) = 0 \tag{4.6}$$

$$y_1(0) = y_2(0), \quad \frac{dy_1(0)}{dx} = \lambda \frac{dy_2(0)}{dx} \tag{4.7}$$

This problem might, in particular, be interpreted as a model for steady-state heat conduction in a compound bar consisting of two physically homogeneous segments built of different materials. Parameter λ represents here the ratio λ_2/λ_1 of the heat conductivities of the materials of which the bar is composed.

The homogeneous problem corresponding to that of eqns (4.4)–(4.7) has only the trivial solution (see exercise 4.1(a)). Following Lagrange's method of variation of parameters, we represent the general solution of eqn (4.4) as

$$y_1(x) = C_1(x) + xC_2(x) \qquad (4.8)$$

This yields the following system of linear algebraic equations

$$\begin{pmatrix} 1 & x \\ 0 & 1 \end{pmatrix} \times \begin{pmatrix} C_1'(x) \\ C_2'(x) \end{pmatrix} = \begin{pmatrix} 0 \\ f_1(x) \end{pmatrix}$$

with a well-posed coefficient matrix. From this, it follows that

$$C_1'(x) = -xf_1(x), \quad C_2'(x) = f_1(x)$$

Expressions for $C_1(x)$ and $C_2(x)$ are obtained by a straightforward integration as

$$C_1(x) = -\int_{-1}^{x} sf_1(s)ds + M_1, \quad C_2(x) = \int_{-1}^{x} f_1(s)ds + M_2$$

Substituting these expressions of $C_1(x)$ and $C_2(x)$ into eqn (4.8) and combining the integral terms, we obtain

$$y_1(x) = \int_{-1}^{x} (x-s)f_1(s)ds + M_1 + M_2 x, \quad x \in [-1, 0] \qquad (4.9)$$

Analogously for $y_2(x)$, we obtain

$$y_2(x) = \int_{0}^{x} (x-s)f_2(s)ds + N_1 + N_2 x, \quad x \in [0, 1] \qquad (4.10)$$

The boundary and contact conditions in eqns (4.6) and (4.7) applied to $y_1(x)$ and $y_2(x)$ will be used to compute the values of $M_1, M_2, N_1,$ and N_2. The boundary conditions in eqn (4.6) yield

$$M_1 - M_2 = 0,$$

$$N_1 + N_2 = \int_{0}^{1} (s-1)f_2(s)ds.$$

while the contact conditions in eqn (4.7) result in

$$M_1 - N_1 = \int_{-1}^{0} sf_1(s)ds,$$

4.1. MULTI-POINT POSED PROBLEMS

$$\lambda N_2 - M_2 = \int_{-1}^{0} f_1(s)ds.$$

These relations form a well-posed system of linear algebraic equations in $M_1, M_2, N_1,$ and N_2, whose solution is

$$M_1 = M_2 = \frac{1}{1+\lambda}\int_{-1}^{0}(\lambda s - 1)f_1(s)ds + \frac{\lambda}{1+\lambda}\int_{0}^{1}(s-1)f_2(s)ds,$$

$$N_1 = -\frac{1}{1+\lambda}\int_{-1}^{0}(s+1)f_1(s)ds + \frac{\lambda}{1+\lambda}\int_{0}^{1}(s-1)f_2(s)ds,$$

and

$$N_2 = \frac{1}{1+\lambda}\int_{-1}^{0}(s+1)f_1(s)ds + \frac{1}{1+\lambda}\int_{0}^{1}(s-1)f_2(s)ds.$$

Hence, substituting these values of the coefficients into eqns (4.9) and (4.10), after elementary transformations, one obtains

$$y_1(x) = \int_{-1}^{0}\frac{(x+1)(\lambda s - 1)}{1+\lambda}f_1(s)ds$$

$$+ \int_{-1}^{x}(x-s)f_1(s)ds + \int_{0}^{1}\frac{\lambda(x+1)(s-1)}{1+\lambda}f_2(s)ds \quad (4.11)$$

and

$$y_2(x) = \int_{-1}^{0}\frac{(x-1)(s+1)}{1+\lambda}f_1(s)ds$$

$$+ \int_{0}^{x}(x-s)f_2(s)ds + \int_{0}^{1}\frac{(x+\lambda)(s-1)}{1+\lambda}f_2(s)ds \quad (4.12)$$

Combining the first and the second integrals from eqn (4.11), we obtain

$$y_1(x) = -\int_{-1}^{0}g_{11}(x,s)f_1(s)ds - \int_{0}^{1}g_{12}(x,s)f_2(s)ds \quad (4.13)$$

while combining the second and the third integrals in eqn (4.12), we have

$$y_2(x) = -\int_{-1}^{0}g_{21}(x,s)f_1(s)ds - \int_{0}^{1}g_{22}(x,s)f_2(s)ds \quad (4.14)$$

where the kernel functions $g_{ij}(x,s)$ are obtained as

$$g_{11}(x,s) = \begin{cases} \beta(x+1)(1-\lambda s), & \text{for } -1 \leq x \leq s < 0 \\ \beta(s+1)(1-\lambda x), & \text{for } -1 < s \leq x \leq 0, \end{cases}$$

$$g_{12}(x,s) = \lambda\beta(x+1)(1-s), \quad \text{for } -1 \leq x \leq 0 < s < 1,$$

$$g_{21}(x,s) = \beta(1-x)(s+1), \quad \text{for } -1 < s < 0 \leq x \leq 1,$$

and

$$g_{22}(x,s) = \begin{cases} \beta(x+\lambda)(1-s), & \text{for } 0 \leq x \leq s < 1 \\ \beta(s+\lambda)(1-x), & \text{for } 0 < s \leq x \leq 1 \end{cases}$$

with $\beta = (1+\lambda)^{-1}$.

In accordance with the approach suggested earlier in this section, we introduce the vector $\mathbf{Y}(x)$ whose components are defined as

$$Y_1(x) = \begin{cases} y_1(x), & \text{for } x \in (-1,0) \\ 0, & \text{for } x \in (0,1) \end{cases} \qquad Y_2(x) = \begin{cases} 0, & \text{for } x \in (-1,0) \\ y_2(x), & \text{for } x \in (0,1) \end{cases}$$

and the vector $\mathbf{F}(x)$ with components defined as

$$F_1(x) = \begin{cases} f_1(x), & \text{for } x \in (-1,0) \\ 0, & \text{for } x \in (0,1) \end{cases} \qquad F_2(x) = \begin{cases} 0, & \text{for } x \in (-1,0) \\ f_2(x), & \text{for } x \in (0,1) \end{cases}$$

In terms of the vectors $\mathbf{Y}(x)$ and $\mathbf{F}(x)$, the integrals from eqns (4.13) and (4.14) can be rewritten as a single integral

$$\mathbf{Y}(x) = -\int_{-1}^{1} G(x,s)\mathbf{F}(s)ds$$

Thus, from the definition introduced in this section, it follows that the expressions of $g_{ij}(x,s)$ just derived can be referred to as the entries of the matrix of Green's type $G(x,s)$ for the three-point posed homogeneous problem associated with that of eqns (4.4)–(4.7).

EXAMPLE 2 Consider a system of Cauchy–Euler equations

$$\frac{d}{dx}\left(x\frac{dy_1(x)}{dx}\right) - \frac{1}{x}y_1(x) = f_1(x), \quad x \in (0, a) \tag{4.15}$$

$$\frac{d}{dx}\left(x\frac{dy_2(x)}{dx}\right) - \frac{1}{x}y_2(x) = f_2(x), \quad x \in (a, \infty) \tag{4.16}$$

with the boundary and contact conditions imposed as

$$|y_1(0)| < \infty, \quad |y_2(\infty)| < \infty \tag{4.17}$$

$$y_1(a) = y_2(a), \quad \frac{dy_1(a)}{dx} = \lambda\frac{dy_2(a)}{dx} \tag{4.18}$$

4.1. MULTI-POINT POSED PROBLEMS

This formulation is presented to show how Lagrange's method works for boundary value problems with singular points for governing equations and stated over semi-infinite domains.

Exercise 4.1(b) shows that the homogeneous problem associated with that in eqns (4.15)–(4.18) has only the trivial solution, justifying, consequently, the existence and uniqueness of its matrix of Green's type.

Clearly, the functions $y_1(x) \equiv x$ and $y_2(x) \equiv x^{-1}$ constitute a fundamental solution set for the homogeneous equation corresponding to that in eqn (4.15). Therefore, tracing out our procedure, we seek the general solution of eqn (4.15) as

$$y(x) = C_1(x)\,x + C_2(x)\,x^{-1} \tag{4.19}$$

This yields the following well-posed system of linear algebraic equations

$$\begin{pmatrix} x & x^{-1} \\ 1 & -x^{-2} \end{pmatrix} \times \begin{pmatrix} C_1'(x) \\ C_2'(x) \end{pmatrix} = \begin{pmatrix} 0 \\ f_1(x)/x \end{pmatrix}$$

in $C_1'(x)$ and $C_2'(x)$, whose solution is

$$C_1'(x) = \frac{f_1(x)}{2x}, \quad C_2'(x) = -\frac{x\,f_1(x)}{2}$$

Integrating these relations and substituting the values of $C_1(x)$ and $C_2(x)$ into eqn (4.19) provides

$$y_1(x) = \int_0^x \frac{x^2 - s^2}{2xs} f_1(s)\,ds + D_{11}x + D_{12}x^{-1} \tag{4.20}$$

Analogously, for $y_2(x)$ one obtains

$$y_2(x) = \int_a^x \frac{x^2 - s^2}{2xs} f_2(s)\,ds + D_{21}x + D_{22}x^{-1} \tag{4.21}$$

Notice that the lower limits of the above two integrals are different representing the left-end points of the intervals $(0, a)$ and (a, ∞), respectively.

The constants of integration in eqns (4.20) and (4.21) are to be obtained by the boundary and contact conditions from eqns (4.17) and (4.18). The first condition in eqn (4.17) requires $D_{12} = 0$, since x^{-1} is unbounded as x approaches zero. To satisfy the second condition in eqn (4.17), we regroup the terms in eqn (4.21) in the following manner

$$y_2(x) = \left(\int_a^x \frac{1}{2s} f_2(s)\,ds + D_{21}\right) x + \left(-\int_a^x \frac{s}{2} f_2(s)\,ds + D_{22}\right) x^{-1} \tag{4.22}$$

By observation, it can easily be seen that, as x goes to infinity, the coefficient of x in eqn (4.32) has to equal zero, resulting in

$$D_{21} = -\int_a^\infty \frac{1}{2s} f_2(s)\,ds$$

Recalling the values of D_{12} and D_{21} just found, we write the first condition in eqn (4.18) in the form

$$D_{11}a - D_{22}a^{-1} = -\int_0^a \frac{a^2 - s^2}{2as} f_1(s)\,ds - \int_a^\infty \frac{a}{2s} f_2(s)\,ds \qquad (4.23)$$

To properly treat the second condition in eqn (4.18), we first differentiate $y_1(x)$ and $y_2(x)$, providing

$$y_1'(x) = \int_0^x \frac{x^2 + s^2}{2sx^2} f_1(s)\,ds + D_{11}$$

and

$$y_2'(x) = \int_a^x \frac{x^2 + s^2}{2sx^2} f_2(s)\,ds + D_{21} - D_{22}\,x^{-2}$$

Hence, the second condition in eqn (4.18) yields

$$D_{11} + \lambda D_{22}\,a^{-2} = -\int_0^a \frac{a^2 + s^2}{2sa^2} f_1(s)\,ds - \int_a^\infty \frac{\lambda}{2s} f_2(s)\,ds \qquad (4.24)$$

Equations (4.23) and (4.24) form a well-posed system of linear algebraic equations in D_{11} and D_{22}, whose solution is

$$D_{11} = -\int_a^\infty \frac{\lambda}{(1+\lambda)s} f_2(s)\,ds - \int_0^a \frac{(a^2 + s^2) + \lambda(a^2 - s^2)}{2(1+\lambda)a^2 s} f_1(s)\,ds$$

and

$$D_{22} = -\int_0^a \frac{(a^2 + s^2) - \lambda(a^2 - s^2)}{4\lambda s} f_1(s)\,ds$$

Substituting the values of $D_{ij}, (i, j = 1, 2)$ just computed into eqns (4.20) and (4.21), we obtain the solution of the problem posed by eqns (4.15)–(4.18) as

$$y_1(x) = -\int_0^a \frac{x[(a^2 + s^2) + \lambda(a^2 - s^2)]}{2(1+\lambda)a^2 s} f_1(s)\,ds$$
$$+ \int_0^x \frac{x^2 - s^2}{2xs} f_1(s)\,ds - \int_a^\infty \frac{\lambda x}{(1+\lambda)s} f_2(s)\,ds$$

and

$$y_2(x) = -\int_0^a \frac{(a^2 + s^2) - \lambda(a^2 - s^2)}{4\lambda xs} f_1(s)\,ds$$

4.1. MULTI-POINT POSED PROBLEMS

$$+ \int_a^x \frac{x^2 - s^2}{2xs} f_2(s) ds - \int_a^\infty \frac{x}{2s} f_2(s) ds$$

From these integral representations for $y_1(x)$ and $y_2(x)$, in accordance with the definition of a matrix of Green's type given at the beginning of this section, the entries $g_{ij}(x,s)$ of the matrix of Green's type to the homogeneous boundary value problem associated with that occurring in eqns (4.15)–(4.18) are finally found to be as follows:

$$g_{11}(x,s) = \begin{cases} x[(a^2+s^2) + \lambda(a^2-s^2)][2(1+\lambda)a^2 s]^{-1}, & \text{for } 0 \le x \le s < a \\ s[(a^2+x^2) + \lambda(a^2-x^2)][2(1+\lambda)a^2 x]^{-1}, & \text{for } 0 < s \le x \le a, \end{cases}$$

$$g_{12}(x,s) = \lambda x[(1+\lambda)s]^{-1}, \quad \text{for } 0 \le x \le a < s < \infty,$$

$$g_{21}(x,s) = [(a^2+s^2) - \lambda(a^2-s^2)](4\lambda x s)^{-1}, \quad \text{for } 0 < s < a \le x < \infty,$$

$$g_{22}(x,s) = \begin{cases} x(2s)^{-1}, & \text{for } a \le x \le s < \infty \\ s(2x)^{-1}, & \text{for } a < s \le x < \infty. \end{cases}$$

Before we start with the next example, it is worth noting that in the formulations that have been discussed so far in this section, we considered multi-point posed boundary value problems where domains of independent variables consist of series of segments. The four-point posed problem to be considered in *EXAMPLE 3* that follows, is different. Three segments are joined in an assembly by allowing their left-end points to contact in a way shown in Figure 4.1.

EXAMPLE 3 Let us consider the following problem

$$\frac{d^2 y_i(x)}{dx^2} = -f_i(x), \quad x \in (0,1), \quad i = 1,2,3 \tag{4.25}$$

$$y_1(0) = y_2(0) = y_3(0), \quad h_1 y_1'(0) + h_2 y_2'(0) + h_3 y_3'(0) = 0 \tag{4.26}$$

$$y_1(1) = 0, \quad y_2(1) = 0, \quad y_3(1) = 0 \tag{4.27}$$

This formulation can, for example, be associated with a steady-state heat conduction in an assembly of three rods each of unit length as shown in Figure 4.1. The rods are assumed to be made of conductive materials whose heat conductivities are h_1, h_2, and h_3. The relations in eqn (4.26) in a case of such interpretation can be referred to as conditions of the ideal thermal contact.

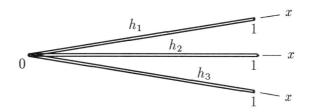

Figure 4.1 Heat conduction in an assembly of rods

In what follows, we will show how the technique described earlier in this section can be applied to the construction of matrices of Green's type for problems of the kind in eqns (4.25)–(4.27).

Exercise 4.1(c) shows that the homogeneous problem associated with that in eqns (4.25)–(4.27) has only the trivial solution, justifying, consequently, the existence and uniqueness of its matrix of Green's type.

Clearly, the general solution of eqn (4.25) can be expressed as

$$y_i(x) = C_i(x) + D_i(x)\,x, \quad i = 1, 2, 3$$

By the method of variation of parameters, this can be reduced to the following integral form

$$y_i(x) = \int_0^x (s-x) f_i(s)\,ds + M_i + N_i\,x, \quad i = 1, 2, 3 \qquad (4.28)$$

Satisfying the first group of condition $y_1(0) = y_2(0) = y_3(0)$ in eqn (4.26), we derive the following two equations in M_1, M_2 and M_3

$$M_1 = M_2 = M_3 \qquad (4.29)$$

The last contact condition from eqn (4.26) yields

$$h_1 N_1 + h_2 N_2 + h_3 N_3 = 0 \qquad (4.30)$$

The boundary conditions in eqn (4.27) finally provide three additional relations for M_i and N_i

$$M_i + N_i = \int_0^1 (1-s) f_i(s)\,ds, \quad i = 1, 2, 3 \qquad (4.31)$$

Relations (4.29)–(4.31) form a well-posed system of six linear algebraic equations in six unknowns, whose solution is

$$M_1 = M_2 = M_3 = H \int_0^1 (1-s)[h_1 f_1(s) + h_2 f_2(s) + h_3 f_3(s)]\,ds,$$

4.1. MULTI-POINT POSED PROBLEMS

$$N_1 = H\int_0^1 (1-s)[(h_2+h_3)f_1(s) - h_2 f_2(s) - h_3 f_3(s)]ds,$$

$$N_2 = H\int_0^1 (1-s)[(h_1+h_3)f_2(s) - h_1 f_1(s) - h_3 f_3(s)]ds,$$

and

$$N_3 = H\int_0^1 (1-s)[(h_1+h_2)f_3(s) - h_1 f_1(s) - h_2 f_2(s)]ds$$

where $H = (h_1 + h_2 + h_3)^{-1}$.

Substituting these values into eqn (4.28), one obtains the solution of the problem formulated by eqns (4.25)–(4.27) in the matrix form

$$\begin{pmatrix} y_1(x) \\ y_2(x) \\ y_3(x) \end{pmatrix} = \int_0^1 \begin{pmatrix} g_{11}(x,s) & g_{12}(x,s) & g_{13}(x,s) \\ g_{21}(x,s) & g_{22}(x,s) & g_{23}(x,s) \\ g_{31}(x,s) & g_{32}(x,s) & g_{33}(x,s) \end{pmatrix} \begin{pmatrix} f_1(s) \\ f_2(s) \\ f_3(s) \end{pmatrix} ds \quad (4.32)$$

The entries $g_{ij}(x,s)$ of the kernel-matrix in the above integral representation are expressed as follows

$$g_{11}(x,s) = \begin{cases} H(1-s)[h_1 + x(h_2+h_3)], & \text{for } x \le s \\ H(1-x)[h_1 + s(h_2+h_3)], & \text{for } s \le x, \end{cases}$$

$$g_{12}(x,s) = Hh_2(1-s)(1-x), \quad g_{13}(x,s) = Hh_3(1-s)(1-x),$$

$$g_{22}(x,s) = \begin{cases} H(1-s)[h_2 + x(h_1+h_3)], & \text{for } x \le s \\ H(1-x)[h_2 + s(h_1+h_3)], & \text{for } s \le x, \end{cases}$$

$$g_{21}(x,s) = Hh_1(1-s)(1-x), \quad g_{23}(x,s) = Hh_3(1-s)(1-x),$$

$$g_{31}(x,s) = Hh_1(1-s)(1-x), \quad g_{32}(x,s) = Hh_2(1-s)(1-x),$$

and

$$g_{33}(x,s) = \begin{cases} H(1-s)[h_3 + x(h_1+h_2)], & \text{for } x \le s \\ H(1-x)[h_3 + s(h_1+h_2)], & \text{for } s \le x. \end{cases}$$

From the definition introduced in the opening part of this section, it follows that the kernel-matrix in eqn (4.32) represents the matrix of Green's type to the homogeneous problem corresponding to that in eqns (4.25)–(4.27). That is, this matrix can be interpreted as the influence function of a point source for the entire assembly.

Matrices of Green's type, as well as Green's functions in the case of a single equation, can naturally be utilized for solutions of boundary value problems for nonhomogeneous systems of equations subject to homogeneous boundary conditions (see, for instance, exercise 4.3).

4.2 Bending of multi-spanned beams

The mathematics in this section have a bearing upon the theory of the multi-point posed boundary value problems stated for a specific type of systems of ordinary linear differential equations that has already been discussed in the previous section, where we have appropriately extended the Green's function formalism to piecewise homogeneous media. Matrices of Green's type were introduced as a natural extension of the Green's function notion. Here we will deeper develop this subject towards its application to the static equilibrium of Kirchhoff beams comprised of more than one span, each one of which might have a different flexural rigidity.

We open this discussion by considering a compound cantilever beam overhanging an intermediate simple support. The beam is comprised of two spans having uniform flexural rigidities EI_1 and EI_2 as shown in Figure 4.2.

To determine the influence function of a transverse unit force (applied to either span) for the above beam, let us formulate the following three-point posed boundary value problem

$$\frac{d^4 w_1(x)}{dx^4} = -\frac{q_1(x)}{EI_1} = -f_1(x), \quad x \in (0, b) \tag{4.33}$$

$$\frac{d^4 w_2(x)}{dx^4} = -\frac{q_2(x)}{EI_2} = -f_2(x), \quad x \in (b, a) \tag{4.34}$$

$$w_1(0) = \frac{dw_1(0)}{dx} = 0, \quad \frac{d^2 w_2(a)}{dx^2} = \frac{d^3 w_2(a)}{dx^3} = 0 \tag{4.35}$$

$$w_1(b) = w_2(b) = 0, \quad \frac{dw_1(b)}{dx} = \frac{dw_2(b)}{dx} \tag{4.36}$$

$$EI_1 \frac{d^2 w_1(b)}{dx^2} = EI_2 \frac{d^2 w_2(b)}{dx^2} \tag{4.37}$$

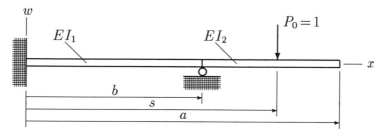

Figure 4.2 A compound beam overhanging one support

4.2. BENDING OF MULTI-SPANNED BEAMS

where $q_1(x)$ and $q_2(x)$ in eqns (4.33) and (4.34) represent arbitrary transverse continuously distributed loads applied to the left-hand and right-hand span, respectively. From our preceding discussions on Lagrange's method of variation of parameters, the reader may recall that $q_1(x)$ and $q_2(x)$ do not need to be specified. However, at a certain stage of the procedure, they aid us in determining the influence matrix that we are seeking. The functions $f_1(x)$ and $f_2(x)$, in turn, are introduced here for the sake of notational convenience in the development that follows.

To comprehend the forthcoming material, the discussion that we have had in Section 4.1 is essential, with a specific emphasis on *EXAMPLE 2*. The only difference of the statement in this section compared to the problems analyzed in Section 4.1 is that we now pose a problem for equations of higher order. This, of course, will result in a more cumbersome computation.

Let $G(x,s)$ represent the matrix of Green's type for the homogeneous boundary value problem corresponding to that in eqns (4.33)–(4.37). Clearly, $G(x,s)$ represents the influence function that we are looking for. Namely, the entries $g_{11}(x,s)$ and $g_{12}(x,s)$ of $G(x,s)$ represent the deflection in the left-hand span $(0 \le x \le b)$ of the beam, caused by a unit force applied within the left-hand span $(s \in (0,b))$ and the right-hand span $(s \in (b,a))$, respectively. The entries $g_{21}(x,s)$ and $g_{22}(x,s)$, in turn, show how the right-hand span $(b \le x \le a)$ responds to a unit force applied within the left-hand and right-hand span, respectively.

In order to solve the boundary value problem in eqns (4.33)–(4.37), we recall the standard technique of Lagrange's method of variation of parameters, repeatedly used in Chapters 1 and 2. In doing so, one represents the general solutions of eqns (4.33) and (4.34) as

$$w_i(x) = A_i(x) + B_i(x)x + C_i(x)x^2 + D_i(x)x^3, \quad i=1,2 \qquad (4.38)$$

The system of linear equations in $A'_i(x), B'_i(x), C'_i(x)$, and $D'_i(x)$ ($i=1,2$), which results from the standard procedure of Lagrange's method, is found to be in this case in the form

$$\begin{pmatrix} 1 & x & x^2 & x^3 \\ 0 & 1 & 2x & 3x^2 \\ 0 & 0 & 2 & 6x \\ 0 & 0 & 0 & 6 \end{pmatrix} \times \begin{pmatrix} A'_i(x) \\ B'_i(x) \\ C'_i(x) \\ D'_i(x) \end{pmatrix} = \begin{pmatrix} 0 \\ 0 \\ 0 \\ -f_i(x) \end{pmatrix}$$

The upper triangular form of the coefficient matrix makes the solution of the system a simple backward substitution. Thus, we have

$$A'_i(x) = \frac{x^3}{6} f_i(x), \quad B'_i(x) = -\frac{x^2}{2} f_i(x)$$

$$C'_i(x) = \frac{x}{2} f_i(x), \quad D'_i(x) = -\frac{1}{6} f_i(x)$$

Hence, the values of the coefficients for $w_1(x)$ in eqn (4.38) can be obtained by integrating their derivatives just obtained over $[0, x]$. This yields

$$A_1(x) = \int_0^x \frac{s^3}{6} f_1(s) ds + H_1, \quad B_1(x) = -\int_0^x \frac{s^2}{2} f_1(s) ds + K_1$$

$$C_1(x) = \int_0^x \frac{s}{2} f_1(s) ds + L_1, \quad D_1(x) = -\int_0^x \frac{1}{6} f_1(s) ds + M_1$$

The coefficients for $w_2(x)$ can accordingly be found as integrals over $[b, x]$

$$A_2(x) = \int_b^x \frac{s^3}{6} f_2(s) ds + H_2, \quad B_2(x) = -\int_b^x \frac{s^2}{2} f_2(s) ds + K_2$$

$$C_2(x) = \int_b^x \frac{s}{2} f_2(s) ds + L_2, \quad D_2(x) = -\int_b^x \frac{1}{6} f_2(s) ds + M_2$$

Upon substituting these expressions into eqn (4.38) and grouping all the integral terms together, one obtains the following expressions for the deflection functions $w_1(x)$ and $w_2(x)$

$$w_1(x) = \int_0^x \frac{(s-x)^3}{6} f_1(s) ds + H_1 + K_1 x + L_1 x^2 + M_1 x^3 \qquad (4.39)$$

$$w_2(x) = \int_b^x \frac{(s-x)^3}{6} f_2(s) ds + H_2 + K_2 x + L_2 x^2 + M_2 x^3 \qquad (4.40)$$

To compute the constant coefficients in the above representations, we take advantage of the boundary and contact conditions from eqns (4.35)–(4.37). In doing so, the first condition $w_1(0) = 0$ of a clamped edge in eqn (4.35) yields $H_1 = 0$. While the second condition $w'(0) = 0$ of a clamped edge in eqn (4.35) analogously results in $K_1 = 0$.

Satisfying then the first condition of a free edge at $x = a$ in eqn (4.35), which is associated with vanishing of the bending moment, one obtains

$$2L_2 + 6M_2 a = -\int_b^a (s-a) f_2(s) ds \qquad (4.41)$$

while the second condition at $x = a$, which is associated with vanishing of the shear force, yields

$$M_2 = \int_b^a \frac{1}{6} f_2(s) ds$$

4.2. BENDING OF MULTI-SPANNED BEAMS

Substituting the value of M_2 just found into eqn (4.41), one obtains

$$L_2 = -\int_b^a \frac{s}{2} f_2(s)\,ds$$

The first contact condition $w_1(b)=0$ in eqn (4.36) yields

$$L_1 b^2 + M_1 b^3 = -\int_0^b \frac{(s-b)^3}{6} f_1(s)\,ds \qquad (4.42)$$

while the second contact condition $w_2(b)=0$ in eqn (4.36) provides

$$H_2 + K_2 b + L_2 b^2 + M_2 b^3 = 0 \qquad (4.43)$$

Taking into account the values of L_2 and M_2 found earlier, we rewrite the relation in eqn (4.43) in the form

$$H_2 + K_2 b = -\int_b^a \frac{b^2(b-3s)}{6} f_2(s)\,ds \qquad (4.44)$$

The third contact condition $w_1'(b) = w_2'(b)$ in eqn (4.36), prescribing the equality of the right-hand and the left-hand limits of the slope of the deflection function at $x=b$, results in

$$2L_1 b + 3M_1 b^2 - \int_0^b \frac{(s-b)^2}{2} f_1(s)\,ds =$$

$$K_2 + \int_b^a \frac{b(b-2s)}{2} f_2(s)\,ds \qquad (4.45)$$

The last contact condition in eqn (4.37), which prescribes the equality of the right-hand and the left-hand limits of the bending moment at $x=b$, yields

$$EI_1 \left[2L_1 + 6M_1 b + \int_0^b (s-b) f_1(s)\,ds \right] = EI_2 \int_b^a (b-s) f_2(s)\,ds \qquad (4.46)$$

The relations in eqns (4.42), (4.44), (4.45), and (4.46) form a well-posed system of linear algebraic equations in the four unknowns $L_1, M_1, H_2,$ and K_2. Its solution can be obtained in a special way. Indeed, equations (4.42) and (4.46) involve only two of the unknowns L_1 and M_1. Hence, these equations can be considered as a system independent from the ones in eqns (4.44) and (4.45). Their solution can be readily found as

$$L_1 = \int_0^b \frac{1}{4b^2}[s(b-s)(s-2b)] f_1(s)\,ds - \int_b^a \frac{\lambda}{4}(b-s) f_2(s)\,ds$$

and
$$M_1 = \int_0^b \frac{1}{12b^3}[(s-b)(s^2-2bs-2b^2)]ds + \int_b^a \frac{\lambda}{4b}(b-s)f_2(s)ds$$

where λ represents the ratio EI_2/EI_1 of the flexural rigidities.

Since values of the coefficients H_1, K_1, L_1, and M_1 are already available at this moment, one can write an explicit expression for the deflection function $w_1(x)$ in the left-hand span of the beam, caused by a combination of the two continuously distributed transverse loads $q_1(x)$ and $q_2(x)$ applied to the left-hand and right-hand span respectively. Substituting these coefficients into eqn (4.39) and performing some elementary algebra, one obtains

$$w_1(x) = \int_0^b \frac{x^2}{12b^3}(s-b)[s(3b-x)(2b-s)-2b^2x]f_1(s)ds$$

$$+ \int_b^a \frac{\lambda x^2}{4b}(b-s)(x-b)f_2(s)ds + \int_0^x \frac{(s-x)^3}{6}f_1(s)ds \quad (4.47)$$

While obtaining the deflection function $w_2(x)$ in the right-hand span, notice that, since the values of L_1 and M_1 are already found, eqn (4.45) provides the value of K_2

$$K_2 = \int_0^b \frac{1}{4b}s^2(b-s)f_1(s)ds + \int_b^a \frac{b}{4}(b-s)(\lambda-2)f_2(s)ds$$

Based on this, eqn (4.44) can be solved to provide the value of H_2. This yields

$$H_2 = -\int_0^b \frac{1}{4}s^2(b-s)f_1(s)ds + \int_b^a \frac{b^2}{12}[3\lambda(s-b)-2(3s-2b)]f_2(s)ds$$

Substituting values of the coefficients H_2, K_2, L_2, and M_2 into eqn (4.40) and performing trivial algebra, one obtains an explicit expression for the deflection $w_2(x)$ in the right-hand span of the beam, caused by the combination of the two continuously distributed loads $q_1(x)$ and $q_2(x)$, in the form

$$w_2(x) = \int_0^b \frac{1}{4b}s^2(b-s)(b-x)f_1(s)ds + \int_b^x \frac{(s-x)^3}{6}f_2(s)ds$$

$$+ \int_b^a \frac{1}{12}(x-b)[2(x-b)^2 - 3(s-b)(b(\lambda-2)+2x)]f_2(s)ds \quad (4.48)$$

Combining now the first and the last integrals in eqn (4.47) and recalling the relations between the functions $f_1(x)$ and $f_2(x)$ and the loading functions $q_1(x)$ and $q_2(x)$ (see eqns (4.33) and (4.34)), we can formally rewrite $w_1(x)$ in the form

$$w_1(x) = \int_0^b g_{11}(x,s)q_1(s)ds + \int_b^a g_{12}(x,s)q_2(s)ds$$

4.2. BENDING OF MULTI-SPANNED BEAMS

where

$$g_{11}(x,s) = \frac{1}{12b^3 EI_1} \begin{cases} x^2(s-b)[s(3b-x)(2b-s)-2b^2 x], & \text{for } x \leq s \\ s^2(x-b)[x(3b-s)(2b-x)-2b^2 s], & \text{for } s \leq x \end{cases} \quad (4.49)$$

with both variables x and s ranging between 0 and b. Whereas, for $g_{12}(x,s)$, with $0 \leq x \leq b$ and $b \leq s \leq a$, we have

$$g_{12}(x,s) = \frac{\lambda x^2}{4bEI_2}(b-s)(x-b) \quad (4.50)$$

Analogously, combining the second and the last integrals in eqn (4.48), we formally rewrite $w_2(x)$ in the form

$$w_2(x) = \int_0^b g_{21}(x,s) q_1(s) ds + \int_b^a g_{22}(x,s) q_2(s) ds$$

where

$$g_{21}(x,s) = \frac{s^2}{4bEI_1}(b-s)(x-b) \quad (4.51)$$

with $b \leq x \leq a$ and $0 \leq s \leq b$. For $g_{22}(x,s)$, with both variables x and s ranging between b and a, we obtain

$$g_{22}(x,s) = \frac{1}{12EI_2} \begin{cases} (x-b)[2(x-b)^2 + 3(b-s)(2x-b(2-\lambda))], & x \leq s \\ (s-b)[2(s-b)^2 + 3(b-x)(2s-b(2-\lambda))], & s \leq x \end{cases} \quad (4.52)$$

Thus, by virtue of the definition introduced in Section 4.1 of this chapter, it follows that the functions $g_{ij}(x,s)$ derived in eqns (4.49)–(4.52) represent the entries of the matrix of Green's type, $G(x,s)$, for the homogeneous boundary value problem corresponding to that in eqns (4.33)–(4.37). That is, $G(x,s)$ represents the influence matrix (in the sense described earlier in this section) to a transverse concentrated unit force for the compound beam shown in Figure 4.2.

Any component of the stress-strain state, caused by a reasonable combination of conventional loads applied to this beam, can readily be computed based on the influence matrix just obtained by using the influence function method. We will address this issue by considering examples below.

EXAMPLE 1 Compute components of the stress-strain state for the beam depicted in Figure 4.3 loaded with a concentrated bending moment of magnitude M and transverse force of magnitude P as shown.

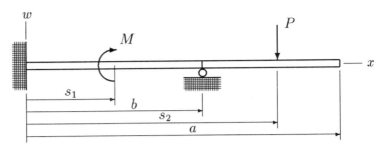

Figure 4.3 A beam subjected to a combination of loads

Clearly, the deflection $w_{1,p}(x)$ in the left-hand span of the beam, caused by the force P, can be computed as

$$w_{1,p}(x) = -Pg_{12}(x,s_2) = -P\frac{\lambda x^2}{4b}(b-s_2)(x-b) \qquad (4.53)$$

while the deflection $w_{1,m}(x)$ in the left-hand span, caused by the bending moment M, equals

$$w_{1,m}(x) = -M\frac{\partial g_{11}(x,s_1)}{\partial s}$$

Due to the piecewise nature of $g_{11}(x,s)$ (see eqn (4.49)), the deflection $w_{1,m}(x)$ is to be computed by different formulae for $x \leq s_1$ (on the left of s_1) and for $x \geq s_1$ (on the right of that point). This yields

$$w_{1,m}(x) = -\frac{M}{4b^2 EI_1}\begin{cases} x^2[s_1(3b-x)(2b-s_1)-2b^3], & x \leq s_1 \\ s_1[x^2(3b-x)(2b-s_1)-2b^3(2x-s_1)], & s_1 \leq x \end{cases} \qquad (4.54)$$

The deflection $w_{2,p}(x)$ in the right-hand span, caused by P, is expressed in terms of the entry $g_{22}(x,s_2)$ as

$$w_{2,p}(x) = \frac{-P}{12EI_2}\begin{cases} (x-b)[2(x-b)^2+3(b-s_2)(2x-b(2-\lambda))], & x \leq s_2 \\ (s_2-b)[2(s_2-b)^2+3(b-x)(2s_2-b(2-\lambda))], & s_2 \leq x \end{cases} \qquad (4.55)$$

Finally, the deflection function $w_{2,m}(x)$ in the right-hand span, caused by M, is computed as

$$w_{2,m}(x) = -M\frac{\partial g_{21}(x,s_2)}{\partial s} = -\frac{Ms}{4b}(x-b)(2b-s_2) \qquad (4.56)$$

Thus, the resultant deflection $w_1(x)$ in the left-hand span of the beam under consideration is obtained as a sum of the terms in eqns (4.53) and

4.2. BENDING OF MULTI-SPANNED BEAMS

(4.54), while summing up the deflections in eqns (4.55) and (4.56), one computes the resultant deflection $w_2(x)$ in the right-hand span.

To obtain explicit expressions for the bending moment $M(x)$ and the shear force $Q(x)$ at any cross-section of the beam in terms of the deflection function $w(x)$, one can directly utilize the relations immediately preceding eqn (2.49) in Chapter 2. Hence, obtaining $M(x)$ and $Q(x)$ is just a matter of taking appropriate derivatives of the expressions for the deflection that occurred in eqns (4.53)–(4.56).

EXAMPLE 2 For the next example, let us obtain components of the stress-strain state of the double-spanned simply supported continuous compound (EI_1 and EI_2) beam subjected to a combination of loads as depicted in Figure 4.4.

Following the procedure described earlier in this section, one obtains the entries $g_{ij}(x,s)$ of the influence matrix $G(x,s)$ of a transverse concentrated unit force for this beam. For $g_{11}^+(x,s)$ (that is the branch of $g_{11}(x,s)$ which is valid for $-a \leq x \leq s \leq 0$) we have

$$g_{11}^+(x,s) = \frac{s(x+a)}{p}\{2a^2[(x+a)^2+(s^2-a^2)]$$

$$+ \lambda s[2a^2 s - x(s+3a)(x+2a)]\} \quad (4.57)$$

where we denote $p = 12a^3(EI_1 + EI_2)$ and $\lambda = EI_2/EI_1$.

For the branch $g_{11}^-(x,s)$, with $-a \leq s \leq x \leq 0$, we obtain

$$g_{11}^-(x,s) = \frac{x(s+a)}{p}\{2a^2[(s+a)^2+(x^2-a^2)]$$

$$+ \lambda x[2a^2 x - s(x+3a)(s+2a)]\} \quad (4.58)$$

The entry $g_{12}(x,s)$, with $-a \leq x \leq 0$ and $0 \leq s \leq a$, is obtained as

$$g_{12}(x,s) = \frac{xs}{p}(s-a)(x+a)(x+2a)(s-2a)$$

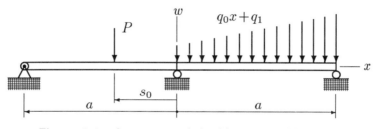

Figure 4.4 A compound double-spanned beam

For the entry $g_{21}(x,s)$, with $-a \le s \le 0$ and $0 \le x \le a$, we obtain

$$g_{21}(x,s) = \frac{xs}{p}(x-a)(s+a)(s+2a)(x-2a)$$

Finally, the branch of $g_{22}(x,s)$, which is valid for $0 \le x \le s \le a$, is presented in the form

$$g_{22}^+(x,s) = \frac{x(s-a)}{\lambda p}\{2\lambda a^2[(s-a)^2 + (x^2-a^2)]$$

$$+ x[2a^2 x - s(2a-s)(3a-x)]\} \tag{4.59}$$

while $g_{22}^-(x,s)$, with $0 \le s \le x \le a$, is obtained as

$$g_{22}^-(x,s) = \frac{s(x-a)}{\lambda p}\{2\lambda a^2[(x-a)^2 + (s^2-a^2)]$$

$$+ s[2a^2 s - x(2a-x)(3a-s)]\} \tag{4.60}$$

To compute the resultant deflection of the beam that is caused by both the distributed and the concentrated load in the statement, one has to account for a piecewise form of the entries $g_{11}(x,s)$ and $g_{22}(x,s)$ of the influence matrix (see eqns (4.57) and (4.60)). For the resultant deflection $w_1(x)$ in the left-hand span, on the left of s_0, we eventually obtain

$$w_1(x) = Pg_{11}^+(x,s_0) + \int_0^a g_{12}(x,s)(q_0 s + q_1)ds$$

where $g_{11}^+(x,s_0)$ is presented in eqn (4.57). While on the right of s_0, we have

$$w_1(x) = Pg_{11}^-(x,s_0) + \int_0^a g_{12}(x,s)(q_0 s + q_1)ds$$

where $g_{11}^-(x,s_0)$ can be found in eqn (4.58).

The resultant deflection $w_2(x)$ caused by the concentrated force and the distributed load in the right-hand span can be computed as

$$w_2(x) = Pg_{21}(x,s_0) + \int_0^x g_{22}^-(x,s)(q_0 s + q_1)ds + \int_x^a g_{22}^+(x,s)(q_0 s + q_1)ds$$

where expressions for $g_{22}^-(x,s)$ and $g_{22}^+(x,s)$ can be found in eqns (4.60) and (4.59), respectively.

Upon accomplishing the above integration and performing routine algebra, one obtains explicit analytic expressions for $w_1(x)$ and $w_2(x)$. These

4.2. BENDING OF MULTI-SPANNED BEAMS

in turn can be utilized to compute values of the bending moment and shear force at any cross-section of the beam.

So far in this section, we have considered beams with a simple intermediate support (see Figures 4.3 and 4.4). Let us show how different kinds of an intermediate support can also be treated in the scope of the influence function method. With this in mind, we consider the following illustrative example.

EXAMPLE 3 Compute the influence matrix (matrix of Green's type) of a transverse unit concentrated force for the compound (EI_1 and EI_2) cantilever beam overhanging an intermediate elastic support with the spring constant k^* as shown in Figure 4.5.

To come forth with the construction procedure, we formulate the following three-point posed boundary value problem

$$\frac{d^4 w_1(x)}{dx^4} = -\frac{q_1(x)}{EI_1} = -f_1(x), \quad x \in (-a, 0) \tag{4.61}$$

$$\frac{d^4 w_2(x)}{dx^4} = -\frac{q_2(x)}{EI_2} = -f_2(x), \quad x \in (0, a) \tag{4.62}$$

$$w_1(-a) = \frac{dw_1(-a)}{dx} = 0, \quad \frac{d^2 w_2(a)}{dx^2} = \frac{d^3 w_2(a)}{dx^3} = 0 \tag{4.63}$$

$$w_1(0) = w_2(0), \quad \frac{dw_1(0)}{dx} = \frac{dw_2(0)}{dx} \tag{4.64}$$

$$EI_1 \frac{d^2 w_1(0)}{dx^2} = EI_2 \frac{d^2 w_2(0)}{dx^2} \tag{4.65}$$

$$EI_1 \frac{d^3 w_1(0)}{dx^3} - kw_1(0) = EI_2 \frac{d^3 w_2(0)}{dx^3} + kw_2(0), \quad k = 2k^* \tag{4.66}$$

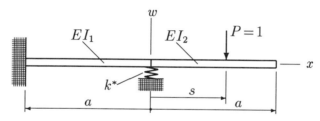

Figure 4.5 A beam overhanging an elastic support

for the deflection functions $w_1(x)$ and $w_2(x)$ to be determined on the intervals $[-a, 0]$ and $[0, a]$, respectively.

Following the technique described in detail earlier in this section, we present $w_1(x)$ and $w_2(x)$ in the form

$$w_1(x) = \int_{-a}^{x} \frac{(s-x)^3}{6} f_1(s)ds + H_1 + K_1 x + L_1 x^2 + M_1 x^3$$

and

$$w_2(x) = \int_{0}^{x} \frac{(s-x)^3}{6} f_2(s)ds + H_2 + K_2 x + L_2 x^2 + M_2 x^3$$

In computing values of the constant coefficients H_i, K_i, L_i and M_i, we take advantage of the set of boundary and contact conditions imposed by eqns (4.63) through (4.66). After these coefficients are obtained and substituted into the above expressions for $w_1(x)$ and $w_2(x)$, the latter are found in the form

$$w_1(x) = \int_{-a}^{x} \frac{(s-x)^3}{6} f_1(s)ds$$

$$+ \int_{-a}^{0} \frac{(a+x)^2}{6p} \{ks[x(s^2-a^2) - 2a(s^2+ax)] + 3EI_1[(x+a) - 3(s+a)]\} f_1(s)ds$$

$$+ \lambda \int_{0}^{a} \frac{(a+x)^2}{2p} \{EI_1[(a+x) - 3(a+s)] - ka^2 xs\} f_2(s)ds \qquad (4.67)$$

and

$$w_2(x) = \int_{-a}^{0} \frac{(a+s)^2}{2p} \{EI_1[(a+s) - 3(x+a)] - ka^2 sx\} f_1(s)ds$$

$$+ \int_{0}^{x} \frac{(s-x)^3}{6} f_2(s)ds + \int_{0}^{a} \frac{1}{6p} \{px^2(x-3s) - 3\lambda ka^4 xs$$

$$- 3EI_2 a[a(2a+3s) + 3x(a+2s)]\} f_2(s)ds, \qquad (4.68)$$

where $p = (2a^3 k + 3EI_1)$ and λ again represents the ratio EI_2/EI_1.

In view of the definition in Section 4.1 of this chapter, it follows from the representations in eqns (4.67) and (4.68), that the branch $g_{11}^+(x, s)$ of $g_{11}(x, s)$, which is valid for $-a \leq x \leq s \leq 0$, is presented as

$$g_{11}^+(x,s) = \frac{(a+x)^2}{6pEI_1} \{ks[x(s^2-a^2) - 2a(s^2+ax)] + 3EI_1[(x+a) - 3(s+a)]\},$$

while for $-a \leq s \leq x \leq 0$, we have

$$g_{11}^-(x,s) = \frac{(a+s)^2}{6pEI_1} \{kx[s(x^2-a^2) - 2a(x^2+as)] + 3EI_1[(s+a) - 3(x+a)]\}.$$

4.2. BENDING OF MULTI-SPANNED BEAMS

For the entry $g_{12}(x,s)$, as $x \in [-a, 0]$ and $s \in [0, a]$, we obtain

$$g_{12}(x,s) = \frac{(a+x)^2}{2pEI_1}\{EI_1[(a+x)-3(a+s)] - ka^2 xs\}$$

and for $g_{21}(x,s)$, as $s \in [-a, 0]$ and $x \in [0, a]$, we have

$$g_{21}(x,s) = \frac{(a+s)^2}{2pEI_1}\{EI_1[(a+s)-3(x+a)] - ka^2 sx\}.$$

Finally, for $g_{22}^+(x,s)$, with both variables x and s belonging to the interval [0,a] and $x \leq s$, we obtain

$$g_{22}^+(x,s) = \frac{1}{6pEI_2}\{px^2(x-3s) - 3\lambda ka^4 xs$$

$$-3EI_2 a[a(2a+3s)+3x(a+2s)]\}$$

while for $g_{22}^-(x,s)$ with $x \geq s$, we have

$$g_{22}^-(x,s) = \frac{1}{6pEI_2}\{ps^2(s-3x) - 3\lambda ka^4 xs$$

$$-3EI_2 a[a(2a+3x)+3s(a+2x)]\}.$$

Since the entries of the influence matrix of a transverse concentrated unit force applied to an arbitrary point in any span of the beam are already available, one is welcome to utilize them in computing the response of the beam to any reasonable combination of transverse and bending loads.

EXAMPLE 4 For our last example in this section, we determine the response of the compound multi-spanned beam (depicted in Figure 4.6) to a transverse concentrated force of magnitude P applied at a point s_0 in the left-hand span.

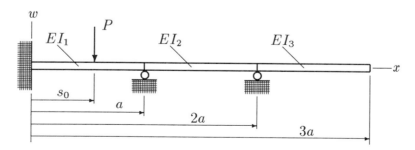

Figure 4.6 A compound beam overhanging two supports

To obtain the influence matrix for such a beam, we formulate the following multi-point posed boundary value problem

$$\frac{d^4 w_1(x)}{dx^4} = -\frac{q_1(x)}{EI_1}, \quad x \in (0, a) \tag{4.69}$$

$$\frac{d^4 w_2(x)}{dx^4} = -\frac{q_2(x)}{EI_2}, \quad x \in (a, 2a) \tag{4.70}$$

$$\frac{d^4 w_3(x)}{dx^4} = -\frac{q_3(x)}{EI_3}, \quad x \in (2a, 3a) \tag{4.71}$$

$$w_1(0) = \frac{dw_1(0)}{dx} = 0, \quad \frac{d^2 w_3(3a)}{dx^2} = \frac{d^3 w_3(3a)}{dx^3} = 0 \tag{4.72}$$

$$w_1(a) = w_2(a) = 0, \quad \frac{dw_1(a)}{dx} = \frac{dw_2(a)}{dx} \tag{4.73}$$

$$EI_1 \frac{d^2 w_1(a)}{dx^2} = EI_2 \frac{d^2 w_2(a)}{dx^2} \tag{4.74}$$

$$w_2(2a) = w_3(2a) = 0, \quad \frac{dw_2(2a)}{dx} = \frac{dw_3(2a)}{dx} \tag{4.75}$$

$$EI_2 \frac{d^2 w_2(2a)}{dx^2} = EI_3 \frac{d^2 w_3(2a)}{dx^2} \tag{4.76}$$

in the deflection functions $w_1(x)$ defined for $x \in (0, a)$, $w_2(x)$ defined for $x \in (a, 2a)$, and $w_3(x)$ defined for $x \in (2a, 3a)$.

Tracing out again the standard procedure of the method of variation of parameters in solving the boundary value problem posed by eqns (4.69) through (4.76), we obtain a 3×3 matrix of Green's type $G(x, s)$ (that is, the influence matrix of a transverse unit concentrated force for the beam under consideration). Since the procedure in this case is quite cumbersome, we omit its details and present just the final result. For the branch $g_{11}^+(x, s)$ of $g_{11}(x, s)$, which is valid for $0 \le x \le s \le a$, we obtain

$$g_{11}^+(x, s) = \frac{x^2(a-s)}{6pa^3} \{2[s(2a-s)(x-3a) + 2a^2 x]$$

$$+ 3\lambda_1(a-s)[x(a+s) + s(x-3a)]\},$$

4.2. BENDING OF MULTI-SPANNED BEAMS

while the other branch $g_{11}^-(x,s)$ of $g_{11}(x,s)$, which is valid for $0 \le s \le x \le a$, is found to be in the form

$$g_{11}^-(x,s) = \frac{s^2(a-x)}{6pa^3}\{2[x(2a-x)(s-3a)+2a^2s]$$

$$+3\lambda_1(a-x)[s(a+x)+x(s-3a)]\},$$

where $p = 4EI_1 + 3EI_2$ and $\lambda_1 = EI_2/EI_1$.

For the entry $g_{21}(x,s)$ whose arguments have different domains, namely $x \in [a,2a]$ and $s \in [0,a]$, we obtain

$$g_{21}(x,s) = \frac{1}{2pa^3}s^2(s-a)(3a-x)(a-x)(2a-x)$$

and for $g_{31}(x,s)$, with $x \in [2a,3a]$ and $s \in [0,a]$, we have

$$g_{31}(x,s) = \frac{1}{2pa}s^2(a-s)(2a-x).$$

Clearly, scalar multiples $Pg_{11}(x,s_0)$, $Pg_{21}(x,s_0)$, and $Pg_{31}(x,s_0)$ represent the deflection functions $w_1(x)$, $w_2(x)$, and $w_3(x)$, in the left-hand, intermediate, and right-hand span of the beam, respectively, caused by the transverse force of magnitude P concentrated at the point s_0 in the left-hand span. Based on these relations, one can readily obtain explicit expressions for the bending moments and shear forces occurring in this beam, by appropriately differentiating the deflection functions.

Thus, the particular problem posed in *EXAMPLE 4* is formally solved. Indeed, the response of this beam to the force P applied at an arbitrary point in the left-hand span is already found. If, however, an external load would also be applied to other span (or spans) of the beam under consideration, then the rest of the entries of the influence matrix ought to be available. Therefore, we present all of them.

The entry $g_{12}(x,s)$, domains for the arguments in which are $x \in [0,a]$ and $s \in [a,2a]$, is obtained in the form

$$g_{12}(x,s) = \frac{1}{2pa^3}x^2(a-s)(3a-s)(x-a)(2a-s)$$

The branch $g_{22}^+(x,s)$ of $g_{22}(x,s)$, with both variables x and s belonging to $[a,2a]$, as $x \le s$, is expressed as

$$g_{22}^+(x,s) = \frac{1}{6\lambda_1 pa^3}(2a-s)(a-x)\{2(x-a)[s(s-4a)(x-4a)$$

$$+a^2(x-10a)]-3\lambda_1 a^2[(s-3a)(s-a)+(x-a)^2]\},$$

while the other branch of $g_{22}(x,s)$, as $s \le x$, is found to be in the form

$$g_{22}^-(x,s) = \frac{1}{6\lambda_1 pa^3}(2a-x)(a-s)\{2(s-a)[x(x-4a)(s-4a)$$

$$+ a^2(s-10a)] - 3\lambda_1 a^2[(x-3a)(x-a)+(s-a)^2]\}.$$

The entry $g_{32}(x,s)$, with $x \in [2a, 3a]$ and $s \in [a, 2a]$, is found to be in the following form

$$g_{32}(x,s) = \frac{1}{2\lambda_1 pa}(s-a)(x-2a)(s-2a)[2(a-s) - \lambda_1 s].$$

For $g_{13}(x,s)$, with $x \in [0, a]$ and $s \in [2a, 3a]$, we obtain

$$g_{13}(x,s) = \frac{1}{2pa}x^2(a-x)(2a-s).$$

The entry $g_{23}(x,s)$, with $x \in [a, 2a]$ and $s \in [2a, 3a]$, is expressed as

$$g_{23}(x,s) = \frac{1}{2\lambda_1 pa}(x-2a)(x-a)(s-2a)[2(a-x) - \lambda_1 x].$$

Finally, for the branch $g_{33}^+(x,s)$ of $g_{33}(x,s)$, with $2a \le x \le s \le 3a$, we obtain

$$g_{33}^+(x,s) = (x-2a)\left\{\frac{1+\lambda_1}{\lambda_1 p}a(2a-s) + \frac{(x-2a)}{6EI_3}[(x+a)+3(a-s)]\right\},$$

while for the other branch $g_{33}^-(x,s)$ of $g_{33}(x,s)$, which is defined for $2a \le s \le x \le 3a$, we have

$$g_{33}^-(x,s) = (s-2a)\left\{\frac{1+\lambda_1}{\lambda_1 p}a(2a-x) + \frac{(s-2a)}{6EI_3}[(s+a)+3(a-x)]\right\}.$$

The influence matrix just presented allows one to obtain components of a stress-strain state caused by any reasonable combination of transverse and bending loads applied to the beam. For example, if the beam is subject to continuous transverse loads $q_1(x)$, $q_2(s)$, and $q_3(x)$ applied to the left-hand, intermediate, and right-hand span, respectively, then the deflection function caused by these loads can be expressed by means of the influence function method as

$$w_j(x) = \int_0^a g_{j1}(x,s)q_1(s)ds + \int_a^{2a} g_{j2}(x,s)q_2(s)ds +$$

$$\int_{2a}^{3a} g_{j3}(x,s)q_3(s)ds, \quad (j=1,2,3)$$

where the subscript j represents the span number.

Clearly, when the deflection function is explicitly available, one can routinely compute bending moments and shear forces in any cross-section of the beam by correspondingly differentiating the deflection function. If the loading functions $q_1(x)$, $q_2(x)$, and $q_3(x)$ have simple form (polynomial, exponential, trigonometric, or their elementary combinations), then the above integration can be conducted analytically, otherwise it can be accomplished approximately by applying appropriate quadrature formulas.

The examples that have been discussed in this section are very helpful in comprehending the material but they cannot clear up all possible peculiarities of the influence function method as applied to multi-spanned compound beams. The reader is therefore encouraged to go through the exercises in this chapter to gain experience in obtaining influence matrices for compound multi-spanned beams as well as in utilizing influence functions for computing components of stress-strain states in beams undergoing various combinations of loads.

4.3 Problems posed on graphs

As follows from Section 4.1, the applicability of the influence matrix formalism developed there is limited to a 'sandwich' type of a piecewise homogeneity of a material of which an assembly is composed. To make the range of possible applications of this formalism broader, we extend it to multi-point posed boundary value problems of a more complex type. The framework of graph theory is used. Sets of linear ordinary differential equations are considered, ones formulated on finite weighted graphs in such a way that every equation governs a single unknown function and is stated on a single edge of the graph. The individual equations are put into a system form by subjecting contact and boundary conditions at the vertices and endpoints of the graph.

Based on such a statement, a new definition of the matrix of Green's type is introduced. Existence and uniqueness of such matrices are discussed and two methods for their practical construction are proposed. Several particular examples from mechanics are considered.

Many authors (see, for example, [6, 7, 12, 15, 22, 26, 28, 30–34, 42–60, 66]) recommend the computational utilization of the Green's function approach to problems of applied mathematical physics. However, as we have already mentioned in this text, the practical use of these functions for actual computations in engineering and science is substantially limited because of a lack of their appropriate representations available in the literature.

The Green's function formalism is traditionally utilized in situations where governing differential equations have continuous coefficients. However, in [43-46, 50, 55, 57, 70] the attempts were undertaken to extend this formalism to boundary value problems of continuum mechanics, formulated throughout piecewise homogeneous regions, yielding discontinuity of the coefficients in governing differential equations. In [43, 44, 53], an effort has been put forth to implement this formalism for treating the so-called multipoint posed boundary value problems which model various situations in continuum mechanics for piecewise homogeneous media.

The first attempts have been undertaken in [45-47, 55, 57] to introduce the notion of a matrix of Green's type. In Section 4.1, we followed the concept which was proposed in [44]. It is worth noting, however, that the range of fruitful implementations of that notion is limited to the sandwich type of the material inhomogeneity. Our intention in the present section is to introduce the notion of a matrix of Green's type in a different way. The objective is to provide the extension of the Green's function formalism to multi-point posed boundary value problems occurring in complex assemblies of different homogeneous elements.

For the notational convenience in what follows, boundary value problems for governing systems of differential equations are set up on finite weighted graphs. This allows a considerably systematic analysis of a variety of problems stated in assemblies of 1-D elements.

A finite weighted graph R is considered, one containing m endpoints, $E_h, (h = \overline{1,m})$ and r vertices, $V_k, (k = \overline{1,r})$ joined by n edges, $e_i, (i = \overline{1,n})$ (see Figure 4.7). Let the positive real numbers $l_i, (i = \overline{1,n})$, representing the lengths of the corresponding edges e_i, be regarded as their weights.

Suppose also that every edge e_i of R (every element of the assembly) is occupied with a conductive material (of either thermal or electrical or any other relevant nature) whose conductivity $p_i(x)$ is a continuously differentiable function of a local for every edge longitudinal coordinate x.

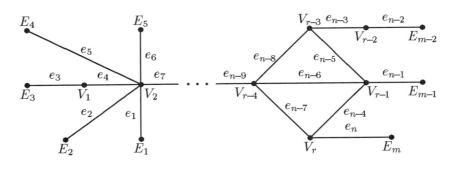

Figure 4.7 Graph R hosting a system of equations

4.3. PROBLEMS POSED ON GRAPHS

Let $u_i(x)$ represent the unknown function (temperature, electric potential, etc.) to be determined throughout the edge e_i of R. We will determine the set of these functions by the following set of differential equations

$$\frac{d}{dx}\left(p_i(x)\frac{du_i(x)}{dx}\right)+q_i(x)u_i(x)=-f_i(x), \quad x\in(0,l_i),\ (i=\overline{1,n}) \qquad (4.77)$$

These individual equations are put into a system format by assigning the set of contact conditions

$$u_1(V_k)=\ldots=u_{d_k}(V_k),\quad \sum_{h=1}^{d_k} p_h(V_k)\frac{du_h(V_k)}{dx}=0,\quad (k=\overline{1,r}) \qquad (4.78)$$

at each of the vertices V_k, with d_k being their degrees. Notice that for the notational convenience, in formulating these conditions, we use a 'local' numbering of the edges incident to the vertex V_k. It can easily be seen that the number of the contact conditions assigned at each of the vertices equals the degree of the vertex. Clearly, the above contact conditions describe conservation of energy at every vertex V_k of R. In addition, the boundary conditions

$$\alpha_h\frac{du_i(E_h)}{dx}+\beta_h u_i(E_h)=0,\quad (h=\overline{1,m}) \qquad (4.79)$$

are subjected at each of the endpoints E_h of R. This implies that $u_i(x)$ in the above equation is defined on the end edges e_i incident to E_h.

It can be easily seen that the total number of the contact and end conditions imposed by eqns (4.78) and (4.79) equals $2n$.

In this study, we will be focusing on the influence function (matrix) which represents the response of the entire assembly to a unit energy source acting at an arbitrary point s within an arbitrary edge of R.

Notice that the emphasis in this section is on boundary value problems of the type in eqns (4.77)–(4.79). However, the results of this work can readily be extended to problems formulated for differential equations of higher order.

We are now in a position to extend the conventional definition of the Green's function so as to make it valid for the multi-point posed boundary value problems of the type in eqns (4.77)–(4.79).

Definition: An $n\times n$ matrix $G(x,s)$, whose entries $g_{ij}(x,s)$ are defined for $x\in e_i$ and $s\in e_j$ on R, is referred to as the *matrix of Green's type* of the homogeneous boundary value problem corresponding to that posed by eqns (4.77)–(4.79), if for any fixed value of s, the entries $g_{ij}(x,s)$ hold the following properties:

178   CHAPTER 4. ASSEMBLIES OF ELEMENTS

1. As $x \neq s$, the entries $g_{ii}(x,s)$ of the principal diagonal $(i=j)$ represent continuous functions of x on e_i, they have continuous partial derivatives with respect to x up to the second order included, and satisfy the homogeneous equations corresponding to those in eqn (4.77);

2. As $x = s$, the entries $g_{ii}(x,s)$ of the principal diagonal are continuous functions of x, whereas their partial derivatives of the first order with respect to x are discontinuous functions, providing

$$\frac{\partial g_{ii}(s+0,s)}{\partial x} - \frac{\partial g_{ii}(s-0,s)}{\partial x} = -\frac{1}{p_i(s)}$$

and

$$\frac{\partial g_{ii}(s,s+0)}{\partial x} - \frac{\partial g_{ii}(s,s-0)}{\partial x} = \frac{1}{p_i(s)}$$

3. The peripheral $(i \neq j)$ entries $g_{ij}(x,s)$ of $G(x,s)$ are continuous functions of x for any value of $s \in e_j$, they have continuous partial derivatives with respect to x up to the second order included, and satisfy the homogeneous equations corresponding to those in eqn (4.77);

4. All of the entries $g_{ij}(x,s)$ of $G(x,s)$ satisfy the contact and the end conditions (which they are involved in) in eqns (4.78) and (4.79), in the sense that each of these conditions is satisfied for s belonging to any of the edges $e_j, (j=\overline{1,n})$.

In the discussion that follows, the arguments x and s of the matrix of Green's type are conventionally referred to as the *observation point* and the *source point*, respectively.

The following theorem can be formulated to stipulate the existence and uniqueness of the matrix of Green's type for the homogeneous boundary value problem corresponding to that posed by eqns (4.77)–(4.79).

Theorem 4.1: If the multi-point posed boundary value problem stated by eqns (4.77)–(4.79) has a unique solution (that is, the corresponding homogeneous problem has only the trivial solution), then there exists a unique matrix of Green's type $G(x,s)$ of the corresponding homogeneous problem.

Proof: Let $u_{i1}(x)$ and $u_{i2}(x)$, $(i=\overline{1,n})$ represent pairs of linearly independent on e_i particular solutions (fundamental solution sets) of the homogeneous equations corresponding to those in eqn (4.77). If so then, by virtue of the defining property 1, the diagonal entries $g_{ii}(x,s)$ of $G(x,s)$ can be sought in the form

$$g_{ii}(x,s) = \begin{cases} a_{i1}(s)u_{i1}(x) + a_{i2}(s)u_{i2}(x), & \text{for } x \leq s \\ b_{i1}(s)u_{i1}(x) + b_{i2}(s)u_{i2}(x), & \text{for } x \geq s \end{cases} \quad (4.80)$$

4.3. PROBLEMS POSED ON GRAPHS

whereas, in compliance with the defining property 3, the peripheral $(i \neq j)$ entries $g_{ij}(x,s)$ of $G(x,s)$ can be written as

$$g_{ij}(x,s) = c_{ij}(s)u_{i1}(x) + d_{ij}(s)u_{i2}(x) \qquad (4.81)$$

The coefficients $a_{i1}(s), a_{i2}(s), b_{i1}(s), b_{i2}(s), c_{ij}(s)$, and $d_{ij}(s)$ in the above representations are to be determined upon applying the remaining defining properties of the matrix of Green's type. Notice that the total number of these coefficients equals $2n(n+1)$ while the total number of the relations provided by properties 2 and 4 for their obtaining equals also $2n(n+1)$.

Indeed, by virtue of property 2, one obtains the n well-posed systems of linear algebraic equations

$$\begin{pmatrix} u_{i1}(s) & u_{i2}(s) \\ u'_{i1}(s) & u'_{i2}(s) \end{pmatrix} \times \begin{pmatrix} C_{i1}(s) \\ C_{i2}(s) \end{pmatrix} = \begin{pmatrix} 0 \\ p_i^{-1}(s) \end{pmatrix}, \quad (i = \overline{1,n}) \qquad (4.82)$$

in two unknowns each, of the total amount of $2n$ equations in $2n$ unknowns $C_{i1}(s)$ and $C_{i2}(s), (i = \overline{1,n})$. These unknowns are expressed in terms of the coefficients of $g_{ii}(x,s)$ in eqn (4.80) as

$$C_{ik}(s) = b_{ik}(s) - a_{ik}(s), \quad k=1,2 \qquad (4.83)$$

The well-posedness of the systems in eqn (4.82) follows from the fact that the determinants of their coefficient matrices represent Wronskians of the linearly independent functions $u_{i1}(x)$ and $u_{i2}(x)$. Hence, the unique values of $C_{i1}(s)$ and $C_{i2}(s)$ can readily be obtained. Subsequently, in compliance with eqn (4.83), the coefficients $a_{i1}(s)$ and $a_{i2}(s)$ can uniquely be expressed in terms of $b_{i1}(s)$ and $b_{i2}(s)$ and vice versa.

Thus, the number of undetermined coefficients in eqns (4.80) and (4.81) reduces to $2n^2$. And they can ultimately be found by applying the defining property 4. Indeed, by satisfying the entire set of the boundary and contact conditions posed by eqns (4.78) and (4.79) n times (once for each location of the source point $s \in e_j, j = \overline{1,n}$), we finally obtain a nonhomogeneous system of $2n^2$ linear algebraic equations in $2n^2$ unknowns. The coefficient matrix of this system can be reduced to the following partitioned diagonal form

$$M = \begin{pmatrix} A_{11} & 0 & \cdots & 0 \\ 0 & A_{22} & \cdots & 0 \\ \cdot & \cdot & \cdots & \cdot \\ 0 & 0 & \cdots & A_{nn} \end{pmatrix}$$

in which $A_{ii}(i = \overline{1,n})$ represent $2n \times 2n$ submatrices whose nonsingularity follows from the well-posedness of the original boundary value problem in eqns (4.77)–(4.79). The peripheral submatrices of M represent the null $2n \times 2n$ matrices. Thus, M represents a nonsingular matrix provided all of the coefficients of the representations in eqns (4.80) and (4.81) can be uniquely found.

This completes the proof of Theorem 4.1 because, once the values of the coefficients $a_{i1}(s), a_{i2}(s), b_{i1}(s), b_{i2}(s), c_{ij}(s)$, and $d_{ij}(s)$ are found, one can immediately obtain explicit representations of the entries of $G(x, s)$ by substituting those values into eqns (4.80) and (4.81). Notice that this proof suggests the procedure for the actual construction of matrices of Green's type for boundary value problems posed on graphs.

Another effective procedure of obtaining matrices of Green's type for homogeneous boundary value problems of the type posed on graphs by eqns (4.77)–(4.79) is also brought forward in this section. To describe it let us introduce a vector-function $\mathbf{U}(\mathbf{x})$ whose components $U_i(x), (i = \overline{1,n})$ are defined in terms of the solutions $u_i(x)$ of eqn (4.77) as

$$U_i(x) = \begin{cases} u_i(x), & \text{for } x \in e_i \\ 0, & \text{for } x \in R \setminus e_i \end{cases} \quad (4.84)$$

We also introduce a vector-function $\mathbf{F}(\mathbf{x})$ whose components $F_i(x)$ are defined in terms of the right-hand side functions $f_i(x)$ of eqn (4.77) in the form

$$F_i(x) = \begin{cases} f_i(x), & \text{for } x \in e_i \\ 0, & \text{for } x \in R \setminus e_i \end{cases} \quad (4.85)$$

The following theorem is formulated and proved to determine the solution of the boundary value problem posed by eqns (4.77)–(4.79) in terms of the matrix of Green's type of the corresponding homogeneous problem.

Theorem 4.2: If $G(x, s)$ represents the matrix of Green's type of the homogeneous boundary value problem corresponding to that in eqns (4.77)–(4.79), then the solution of the problem posed by eqns (4.77)–(4.79) on R can be written as

$$\mathbf{U}(x) = \int_R G(x, s)\mathbf{F}(s) dR(s), \quad x \in R \quad (4.86)$$

where the integration is carried out over the entire graph R. The converse is also true. That is, if the solution of the problem posed by eqns (4.77)–(4.79) on R is obtained in the form of the integral in eqn (4.86), then the kernel $G(x, s)$ of that integral represents the matrix of Green's type for the homogeneous boundary value problem corresponding to that in eqns (4.77)–(4.79).

4.3. PROBLEMS POSED ON GRAPHS

Proof: By virtue of the relations in eqns (4.84) and (4.85), the integral in eqn (4.86) can be read off in the scalar form as

$$u_i(x) = \sum_{j=1}^{n} \int_{e_j} g_{ij}(x,s) f_j(s) de_j(s), \quad i = \overline{1,n}$$

which can be rewritten in terms of the local coordinates as

$$u_i(x) = \sum_{j=1}^{n} \int_{0}^{l_j} g_{ij}(x,s) f_j(s) ds, \quad x \in [0, l_i], \quad i = \overline{1,n} \qquad (4.87)$$

Since the diagonal $g_{ii}(x,s)$ and the peripheral $g_{ij}(x,s)$ entries of the matrix of Green's type are defined in different manner (see eqns (4.80) and (4.81)), we isolate the i-th term of the finite sum in eqn (4.87)

$$u_i(x) = \sum_{j=1}^{i-1} \int_{0}^{l_j} g_{ij}(x,s) f_j(s) ds + \int_{0}^{l_i} g_{ii}(x,s) f_i(s) ds$$

$$+ \sum_{j=i+1}^{n} \int_{0}^{l_j} g_{ij}(x,s) f_j(s) ds, \quad x \in [0, l_i], \quad i = \overline{1,n}$$

Since the diagonal entries of $G(x,s)$ are defined in pieces, we break down the integral containing $g_{ii}(x,s)$ into two integrals as shown

$$u_i(x) = \sum_{j=1}^{i-1} \int_{0}^{l_j} g_{ij}(x,s) f_j(s) ds + \int_{0}^{x} g_{ii}^{-}(x,s) f_i(s) ds$$

$$+ \int_{x}^{l_i} g_{ii}^{+}(x,s) f_i(s) ds + \sum_{j=i+1}^{n} \int_{0}^{l_j} g_{ij}(x,s) f_j(s) ds, \quad x \in [0, l_i], \quad i = \overline{1,n}$$

where $g_{ii}^{-}(x,s)$ and $g_{ii}^{+}(x,s)$ represent the lower and the upper branches of the diagonal entries of $G(x,s)$, which are valid for $x \geq s$ and $x \leq s$, respectively (see eqn (4.80)).

To properly differentiate the functions $u_i(x)$ in the above equation, we recall the defining properties of the entries of $G(x,s)$ and notice also that the above expression contains integrals involving parameter and having variable limits. With this in mind, one obtains

$$\frac{du_i(x)}{dx} = \sum_{j=1}^{i-1} \int_{0}^{l_j} \frac{\partial g_{ij}(x,s)}{\partial x} f_j(s) ds + \int_{0}^{x} \frac{\partial g_{ii}^{-}(x,s)}{\partial x} f_i(s) ds$$

$$+ g_{ii}(x, x-0) f_i(x) + \int_{x}^{l_i} \frac{\partial g_{ii}^{+}(x,s)}{\partial x} f_i(s) ds - g_{ii}(x, x+0) f_i(x)$$

$$+ \sum_{j=i+1}^{n} \int_{0}^{l_j} \frac{\partial g_{ij}(x,s)}{\partial x} f_j(s) ds, \quad x \in [0, l_i], \quad i = \overline{1, n}$$

The sum of the two nonintegral terms

$$g_{ii}(x, x-0) f_i(x) - g_{ii}(x, x+0) f_i(x)$$

equals zero because, according to the definition of the matrix of Green's type, its diagonal entries are continuous as $x = s$. This yields

$$\frac{du_i(x)}{dx} = \sum_{j=1}^{i-1} \int_{0}^{l_j} \frac{\partial g_{ij}(x,s)}{\partial x} f_j(s) ds + \int_{0}^{x} \frac{\partial g_{ii}^{-}(x,s)}{\partial x} f_i(s) ds$$

$$+ \int_{x}^{l_i} \frac{\partial g_{ii}^{+}(x,s)}{\partial x} f_i(s) ds + \sum_{j=i+1}^{n} \int_{0}^{l_j} \frac{\partial g_{ij}(x,s)}{\partial x} f_j(s) ds, \quad x \in [0, l_i], \quad i = \overline{1, n}$$

or in a compact form

$$\frac{du_i(x)}{dx} = \sum_{j=1}^{n} \int_{0}^{l_j} \frac{\partial g_{ij}(x,s)}{\partial x} f_j(s) ds, \quad x \in [0, l_i], \quad i = \overline{1, n} \quad (4.88)$$

Hence, the first order derivatives of the integral representations of $u_i(x)$ can be obtained by a straightforward differentiation of their integrands. Consequently, these representations of $u_i(x)$ satisfy the boundary conditions in eqns (4.78) and (4.79) because so do the entries of $G(x, s)$.

To find out whether $u_i(x)$, as those shown in eqn (4.87), satisfy the governing differential equations, we obtain the second derivatives of $u_i(x)$

$$\frac{d^2 u_i(x)}{dx^2} = \sum_{j=1}^{i-1} \int_{0}^{l_j} \frac{\partial^2 g_{ij}(x,s)}{\partial x^2} f_j(s) ds + \int_{0}^{x} \frac{\partial^2 g_{ii}^{-}(x,s)}{\partial x^2} f_i(s) ds$$

$$+ \frac{\partial g_{ii}(x, x-0)}{\partial x} f_i(x) + \int_{x}^{l_i} \frac{\partial^2 g_{ii}^{+}(x,s)}{\partial x^2} f_i(s) ds - \frac{\partial g_{ii}(x, x+0)}{\partial x} f_i(x)$$

$$+ \sum_{j=i+1}^{n} \int_{0}^{l_j} \frac{\partial^2 g_{ij}(x,s)}{\partial x^2} f_j(s) ds, \quad x \in [0, l_i], \quad i = \overline{1, n}$$

In compliance with property 2 of the definition of $G(x, s)$, it follows that

$$\frac{\partial g_{jj}(x, x-0)}{\partial x} f_i(x) - \frac{\partial g_{jj}(x, x+0)}{\partial x} f_i(x) = -\frac{f_i(x)}{p_i(s)}$$

And for the second derivative of $u_i(x)$, we finally obtain its compact representation written as

$$\frac{d^2 u_i(x)}{dx^2} = \sum_{j=1}^{n} \int_{0}^{l_j} \frac{\partial^2 g_{ij}(x,s)}{\partial x^2} f_j(s) ds - \frac{f_i(x)}{p_i(x)}, \quad x \in [0, l_i], \quad i = \overline{1, n} \quad (4.89)$$

4.3. PROBLEMS POSED ON GRAPHS

Upon substituting the values of $u_i(x)$ and their derivatives from eqns (4.87)–(4.89) into eqn (4.77), we finally obtain

$$\sum_{j=1}^{n} \int_0^{l_j} L[g_{ij}(x,s)]f_j(s)ds - f_i(x) = -f_i(x), \quad x \in (0, l_i)$$

where L represents the differential operator of eqn (4.77).

Thus, the integral representations in eqn (4.87) satisfy the governing differential equations because the entries of the matrix of Green's type satisfy the homogeneous equations corresponding to those in eqn (4.77). That is, $L[g_{ij}(x,s)] = 0$, vanishing the integral terms in the above equation. Hence, the theorem has been proven.

This theorem clearly suggests that, once the solution of the original problem posed by eqns (4.77)–(4.79) is expressed in terms of the integral in eqn (4.86), the kernel $G(x,s)$ of such an integral represents the matrix of Green's type of the corresponding homogeneous problem.

A version of the method of variation of parameters is proposed below to obtain an integral representation of the form in eqn (4.87) for the solution of the nonhomogeneous boundary value problem posed by eqn (4.77)–(4.79).

In doing so, we again recall the fundamental solution sets $u_{i1}(x)$ and $u_{i2}(x)$ of the homogeneous equations corresponding to those in eqn (4.77). The general solution $u_i(x)$ of eqn (4.77) is sought as follows

$$u_i(x) = D_{i1}(x)u_{i1}(x) + D_{i2}(x)u_{i2}(x), \quad i = \overline{1,n} \qquad (4.90)$$

Based on this and following the standard procedure of the method of variation of parameters, one obtains the well-posed systems of linear algebraic equations

$$\begin{pmatrix} u_{i1}(x) & u_{i2}(x) \\ u'_{i1}(x) & u'_{i2}(x) \end{pmatrix} \times \begin{pmatrix} D'_{i1}(x) \\ D'_{i2}(x) \end{pmatrix} = \begin{pmatrix} 0 \\ -f_i(x)/p_i(x) \end{pmatrix}, \quad i = \overline{1,n}$$

in the derivatives of the coefficients $D_{i1}(x)$ and $D_{i2}(x)$ of $u_i(x)$. From this system it follows that

$$D'_{i1}(x) = \frac{u_{i2}(x)f_i(x)}{p_i(x)W_i(x)}, \quad D'_{i2}(x) = -\frac{u_{i1}(x)f_i(x)}{p_i(x)W_i(x)}, \quad i = \overline{1,n}$$

where $W_i(x) = u_{i1}(x)u'_{i2}(x) - u_{i2}(x)u'_{i1}(x)$ represent the Wronskians of the fundamental solution sets $u_{i1}(x)$ and $u_{i2}(x)$.

Integration of the derivatives $D'_{i1}(x)$ and $D'_{i2}(x)$ yields

$$D_{i1}(x) = \int_0^x \frac{u_{i2}(s)f_i(s)}{p_i(s)W_i(s)}ds + E_{i1}, \quad i = \overline{1,n}$$

and
$$D_{i2}(x) = -\int_0^x \frac{u_{i1}(s)f_i(s)}{p_i(s)W_i(s)}ds + E_{i2}, \quad i=\overline{1,n}$$

where E_{i1} and E_{i2} represent undetermined coefficients. Upon substituting $D_{i1}(x)$ and $D_{i2}(x)$ just found into eqn (4.90), the latter can be rewritten as

$$u_i(x) = u_{i1}(x)\int_0^x \frac{u_{i2}(s)f_i(s)}{p_i(s)W_i(s)}ds - u_{i2}(x)\int_0^x \frac{u_{i1}(s)f_i(s)}{p_i(s)W_i(s)}ds$$

$$+E_{i1}u_{i1}(x) + E_{i2}u_{i2}(x), \quad i=\overline{1,n}$$

By combining the integral terms in the above equation, the general solution of eqn (4.77) is finally obtained in the form

$$u_i(x) = \int_0^x \frac{u_{i1}(x)u_{i2}(s) - u_{i2}(x)u_{i1}(s)}{p_i(s)W_i(s)}f_i(s)ds$$

$$+E_{i1}u_{i1}(x) + E_{i2}u_{i2}(x), \quad x\in(0,l_i), \quad i=\overline{1,n} \tag{4.91}$$

The undetermined coefficients E_{i1} and E_{i2}, of a total number of $2n$, can be obtained upon satisfying the contact and boundary conditions in eqns (4.78) and (4.79), whose total number equals also $2n$. This yields a well-posed system of linear algebraic equations which leads finally to the integral representation of the form in eqn (4.86) for the solution of the problem under consideration. The kernel of that integral represents the matrix of Green's type of the problem.

We will here apply the matrix of Green's type formalism to a problem formulated for the medium whose property is a discontinuous function of the spatial variable. The influence matrix will be constructed for the steady-state heat conduction in an assembly of rods (see Figure 4.8) each of which is composed of a homogeneous material whose heat conductivity is p_i.

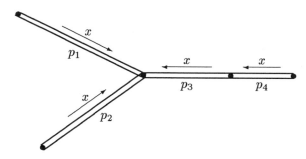

Figure 4.8 An assembly of heat conductive rods

4.3. PROBLEMS POSED ON GRAPHS

On the weighted graph associated with the above assembly, we formulate the following multi-point posed boundary value problem

$$p_i \frac{d^2 u_i(x)}{dx^2} = -f_i(x), \quad x \in (0, l_i), \quad i = \overline{1,4} \qquad (4.92)$$

$$u_1(l_1) = u_2(l_2) = u_3(l_3) \qquad (4.93)$$

$$p_1 \frac{du_1(l_1)}{dx} + p_2 \frac{du_2(l_2)}{dx} + p_3 \frac{du_3(l_3)}{dx} = 0 \qquad (4.94)$$

$$u_3(0) = u_4(l_4) \qquad (4.95)$$

$$p_3 \frac{du_3(0)}{dx} - p_4 \frac{du_4(l_4)}{dx} = 0 \qquad (4.96)$$

$$u_1(0) = u_2(0) = u_4(0) = 0 \qquad (4.97)$$

that describes the steady-state heat conduction phenomenon in the assembly. Here l_i, $(i = \overline{1,4})$ represents the lengths of the rods.

In compliance with the procedure of the method of variation of parameters, we seek the general solution of eqn (4.92) in the form

$$u_{i,g}(x) = D_{i1}(x) + D_{i2}(x)x, \quad i = \overline{1,4}$$

that ultimately reduces in this case (see eqn (4.91)) to

$$u_i(x) = \int_0^x \frac{s-x}{p_i} f_i(s)ds + E_{i1} + E_{i2}x, \quad x \in (0, l_i), \quad i = \overline{1,4} \qquad (4.98)$$

The undetermined coefficients E_{i1} and E_{i2}, $(i = \overline{1,4})$ in eqn (4.98) are to be determined upon satisfying the contact and boundary conditions posed by eqns (4.93)–(4.97). The conditions in eqn (4.97) yield, in particular, $E_{11} = E_{21} = E_{41} = 0$. For the rest of the coefficients, one obtains the following well-posed system of linear algebraic equations written as

$$\begin{pmatrix} l_1 & -l_2 & 0 & 0 & 0 \\ l_1 & 0 & -1 & -l_3 & 0 \\ p_1 & p_2 & 0 & p_3 & 0 \\ 0 & 0 & 1 & 0 & -l_4 \\ 0 & 0 & 0 & p_3 & -p_4 \end{pmatrix} \times \begin{pmatrix} E_{12} \\ E_{22} \\ E_{31} \\ E_{32} \\ E_{42} \end{pmatrix} = \begin{pmatrix} A_2 - A_1 \\ A_3 - A_1 \\ B_1 + B_2 + B_3 \\ A_4 \\ -B_4 \end{pmatrix} \qquad (4.99)$$

where
$$A_i = \int_0^{l_i} \frac{s-l_i}{p_i} f_i(s)ds, \quad B_i = \int_0^{l_i} f_i(s)ds, \quad i = \overline{1,4}$$

For the sake of simplicity, we assume in what follows that the edges of the graph have equal lengths, that is $l_1 = l_2 = l_3 = l_4 = l$. The determinant Δ of the coefficient matrix of the system in eqn (4.99) is found, in such a case, to be in the form

$$\Delta = l^2[(p_1 + p_2)(p_3 + p_4) + p_3 p_4]$$

When solving the system in eqn (4.99) and substituting thereupon values of the coefficients E_{i1} and E_{i2} found into eqn (4.98), we finally obtain

$$u_1(x) = \int_0^l \frac{x}{\Delta^* p_1} \{\Delta^* - s[p_2(p_3+p_4) + p_3 p_4]\} f_1(s) ds$$

$$+ \int_0^x \frac{s-x}{p_1} f_1(s)ds + + \int_0^l \frac{xs}{\Delta^*}(p_3+p_4) f_2(s) ds$$

$$+ \int_0^l \frac{x}{\Delta^*}(lp_3 + sp_4) f_3(s) ds + \int_0^l \frac{xs}{\Delta^*} p_3 f_4(s) ds, \qquad (4.100)$$

$$u_2(x) = \int_0^l \frac{xs}{\Delta^*}(p_3+p_4) f_1(s) ds$$

$$+ \int_0^x \frac{s-x}{p_2} f_2(s)ds + \int_0^l \frac{x}{\Delta^* p_2} \{\Delta^* - s[p_1(p_3+p_4) + p_3 p_4]\} f_2(s) ds$$

$$+ \int_0^l \frac{x}{\Delta^*}(lp_3 + sp_4) f_3(s) ds + \int_0^l \frac{xs}{\Delta^*} p_3 f_4(s) ds, \qquad (4.101)$$

$$u_3(x) = \int_0^l \frac{s}{\Delta^*}(lp_3 + xp_4) f_1(s) ds + \int_0^l \frac{s}{\Delta^*}(lp_3 + xp_4) f_2(s) ds$$

$$+ \int_0^l \frac{1}{\Delta^* p_3}[l(p_1+p_2+p_3) - s(p_1+p_2)](lp_3 + xp_4) f_3(s) ds$$

$$+ \int_0^x \frac{s-x}{p_3} f_3(s)ds + \int_0^l \frac{s}{\Delta^*}[l(p_1+p_2+p_3) - x(p_1+p_2)] f_4(s) ds, \qquad (4.102)$$

and
$$u_4(x) = \int_0^l \frac{xs}{\Delta^*} p_3 f_1(s) ds + \int_0^l \frac{xs}{\Delta^*} p_3 f_2(s) ds$$

$$+ \int_0^l \frac{x}{\Delta^*}[l(p_1+p_2+p_3)-s(p_1+p_2)]f_3(s)ds$$

$$+ \int_0^x \frac{s-x}{p_4}f_4(s)ds + \int_0^l \frac{x}{\Delta^* p_4}[\Delta^* - sp_3(p_1+p_2)]f_4(s)ds, \qquad (4.103)$$

where $\Delta^* = \Delta/l$.

Thus, the solution of the boundary value problem posed by eqns (4.92)–(4.97) is finally expressed in the form of the integral in eqn (4.86). This allows the entries $g_{ij}(x,s)$ of the matrix of Green's type $G(x,s)$ of the corresponding homogeneous problem to be read off from the integral representations in eqns (4.100)–(4.103). We exhibit below the entries of the first column of $G(x,s)$

$$g_{11}(x,s) = \frac{1}{\Delta^* p_1} \begin{cases} x\{\Delta^* - s[p_2(p_3+p_4)+p_3 p_4]\}, & \text{for } x \leq s \\ s\{\Delta^* - x[p_2(p_3+p_4)+p_3 p_4]\}, & \text{for } x \geq s \end{cases}$$

$$g_{21}(x,s) = \frac{xs}{\Delta^*}(p_3+p_4), \quad g_{31}(x,s) = \frac{s}{\Delta^*}(lp_3+xp_4), \quad g_{41}(x,s) = \frac{xs}{\Delta^*}p_3$$

which represent the response of the entire assembly to a unit source acting at the source point s located in the first rod. Other entries of $G(x,s)$, if required, can also be directly obtained from the integral representations in eqns (4.100)–(4.103).

4.4 Review Exercises

4.1 Determine whether the following multi-point posed boundary value problems have only the trivial solution:

  (a) Homogeneous problem corresponding to that in eqns (4.4)–(4.7);
  (b) Homogeneous problem corresponding to that in eqns (4.15)–(4.18);
  (c) Homogeneous problem corresponding to that in eqns (4.25)–(4.27).

4.2 Construct matrices of Green's type for the following multi-point posed boundary value problems:

  (a) $y_1''(x) = 0$ for $x \in (-a, 0)$ and $y_2''(x) - k^2 y_2(x) = 0$ for $x \in (0, \infty)$
    with $y_1(-a) = 0$, $|y_2(\infty)| < \infty$, $y_1(0) = y_2(0)$, $y_1'(0) = \lambda y_2'(0)$;
  (b) $y_1''(x) - k^2 y_1(x) = 0$, $x \in (-a, 0)$ and $y_2''(x) - k^2 y_2(x) = 0$, $x \in (0, \infty)$
    with $y_1(-a) = 0$, $|y_2(\infty)| < \infty$, $y_1(0) = y_2(0)$, $y_1'(0) = \lambda y_2'(0)$;

(c) $y_1''(x)+k^2y_1(x)=0$ for $x\in(-a,0)$ and $y_2''(x)+k^2y_2(x)=0$ for $x\in(0,a)$ with $y_1(-a)=0$, $y_2(a)=0$, $y_1(0)=y_2(0)$, $y_1'(0)=\lambda y_2'(0)$;

(d) $y_i''(x)-k^2y_i(x)=0$, $(i=1,2,3)$ for $x\in(0,\infty)$ with $y_1(0)=y_2(0)=y_3(0)$, $h_1y_1'(0)+h_2y_2'(0)+h_3y_3'(0)=0$, $|y_i(\infty)|<\infty$, $(i=1,2,3)$.

4.3 Solve the four-point posed boundary value problem posed by eqns (4.25)–(4.27) for $f_1(x)\equiv\sin(\pi x)$, $f_2(x)=f_3(x)\equiv 0$.

4.4 Utilize the influence matrix, derived in Section 4.2 (eqns (4.49)–(4.52), with $EI_1=EI_2=EI$), and compute the deflections, bending moment, and shear force caused by the loads shown below:

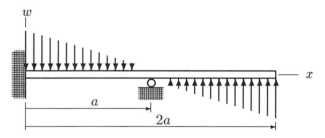

(a) Continuously distributed load, $q(x)=q_\circ(x-a)/a$;

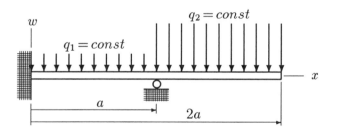

(b) Piecewise uniformly distributed load;

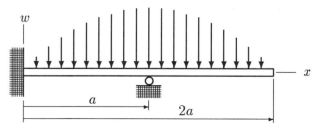

(c) Continuously distributed load, $q(x)=q_\circ x(2a-x)/a^2$;

4.4. REVIEW EXERCISES

4.5 Utilize the entries of the influence matrix presented in Section 4.2 (see eqns (4.57) through (4.60)) to determine the basic components (deflection, bending moment, and shear force) of the stress-strain state for the double-spanned beam depicted below. Let EI_1 and EI_2 represent the flexural rigidities of the left-hand and right-hand span, respectively, and let the loads be applied as shown:

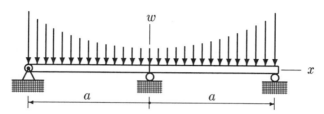

(a) Continuously distributed load, $q(x) = q_0 x^2 + q_1 x + q_2$;

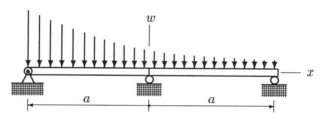

(b) Continuously distributed load, $q(x) = q_0 \exp(-mx)$.

4.6 Construct the influence matrix of a transverse unit concentrated force for the double-spanned clamped compound (EI_1 and EI_2) beam having an intermediate simple support right at the midpoint as depicted in Figure 4.9.

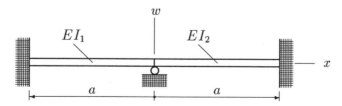

Figure 4.9 A beam having an intermediate support

4.7 Utilize the influence matrix constructed in exercise 4.6 and determine

190   CHAPTER 4. ASSEMBLIES OF ELEMENTS

the basic components (deflection, bending moment, and shear force) of the stress-strain state for the double-spanned compound beam having an intermediate support and loaded as shown below:

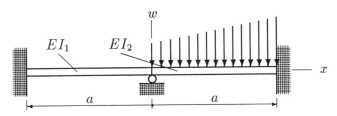

(a)   Load $q(x) = q_\circ x + q_1$ applied to the right-hand span;

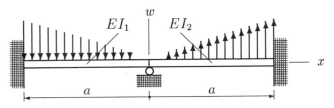

(b)   Linearly distributed load, $q(x) = q_\circ x + q_1$.

4.8 Construct the influence matrix of a transverse unit concentrated force for the double-spanned clamped compound (EI_1 and EI_2) beam having an intermediate simple support right at the midpoint as depicted in Figure 4.10.

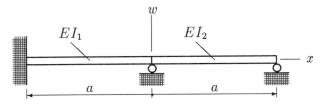

Figure 4.10   A beam with an intermediate support

4.9 Utilize the influence matrix constructed in exercise 4.8 and determine the basic components (deflection, bending moment, and shear force) of the stress-strain state for the clamped – simply supported compound beam having an intermediate support and loaded as shown below:

4.4. REVIEW EXERCISES

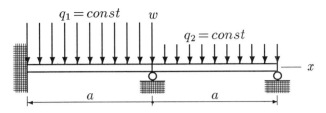

(a) Piecewise uniformly distributed load

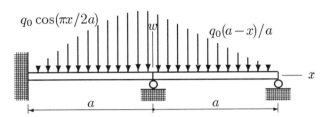

(b) Continuous load applied to the entire beam length

In exercises 4.10 and 4.11, we suggest the reader go through quite cumbersome computations associated with the extension of our procedure for the construction of influence functions to triple-spanned beams.

4.10 Construct the influence matrix of a transverse unit concentrated force for the triple-spanned cantilever compound beam overhanging a simple support as depicted in Figure 4.11.

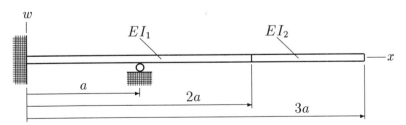

Figure 4.11 A compound beam overhanging one support

4.11 Construct the influence matrix of a transverse unit concentrated force for the triple-spanned beam of a uniform flexural rigidity EI, having two simple supports as depicted in Figure 4.12.

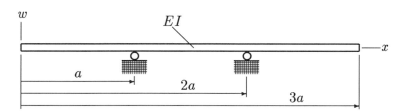

Figure 4.12 A triple-spanned beam having two supports

It is worth noting that to accomplish computation in exercises 4.4, 4.5, 4.7, and 4.9, the reader will be required to utilize the influence matrices that are either presented earlier in the text or are assumed to be constructed in the corresponding exercises of this set.

Chapter 5

Potential and Related Fields

So far in this text, we have been involved with problems of applied mechanics that are modeled with ordinary differential equations. We now turn to problems described by partial differential equations for which the list of available Green's (influence) functions and matrices is very limited.

In this chapter, the detailed description is presented of the technique that was originally developed (see, for example, [24, 42, 44, 45, 53]) for the construction of Green's functions and matrices for elliptic equations in two dimensions. The technique is based on the so-called method of eigenfunction expansion [63] and has proven to be especially effective for a variety of problems in computational continuum mechanics, which reduce to Laplace's, Klein-Gordon, and biharmonic equations as well as to Lame's system for the displacement formulation of the plane problem in the theory of elasticity.

Two stages are distinguished within the scope of the suggested technique that represents Green's functions in terms of their Fourier expansions with respect to one of the independent variables. This consequently results in the construction of Green's functions for ordinary differential equations in the coefficients of the Fourier expansions (the first stage of the technique), which can be managed by using either the method based on the defining properties of Green's functions (see Section 1.1 for details), or by the method of variation of parameters, described in Section 1.3. The second stage of the technique concerns either complete or partial summation of the Fourier series representing the Green's function to be found.

Recall that, in the theory of potential, the notion of an influence function of a point source is associated with the notion of a Green's function of the corresponding boundary value problem for Laplace's equation.

A number of compact representations of Green's functions for Laplace's and Klein-Gordon equations expressed in various coordinate systems are presented within the scope of this chapter. Various configurations of domains and types of boundary conditions are considered. Some mathematical issues are addressed here as they are closely related to our discussion (such as improvement of the convergence of the above mentioned series, estimation of their remainders, splitting off singular terms of Green's functions, expressing their regular components in terms of uniformly converging Fourier series, and so on).

Before presenting the construction procedure, the notion of the Green's function $G(P,Q)$ for Laplace's equation has to be introduced. In doing so, we set up the following boundary value problem

$$\Delta u(P) = -f(P), \quad P \in \Omega \tag{5.1}$$

$$M_i u(P) \equiv \alpha_i(P) \frac{\partial u(P)}{\partial n_i} + \beta_i(P) u(P) = 0, \quad i = \overline{1,m}, \quad P \in \Gamma_i \tag{5.2}$$

where Δ represents Laplace's operator, Ω is a simply connected region to be considered in two-dimensional Euclidean space, $\Gamma = \bigcup_{i=1}^{m} \Gamma_i$ denotes the piecewise smooth contour of Ω, $\alpha_i(P)$ and $\beta_i(P)$ are given functions defined on Γ in such a way that at least one of them is nonzero for every piece Γ_i of Γ, and n_i represents the normal direction to Γ_i at the point P. The points P and Q are commonly referred to as the *observation* and the *source* point, respectively.

Assume that the boundary value problem posed by eqns (5.1) and (5.2) has a unique solution. That is, the corresponding homogeneous problem

$$\Delta u(P) = 0, \quad P \in \Omega$$

$$M_i u(P) = 0, \quad i = \overline{1,m}, \quad P \in \Gamma_i$$

has only the trivial solution.

For the purpose of this study, the Green's function for the above homogeneous boundary value problem is defined in the following fashion.

DEFINITION: If, for any allowable right-hand term $f(P)$ in eqn (5.1), the solution of the boundary value problem posed by eqns (5.1) and (5.2) is expressible in the form

$$u(P) = \iint_\Omega G(P,Q) f(Q) d\Omega(Q) \tag{5.3}$$

then the kernel $G(P,Q)$ of this representation is said to be the *Green's function* for the homogeneous problem associated with that of eqns (5.1) and (5.2).

For any location of the source point $Q \in \Omega$, $G(P,Q)$ as a function of the coordinates of the observation point P holds the following properties:

5.1. FORMULATIONS IN CARTESIAN COORDINATES

1. at any $P \in \Omega$ except at $P = Q$, $G(P,Q)$ is a harmonic function, that is
$$\Delta G(P,Q) = 0, \quad P \neq Q$$

2. for $P = Q$, $G(P,Q)$ possesses the logarithmic singularity of the type
$$\frac{1}{2\pi} \ln \frac{1}{|P-Q|}$$

3. $G(P,Q)$ satisfies the boundary conditions in eqn (5.2), that is
$$M_i G(P,Q) = 0, \quad P \in \Gamma_i, \quad i = \overline{1,m}$$

From this definition, it clearly follows that the Green's function of the homogeneous boundary value problem associated with that posed by eqns (5.1) and (5.2) can be viewed as

$$G(P,Q) = \frac{1}{2\pi} \ln \frac{1}{|P-Q|} + R(P,Q)$$

where $R(P,Q)$ is a function harmonic everywhere on Ω, regardless of a mutual location of P and Q. $R(P,Q)$ is called the *regular component* of the Green's function.

It is worth reminding that the Green's function of a boundary value problem of Laplace's equation, set up over a certain region, can be identified with the influence function of a point source of the potential field for which that boundary value problem is a mathematical model.

5.1 Formulations in Cartesian coordinates

We begin the discussion on the construction procedure for influence (Green's) functions for Laplace's equation by focusing on Dirichlet problem for a semi-strip $\Omega_{ss} = \{(x,y): x > 0, 0 < y < b\}$ of width b (see Figure 5.1). This can be taken as a test problem, since its Green's function

$$G(x,y;s,t) = \frac{1}{4\pi} \ln \frac{\cosh(\omega(x+s)) - \cos(\omega(y-t))}{\cosh(\omega(x-s)) - \cos(\omega(y-t))}$$

$$+ \frac{1}{4\pi} \ln \frac{\cosh(\omega(x-s)) - \cos(\omega(y+t))}{\cosh(\omega(x+s)) - \cos(\omega(y+t))}, \quad \omega = \frac{\pi}{b} \quad (5.4)$$

is available in the literature (see, for example, [38], where it is shown for the semi-strip of width π, that is $\omega = 1$ in eqn (5.4)). In this representation, (x,y) and (s,t) are the observation and the source point, respectively.

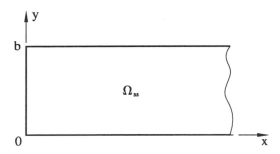

Figure 5.1 Semi-strip-shaped region

The expression for $G(x,y;s,t)$ shown in eqn (5.4) is usually derived by means of the classical method based on the Riemann's mapping theorem. We, however, will derive it by using a different approach. Our construction procedure stems from another classical method, that is the so-called *method of eigenfunction expansion* (see, for example, [63]).

Before going any further with the description of our construction procedure, we reduce the expression in eqn (5.4) to a more compact form. In doing so, each of the four components in the argument of the logarithmic function is viewed as $\cosh(\alpha)-\cos(\beta)$ and is transformed as follows

$$\cosh(\alpha)-\cos(\beta)=\frac{1}{2}[\exp(\alpha)+\exp(-\alpha)]-\cos(\beta)$$

$$=\frac{1}{2}\exp(-\alpha)[1-2\exp(\alpha)\cos(\beta)+\exp(2\alpha)]$$

The expression in brackets can be interpreted as a square of the modulus of the function $1-\exp(\alpha+i\beta)$ of a complex variable $\alpha+i\beta$. That is,

$$|1-\exp(\alpha+i\beta)|^2 = |1-\exp(\alpha)[\cos(\beta)+i\sin(\beta)]|^2$$

$$= [1-\exp(\alpha)\cos(\beta)]^2+[\exp(\alpha)\sin(\beta)]^2$$

$$= 1-2\exp(\alpha)\cos(\beta)+\exp(2\alpha)$$

Hence, the expression $\cosh(\alpha)-\cos(\beta)$ transforms to

$$\cosh(\alpha)-\cos(\beta) = \frac{1}{2}\exp(-\alpha)|1-\exp(\alpha+i\beta)|^2$$

and the components in the argument of the logarithmic function in eqn (5.4) can consequently be transformed to

$$\cosh(\mu(x+s))-\cos(\mu(y-t))$$

5.1. FORMULATIONS IN CARTESIAN COORDINATES

$$= \frac{1}{2}\exp(-\mu(x+s))|1-\exp(\mu((x+s)+i(y-t)))|^2$$

$$= \frac{1}{2}\exp(-\mu(x+s))\left|1-\exp(\mu(z+\overline{\zeta}))\right|^2,$$

$$\cosh(\mu(x-s))-\cos(\mu(y+t))$$

$$= \frac{1}{2}\exp(-\mu(x-s))|1-\exp(\mu((x-s)+i(y+t)))|^2$$

$$= \frac{1}{2}\exp(-\mu(x-s))\left|1-\exp(\mu(z-\overline{\zeta}))\right|^2,$$

$$\cosh(\mu(x-s))-\cos(\mu(y-t))$$

$$= \frac{1}{2}\exp(-\mu(x-s))|1-\exp(\mu((x-s)+i(y-t)))|^2$$

$$= \frac{1}{2}\exp(-\mu(x-s))\left|1-\exp(\mu(z-\zeta))\right|^2,$$

and

$$\cosh(\mu(x+s))-\cos(\mu(y+t))$$

$$= \frac{1}{2}\exp(-\mu(x+s))|1-\exp(\mu((x+s)+i(y+t)))|^2$$

$$= \frac{1}{2}\exp(-\mu(x+s))\left|1-\exp(\mu(z+\zeta))\right|^2$$

where the complex variables z and ζ denote the observation and the source point, respectively, and are defined in terms of their coordinates x, y and s, t as $z = x+iy$ and $\zeta = s+it$. The bar on ζ indicates complex conjugate.

After the above expressions for the components of the logarithmic function are substituted into eqn (5.4), the latter reduces to

$$G(x,y;s,t) = \frac{1}{2\pi} \ln \frac{\left|1-\exp(\mu(z+\overline{\zeta}))\right|\left|1-\exp(\mu(z-\overline{\zeta}))\right|}{\left|1-\exp(\mu(z-\zeta))\right|\left|1-\exp(\mu(z+\zeta))\right|}$$

This expression reduces finally to the compact form

$$G(x,y;s,t) = \frac{1}{2\pi} \ln \frac{E(z+\overline{\zeta})E(z-\overline{\zeta})}{E(z-\zeta)E(z+\zeta)} \tag{5.5}$$

where the real-valued function $E(p)$ of a complex variable p is defined as

$$E(p) = |1-\exp(\mu p)| \tag{5.6}$$

To present the detailed description of the technique which we proposed in [24, 42] for the construction of Green's functions of boundary value problems for elliptic equations, let us consider the Poisson equation

$$\frac{\partial^2 u(x,y)}{\partial x^2} + \frac{\partial^2 u(x,y)}{\partial y^2} = -f(x,y), \quad (x,y) \in \Omega_{ss} \tag{5.7}$$

throughout Ω_{ss} subject to the Dirichlet conditions

$$u(0,y)=0, \quad u(x,0)=u(x,b)=0 \qquad (5.8)$$

imposed on the boundary of Ω_{ss}.

A reasonable question may occur in conjunction with the above formulation. Indeed, why in an attempt to construct a Green's function for Laplace's equation, do we set up a boundary value problem for Poisson equation? Addressing this question, notice that, according to the definition introduced in the opening part of this chapter, if the solution of the problem posed by eqns (5.7) and (5.8) is found to be in a form of the integral

$$u(x,y) = \int_0^\infty \int_0^b K(x,y;s,t)f(s,t)dtds$$

then the kernel function $K(x,y;s,t)$ of this integral represents the Green's function of Dirichlet boundary value problem for Laplace's equation over the semi-strip Ω_{ss}.

Hence, in solving the boundary value problem posed by eqns (5.7) and (5.8), our goal is not to just somehow find its solution, but rather to obtain it in the form of the above integral, because this yields an explicit expression for the Green's function of our interest. This determines a certain strategy in solving the problem. We will always follow that strategy within the remaining part of the text when Green's (influence) functions and matrices are to be constructed.

Let us turn to the formulation in eqns (5.7) and (5.8). In addition to the boundary conditions imposed by eqn (5.8), we require for $u(x,y)$ to remain bounded as x approaches infinity. Assume also that the right-hand part $f(x,y)$ of eqn (5.7) is a suitably well-behaved function on Ω_{ss} in the sense that its integral over Ω_{ss} is bounded, that is

$$\left| \iint_{\Omega_{ss}} f(x,y)d\Omega_{ss} \right| < \infty$$

Assume, for the solution $u(x,y)$ of the problem set up by eqns (5.7) and (5.8), the following expansion

$$u(x,y) = \sum_{n=1}^\infty u_n(x)\sin(\nu y), \quad \nu = \frac{n\pi}{b} \qquad (5.9)$$

that follows from the standard procedure of the separation of variables being applied to the original formulation in eqns (5.7) and (5.8).

Expand also the right-hand side of eqn (5.7) in the Fourier sine-series

$$f(x,y) = \sum_{n=1}^\infty f_n(x)\sin(\nu y) \qquad (5.10)$$

5.1. FORMULATIONS IN CARTESIAN COORDINATES

Clearly, the expansion in eqn (5.9) satisfies the second and the third of the boundary conditions of eqn (5.8). A straightforward substitution of the quantities from eqns (5.9) and (5.10) into eqn (5.7) yields

$$\sum_{n=1}^{\infty} \left(\frac{d^2 u_n(x)}{dx^2} - \nu^2 u_n(x) \right) \sin(\nu y) = - \sum_{n=1}^{\infty} f_n(x) \sin(\nu y)$$

Equating the corresponding coefficients of the above series, one obtains a set of the following ordinary differential equations

$$\frac{d^2 u_n(x)}{dx^2} - \nu^2 u_n(x) = -f_n(x), \quad n = 1, 2, 3, \ldots \quad (5.11)$$

in the Fourier coefficients $u_n(x)$ of the expansion in eqn (5.9).

The solution of the above equation must be subjected to the boundary conditions written as

$$u_n(0) = 0, \quad |u_n(\infty)| < \infty \quad (5.12)$$

The Green's function $g_n(x,s)$ for the homogeneous ($f_n(x) \equiv 0$) boundary value problem corresponding to that posed by eqns (5.11) and (5.12) can be found in *EXAMPLE 4* of Section 1.1 (Chapter 1). Recalling it and rewriting in terms of our current notations, we obtain

$$g_n(x,s) = \frac{1}{2\nu} \begin{cases} \exp(\nu(x-s)) - \exp(-\nu(x+s)), & \text{for } x \le s \\ \exp(\nu(s-x)) - \exp(-\nu(s+x)), & \text{for } s \le x \end{cases} \quad (5.13)$$

Thus, in light of Theorem 1.3 of Section 1.3, the solution of the nonhomogeneous boundary value problem posed by eqns (5.11) and (5.12) can be written by means of the following integral

$$u_n(x) = \int_0^{\infty} g_n(x,s) f_n(s) ds \quad (5.14)$$

In conformity with the series expansion of eqn (5.10), the Fourier coefficients $f_n(x)$ of the right-hand term $f(x,y)$ of eqn (5.7) can be written in terms of Fourier–Euler formula as follows

$$f_n(s) = \frac{2}{b} \int_0^b f(s,t) \sin(\nu t) dt$$

providing for $u_n(x)$ the following representation

$$u_n(x) = \frac{2}{b} \int_0^b \int_0^{\infty} g_n(x,s) \sin(\nu t) f(s,t) ds dt$$

Substituting this into eqn (5.9) and assuming a possibility for interplacing the order of operations in it, we obtain

$$u(x,y) = \int_0^b \int_0^\infty \left[\frac{2}{b} \sum_{n=1}^\infty g_n(x,s) \sin(\nu y) \sin(\nu t) \right] f(s,t) \, ds \, dt \qquad (5.15)$$

Thus, since the solution of the boundary value problem posed by eqns (5.7) and (5.8) is expressed in the form of the integral in eqn (5.3), then, according to the definition presented in the opening part of this chapter, the kernel function

$$G(x,y;s,t) = \frac{2}{b} \sum_{n=1}^\infty g_n(x,s) \sin(\nu y) \sin(\nu t) \qquad (5.16)$$

of the latter integral representation is nothing but the Green's function of the Dirichlet problem for Laplace's equation on the semi-strip Ω_{ss}.

Later in this section, we will show that the series appearing in eqn (5.16) does not uniformly converge on Ω_{ss}, because the Green's function contains the singular component which by no means can be expressed by a uniformly convergent series. This circumstance dramatically affects the practicality of the series representations of Green's functions, since it notably narrows down the sphere of possible direct numerical implementations of expressions of the type in eqn (5.16).

Indeed, the finalization of such series by their truncation yields an unacceptable rise of computational errors. And, consequently, the range of successful direct numerical implementations of the series of eqn (5.16) could not be broad enough. The situation, nevertheless, can be radically improved by the analytical summation of the series, which decomposes the regular and singular parts of $G(x,y;s,t)$. In the discussion that follows, the emphasis will also be on the improvement of the convergence of the series in eqn (5.16).

To accomplish the analytical summation of the series of eqn (5.16), we convert the product of sines into the difference of cosines as shown

$$G(x,y;s,t) = \frac{1}{b} \sum_{n=1}^\infty g_n(x,s)[\cos(\nu(y-t)) - \cos(\nu(y+t))]$$

and substitute then the expression for $g_n(x,s)$ from eqn (5.13). Notice that either of the branches of $g_n(x,s)$ is appropriate for this. Upon substituting the upper branch, that is valid for $x \le s$, one obtains

$$G(x,y;s,t) = \frac{1}{2\pi} \sum_{n=1}^\infty \frac{1}{n} [\exp(\nu(x-s)) - \exp(\nu(x+s))]$$

$$\times [\cos(\nu(y-t)) - \cos(\nu(y+t))]$$

5.1. FORMULATIONS IN CARTESIAN COORDINATES

Removing the brackets in this expression, we obtain four similar series of the following type

$$\sum_{n=1}^{\infty} \frac{p^n}{n} \cos(n\alpha)$$

where $p<1$ and $0 \leq \alpha < 2\pi$. As is known (see, for example, [1, 29]), the above series converges providing

$$\sum_{n=1}^{\infty} \frac{p^n}{n} \cos(n\alpha) = -\ln\sqrt{1 - 2p\cos\alpha + p^2} \qquad (5.17)$$

Hence, in view of this relation, the series representation of the Green's function from eqn (5.16) can be written in the following closed form

$$G(x, y; s, t) =$$

$$\frac{1}{2\pi}\left\{ \ln\sqrt{1 - 2\exp(-\frac{\pi}{b}(x+s))\cos(\frac{\pi}{b}(y-t)) + \exp(-\frac{2\pi}{b}(x+s))} \right.$$

$$-\ln\sqrt{1 - 2\exp(\frac{\pi}{b}(x-s))\cos(\frac{\pi}{b}(y-t)) + \exp(\frac{2\pi}{b}(x-s))}$$

$$+\ln\sqrt{1 - 2\exp(\frac{\pi}{b}(x-s))\cos(\frac{\pi}{b}(y+t)) + \exp(\frac{2\pi}{b}(x-s))}$$

$$\left. -\ln\sqrt{1 - 2\exp(-\frac{\pi}{b}(x+s))\cos(\frac{\pi}{b}(y+t)) + \exp(-\frac{2\pi}{b}(x+s))} \right\}$$

The negative exponents, available in the first and the last logarithms, can be eliminated by adding and subtracting the term

$$\ln\sqrt{\exp(\frac{2\pi}{b}(x+s))}$$

to the first and from the last logarithms, respectively. This ultimately yields the equivalent form of the Green's function

$$G(x, y; s, t) =$$

$$\frac{1}{2\pi}\left\{ \ln\sqrt{1 - 2\exp(\frac{\pi}{b}(x+s))\cos(\frac{\pi}{b}(y-t)) + \exp(\frac{2\pi}{b}(x+s))} \right.$$

$$-\ln\sqrt{1 - 2\exp(\frac{\pi}{b}(x-s))\cos(\frac{\pi}{b}(y-t)) + \exp(\frac{2\pi}{b}(x-s))}$$

$$+\ln\sqrt{1 - 2\exp(\frac{\pi}{b}(x-s))\cos(\frac{\pi}{b}(y+t)) + \exp(\frac{2\pi}{b}(x-s))}$$

$$-\ln\sqrt{1-2\exp(\frac{\pi}{b}(x+s))\cos(\frac{\pi}{b}(y+t))+\exp(\frac{2\pi}{b}(x+s))}\Bigg\} \qquad (5.18)$$

This expression reduces to a more compact form upon introducing the complex variables $z=x+iy$ and $\zeta=s+it$. Indeed, since $z+\overline{\zeta}=(x+s)+i(y-t)$ and

$$\exp(\frac{\pi}{b}(z+\overline{\zeta})) = \exp(\frac{\pi}{b}(x+s))\left[\cos(\frac{\pi}{b}(y-t))+i\sin(\frac{\pi}{b}(y-t))\right],$$

the real $\Re$ and imaginary $\Im$ parts of the function

$$\exp(\frac{\pi}{b}(z+\overline{\zeta}))-1$$

can be expressed as

$$\Re\left(\exp(\frac{\pi}{b}(z+\overline{\zeta}))-1\right) = \exp(\frac{\pi}{b}(x+s))\cos(\frac{\pi}{b}(y-t))-1$$

and

$$\Im\left(\exp(\frac{\pi}{b}(z+\overline{\zeta}))-1\right) = \exp(\frac{\pi}{b}(x+s))\sin(\frac{\pi}{b}(y-t))$$

and its modulus consequently is

$$\left|\exp(\frac{\pi}{b}(z+\overline{\zeta}))-1\right|$$

$$=\sqrt{1-2\exp(\frac{\pi}{b}(x+s))\cos(\frac{\pi}{b}(y-t))+\exp(\frac{2\pi}{b}(x+s))}$$

Hence, the first logarithmic term of eqn (5.18) can be rewritten in terms of the earlier introduced function $E(p)$ (see eqn (5.6)) in the form

$$\ln\left|\exp(\frac{\pi}{b}(z+\overline{\zeta}))-1\right| = \ln\left(E(z+\overline{\zeta})\right)$$

The rest of the additive components from eqn (5.18) can be correspondingly transformed into

$$\ln\left|\exp(\frac{\pi}{b}(z-\zeta))-1\right| = \ln\left(E(z-\zeta)\right)$$

$$\ln\left|\exp(\frac{\pi}{b}(z-\overline{\zeta}))-1\right| = \ln\left(E(z-\overline{\zeta})\right)$$

and

$$\ln\left|\exp(\frac{\pi}{b}(z+\zeta))-1\right| = \ln\left(E(z+\zeta)\right)$$

5.1. FORMULATIONS IN CARTESIAN COORDINATES

respectively. Hence, the function obtained in eqn (5.18) finally reduces to the classical representation of the Green's function of Laplace's equation for Dirichlet problem on a semi-strip, shown in eqn (5.5).

Clearly, the singularity of the obtained representation is associated with that logarithm whose argument is $E(z-\zeta)$.

Thus, the suggested technique can be used as an alternative method for obtaining the Green's function of the Dirichlet problem on the semi-strip. This is promising and may encourage the users of the technique. However, the Green's function just obtained can be also constructed by using various classical approaches. For example, it can be obtained by the method based on the Riemann mapping theorem, with the semi-strip $\Omega_{ss} = \{(x,y): x>0, \; 0<y<b\}$ being mapped in a conformal manner on the interior of the unit circle, or by the reflection method.

But as we know, the standard methods fail when applied to the construction of Green's functions for more complex formulations, such as mixed boundary value problems, for example. Therefore, despite the fact that the suggested technique is capable of obtaining the Green's function for Dirichlet problem for a semi-strip, a reasonable question may arise regarding to whether the technique is capable when applied to more intricate problems. And what is even more important, whether it is capable of tackling those problems whose Green's functions are not affordable by means of the classical approaches.

The remaining part of this text is designed to provide answers to the questions just posed and to convince the reader of the great potential of the suggested technique. In the discussion that follows in this chapter, in particular, several rather complicated statements are considered. It is shown that the suggested technique is very productive when applied to a certain class of boundary value problems for which Green's functions are either not available yet in the literature or their known representations are not compact enough, limiting their computational implementations.

As our next example, we set up a mixed boundary value problem in which different portions of the region's boundary are subjected to different types of boundary conditions. That is

$$\left.\left(\frac{\partial u}{\partial x} - \beta u\right)\right|_{x=0} = u\big|_{y=0,b} = 0, \quad |u|_{x=\infty} < \infty, \quad \beta \geq 0 \tag{5.19}$$

for eqn (5.7) on the semi-strip $\Omega_{ss} = \{(x,y): x>0, \; 0<y<b\}$. To the best of the author's knowledge, the Green's function of this problem was published for the first time in [53] where it was obtained by the same technique.

If the solution $u(x,y)$ and the right-hand term $f(x,y)$ of eqn (5.7) are expressed by the Fourier expansions given by eqn (5.9) and (5.10) respectively, then the Green's function of the homogeneous problem corresponding to that posed by eqns (5.7) and (5.19) on Ω_{ss} is again given by the series of eqn (5.16), whose Fourier coefficient $g_n(x,s)$ represents in this case the Green's function of the homogeneous boundary value problem

$$\frac{d^2 u_n(x)}{dx^2} - \nu^2 u_n(x) = 0, \quad n=1,2,3,\ldots \tag{5.20}$$

$$\frac{du_n(0)}{dx} - \beta u_n(0) = 0, \quad |u_n(\infty)| < \infty \tag{5.21}$$

This Green's function is available in the text (see *EXAMPLE 4* of Section 1.3). Recalling it and rewriting in terms of our current notations, we obtain

$$g_n(x,s) = \frac{1}{2\nu} \begin{cases} \exp(-\nu s)\,[\exp(\nu x) + \beta^* \exp(-\nu x)], & \text{for } x \le s \\ \exp(-\nu x)\,[\exp(\nu s) + \beta^* \exp(-\nu s)], & \text{for } s \le x \end{cases} \tag{5.22}$$

where $\nu = n\pi/b$ and $\beta^* = (\nu - \beta)/(\nu + \beta)$.

The series of eqn (5.16), with the coefficients given by eqn (5.22), does not converge uniformly. Similarly to the case of Dirichlet problem, the rate of its convergence is low (the 'logarithmic' convergence takes place here of the order of $1/n$). But in the present case, on the contrary to the 'Dirichlet' case, the series of eqn (5.16) cannot be analytically summed up. Its convergence, however, can be significantly improved by breaking the series down into two parts, one of which is a slowly convergent series (but happens to be summable), whereas the other part is a series which converges notably faster allowing a reasonable truncation without losing the accuracy.

To carry out the breakage, let us consider the upper branch $g_n^+(x,s)$ of $g_n(x,s)$ from eqn (5.22)

$$g_n^+(x,s) = \frac{1}{2\nu} \left[\exp(\nu(x-s)) + \frac{\nu-\beta}{\nu+\beta} \exp(-\nu(x+s)) \right], \quad \text{for } x \le s \tag{5.23}$$

and present the coefficient of the second exponential term in it in the form

$$\frac{\nu-\beta}{\nu+\beta} = 1 - \frac{2\beta}{\nu+\beta}$$

This yields for $g_n^+(x,s)$

$$g_n^+(x,s) = \frac{1}{2\nu}\{[\exp(\nu(x-s)) + \exp(-\nu(x+s))]$$

$$- \frac{2\beta}{\nu+\beta}\exp(-\nu(x+s))\}$$

5.1. FORMULATIONS IN CARTESIAN COORDINATES

Thus, the Green's function under consideration can now be rewritten as

$$G(x,y;s,t) =$$

$$\frac{1}{b}\sum_{n=1}^{\infty}\frac{1}{\nu}[\exp(\nu(x-s))+\exp(-\nu(x+s))]\sin(\nu y)\sin(\nu t)$$

$$-\frac{2\beta}{b}\sum_{n=1}^{\infty}\frac{\exp(-\nu(x+s))}{\nu(\nu+\beta)}\sin(\nu y)\sin(\nu t), \quad \nu=\frac{n\pi}{b}$$

The first of the above series can be summed up in the manner similar to that used earlier for the Dirichlet problem. This finally yields the following representation

$$G(x,y;s,t) = \frac{1}{2\pi}\ln\frac{E(z+\zeta)E(z-\overline{\zeta})}{E(z-\zeta)E(z+\overline{\zeta})}$$

$$-\frac{2\beta}{b}\sum_{n=1}^{\infty}\frac{\exp(-\nu(x+s))}{\nu(\nu+\beta)}\sin(\nu y)\sin(\nu t) \quad (5.24)$$

for the Green's function of the homogeneous boundary value problem corresponding to that posed by eqns (5.7) and (5.19). Recall that the expression for E was earlier introduced by eqn (5.6).

Notice that if $\beta=0$ in the above expression, then the latter reduces to the Green's function

$$G(x,y;s,t) = \frac{1}{2\pi}\ln\frac{E(z+\zeta)E(z-\overline{\zeta})}{E(z-\zeta)E(z+\overline{\zeta})}$$

of the boundary value problem for the semi-strip Ω_{ss}, with boundary conditions imposed as follows

$$\left.\frac{\partial u}{\partial x}\right|_{x=0} = u\bigm|_{y=0,b} = 0, \quad |u|_{x=\infty} < \infty$$

Recall again the series of eqn (5.24). It converges uniformly. Moreover, its convergence of the order of $1/n^2$ is much 'faster' than that of the original series appeared in eqn (5.16). The truncation, therefore, could be relatively more accurate. To address this issue in detail, we recall the remainder

$$R_N(x,y;s,t) = \sum_{n=N+1}^{\infty}\frac{\exp(-\nu(x+s))}{\nu(\nu+\beta)}\sin(\nu y)\sin(\nu t) \quad (5.25)$$

of the N-th partial sum

$$S_N(x,y;s,t) = \sum_{n=1}^{N}\frac{\exp(-\nu(x+s))}{\nu(\nu+\beta)}\sin(\nu y)\sin(\nu t)$$

Table 5.1 Efficiency of the estimate obtained in eqn (5.26)

Parameter N	5	10	20	50	100	∞
S_N^*	1.4636	1.5498	1.5962	1.6251	1.6350	1.6449
error, %	11.02	5.78	2.96	1.21	0.59	—

of the series occurring in eqn (5.24) and proceed to the estimation of the remainder $R_N(x,y;s,t)$. Clearly, the exponential and trigonometric factors in $R_N(x,y;s,t)$ never exceed unity. Taking into account in addition that the parameter β is nonnegative (see eqn (5.19)), one obtains the following estimate

$$|R_N(x,y;s,t)| \leq \sum_{n=N+1}^{\infty} \frac{1}{\nu(\nu+\beta)} \leq \sum_{n=N+1}^{\infty} \frac{1}{\nu^2}$$

$$= \frac{b^2}{\pi^2} \sum_{n=N+1}^{\infty} \frac{1}{n^2} = \frac{b^2}{\pi^2} \left(\sum_{n=1}^{\infty} \frac{1}{n^2} - \sum_{n=1}^{N} \frac{1}{n^2} \right)$$

The infinite series in the above equation is usually referred to as the generalized harmonic series (see, for instance, [1, 29, 68]). On the contrary to the standard divergent harmonic series

$$\sum_{n=1}^{\infty} \frac{1}{n},$$

the series

$$\sum_{n=1}^{\infty} \frac{1}{n^2}$$

converges and its sum is $\pi^2/6$. Hence, the estimate just derived can be rewritten as

$$|R_N(x,y;s,t)| \leq \frac{b^2}{\pi^2} \left(\frac{\pi^2}{6} - \sum_{n=1}^{N} \frac{1}{n^2} \right) \tag{5.26}$$

This estimate is very compact. It could therefore be directly used in actual computations involving the Green's function from eqn (5.24). Table 5.1 exhibits some data illustrating the efficiency of the estimate.

It is worth noting, however, that, as follows from the method by which the estimate in eqn (5.26) was derived, the data of Table 5.1 must significantly overestimate possible errors of actual computations.

In what follows, we show that by using a different approach, a more accurate estimate of the series in eqn (5.24) can be derived. That is, the estimate found in eqn (5.26) can notably be improved. Indeed, replacing the

5.1. FORMULATIONS IN CARTESIAN COORDINATES

trigonometric factors in $R_N(x,y;s,t)$ of eqn (5.25) with unity and expressing the parameter ν in its denominator in terms of n, we obtain

$$|R_N(x,y;s,t)| \leq \sum_{n=N+1}^{\infty} \frac{\exp(-\nu(x+s))}{\nu(\nu+\beta)}$$

$$= \frac{b^2}{\pi^2} \sum_{n=N+1}^{\infty} \frac{\exp(-\nu(x+s))}{n(n+\beta_0)}$$

where $\beta_0 = \beta b/\pi$. In a special case, if $\beta_0 \geq 1$, the last estimate can be improved in the following way

$$|R_N(x,y;s,t)| \leq \frac{b^2}{\pi^2} \sum_{n=N+1}^{\infty} \frac{\exp(-\nu(x+s))}{n(n+1)}$$

$$= \frac{b^2}{\pi^2} \left[\sum_{n=1}^{\infty} \frac{\exp(-\nu(x+s))}{n(n+1)} - \sum_{n=1}^{N} \frac{\exp(-\nu(x+s))}{n(n+1)} \right]$$

$$= \frac{b^2}{\pi^2} \left\{ 1 + [\exp(\frac{\pi}{b}(x+s))-1] \ln\left(1-\exp(-\frac{\pi}{b}(x+s))\right) \right.$$

$$\left. - \sum_{n=1}^{N} \frac{\exp(-\nu(x+s))}{n(n+1)} \right\} \quad (5.27)$$

The finite term

$$1 + [\exp(\frac{\pi}{b}(x+s))-1] \ln\left(1-\exp(-\frac{\pi}{b}(x+s))\right)$$

occurring in eqn (5.27) represents the sum of the functional series

$$\sum_{n=1}^{\infty} \frac{\exp(-\nu(x+s))}{n(n+1)}$$

This series was summed up by recalling the standard summation formula (see, for example, [29])

$$\sum_{n=1}^{\infty} \frac{p^n}{n(n+1)} = 1 - \frac{1-p}{p} \ln \frac{1}{1-p}, \quad p^2 < 1$$

As can easily be seen, the estimate obtained in eqn (5.27) is not uniform. Indeed, it depends on the positions of the field and source points for which the estimate is applied. This makes it possible, therefore, to use different truncations of the series derived in eqn (5.24) in different zones of Ω_{ss} in order to keep a certain fixed level of the desired accuracy for the entire region.

Table 5.2 Efficiency of the estimate obtained in eqn (5.27)

Parameter	Truncating parameter, N					
$x+s$	5	10	20	50	100	∞
1.000	.21178	.21187	.21187	.21187	.21187	.21187
0.500	.39268	.39484	.39490	.39490	.39490	.39490
0.200	.60155	.61900	.62176	.62189	.62189	.62189
0.100	.70434	.74038	.75097	.75260	.75262	.75262
0.010	.81901	.88928	.92676	.94734	.95231	.95367
0.001	.83189	.90707	.94974	.97690	.98595	.99309

In Table 5.2, some data are displayed that can be used in making a decision concerning the number N of terms of the series in eqn (5.24) required for attaining a desired level of accuracy in different zones of the semi-strip Ω_{ss} of width $b = \pi$.

These data suggest, in particular, that the closer both the observation and the source points are to the edge $x = 0$ of the region (the smaller the parameter $x+s$ is), the bigger the truncating parameter N must be to attain the required level of accuracy in computing values of the Green's function by truncating the series of eqn (5.24).

From comparison of the data available in Tables 5.1 and 5.2, it follows that the second of the two derived estimates is much more efficient. Its use could significantly cut off expenses of the computational implementations of the expression obtained in eqn (5.24) for the Green's function of the problem stated by eqns (5.7) and (5.19) on the semi-strip Ω_{ss}.

The next problem whose Green's function will be presented here is also stated on the semi-strip Ω_{ss}. That is

$$\frac{\partial^2 u(x,y)}{\partial x^2} + \frac{\partial^2 u(x,y)}{\partial y^2} = 0, \quad (x,y) \in \Omega_{ss}$$

$$\left(\frac{\partial u}{\partial x} - \beta u\right)\bigg|_{x=0} = u\big|_{y=0} = \frac{\partial u}{\partial y}\bigg|_{y=b} = 0, \quad |u|_{x=\infty} < \infty, \quad \beta \geq 0$$

The distinguishing feature of the above formulation is that different kinds of boundary conditions (Dirichlet, Neumann, and mixed) are imposed on each single portion of the boundary.

Implementing the suggested technique and utilizing the standard [1, 29] summation formula

$$\sum_{n=1}^{\infty} \frac{p^{2n-1}}{2n-1} \cos((2n-1)x) = \frac{1}{4} \ln \frac{1+2p\cos(x)+p^2}{1-2p\cos(x)+p^2} \qquad (5.28)$$

5.1. FORMULATIONS IN CARTESIAN COORDINATES

that holds for $p^2 < 1$ and $0 \leq x < 2\pi$, one can readily obtain the Green's function of the last formulation in the form

$$G(x,y;s,t) = \frac{1}{2\pi} \ln \frac{E_1(z-\zeta)E_1(z+\overline{\zeta})E_2(z+\zeta)E_2(z-\overline{\zeta})}{E_1(z+\zeta)E_1(z-\overline{\zeta})E_2(z-\zeta)E_2(z+\overline{\zeta})}$$

$$-\frac{2\beta}{b}\sum_{n=1}^{\infty}\frac{\exp(-\nu(x+s))}{\nu(\nu+\beta)}\sin(\nu y)\sin(\nu t), \quad \nu = \frac{(2n-1)\pi}{2b}$$

where we have used notations as follows

$$E_1(p) = \left|\exp\left(\frac{\pi p}{2b}\right)+1\right|, \quad E_2(p) = \left|\exp\left(\frac{\pi p}{2b}\right)-1\right|$$

Clearly, the rate of convergence of the above series is the same as of the series in eqn (5.24). Following the approach utilized in the previous example, one can obtain the estimate of the remainder $R_N(x,y;s,t)$ of the above series in the form

$$|R_N(x,y;s,t)| \leq \frac{4b^2}{\pi^2}\left(\frac{\pi^2}{8} - \sum_{n=1}^{N}\frac{1}{(2n-1)^2}\right)$$

similar to that earlier obtained in eqn (5.26).

We will now turn the reader's attention to a rectangular region. Evidently, this shape brings a case of a great practical importance, because it has a broad range of applications in science and engineering. It is worth noting, however, that the Dirichlet problem provides the only case for the rectangle, for which representations of the Green's function are available in the literature. Namely, there is no doubt that the reader is familiar with the following double Fourier series expansion of the Green's function

$$G(x,y;s,t) = -\frac{4ab}{\pi^2}\sum_{m,n=1}^{\infty}\frac{\sin(\mu x)\sin(\mu s)\sin(\nu y)\sin(\nu t)}{(na)^2+(mb)^2}$$

for the Dirichlet problem on the rectangle $\Omega_r = \{(x,y): 0<x<a, \, 0<y<b\}$ shown in Figure 5.2. Here, $\mu = m\pi/a$ and $\nu = n\pi/b$.

The above representation is available in most texts on the subject matter (see, for example, [14, 26, 31, 36, 38, 63, 70]). Another form of this Green's function has been obtained by means of the Riemann mapping theorem, where the rectangle is mapped in a conformal manner on the interior of the unit circle with the aid of the so-called Weierstrass functions (see [68]).

However, both of the above mentioned representations are inconvenient for computational implementations. Indeed, the above double series converges non-uniformly due to the logarithmic singularity of the Green's function which represents the sum of the series. Standard procedures for computing the Weierstrass functions, in turn, are yet unaccessible in existing

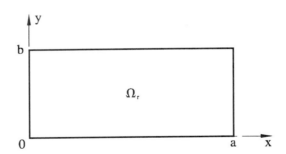

Figure 5.2 Rectangular region

software and their known expressions in the form of infinite products are cumbersome and difficult to accurately compute. This situation strongly motivates the implementation of the suggested technique for the construction of Green's functions on a rectangular region.

In what follows, we demonstrate the productivity of the technique as applied to various kinds of boundary value problems. The discussion begins with the Dirichlet problem

$$u(0,y) = u(a,y) = u(x,0) = u(x,b) = 0$$

set up for the rectangle Ω_r. Upon utilizing the suggested procedure in this case, one obtains the Green's function to be found as expressed by the expansion occurred in eqn (5.16)

$$G(x,y;s,t) = \frac{2}{b} \sum_{n=1}^{\infty} g_n(x,s) \sin(\nu y) \sin(\nu t), \quad \nu = \frac{n\pi}{b}$$

whose coefficient $g_n(x,s)$ represents in this case the Green's function of the following self-adjoint boundary value problem

$$\frac{d^2 u_n(x)}{dx^2} - \nu^2 u_n(x) = 0, \quad n = 1, 2, 3, \ldots$$

$$u_n(0) = 0, \quad u_n(a) = 0$$

For $x \leq s$, $g_n(x,s)$ is obtained in the form

$$g_n(x,s) = \frac{1}{2\nu} [\exp(\nu(x-s)) - \exp(\nu(x+s))]$$

$$- \frac{\sinh(\nu x) \sinh(\nu s)}{\nu \exp(\nu a) \sinh(\nu a)}.$$

5.1. FORMULATIONS IN CARTESIAN COORDINATES

Due to the self-adjointness of the latter boundary value problem, the expression of $g_n(x,s)$ for $x \geq s$ can be obtained from that above by interchanging of x with s.

Unfortunately, the expansion for $G(x,y;s,t)$ that we just came up with, cannot be completely summed up. However, one can accomplish its partial summation in the manner similar to that used earlier in this section for the semi-strip. This finally yields

$$G(x,y;s,t) = \frac{1}{2\pi} \ln \frac{E(z+\overline{\zeta})E(z-\overline{\zeta})}{E(z-\zeta)E(z+\zeta)}$$

$$-\frac{2}{b}\sum_{n=1}^{\infty} \frac{\sinh(\nu x)\sinh(\nu s)}{\nu \exp(\nu a)\sinh(\nu a)} \sin(\nu y)\sin(\nu t) \qquad (5.29)$$

It should be pointed out that the above trigonometric series rapidly converges on the rectangle Ω_r, unless both the source (s,t) and observation (x,y) points approach the edge $x=a$ of Ω. In such a situation, the convergence of the series dramatically 'slows down' and the direct computation of values of $G(x,y;s,t)$ by truncating the series results in rather notable errors and becomes practically unacceptable.

The point is that there exists the singularity in the series appearing in eqn (5.29), which is 'responsible' for slowing down the convergence of the series when x and s approach a. The development that follows is undertaken to improve the convergence of the series by isolating that singularity from the regular components of the series. For this purpose, we transform the coefficient of the series from eqn (5.29)

$$\frac{\sinh(\nu x)\sinh(\nu s)}{\nu \exp(\nu a)\sinh(\nu a)} = \frac{\cosh(\nu(x+s)) - \cosh(\nu(x-s))}{2\nu \exp(\nu a)\sinh(\nu a)}$$

$$= \frac{\cosh(\nu(x+s)) - \cosh(\nu(x-s))}{\nu[\exp(2\nu a) - 1]}$$

$$= \frac{\cosh(\nu(x+s))}{\nu[\exp(2\nu a) - 1]} - \frac{\cosh(\nu(x-s))}{\nu[\exp(2\nu a) - 1]}$$

Clearly, the first of the above two terms

$$\frac{\cosh(\nu(x+s))}{\nu[\exp(2\nu a) - 1]}$$

entails the non-uniform convergence of the series in eqn (5.29). With this in mind, in order to eliminate the influence of the above term, we add and subtract the term as follows

$$\frac{\cosh(\nu(x+s))}{\nu \exp(2\nu a)}$$

This yields the transformation

$$\frac{\sinh(\nu x)\sinh(\nu s)}{\nu \exp(\nu a)\sinh(\nu a)} = \frac{\cosh(\nu(x+s))}{\nu[\exp(2\nu a)-1]} - \frac{\cosh(\nu(x+s))}{\nu \exp(2\nu a)}$$

$$+ \frac{\cosh(\nu(x+s))}{\nu \exp(2\nu a)} - \frac{\cosh(\nu(x-s))}{\nu[\exp(2\nu a)-1]}$$

$$= \frac{\cosh(\nu(x+s))}{\nu \exp(2\nu a)[\exp(2\nu a)-1]} + \frac{\cosh(\nu(x+s))}{\nu \exp(2\nu a)} - \frac{\cosh(\nu(x-s))}{\nu[\exp(2\nu a)-1]}$$

It can readily be seen that the second term in the last line of the above development allows summation of the series of eqn (5.29) providing consequently some logarithmic terms additional to those already available in $G(x,y;s,t)$ (see eqn (5.29)). The first and the third terms in the last line of the above development make no contribution to the singularity. Therefore, for the notational convenience in what follows, we denote the sum of those two terms with $S_n(x,s)$. A trivial algebra yields

$$S_n(x,s) = \frac{\exp(\nu x)\sinh(\nu(a-s)) - \exp(-\nu x)\sinh(\nu(a+s))}{2\nu \exp(2\nu a)\sinh(\nu a)} \quad (5.30)$$

Hence, the series

$$\frac{2}{b}\sum_{n=1}^{\infty} \frac{\sinh(\nu x)\sinh(\nu s)}{\nu \exp(\nu a)\sinh(\nu a)} \sin(\nu y)\sin(\nu t)$$

occurring in eqn (5.29) can now be rewritten as

$$\frac{2}{b}\sum_{n=1}^{\infty} \frac{\cosh(\nu(x+s))}{\nu \exp(2\nu a)} \sin(\nu y)\sin(\nu t)$$

$$+ \frac{2}{b}\sum_{n=1}^{\infty} S_n(x,s)\sin(\nu y)\sin(\nu t)$$

So the isolation of the singularity in the series of eqn (5.29) is already accomplished. The second of the above two series converges uniformly, whereas the first is not uniformly convergent, but it can be summed up. In doing so, we rewrite it in the form

$$\frac{2}{b}\sum_{n=1}^{\infty} \frac{\cosh(\nu(x+s))}{\nu \exp(2\nu a)}\sin(\nu y)\sin(\nu t)$$

$$= \frac{2}{b}\sum_{n=1}^{\infty} \frac{\exp(\nu(x+s)) + \exp(-\nu(x+s))}{2\nu \exp(2\nu a)}\sin(\nu y)\sin(\nu t)$$

5.1. FORMULATIONS IN CARTESIAN COORDINATES

$$= \frac{1}{2\pi} \sum_{n=1}^{\infty} \frac{1}{n} [\exp(\nu((x+s)-2a)) + \exp(-\nu(x+s)+2a))]$$

$$\times [\cos(\nu(y-t)) - \cos(\nu(y+t))]$$

This can be treated analogously to the series appearing earlier in the problems stated on a semi-strip. Only the arguments of the exponential functions are here slightly more cumbersome. But this does not affect the summation itself, which yields

$$\frac{2}{b} \sum_{n=1}^{\infty} \frac{\cosh(\nu(x+s))}{\nu \exp(2\nu a)} \sin(\nu y) \sin(\nu t) =$$

$$\frac{1}{2\pi} \left\{ \ln \sqrt{1 - 2\exp(\frac{\pi}{b}(x+s-2a))\cos(\frac{\pi}{b}(y+t)) + \exp(\frac{2\pi}{b}(x+s-2a))} \right.$$

$$- \ln \sqrt{1 - 2\exp(\frac{\pi}{b}(x+s-2a))\cos(\frac{\pi}{b}(y-t)) + \exp(\frac{2\pi}{b}(x+s-2a))}$$

$$+ \ln \sqrt{1 - 2\exp(-\frac{\pi}{b}(x+s+2a))\cos(\frac{\pi}{b}(y+t)) + \exp(-\frac{2\pi}{b}(x-s+2a))}$$

$$\left. - \ln \sqrt{1 - 2\exp(-\frac{\pi}{b}(x+s+2a))\cos(\frac{\pi}{b}(y-t)) + \exp(-\frac{2\pi}{b}(x+s+2a))} \right\}$$

To eliminate the negative exponents in the last two logarithms, we again apply the 'trick' that was helpful when the analogous situation has been resolved for the Dirichlet problem on the semi-strip. Namely, we multiply the expressions under the radical signs of each of those logarithms by the positive exponential function

$$\exp(\frac{2\pi}{b}(x+s+2a))$$

Obviously, this provides an equivalent transformation and yields

$$\frac{2}{b} \sum_{n=1}^{\infty} \frac{\cosh(\nu(x+s))}{\nu \exp(2\nu a)} \sin(\nu y) \sin(\nu t) =$$

$$\frac{1}{2\pi} \left\{ \ln \sqrt{1 - 2\exp(\frac{\pi}{b}(x+s-2a))\cos(\frac{\pi}{b}(y+t)) + \exp(\frac{2\pi}{b}(x+s-2a))} \right.$$

$$- \ln \sqrt{1 - 2\exp(\frac{\pi}{b}(x+s-2a))\cos(\frac{\pi}{b}(y-t)) + \exp(\frac{2\pi}{b}(x+s-2a))}$$

$$+ \ln \sqrt{1 - 2\exp(\frac{\pi}{b}(x+s+2a))\cos(\frac{\pi}{b}(y+t)) + \exp(\frac{2\pi}{b}(x-s+2a))}$$

$$-\ln\sqrt{1-2\exp(\frac{\pi}{b}(x+s+2a))\cos(\frac{\pi}{b}(y-t))+\exp(\frac{2\pi}{b}(x+s+2a))}\Bigg\}$$

Upon introducing the complex variables as shown

$$z_1=(x+a)+iy,\quad z_2=(x-a)+iy$$

$$\zeta_1=(s+a)+it,\quad \zeta_2=(s-a)+it$$

one obtains the following compact form of the sum of the series under consideration

$$\frac{2}{b}\sum_{n=1}^{\infty}\frac{\cosh(\nu(x+s))}{\nu\exp(2\nu a)}\sin(\nu y)\sin(\nu t)=\frac{1}{2\pi}\ln\frac{E(z_1+\zeta_1)E(z_2+\zeta_2)}{E(z_1+\overline{\zeta_1})E(z_2+\overline{\zeta_2})}$$

where we again use the real-valued function E defined earlier by eqn (5.6).

Summarizing all the preparatory work, one finally obtains the Green's function for the Dirichlet problem on the rectangle Ω_r in the form

$$G(x,y;s,t)=-\frac{2}{b}\sum_{n=1}^{\infty}S_n(x,s)\sin(\nu y)\sin(\nu t)$$

$$+\frac{1}{2\pi}\ln\frac{E(z-\overline{\zeta})E(z+\overline{\zeta})E(z_1+\overline{\zeta_1})E(z_2+\overline{\zeta_2})}{E(z-\zeta)E(z+\zeta)E(z_1+\zeta_1)E(z_2+\zeta_2)} \quad (5.31)$$

where the coefficient $S_n(x,s)$ of the above series can be found in eqn (5.30).

To analyze the convergence of the above series, let us estimate the modulus of its remainder $R_N(x,y;s,t)$

$$|R_N(x,y;s,t)|=\left|\sum_{n=N+1}^{\infty}S_n(x,s)\sin(\nu y)\sin(\nu t)\right|$$

$$\leq\sum_{n=N+1}^{\infty}\left|\frac{\exp(\nu x)\sinh(\nu(a-s))-\exp(-\nu x)\sinh(\nu(a+s))}{2\nu\exp(2\nu a)\sinh(\nu a)}\right|$$

Since the second term

$$\exp(-\nu x)\sinh(\nu(a+s))$$

in the numerator of the right-hand term in the above inequality is nonnegative on Ω_r, the last estimate can be rewritten as

$$|R_N(x,y;s,t)|\leq\sum_{n=N+1}^{\infty}\frac{\exp(\nu x)\sinh(\nu(a-s))}{2\nu\exp(2\nu a)\sinh(\nu a)}$$

5.1. FORMULATIONS IN CARTESIAN COORDINATES

By virtue of the following inequalities
$$\exp(\nu x) < \exp(\nu a) \quad \text{and} \quad \sinh(\nu(a-s)) < \sinh(\nu a)$$
that obviously hold on Ω_r, we further obtain
$$|R_N(x,y;s,t)| \le \sum_{n=N+1}^{\infty} \frac{1}{2\nu \exp(\nu a)}$$
$$= \frac{b}{2\pi} \left\{ \sum_{n=1}^{\infty} \frac{1}{n} [\exp(-\frac{\pi a}{b})]^n - \sum_{n=1}^{N} \frac{1}{n} [\exp(-\frac{\pi a}{b})]^n \right\}$$

Recalling the known [1, 29] relation
$$\sum_{n=1}^{\infty} \frac{p^n}{n} = \ln \frac{1}{1-p}$$
which is valid for $p^2 < 1$, the remainder of the series in eqn (5.31) is finally estimated as
$$|R_N(x,y;s,t)| \le \frac{b}{2\pi} \left\{ -\ln(1-\exp(-\frac{\pi a}{b})) - \sum_{n=1}^{N} \frac{1}{n} [\exp(-\frac{\pi a}{b})]^n \right\}$$

This estimate reveals an extremely high rate of convergence of the series in eqn (5.31). It can be used in practical computations of values of the Green's function given by eqn (5.31) by a straightforward truncation of the series in it.

The next example is designed to reveal the potential capability of the suggested technique in dealing with problems whose Green's functions are not available in the classical sources.

The mixed boundary value problem
$$u|_{x=0} = (\frac{\partial u}{\partial x} + \beta u)\Big|_{x=a} = u_{y=o,b} = 0, \quad \beta \ge 0$$

for Laplace's equation is set up on the rectangle Ω_r. Following the standard routine of our technique, one discovers the Green's function of this problem to be in the form

$$G(x,y;s,t) = \frac{1}{2\pi} \ln \frac{E(z-\overline{\zeta})E(z+\overline{\zeta})}{E(z-\zeta)E(z+\zeta)}$$

$$-\frac{2}{b} \sum_{n=1}^{\infty} \frac{(\beta-\nu)\exp(-\nu a)\sinh(\nu x)\sinh(\nu s)}{\nu(\beta\sinh(\nu a)+\nu\cosh(\nu a))} \sin(\nu y)\sin(\nu t) \quad (5.32)$$

Computational implementations of this representation are limited because the above series does not converge uniformly and its truncation could result in an unacceptable level of errors. This reflects the existence of an additional singularity in the series as both x and s approach the value of a, that is, as both the observation and the source points are taken to the right-hand part of the boundary of the rectangle.

The convergence of the above series can be improved by decomposing its singular and regular components and by then complete summing the singular component. In doing so, we transform the coefficient of the series in the following manner

$$\frac{(\beta-\nu)\exp(-\nu a)\sinh(\nu x)\sinh(\nu s)}{\nu(\beta\sinh(\nu a)+\nu\cosh(\nu a))} =$$

$$\frac{(\beta-\nu)\sinh(\nu x)\sinh(\nu s)}{\nu\exp(\nu a)} \frac{1}{\nu\cosh(\nu a)+\beta\sinh(\nu a)}$$

$$= \frac{(\beta-\nu)\sinh(\nu x)\sinh(\nu s)}{\nu\exp(\nu a)} \left[\frac{1}{\nu\cosh(\nu a)+\beta\sinh(\nu a)} \right.$$

$$\left. - \frac{1}{\nu\cosh(\nu a)} + \frac{1}{\nu\cosh(\nu a)} \right] = \frac{(\beta-\nu)\sinh(\nu x)\sinh(\nu s)}{\nu\exp(\nu a)}$$

$$\times \left[\frac{1}{\nu\cosh(\nu a)} - \frac{\beta\sinh(\nu a)}{\nu\cosh(\nu a)(\nu\cosh(\nu a)+\beta\sinh(\nu a))} \right]$$

$$= \frac{(\beta-\nu)\sinh(\nu x)\sinh(\nu s)}{\nu^2\exp(\nu a)\cosh(\nu a)}$$

$$- \frac{\beta(\beta-\nu)\sinh(\nu a)\sinh(\nu x)\sinh(\nu s)}{\nu^2\exp(\nu a)\cosh(\nu a)(\nu\cosh(\nu a)+\beta\sinh(\nu a))} \tag{5.33}$$

This analysis shows that the second of the two terms in the above expression has the order of $1/n^2$ providing no hardship while approximately computing the sum of the series from eqn (5.32) by the straightforward truncation, whereas the order of the first term in eqn (5.33) is $1/n$. This preconditions the logarithmic singularity of the series and prompts the idea that should be used to regularize the series. That is, the second of those terms

$$\frac{\beta(\beta-\nu)\sinh(\nu a)\sinh(\nu x)\sinh(\nu s)}{\nu^2\exp(\nu a)\cosh(\nu a)(\nu\cosh(\nu a)+\beta\sinh(\nu a))}$$

should be left in its current form, while the first of them requires a certain transformation in order to split off its singularity in the form that makes it

5.1. FORMULATIONS IN CARTESIAN COORDINATES

possible to analytically sum it up. In doing so, we transform that term in the following manner

$$\frac{(\beta-\nu)\sinh(\nu x)\sinh(\nu s)}{\nu^2 \exp(\nu a)\cosh(\nu a)} = \frac{2(\beta-\nu)\sinh(\nu x)\sinh(\nu s)}{\nu^2(\exp(2\nu a)+1)}$$

$$= \frac{2(\beta-\nu)}{\nu^2}\sinh(\nu x)\sinh(\nu s)\left[\frac{1}{\exp(2\nu a)+1}-\frac{1}{\exp(2\nu a)}+\frac{1}{\exp(2\nu a)}\right]$$

$$= \frac{2(\beta-\nu)}{\nu^2}\sinh(\nu x)\sinh(\nu s)\left[\frac{1}{\exp(2\nu a)}-\frac{1}{\exp(2\nu a)(\exp(2\nu a)+1)}\right]$$

$$= \frac{2(\beta-\nu)\sinh(\nu x)\sinh(\nu s)}{\nu^2 \exp(2\nu a)} - \frac{2(\beta-\nu)\sinh(\nu x)\sinh(\nu s)}{\nu^2 \exp(2\nu a)(\exp(2\nu a)+1)}$$

$$= \frac{2\beta\sinh(\nu x)\sinh(\nu s)}{\nu^2 \exp(2\nu a)} - \frac{2\sinh(\nu x)\sinh(\nu s)}{\nu \exp(2\nu a)}$$

$$- \frac{2(\beta-\nu)\sinh(\nu x)\sinh(\nu s)}{\nu^2 \exp(2\nu a)(\exp(2\nu a)+1)} = \frac{2\beta\sinh(\mu x)\sinh(\nu s)}{\nu^2 \exp(2\nu a)}$$

$$- \frac{\cosh(\nu(x+s))}{\nu \exp(2\nu a)} + \frac{\cosh(\nu(x-s))}{\nu \exp(2\nu a)} - \frac{2(\beta-\nu)\sinh(\nu x)\sinh(\nu s)}{\nu^2 \exp(2\nu a)(\exp(2\nu a)+1)}$$

The second of the above four terms has the order of $1/n$ as both x and s approach the left-hand edge of Ω_r, whereas the rest of the terms have notably higher order of convergence. Hence, the singularity of the series of eqn (5.32) is already isolated and the coefficient of the series can be rewritten as

$$\frac{(\beta-\nu)\exp(-\nu a)\sinh(\nu x)\sinh(\nu s)}{\nu(\beta\sinh(\nu a)+\nu\cosh(\nu a))}$$

$$= -\frac{1}{2\nu}[\exp(\nu(x+s-2a))+\exp(-\nu(x+s+2a))] + T_n(x,s)$$

where the term $T_n(x,s)$ is given as

$$T_n(x,s) = \frac{2\beta\sinh(\nu x)\sinh(\nu s)}{\nu^2 \exp(2\nu a)} + \frac{\cosh(\nu(x-s))}{\nu \exp(2\nu a)}$$

$$- \frac{2(\beta-\nu)\sinh(\nu x)\sinh(\nu s)}{\nu^2 \exp(2\nu a)(\exp(2\nu a)+1)}$$

Thus, the entire series in eqn (5.32) can now be rewritten as

$$\frac{2}{b}\sum_{n=1}^{\infty}\frac{(\beta-\nu)\exp(-\nu a)\sinh(\nu x)\sinh(\nu s)}{\nu(\beta\sinh(\nu a)+\nu\cosh(\nu a))}\sin(\nu y)\sin(\nu t)$$

$$= \frac{1}{2\pi} \sum_{n=1}^{\infty} \frac{1}{n} [\exp(\nu(x+s-2a)) + \exp(-\nu(x+s+2a))]$$

$$\times [\cos(\nu(y+t)) - \cos(\nu(y-t))] + \frac{2}{b} \sum_{n=1}^{\infty} T_n(x,s) \sin(\nu y) \sin(\nu t)$$

This representation can be partially summated. The second of the above two series converges uniformly and accurate computational implementations involving this series are possible by the straightforward truncation. One can readily estimate the remainder of this series following the pattern utilized earlier in this section. So, the accuracy level for partial sums of the series is clearly predictable.

The first of the above two series contains the logarithmic singularity of the Green's function, thus it can be summed up with the aid of the summation formula (5.17). This yields

$$\frac{1}{2\pi} \sum_{n=1}^{\infty} \frac{1}{n} [\exp(\nu(x+s-2a)) + \exp(-\nu(x+s+2a))]$$

$$\times [\cos(\nu(y+t)) - \cos(\nu(y-t))]$$

$$= -\frac{1}{2\pi} \ln \frac{E(z_1+\zeta_1)E(z_2+\zeta_2)}{E(z_1+\overline{\zeta_1})E(z_2+\overline{\zeta_2})}$$

Thus, after accomplishing some elementary algebra, the Green's function of the mixed boundary value problem under consideration, whose representation was given by eqn (5.32) containing a non-uniformly convergent series, is finally found in the following readily computable form

$$G(x,y;s,t) = -\frac{2}{b} \sum_{n=1}^{\infty} T_n(x,s) \sin(\nu y) \sin(\nu t)$$

$$+ \frac{1}{2\pi} \ln \frac{E(z-\overline{\zeta})E(z+\overline{\zeta})E(z_1+\zeta_1)E(z_2+\zeta_2)}{E(z-\zeta)E(z+\zeta)E(z_1+\overline{\zeta_1})E(z_2+\overline{\zeta_2})} \qquad (5.34)$$

where the function $T_n(x,s)$ and the variables z_1, z_2, ζ_1, and ζ_2 were introduced earlier in this section.

In the author's books [44, 45], a number of representations for Green's functions can be found as obtained by the suggested technique. Various types of boundary value problems are considered there.

In the set of exercises related to this section, the reader is advised to construct some other Green's functions for a number of problems (mostly mixed) formulated on a strip, semi-strip, and rectangle. This will stimulate the reader's interest in the further extension of the suggested technique to single problems and problem classes for which it has not yet been but potentially could readily be implemented.

5.2 Formulations in polar coordinates

The present section is concerned with those specific details of the suggested technique that arise within its utilization for the construction of Green's (influence) functions for Laplace's equation set up in polar coordinate system. The reader will soon realize that all the basic aspects of the procedure described in the previous section can readily be adapted to the case of polar coordinates.

Analogously to the development in Section 5.1, we begin here with a validation example. Let us consider the Dirichlet problem

$$\frac{1}{r}\frac{\partial}{\partial r}\left(r\frac{\partial u(r,\phi)}{\partial r}\right) + \frac{1}{r^2}\frac{\partial^2 u(r,\phi)}{\partial \phi^2} = -f(r,\phi), \quad (r,\phi)\in\Omega_c \quad (5.35)$$

$$u(R,\phi) = 0 \quad (5.36)$$

stated in the circle $\Omega_c = \{(r,\phi): 0 < r < R,\ 0 \leq \phi < 2\pi\}$ of radius R.

And again, like in the previous section, to construct a Green's function of Laplace's equation, we formulate a boundary value problem for Poisson equation. But we believe that the reader is not confused any more with this 'trick' which is undertaken for the methodological purpose.

Due to the 2π-periodicity of the problem posed by eqns (5.35) and (5.36) in the direction of ϕ, we assume, for its solution $u(r,\phi)$, the following trigonometric Fourier expansion

$$u(r,\phi) = \frac{1}{2}u_0(r) + \sum_{n=1}^{\infty}\left(u_n^c(r)\cos(n\phi) + u_n^s(r)\sin(n\phi)\right) \quad (5.37)$$

Let also the right-hand term of eqn (5.35) be presented by the Fourier series in a general form

$$f(r,\phi) = \frac{1}{2}f_0(r) + \sum_{n=1}^{\infty}\left(f_n^c(r)\cos(n\phi) + f_n^s(r)\sin(n\phi)\right) \quad (5.38)$$

Upon substituting the expansions just presented into eqn (5.35) and equating the coefficients of the Fourier series on both hand sides of it, one obtains the following set of ordinary differential equations

$$\frac{1}{r}\frac{d}{dr}\left(r\frac{du_n(r)}{dr}\right) - \frac{n^2}{r^2}u_n(r) = -f_n(r), \quad n = 0, 1, 2, \ldots$$

which reduces to the self-adjoint form

$$\frac{d}{dr}\left(r\frac{du_n(r)}{dr}\right) - \frac{n^2}{r}u_n(r) = -rf_n(r), \quad n = 0, 1, 2, \ldots \quad (5.39)$$

by introducing the integrating factor r.

The boundary conditions

$$|u_n(0)|<\infty, \quad u_n(R)=0 \tag{5.40}$$

are imposed where the first one follows from the boundedness at the origin of the solution $u(r,\phi)$ of the original problem posed by eqns (5.35) and (5.36).

For the sake of notational convenience in the opening stage of the development that follows, we omit the superscripts on $u_n(r)$ and $f_n(r)$.

Clearly, in constructing the Green's functions for the homogeneous boundary value problems corresponding to those posed by eqns (5.39) and (5.40), the case of $n=0$ must be considered separately from the general case of $n\geq 1$, because the fundamental solutions set for the general case differs from that for the case of $n=0$.

We will seek the solution to the problem in eqns (5.39) and (5.40) by the Lagrange's method of variation of parameters. Since for $n=0$

$$\frac{d}{dr}\left(r\frac{du_0(r)}{dr}\right) = -rf_0(r) \tag{5.41}$$

the fundamental solution set of the corresponding homogeneous ($f_0(x)\equiv 0$) equation can be expressed by the functions $u\equiv\ln(r)$ and $u\equiv 1$, the general solution of the above equation is given by

$$u_0(r) = C_1(r)\ln(r)+C_2(r) \tag{5.42}$$

Upon substituting this form into eqn (5.41), one obtains

$$C_1'(r)=-rf_0(r) \quad \text{and} \quad C_2'(r)=r\ln(r)f_0(r)$$

and the straightforward integration of these expressions yields

$$C_1(r)=-\int_0^r sf_0(s)ds+D_1, \quad C_2(r)=\int_0^r \ln(s)sf_0(s)ds+D_2$$

Upon substituting the above quantities into eqn (5.42) and combining afterwards the integral terms in it, one obtains the general solution of eqn (5.41) in the form

$$u_0(r) = \int_0^r \ln\left(\frac{s}{r}\right)sf_0(s)ds+D_1\ln(r)+D_2$$

To obtain the values of D_1 and D_2, one takes advantage of the boundary conditions given by eqn (5.40). Clearly, the first of those, $|u_0(0)|<\infty$, can be satisfied if and only if $D_1=0$, while the second yields

$$D_2 = -\int_0^R \ln\left(\frac{s}{R}\right)sf_0(s)ds$$

5.2. FORMULATIONS IN POLAR COORDINATES

Thus, the solution of the boundary value problem posed by eqns (5.41) and (5.40) is found in the form

$$u_0(r) = \int_0^r \ln\left(\frac{s}{r}\right) s f_0(s) ds - \int_0^R \ln\left(\frac{s}{R}\right) s f_0(s) ds$$

or, by combining the above integrals, we finally obtain a more compact form for $u_0(r)$

$$u_0(r) = \int_0^R k_0(r,s) s f_0(s) ds \qquad (5.43)$$

In accordance with Theorem 1.3 of Section 1.3, the kernel

$$k_0(r,s) = \begin{cases} -\ln(s/R), & \text{for } r \leq s \\ -\ln(r/R), & \text{for } s \leq r \end{cases}$$

of the integral in eqn (5.43) represents the Green's function of the homogeneous boundary value problem corresponding to that stated by eqns (5.41) and (5.40). This function can also be found in Section 1.1, where it was constructed by the method based on the defining properties of Green's functions (see the particular case in *EXAMPLE 7*).

Consider now the general case ($n \geq 1$) of the boundary value problem stated by eqns (5.39) and (5.40). A fundamental solution set for the homogeneous equation

$$\frac{d}{dr}\left(r\frac{du_n(r)}{dr}\right) - \frac{n^2}{r} u_n(r) = 0$$

corresponding to that of eqn (5.39) can be represented by r^n and r^{-n}. In compliance with the method of variation of parameters, one expresses the general solution of eqn (5.39) in the form

$$u_n(r) = C_1(r) r^n + C_2(r) r^{-n}$$

and finally derives it as

$$u_n(r) = \int_0^r \frac{1}{2n}\left[\left(\frac{s}{r}\right)^n - \left(\frac{r}{s}\right)^n\right] s f_n(s) ds + D_1 r^n + D_2 r^{-n}$$

where D_1 and D_2 represent undetermined constant coefficients. To find them the boundary conditions in eqn (5.40) must be used. This yields

$$D_2 = 0, \quad D_1 = \int_0^R \frac{1}{2n}\left[\left(\frac{1}{s}\right)^n - \left(\frac{s}{R^2}\right)^n\right] s f_n(s) ds$$

Upon substituting these quantities into the last expression for $u_n(r)$, one obtains

$$u_n(r) = \int_0^r \frac{1}{2n}\left[\left(\frac{s}{r}\right)^n - \left(\frac{r}{s}\right)^n\right] s f_n(s)\,ds$$
$$+ \int_0^R \frac{1}{2n}\left[\left(\frac{r}{s}\right)^n - \left(\frac{rs}{R^2}\right)^n\right] s f_n(s)\,ds$$

or by using a more compact notation, $u_n(r)$ can be rewritten as

$$u_n(r) = \int_0^R k_n(r,s) s f_n(s)\,ds \qquad (5.44)$$

where

$$k_n(r,s) = \frac{1}{2n}\left[\left(\frac{r}{s}\right)^n - \left(\frac{rs}{R^2}\right)^n\right], \quad \text{for } r \leq s$$

represents one of the two branches of the Green's function of the homogeneous boundary value problem corresponding to that of eqns (5.39) and (5.40). The other branch of $k_n(r,s)$, which is valid for $r \geq s$, can be obtained from that above by interchanging of r with s. This clearly reflects the self-adjointness of the formulation in eqns (5.39) and (5.40).

According to the fundamental rule for coefficients of the Fourier series, one writes the following expressions for the coefficients $f_n^c(s)$ and $f_n^s(s)$ of the series that appeared in eqn (5.38)

$$f_n^c(s) = \frac{1}{\pi}\int_0^{2\pi} f(s,t)\cos(nt)\,dt, \quad n=0,1,2,\ldots \qquad (5.45)$$

$$f_n^s(s) = \frac{1}{\pi}\int_0^{2\pi} f(s,t)\sin(nt)\,dt, \quad n=1,2,3,\ldots \qquad (5.46)$$

Clearly, eqn (5.44) can now be rewritten, for both the cosine- and sine-coefficients of the series in eqn (5.37), as

$$u_n^c(r) = \int_0^R k_n(r,s) s f_n^c(s)\,ds \qquad (5.47)$$

$$u_n^s(r) = \int_0^R k_n(r,s) s f_n^s(s)\,ds \qquad (5.48)$$

Upon substituting the expressions for $f_n^c(s)$ and $f_n^s(s)$ from eqns (5.45) and (5.46) into eqns (5.43), (5.47), and (5.48), and then the coefficients $u_0(r)$, $u_n^c(r)$, and $u_n^s(r)$ into eqn (5.37), one obtains the solution of the boundary value problem posed by eqns (5.35) and (5.36) in the form

$$u(r,\phi) = \int_0^{2\pi}\int_0^R \left\{\frac{1}{\pi}\left[\frac{k_0(r,s)}{2} + \sum_{n=1}^{\infty} k_n(r,s)\cos(n\phi)\cos(nt)\right.\right.$$

5.2. FORMULATIONS IN POLAR COORDINATES

$$+ \sum_{n=1}^{\infty} k_n(r,s)\sin(n\phi)\sin(nt)\bigg]\bigg\} f(s,t) s ds dt$$

which can be simplified after both the above series are combined into one. This yields

$$u(r,\phi) = \int_0^{2\pi}\int_0^R \left\{\frac{1}{\pi}\left[\frac{k_0(r,s)}{2} + \sum_{n=1}^{\infty} k_n(r,s)\cos(n(\phi-t))\right]\right\} f(s,t) s ds dt \quad (5.49)$$

Clearly, the expression $sdsdt$ represents the element of area in polar coordinates. So, the integration in eqn (5.49) is accomplished over the entire disk Ω_c. Hence, in view of the definition presented in the opening part of this chapter, the kernel

$$G(r,\phi;s,t) = \frac{1}{2\pi}\left[k_0(r,s) + 2\sum_{n=1}^{\infty} k_n(r,s)\cos(n(\phi-t))\right] \quad (5.50)$$

of the integral in eqn (5.49) represents the Green's function of the Dirichlet problem for Laplace's equation on a disk of radius R.

The trigonometric series of eqn (5.50) converges non-uniformly due to the fact that the function represented by its sum has the logarithmic singularity as $(r,\phi) = (s,t)$. Hence, the direct computational implementations based on the expansion in eqn (5.50) are not recommended. However, one can decompose the regular and the singular terms of the series by summing them up. In doing so, we recall the expressions for $k_0(r,s)$ and $k_n(r,s)$ that were earlier obtained and substitute them into eqn (5.50). Notice that either of the branches of these functions are appropriate for the summation procedure. Taking, for instance, those branches which are valid for $r \leq s$, one obtains

$$G(r,\phi;s,t) = \frac{1}{2\pi}\left\{-\ln\left(\frac{s}{R}\right) + \sum_{n=1}^{\infty}\frac{1}{n}\left[\left(\frac{r}{s}\right)^n - \left(\frac{rs}{R^2}\right)^n\right]\cos(n(\phi-t))\right\}$$

The summation of this series is tackled by recalling the standard formula

$$\sum_{n=1}^{\infty}\frac{p^n}{n}\cos(n\alpha) = -\ln\sqrt{1 - 2p\cos\alpha + p^2}$$

presented earlier in this section (see eqn (5.17)). This results in

$$G(r,\phi;s,t) = \frac{1}{2\pi}\left\{-\ln\left(\frac{s}{R}\right) - \ln\sqrt{1 - 2\left(\frac{r}{s}\right)\cos(\phi-t) + \left(\frac{r}{s}\right)^2}\right.$$

$$\left. + \ln\sqrt{1 - 2\left(\frac{rs}{R^2}\right)\cos(\phi-t) + \left(\frac{rs}{R^2}\right)^2}\right\}$$

$$= \frac{1}{2\pi}\left\{-\ln\left(\frac{s}{R}\right) - \ln\frac{\sqrt{r^2 - 2rs\cos(\phi-t) + s^2}}{s}\right.$$

$$\left. + \ln\frac{\sqrt{R^4 - 2R^2 rs\cos(\phi-t) + r^2 s^2}}{R^2}\right\} \tag{5.51}$$

The above representation can be simplified by introducing the complex variables z and ζ as expressed in trigonometric form in terms of the coordinates of the observation (r, ϕ) and the source (s, t) point, respectively

$$z = r(\cos(\phi) + i\sin(\phi)), \quad \zeta = s(\cos(t) + i\sin(t))$$

In doing so, we derive the expression for $R^2 - z\overline{\zeta}$

$$R^2 - z\overline{\zeta} = R^2 - rs[\cos(\phi) + i\sin(\phi)][\cos(t) - i\sin(t)]$$

Removing the brackets and decomposing the real and imaginary parts, one obtains

$$R^2 - z\overline{\zeta} = [R^2 - rs\cos(\phi-t)] + irs\sin(\phi-t)$$

Hence, the modulus of the above function of complex variables can be finally obtained in the form

$$|R^2 - z\overline{\zeta}| = \sqrt{[R^2 - rs\cos(\phi-t)]^2 + [rs\sin(\phi-t)]^2}$$

$$= \sqrt{R^4 - 2R^2 rs\cos(\phi-t) + r^2 s^2} \tag{5.52}$$

Analogously, one obtains

$$|z - \zeta| = \sqrt{r^2 - 2rs\cos(\phi-t) + s^2} \tag{5.53}$$

Upon replacing the arguments of the second and the third logarithms in eqn (5.51) with the left-hand sides of eqns (5.53) and (5.52), respectively, one obtains the following expression for $G(r, \phi; s, t)$

$$G(r, \phi; s, t) = \frac{1}{2\pi}\left\{-\ln\left(\frac{s}{R}\right) - \ln\frac{|z-\zeta|}{s} + \ln\frac{|R^2 - z\overline{\zeta}|}{R^2}\right\}$$

Combining all of the above logarithms, we finally obtain the Green's function of the Dirichlet problem for the disk of radius R in the form

$$G(r, \phi; s, t) = \frac{1}{2\pi}\ln\frac{|R^2 - z\overline{\zeta}|}{R|z - \zeta|} \tag{5.54}$$

5.2. FORMULATIONS IN POLAR COORDINATES

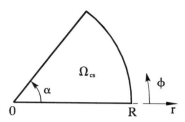

Figure 5.3   Circular sector

This is, in fact, a classical (see, for example, [6, 14, 17, 26, 31, 36, 38, 63, 66, 68, 70]) representation of this Green's function, which is usually obtained by the reflection method. So, in this case the suggested technique can be considered as an alternative to the reflection method.

Notice that all of the analytic representations of Green's functions for two-dimensional boundary value problems for Laplace's equation, which are already available in the existing literature, can be reproduced by the suggested technique. To address this point, we consider one more formulation. That is the Dirichlet problem

$$u(R,\phi)=0, \quad u(r,0)=u(r,\alpha)=0 \qquad (5.55)$$

set up for eqn (5.35) stated on the circular sector $\Omega_{cs} = \{(r,\phi): 0<r<R;\ 0<\phi<\alpha\}$ shown in Figure 5.3.

Assume, for the solution of the problem posed by eqns (5.35) and (5.55), the following trigonometric expansion

$$u(r,\phi) = \sum_{n=1}^{\infty} u_n(r)\sin(\nu\phi), \quad \nu = \frac{n\pi}{\alpha} \qquad (5.56)$$

Represent also the right-hand term of eqn (5.35) with the sine-series

$$f(r,\phi) = \sum_{n=1}^{\infty} f_n(r)\sin(\nu\phi) \qquad (5.57)$$

Clearly, substitution of these series into the original formulation yields the following self-adjoint boundary value problem

$$\frac{d}{dr}\left(r\frac{du_n(r)}{dr}\right) - \frac{\nu^2}{r}u_n(r) = -rf_n(r), \quad n=1,2,3,\ldots \qquad (5.58)$$

$$|u_n(0)|<\infty, \quad u_n(R)=0 \qquad (5.59)$$

which is nearly the same as that posed earlier by eqns (5.39) and (5.40), with only two minor differences in their statements (the coefficients of $u_n(r)$ in the governing equations and the ranges of values of the parameter n are formally different). Therefore, the solution of the problem in eqns (5.58) and (5.59) can also be written in the form of eqn (5.44)

$$u_n(r) = \int_0^R k_n(r,s) s f_n(s) ds$$

where the branch of the kernel function $k_n(r,s)$, which is valid for $r \leq s$ is expressed as

$$k_n(r,s) = \frac{1}{2\nu}\left[\left(\frac{r}{s}\right)^\nu - \left(\frac{rs}{R^2}\right)^\nu\right], \quad \text{for } r \leq s$$

and the corresponding representation of $k_n(r,s)$ for $r \geq s$, due to the self-adjointness of the problem set up by eqns (5.58) and (5.59), can be obtained from that above by the interchanging of r with s.

Hence, in compliance with the suggested technique, the Green's function $G(r, \phi; s, t)$ of the Dirichlet problem for Laplace's equation on the circular sector Ω_{cs} is obtained in the form

$$G(r, \phi; s, t) = \frac{2}{\alpha} \sum_{n=1}^{\infty} k_n(r,s) \sin(\nu\phi) \sin(\nu t)$$

$$= \frac{1}{2\pi} \sum_{n=1}^{\infty} \frac{1}{n}\left[\left(\left(\frac{r}{s}\right)^p\right)^n - \left(\left(\frac{rs}{R^2}\right)^p\right)^n\right]$$

$$\times [\cos(np(\phi-t)) - \cos(np(\phi+t))]$$

where $p = \pi/\alpha$.

By removing the brackets in the above expression, it can be written as a sum of the four summable series of the type of eqn (5.17). After the summation is completed, we obtain

$$G(r, \phi; s, t) = \frac{1}{2\pi}\left\{\ln\sqrt{1 - 2\left(\frac{r}{s}\right)^p \cos(p(\phi+t)) + \left(\frac{r}{s}\right)^{2p}}\right.$$

$$- \ln\sqrt{1 - 2\left(\frac{r}{s}\right)^p \cos(p(\phi-t)) + \left(\frac{r}{s}\right)^{2p}}$$

$$+ \ln\sqrt{1 - 2\left(\frac{rs}{R^2}\right)^p \cos(p(\phi-t)) + \left(\frac{rs}{R^2}\right)^{2p}}$$

$$\left. - \ln\sqrt{1 - 2\left(\frac{rs}{R^2}\right)^p \cos(p(\phi+t)) + \left(\frac{rs}{R^2}\right)^{2p}}\right\}.$$

5.2. FORMULATIONS IN POLAR COORDINATES

Multiplying the arguments of the radicals of the first and the second of the above logarithms by s^{2p} and multiplying the arguments of the radicals of the third and the fourth of the above logarithms by R^{4p}, one transforms $G(r,\phi;s,t)$ to

$$G(r,\phi;s,t) = \frac{1}{2\pi}\left\{\ln\sqrt{r^{2p}-2(rs)^p\cos(p(\phi+t))+s^{2p}}\right.$$

$$-\ln\sqrt{r^{2p}-2(rs)^p\cos(p(\phi-t))+s^{2p}}$$

$$+\ln\sqrt{R^{4p}-2(rsR^2)^p\cos(p(\phi-t))+(rs)^{2p}}$$

$$\left.-\ln\sqrt{R^{4p}-2(rsR^2)^p\cos(p(\phi+t))+(rs)^{2p}}\right\}.$$

The above expression for $G(r,\phi;s,t)$ can be written in a more compact form in terms of the observation z and the source ζ points. This finally yields the Green's function for the Dirichlet problem on a circular sector of radius R in the form

$$G(r,\phi;s,t) = \frac{1}{2\pi}\ln\frac{|z^p-(\overline{\zeta})^p||(R^2)^p-(z\overline{\zeta})^p|}{|z^p-\zeta^p||(R^2)^p-(z\zeta)^p|} \quad (5.60)$$

where z^p is understood as the principal value of a function.

As the central angle α of the sector Ω_{cs} equals π, the expression in eqn (5.60) reduces to the classical representation

$$G(r,\phi;s,t) = \frac{1}{2\pi}\ln\frac{|z-\overline{\zeta}||R^2-z\overline{\zeta}|}{|z-\zeta||R^2-z\zeta|}$$

of the Green's function for a half-disk of radius R. That is, we have found additional evidence of the effectiveness of the suggested technique.

So far in this section, we have considered validation examples. Indeed, the suggested technique has been used for reconstructing existing representations of Green's functions. Our goal in the next example is a bit different. Namely, in the scope of the suggested technique, an alternative representation of a Green's function is derived whose existing form is too complex and is not suitable for computational implementations.

Let us set up the mixed boundary value problem

$$\left.\left(\frac{\partial u}{\partial r}+\beta u\right)\right|_{r=R} = 0, \quad \beta>0 \quad (5.61)$$

on the disk Ω_c of radius R. Notice that in the existing representation of the Green's function for this formulation, which is available, for example, in [68]), the regular component is expressed in terms of a line integral in the complex plane. Consequently, the numerical evaluation of the existing representation results in a cumbersome computation. In the representation which will be obtained here, the regular component of the Green's function is presented in the form of a uniformly convergent Fourier series whose accurate truncation is trivial and can easily be accomplished.

In accordance with our procedure, the solution of the problem in eqns (5.35) and (5.61) is expanded into the Fourier series presented earlier by eqn (5.37). This yields the following self-adjoint boundary value problem

$$|u_n(0)| < \infty, \quad \frac{du_n(R)}{dr} + \beta u_n(R) = 0 \tag{5.62}$$

stated for the ordinary differential equation (5.39).

The Green's function of the homogeneous boundary value problem corresponding to that posed by eqns (5.35) and (5.61) is found to be in the form of the expansion in eqn (5.50), whose coefficient, for $n=0$ is given by

$$k_0(r,s) = \frac{1}{\beta R} - \ln\left(\frac{s}{R}\right), \quad \text{for } r \le s \tag{5.63}$$

while for $n \ge 1$, one obtains

$$k_n(r,s) = \frac{1}{2n}\left[\left(\frac{r}{s}\right)^n + \left(\frac{rs}{R^2}\right)^n\right] - \frac{\beta R}{n(n+\beta R)}\left(\frac{rs}{R^2}\right)^n, \quad \text{for } r \le s \tag{5.64}$$

Clearly, as β approaches infinity, the mixed boundary condition in eqn (5.61) reduces to the Dirichlet condition. We show below that the above expressions for the coefficients $k_0(r,s)$ and $k_n(r,s)$ reduce to the ones which were recently derived for the Dirichlet problem. Indeed, taking a limit of $k_0(r,s)$ from eqn (5.63) as β approaches infinity, we obtain

$$\lim_{\beta \to \infty}\left[\frac{1}{\beta R} - \ln\left(\frac{s}{R}\right)\right] = -\ln\left(\frac{s}{R}\right)$$

(see the expression of $k_0(r,s)$, for $r \le s$, that immediately follows eqn (5.43)). And the expression for $k_n(r,s)$, that immediately follows eqn (5.44), occurs as a limit of $k_n(r,s)$ from eqn (5.64). That is

$$\lim_{\beta \to \infty}\left\{\frac{1}{2n}\left[\left(\frac{r}{s}\right)^n + \left(\frac{rs}{R^2}\right)^n\right] - \frac{\beta R}{n(n+\beta R)}\left(\frac{rs}{R^2}\right)^n\right\}$$

$$= \lim_{\beta \to \infty}\left\{\frac{1}{2n}\left[\left(\frac{r}{s}\right)^n + \left(\frac{rs}{R^2}\right)^n\right] - \left(\frac{1}{n} - \frac{1}{n+\beta R}\right)\left(\frac{rs}{R^2}\right)^n\right\}$$

$$= \frac{1}{2n}\left[\left(\frac{r}{s}\right)^n + \left(\frac{rs}{R^2}\right)^n\right] - \frac{1}{n}\left(\frac{rs}{R^2}\right)^n = \frac{1}{2n}\left[\left(\frac{r}{s}\right)^n - \left(\frac{rs}{R^2}\right)^n\right]$$

5.2. FORMULATIONS IN POLAR COORDINATES

That is, the Green's function for the Dirichlet problem, derived earlier in eqn (5.54), can be obtained as a particular case of that representation which appears below in eqn (5.65), as we take a limit of the latter as β approaches infinity.

We now turn to the series of eqn (5.50), with the coefficients presented by eqn (5.63) and (5.64). This appears to be written as

$$G(r, \phi; s, t) = \frac{1}{2\pi} \left\{ \frac{1}{\beta R} - \ln\left(\frac{s}{R}\right) \right.$$

$$\left. + \sum_{n=1}^{\infty} \left[\frac{1}{n}\left(\left(\frac{r}{s}\right)^n + \left(\frac{rs}{R^2}\right)^n\right) - \frac{2\beta R}{n(n+\beta R)}\left(\frac{rs}{R^2}\right)^n \right] \cos\left(n(\phi-t)\right) \right\}$$

The first two terms in the above series allow a complete summation with the aid of the summation formula from eqn (5.17). This yields

$$G(r, \phi; s, t) = \frac{1}{2\pi} \left\{ \frac{1}{\beta R} - \ln\left(\frac{s}{R}\right) - \ln\frac{|z-\zeta|}{s} - \ln\frac{|R^2 - z\overline{\zeta}|}{R^2} \right.$$

$$\left. - \sum_{n=1}^{\infty} \frac{2\beta R}{n(n+\beta R)}\left(\frac{rs}{R^2}\right)^n \cos\left(n(\phi-t)\right) \right\}$$

Combining all of the above logarithmic terms, we ultimately obtain the Green's function of the mixed boundary value problem posed by eqn (5.61) for Laplace's equation on a disk of radius R in the form

$$G(r, \phi; s, t) = \frac{1}{2\pi} \left[\frac{1}{R\beta} + \ln\frac{R^3}{|z-\zeta||R^2 - z\overline{\zeta}|} \right.$$

$$\left. - \sum_{n=1}^{\infty} \frac{2\beta R}{n(n+\beta R)}\left(\frac{rs}{R^2}\right)^n \cos\left(n(\phi-t)\right) \right] \quad (5.65)$$

In Section 5.1, we have analyzed the uniformly convergent series of the above type and have estimated its remainder. Hence, a required level of accuracy in computing values of the Green's function under consideration can readily be attained by an appropriate truncation of the above series. The form that is just derived is notably less expensive computationally compared to the existing [68] representation of this Green's function.

To highlight the potential of the suggested technique, in the last example of this section, we present the Green's function of Laplace's equation for the mixed boundary value problem written as

$$\left.\left(\frac{\partial u}{\partial r} + \beta u\right)\right|_{r=R} = u|_{\phi=0} = \left.\frac{\partial u}{\partial \phi}\right|_{\phi=\alpha} = 0, \quad \beta > 0$$

for the circular sector Ω_{cs} of central angle α. This problem is particularly interesting because its formulation contains three different kinds of boundary conditions imposed on the three different portions of the region's contour.

In this case we omit a detailed description of the deriving procedure because the reader has already gained the necessary experience. Instead, we just present the final form

$$G(r,\phi;s,t) = -\frac{2\beta R}{\alpha}\sum_{n=1}^{\infty}\frac{1}{\nu(\nu+\beta R)}\left(\frac{rs}{R^2}\right)^{\nu}\sin(\nu\phi)\sin(\nu t)$$

$$+\frac{1}{2\pi}\ln\frac{|z^p-(\overline{\zeta})^p||z^p+\zeta^p||(R^2)^p-(z\zeta)^p||(R^2)^p+(z\overline{\zeta})^p|}{|z^p+(\overline{\zeta})^p||z^p-\zeta^p||(R^2)^p+(z\zeta)^p||(R^2)^p-(z\overline{\zeta})^p|} \quad (5.66)$$

of the Green's function under consideration. The parameters p and ν are given as $p = \pi/2\alpha$ and $\nu = (2n-1)p$. Here, z^p is again understood as the principal value of a function. Hence, the singular and regular components of $G(r,\phi;s,t)$ are decomposed and the latter is presented in the form containing a uniformly convergent series.

To the best of the author's knowledge, the above representation appeared for the first time in the author's work [53].

In the set of exercises related to Sections 5.1 and 5.2, the reader finds a number of statements for which influence (Green's) functions are not available at all in the existing texts and handbooks.

5.3 Potential fields on surfaces

In order to permit a reasonably systematic presentation of the suggested technique, we limited our discussion in the preceding two sections to Cartesian and polar coordinates for which at least some representations of Green's (influence) functions of the Laplace's equation are available in the existing literature.

In the present section, we will extend the sphere of successful use of the suggested technique to the construction of influence functions for potential problems formulated on surfaces of revolution. This problem class is of great importance in engineering and science for determining potential fields of various physical nature (gravitation, electro-statics and magneto-statics, steady-state heat and mass conduction, and so on) in thin plate-shell structures, when the energy flow normal to their face surfaces is small and can therefore be neglected.

Some formulations for spherical and toroidal surfaces are considered herein. This, however, by no means indicates the limitation of applicability

5.3. POTENTIAL FIELDS ON SURFACES

of the technique. In Chapter 7, for example, influence functions are constructed for potential fields in compound plate-shell structures consisting of plate and cylindrical elements as well. Conical, paraboloidal, hyperboloidal, and other surfaces of revolution can also be considered in the scope of the suggested technique.

To begin with, let us assume a spherical surface of radius a to be determined by the following parametric equations

$$x = a\sin(\phi)\cos(\theta), \quad y = a\sin(\phi)\sin(\theta), \quad z = a\cos(\phi) \quad (5.67)$$

That is, a point (x, y, z) on the surface is uniquely defined by the geographic coordinates ϕ and θ representing the latitude and the longitude, respectively.

Due to the change of the independent variables determined by the relations in eqn (5.67), the Cartesian three-dimensional Laplace's operator

$$\Delta(x, y, z) \equiv \frac{\partial^2}{\partial x^2} + \frac{\partial^2}{\partial y^2} + \frac{\partial^2}{\partial z^2}$$

transforms into the two-dimensional differential form written as

$$\Delta(\phi, \theta) \equiv \frac{1}{\sin(\phi)} \frac{\partial}{\partial \phi}\left(\sin(\phi)\frac{\partial}{\partial \phi}\right) + \frac{1}{\sin^2(\phi)} \frac{\partial^2}{\partial \theta^2}.$$

Note that this form can also be obtained directly from the three-dimensional Laplace's operator written in spherical coordinates r, ϕ, and θ

$$\frac{1}{r^2}\frac{\partial}{\partial r}\left(r^2\frac{\partial}{\partial r}\right) + \frac{1}{r^2\sin(\phi)}\frac{\partial}{\partial \phi}\left(\sin(\phi)\frac{\partial}{\partial \phi}\right) + \frac{1}{r^2\sin^2(\phi)}\frac{\partial^2}{\partial \theta^2}$$

by letting r be a constant and neglecting the derivatives with respect to r.

As the first example for the spherical surface, we set up the following mixed boundary value problem

$$\frac{1}{\sin(\phi)\partial\phi}\left(\sin(\phi)\frac{\partial u(\phi,\theta)}{\partial \phi}\right) + \frac{1}{\sin^2(\phi)}\frac{\partial^2}{\partial \theta^2} = -f(\phi,\theta), \quad (\phi,\theta) \in \Omega_{st} \quad (5.68)$$

$$|u|_{\phi=0} < \infty, \quad \left.\frac{\partial u}{\partial \phi}\right|_{\phi=\alpha} = 0, \quad u|_{\theta=0,\beta} = 0 \quad (5.69)$$

on the spherical triangle $\Omega_{st} = \{(\phi, \theta): 0 < \phi < \alpha,\ 0 < \theta < \beta\}$, where $\alpha < \pi$ and $\beta < 2\pi$. Let us assume the following expansion

$$u(\phi, \theta) = \sum_{n=1}^{\infty} u_n(\phi)\sin(\nu\theta), \quad \nu = \frac{n\pi}{\beta} \quad (5.70)$$

for the solution of the problem posed by eqns (5.68) and (5.69). Expand also the right-hand side of eqn (5.68) in the Fourier sine-series

$$f(\phi,\theta) = \sum_{n=1}^{\infty} f_n(\phi)\sin(\nu\theta) \qquad (5.71)$$

Upon substituting these expansions of $u(\phi,\theta)$ and $f(\phi,\theta)$ into the original formulation and equating the corresponding coefficients of the Fourier series appearing on both hand sides of eqn (5.68), one obtains the following linear ordinary differential equation

$$\frac{1}{\sin(\phi)}\frac{d}{d\phi}\left(\sin(\phi)\frac{du_n(\phi)}{d\phi}\right) - \frac{\nu^2}{\sin^2(\phi)}u_n(\phi) = -f_n(\phi,\theta)$$

in the coefficients $u_n(\phi)$ of the series in eqn (5.70). Introducing the integrating factor $\sin(\phi)$, this reduces to the self-adjoint form

$$\frac{d}{d\phi}\left(\sin(\phi)\frac{du_n(\phi)}{d\phi}\right) - \frac{\nu^2}{\sin(\phi)}u_n(\phi) = -f_n(\phi,\theta)\sin(\phi) \qquad (5.72)$$

which is subject to the boundary conditions

$$|u_n(0)|<\infty, \quad \frac{du_n(\alpha)}{d\phi}=0, \quad n=1,2,3,... \qquad (5.73)$$

To determine the fundamental solution set of the homogeneous equation corresponding to that of eqn (5.72), it is worth noting that upon the change of variable given [35] by

$$\omega = \ln\left(\tan(\frac{\phi}{2})\right) \qquad (5.74)$$

one obtains

$$\frac{d\omega}{d\phi} = \frac{1}{2}\frac{\sec^2(\phi/2)}{\tan(\phi/2)} = \frac{1}{2\sin(\phi/2)\cos(\phi/2)} = \frac{1}{\sin(\phi)}$$

and

$$\frac{d}{d\phi} = \frac{d}{d\omega}\frac{d\omega}{d\phi} = \frac{1}{\sin(\phi)}\frac{d}{d\omega}$$

hence

$$\sin(\phi)\frac{d}{d\phi} = \frac{d}{d\omega}$$

and

$$\frac{d}{d\phi}\left(\sin(\phi)\frac{d}{d\phi}\right) = \frac{d^2}{d\omega^2}\frac{d\omega}{d\phi} = \frac{1}{\sin(\phi)}\frac{d^2}{d\omega^2}$$

5.3. POTENTIAL FIELDS ON SURFACES

This suggests that the differential operator in eqn (5.72) can be rewritten in terms of the new variable ω in the form

$$\frac{d}{d\phi}\left(\sin(\phi)\frac{d}{d\phi}\right) - \frac{\nu^2}{\sin(\phi)} \equiv \frac{1}{\sin(\phi)}\left(\frac{d^2}{d\omega^2} - \nu^2\right) \quad (5.75)$$

At this point, it is clearly seen that the fundamental solution set for the operator in the right-hand side of eqn (5.75) can be represented by the exponential functions

$$\exp(\nu\omega) \quad \text{and} \quad \exp(-\nu\omega)$$

In view of the relation in eqn (5.74), these provide the fundamental solution set for the homogeneous equation corresponding to that of eqn (5.72) to be expressed by the functions

$$\tan^\nu\left(\frac{\phi}{2}\right), \quad \cot^\nu\left(\frac{\phi}{2}\right) \quad (5.76)$$

Based on the fundamental solution set just derived, one constructs the Green's function $g_n(\phi, \psi)$ for the homogeneous problem corresponding to that posed by eqns (5.72) and (5.73). The branch of $g_n(\phi, \psi)$ valid for $\phi \leq \psi$ is obtained as

$$g_n(\phi, \psi) = \frac{1}{2\nu}\left\{\left[\tan(\frac{\phi}{2})\tan(\frac{\psi}{2})\cot^2(\frac{\alpha}{2})\right]^\nu + \left[\tan(\frac{\phi}{2})\cot(\frac{\psi}{2})\right]^\nu\right\} \quad (5.77)$$

And from the self-adjointness of the problem in eqns (5.72) and (5.73), it follows that the other branch of $g_n(\phi, \psi)$ valid for $\phi \geq \psi$ can be obtained from that above by interchanging of ϕ with ψ.

Following the suggested procedure, the Green's function of the homogeneous boundary value problem, associated with that posed by eqns (5.68) and (5.69), is expressed by

$$G(\phi, \theta; \psi, \tau) = \frac{1}{\beta}\sum_{n=1}^{\infty} g_n(\phi, \psi)[\cos(\nu(\theta+\tau)) - \cos(\nu(\theta-\tau))]$$

This trigonometric series is completely summable. To carry out the summation, we first rewrite the series in the form

$$G(\phi, \theta; \psi, \tau) = \frac{1}{2\pi}\left\{\sum_{n=1}^{\infty}\frac{1}{n}\left[\left(\frac{\Phi}{\Psi}\right)^n + \left(\frac{\Phi\Psi}{A^2}\right)^n\right]\cos(n\gamma)\right.$$

$$\left. - \sum_{n=1}^{\infty}\frac{1}{n}\left[\left(\frac{\Phi}{\Psi}\right)^n - \left(\frac{\Phi\Psi}{A^2}\right)^n\right]\cos(n\delta)\right\} \quad (5.78)$$

where
$$\Phi = \tan^{\pi/\beta}\left(\frac{\phi}{2}\right), \quad \Psi = \tan^{\pi/\beta}\left(\frac{\psi}{2}\right), \quad A = \tan^{\pi/\beta}\left(\frac{\alpha}{2}\right)$$

and
$$\gamma = \frac{\pi}{\beta}(\theta - \tau), \quad \delta = \frac{\pi}{\beta}(\theta + \tau) \tag{5.79}$$

Notice that the variables ϕ and ψ in eqn (5.78) satisfy the relationship $\phi \leq \psi$ (to ensure this, recall the limitation we assumed for $g_n(\phi, \psi)$ in eqn (5.77)). From this, it follows that, for the series of eqn (5.78), the conditions are met of the applicability of the standard summation formula from eqn (5.17) and, therefore, the series of eqn (5.78) are indeed summable.

Accomplishing the summation in eqn (5.78) and performing some elementary algebra, we finally obtain, for the Green's function to the problem posed by eqns (5.68) and (5.69) the following closed form

$$G(\phi, \theta; \psi, \tau) = \frac{1}{2\pi}\left[\ln\sqrt{\frac{\Phi^2 - 2\Phi\Psi\cos(\delta) + \Psi^2}{\Phi^2 - 2\Phi\Psi\cos(\gamma) + \Psi^2}}\right.$$

$$\left. + \ln\sqrt{\frac{A^4 - 2A^2\Phi\Psi\cos(\delta) + \Phi^2\Psi^2}{A^4 - 2A^2\Phi\Psi\cos(\gamma) + \Phi^2\Psi^2}}\right]$$

which is compact enough for computational implementations.

Thus, from the example just completed, it follows that the suggested technique is applicable to the construction of Green's (influence) functions for mixed boundary value problems of potential, stated on surfaces of revolution. The next example brings to the reader's attention one more problem stated on the spherical surface.

We present a brief description of the construction procedure for the Green's function of the Dirichlet problem stated for Laplace's equation on the spherical segment $\Omega_{ss} = \{(\phi, \theta): 0 < \phi < \alpha, 0 \leq \theta < 2\pi\}$ where $\alpha < \pi$. In accordance with our procedure, the solution $u(\phi, \theta)$ and the right-hand term $f(\phi, \theta)$ of eqn (5.68) are expanded in this case in the general type of Fourier series

$$u(\phi, \theta) = \frac{1}{2}u_0(\phi) + \sum_{n=1}^{\infty}(u_n^c(\phi)\cos(n\theta) + u_n^s(\phi)\sin(n\theta)) \tag{5.80}$$

and
$$f(\phi, \theta) = \frac{1}{2}f_0(\phi) + \sum_{n=1}^{\infty}(f_n^c(\phi)\cos(n\theta) + f_n^s(\phi)\sin(n\theta))$$

5.3. POTENTIAL FIELDS ON SURFACES

This yields the following set of self-adjoint boundary value problems

$$\frac{d}{d\phi}\left(\sin(\phi)\frac{du_n(\phi)}{d\phi}\right) - \frac{n^2}{\sin(\phi)}u_n(\phi) = -f_n(\phi,\theta)\sin(\phi) \qquad (5.81)$$

$$|u_n(0)| < \infty, \quad u_n(\alpha) = 0, \quad n = 0, 1, 2, \ldots \qquad (5.82)$$

in the coefficients $u_n(\phi)$ of the series of eqn (5.80). And the Green's function to be found is ultimately presented as the series

$$G(\phi,\theta;\psi,\tau) = \frac{1}{2\pi}\left[g_0(\phi,\psi) + 2\sum_{n=1}^{\infty} g_n(\phi,\psi)\cos(n(\theta-\tau))\right] \qquad (5.83)$$

whose coefficients $g_0(\phi,\psi)$ and $g_n(\phi,\psi)$ represent the Green's functions for the boundary value problems stated by eqns (5.81) and (5.82). Clearly, in constructing these Green's functions, the case of $n = 0$ should be treated individually, because the fundamental solution set in this case (see [35])

$$\ln\left(\tan(\frac{\phi}{2})\right), \quad \text{and} \quad 1$$

is different from that of the general case of $n \geq 1$ presented earlier by eqn (5.76).

Once $g_0(\phi,\psi)$ and $g_n(\phi,\psi)$ are constructed and substituted into the series of eqn (5.83), the latter is completely summed up providing the influence function of a unit point source of the Dirichlet problem for the spherical segment in the closed form

$$G(\phi,\theta;\psi,\tau) = \frac{1}{2\pi}\ln\left(\frac{A_0\sqrt{\Phi_0^2 - 2\Phi_0\Psi_0\cos(\theta-\tau) + \Psi_0^2}}{\sqrt{A_0^4 - 2A_0^2\Phi_0\Psi_0\cos(\theta-\tau) + \Phi_0^2\Psi_0^2}}\right)$$

where

$$A_0 = \tan\left(\frac{\alpha}{2}\right), \quad \Phi_0 = \tan\left(\frac{\phi}{2}\right), \quad \Psi_0 = \tan\left(\frac{\psi}{2}\right)$$

For the next example, we consider a boundary value problem for the equation of potential, stated over a region on a toroidal surface. Let a and R, with $a < R$, represent the radius of the meridional cross-section of the circular toroidal surface and the distance between the center of the meridional cross-section and the axis of revolution, respectively (see Figure 5.4).

Suppose that a point (x,y,z) on this surface is determined by the geographical coordinates ϕ (the latitude) and θ (the longitude) in accordance with the parameterization

$$x = D(\phi)\cos(\theta), \quad y = D(\phi)\sin(\theta), \quad z = a\cos(\phi)$$

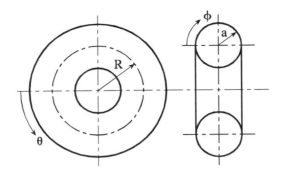

Figure 5.4 A circular toroidal surface

where $D(\phi) = R + a\sin(\phi)$.

The change of variables determined by such a parameterization transforms the three-dimensional Cartesian Laplace's operator to the following two-dimensional form

$$\Delta(\phi,\theta) \equiv \frac{1}{D(\phi)}\frac{\partial}{\partial \phi}\left(D(\phi)\frac{\partial}{\partial \phi}\right) + \frac{a^2}{D^2(\phi)}\frac{\partial^2}{\partial \theta^2}$$

We consider the toroidal sector Ω_{ts} which is closed in the direction of ϕ, whereas it is open in the direction of θ. That is, $\Omega_{ts} = \{(\phi,\theta): 0 < \phi \leq 2\pi,\ 0 < \theta < \beta\}$, where $\beta < 2\pi$. Set up the Dirichlet problem

$$\frac{1}{D(\phi)}\frac{\partial}{\partial \phi}\left(D(\phi)\frac{\partial u(\phi,\theta)}{\partial \phi}\right) + \frac{a^2}{D^2(\phi)}\frac{\partial^2 u(\phi,\theta)}{\partial \theta^2} = -f(\phi,\theta) \quad (5.84)$$

$$u|_{\theta=0} = 0, \quad u|_{\theta=\beta} = 0 \quad (5.85)$$

over Ω_{ts}. Due to the closedness of Ω_{ts} in the direction of ϕ, the above formulation is augmented with the relations

$$u|_{\phi=0} = u|_{\phi=2\pi}, \quad \left.\frac{\partial u}{\partial \phi}\right|_{\phi=0} = \left.\frac{\partial u}{\partial \phi}\right|_{\phi=2\pi} \quad (5.86)$$

that represent the conditions of 2π-periodicity of the solution in the direction of ϕ.

Assume, for the solution $u(\phi,\theta)$ of the above problem, the following trigonometric expansion

$$u(\phi,\theta) = \sum_{n=1}^{\infty} u_n(\phi)\sin(\nu\theta), \quad \nu = \frac{n\pi}{\beta}$$

5.3. POTENTIAL FIELDS ON SURFACES

which allows for $u(\phi,\theta)$ to satisfy the boundary conditions of eqn (5.85).

Represent also the right-hand term $f(\phi,\theta)$ of eqn (5.84) with the series

$$f(\phi,\theta) = \sum_{n=1}^{\infty} f_n(\phi)\sin(\nu\theta)$$

This yields the self-adjoint boundary value problem

$$\frac{d}{d\phi}\left(D(\phi)\frac{du_n(\phi)}{d\phi}\right) - \frac{a^2\nu^2}{D(\phi)}u_n(\phi) = -D(\phi)f_n(\phi) \tag{5.87}$$

$$u_n(0) = u_n(2\pi), \quad \frac{du_n(0)}{d\phi} = \frac{du_n(2\pi)}{d\phi}, \quad n = 1, 2, 3, \ldots \tag{5.88}$$

in the coefficients $u_n(\phi)$ of the above series for $u(\phi,\theta)$.

The fundamental solution set of eqn (5.87), utilized for the construction of the Green's function $g_n(\phi,\psi)$ of the homogeneous boundary value problem corresponding to that posed by eqns (5.87) and (5.88), is found (see [35]) as

$$\exp\left(\frac{2a\nu}{\sqrt{R^2-a^2}}\arctan\left(\frac{a+R\tan(\phi/2)}{\sqrt{R^2-a^2}}\right)\right)$$

and

$$\exp\left(-\frac{2a\nu}{\sqrt{R^2-a^2}}\arctan\left(\frac{a+R\tan(\phi/2)}{\sqrt{R^2-a^2}}\right)\right)$$

Based on $g_n(\phi,\psi)$, one finds a series representation for the Green's function $G(\phi,\theta;\psi,\tau)$ of the Dirichlet problem for the equation of potential posed on the toroidal sector Ω_{ts}. The series is entirely summable with the aid of the summation formula from eqn (5.17). Thus, we finally obtain the following closed form of $G(\phi,\theta;\psi,\tau)$

$$G(\phi,\theta;\psi,\tau) = \frac{1}{2\pi}\ln\sqrt{\frac{\Phi_1^2 - 2\Phi_1\Psi_1\cos(\delta) + \Psi_1^2}{\Phi_1^2 - 2\Phi_1\Psi_1\cos(\gamma) + \Psi_1^2}}$$

where

$$\Phi_1 = \exp\left(\frac{2\pi a}{\beta\sqrt{R^2-a^2}}\arctan\left(\frac{a+R\tan(\phi/2)}{\sqrt{R^2-a^2}}\right)\right)$$

and

$$\Psi_1 = \exp\left(\frac{2\pi a}{\beta\sqrt{R^2-a^2}}\arctan\left(\frac{a+R\tan(\psi/2)}{\sqrt{R^2-a^2}}\right)\right)$$

with δ and γ being earlier defined by eqn (5.79).

Later in this text, while dealing with potential fields on joined surfaces of revolution (see Chapter 7 devoted to influence matrices for compound media), we will take advantage of the experience acquired in this section.

5.4 Klein-Gordon equation

In this section, the technique that was earlier developed for the construction of influence functions for problems of potential, will be extended to a variety of boundary value problems stated for the equation

$$\left(\Delta - k^2\right) u(P) = 0 \qquad (5.89)$$

where Δ represents Laplace's operator and k is a real constant.

From the stand-point of mathematics, this equation is very close to Laplace's equation. In physics it often models some potential related fields. In two dimensions, for example, this equation can be associated with the steady-state heat conduction phenomenon in a thin plate whose surfaces are not perfectly insulated.

It is worth noting that no special name is customarily associated with equation (5.89), though in some sources (see, for example, [66]), it is called the *static Klein-Gordon* equation. This name, however, has rarely been recalled in standard textbooks or handbooks on equations of mathematical physics. Given this, we suspect that this name could be slightly confusing for the reader. Nevertheless, within the scope of the present study, equation (5.89) is always referred to as the *Klein-Gordon* equation, though this may, probably call out a certain criticism from some of the readers.

To introduce the definition of the Green's function $G(P,Q)$ for Klein-Gordon equation (5.89), we set up a boundary value problem written as

$$\Delta u(P) - k^2 u(P) = -f(P), \quad P \in \Omega \qquad (5.90)$$

$$M_i u(P) \equiv \alpha_i(P) \frac{\partial u(P)}{\partial n_i} + \beta_i(P) u(P) = 0, \quad P \in \Gamma_i \qquad (5.91)$$

where Ω represents a simply connected region to be considered in two-dimensional Euclidean space, $\Gamma = \bigcup_{i=1}^{m} \Gamma_i$ denotes the piecewise smooth contour of Ω, $\alpha_i(P)$ and $\beta_i(P)$ are given functions defined on Γ in such a way that at least one of them is nonzero for every piece Γ_i of Γ, and n_i represents the normal direction to Γ_i at the point P.

Assume that the boundary value problem set up by eqns (5.90) and (5.91) has a unique solution. That is, the corresponding homogeneous problem, with $f(P) \equiv 0$, has only the trivial solution.

To define the Green's function for Klein-Gordon equation, we use here the same methodology as we used earlier for Laplace's equation.

DEFINITION: If, for any allowable right-hand term $f(P)$ in eqn (5.90), the solution of the boundary value problem set up by eqns (5.90)

5.4. KLEIN-GORDON EQUATION

and (5.91) is expressible in the form

$$u(P) = \iint_\Omega G(P,Q)f(Q)d\Omega(Q) \tag{5.92}$$

then the kernel $G(P,Q)$ of this representation is said to be the *Green's function* for the homogeneous problem associated with that of eqns (5.90) and (5.91).

For any location of the source point $Q \in \Omega$, $G(P,Q)$ as a function of the coordinates of the observation point P holds the properties written as:

1. at any point $P \in \Omega$, except at $P = Q$, $G(P,Q)$ satisfies the homogeneous Klein-Gordon equation, that is

$$(\Delta - k^2)G(P,Q) = 0, \quad P \neq Q$$

2. for $P \to Q$, $G(P,Q)$ approaches infinity like the modified Bessel (or Macdonald) function $K_0(k|P-Q|)$ of the second kind of order zero. That is [3, 17, 68], $G(P,Q)$ possesses the logarithmic singularity (as well as the Green's function of Laplace's equation).

3. $G(P,Q)$ satisfies the boundary conditions in eqn (5.91), that is

$$M_i G(P,Q) = 0, \quad P \in \Gamma_i, \quad i = \overline{1,m}$$

From this definition, it follows that the Green's function of the homogeneous boundary value problem associated with that posed by eqns (5.90) and (5.91) can be viewed as

$$G(P,Q) = \frac{1}{2\pi} \ln \frac{1}{|P-Q|} + R(P,Q) \tag{5.93}$$

where $R(P,Q)$, as a function of P, satisfies the homogeneous Klein-Gordon equation everywhere on Ω, regardless to a mutual location of P and Q. $R(P,Q)$ is called the *regular component* of the Green's function.

The reader's attention is now turned to the construction of Green's functions. The constructing technique, developed in the preceding sections of this chapter for Laplace's equation, is directly extended here to a variety of boundary value problems for Klein-Gordon equation.

As the first example, we set up the Dirichlet problem

$$(\Delta - k^2)u(x,y) = -f(x,y), \quad (x,y) \in \Omega_s \tag{5.94}$$

$$u(x,0) = u(x,b) = 0 \tag{5.95}$$

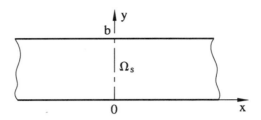

Figure 5.5   Infinite strip of width b

for the infinite strip $\Omega_s = \{(x,y): -\infty < x < \infty,\; 0 < y < b\}$ of width b (see Figure 5.5). In addition, we require, for the solution of the problem posed by eqns (5.94) and (5.95), to remain bounded for $x \to \pm\infty$. Assume also that the right-hand term $f(x,y)$ of eqn (5.94) is a suitably well-behaved function over Ω_s in the sense that

$$\left| \iint_{\Omega_s} f(x,y)\,d\Omega_s \right| < \infty$$

Assume, for the solution $u(x,y)$ of the problem set up by eqns (5.94) and (5.95), the following expansion

$$u(x,y) = \sum_{n=1}^{\infty} u_n(x)\sin(\nu y), \quad \nu = \frac{n\pi}{b}$$

and express also the right-hand term of eqn (5.94) by the Fourier sine-series

$$f(x,y) = \sum_{n=1}^{\infty} f_n(x)\sin(\nu y)$$

This yields the set of self-adjoint boundary value problems written as

$$\frac{d^2 u_n(x)}{dx^2} - (\nu^2 + k^2) u_n(x) = -f_n(x), \quad n = 1, 2, 3, \ldots \qquad (5.96)$$

$$|u_n(-\infty)| < \infty, \quad |u_n(\infty)| < \infty \qquad (5.97)$$

The Green's function $g_n(x,s)$ of the homogeneous problem associated with that posed by eqns (5.96) and (5.97) is found in the form

$$g_n(x,s) = \frac{1}{2\sqrt{\nu^2 + k^2}} \begin{cases} \exp(\sqrt{\nu^2 + k^2}(x-s)), & \text{for } x \le s \\ \exp(\sqrt{\nu^2 + k^2}(s-x)), & \text{for } s \le x \end{cases}$$

5.4. KLEIN-GORDON EQUATION

and the series representation of the Green's function of the Dirichlet problem for the homogeneous Klein-Gordon equation stated over the strip Ω_s, resulting from our technique, is obtained as

$$G(x,y;s,t) = \frac{2}{b} \sum_{n=1}^{\infty} g_n(x,s) \sin(\nu y) \sin(\nu t) \qquad (5.98)$$

The non-uniform convergence of this series is caused by the logarithmic singularity of the Green's function. This dramatically restricts direct computational implementations that imply the truncation of the series. The situation, however, can be radically improved. In doing so, we rewrite that branch of the coefficient $g_n(x,s)$, which is valid for $x \leq s$ (the other branch can readily be used instead), in the form

$$g_n(x,s) = \left[\frac{\exp(h(x-s))}{2h} - \frac{\exp(\nu(x-s))}{2\nu} \right] + \frac{\exp(\nu(x-s))}{2\nu}$$

where for the notational compactness, we denote $h = \sqrt{\nu^2 + k^2}$. This transforms the expression in eqn (5.98) into

$$G(x,y;s,t) = \frac{1}{b} \sum_{n=1}^{\infty} \left[\frac{\exp(h(x-s))}{h} - \frac{\exp(\nu(x-s))}{\nu} \right] \sin(\nu y) \sin(\nu t)$$

$$+ \frac{1}{b} \sum_{n=1}^{\infty} \frac{\exp(\nu(x-s))}{\nu} \sin(\nu y) \sin(\nu t)$$

The second of the above two series allows a complete summation with the aid of the summing formula presented by eqn (5.17). That is

$$G(x,y;s,t) = \frac{1}{2\pi} \ln \frac{E(z-\bar{\zeta})}{E(z-\zeta)} - \frac{1}{b} \sum_{n=1}^{\infty} H_n(x,s) \sin(\nu y) \sin(\nu t) \qquad (5.99)$$

where by E we denote a real-valued function of complex variables earlier introduced in the opening part of Section 5.1 (see eqn (5.6)) and the coefficient $H_n(x,s)$ of the above series is given by

$$H_n(x,s) = \frac{1}{\nu h}[h \exp(\nu(x-s)) - \nu \exp(h(x-s))], \quad \text{for } x \leq s \qquad (5.100)$$

Notice that the expression of $H_n(x,s)$ for $x \geq s$ can be obtained from that above by the interchanging of x with s. Notice also that $H_n(x,s)$ vanishes when $k=0$ and the representation of eqn (5.99) reduces in this case to the well-known closed form

$$G(x,y;s,t) = \frac{1}{2\pi} \ln \frac{E(z-\bar{\zeta})}{E(z-\zeta)}$$

of the Green's function of the Dirichlet problem for Laplace's equation set up on the strip Ω_s (see, for example, [63, 68, 70]). Evidently, this provides an indirect verification of the suggested technique.

Let us now turn to the analysis of the trigonometric series of eqn (5.99). To explore its convergence, we analyze the modulus of its remainder

$$|R_N(x,y;s,t)| =$$

$$\left| \sum_{n=N+1}^{\infty} \frac{1}{\nu h} [h\exp(\nu(x-s)) - \nu\exp(h(x-s))] \sin(\nu y) \sin(\nu t) \right|$$

Upon ignoring the trigonometric factors, one obtains

$$|R_N(x,y;s,t)| \leq \sum_{n=N+1}^{\infty} \left| \frac{1}{\nu h} [h\exp(\nu(x-s)) - \nu\exp(h(x-s))] \right|$$

Clearly, since $k^2 < 0$ and $x \leq s$, the parameter h exceeds the value of ν and, consequently, the first term in the brackets always exceeds the second one. That is why, in the right-hand side of the above inequality, the modulus sign can be omitted. This yields

$$|R_N(x,y;s,t)| \leq \sum_{n=N+1}^{\infty} \frac{1}{\nu h} [h\exp(\nu(x-s)) - \nu\exp(h(x-s))]$$

Accounting for the fact that $\nu < h$, the above estimate is rewritten as

$$|R_N(x,y;s,t)| < \sum_{n=N+1}^{\infty} \frac{1}{\nu^2} [h\exp(\nu(x-s)) - \nu\exp(h(x-s))] \quad (5.101)$$

This estimate can be further developed by using the obvious relation

$$h = \sqrt{\nu^2 + k^2} < \nu + k$$

that may be thought of as the *triangle inequality* for the values of h, ν, and k. One rewrites the inequality derived in eqn (5.101) by increasing the first term in the brackets (based on the fact that $\nu + k > h$) and decreasing the second term (because $x \leq s$ and $\nu + k > h$). That is

$$|R_N(x,y;s,t)| < \sum_{n=N+1}^{\infty} \frac{1}{\nu^2} [(\nu+k)\exp(\nu(x-s)) - \nu\exp((\nu+k)(x-s))]$$

The elementary algebra yields

$$|R_N(x,y;s,t)|$$

5.4. KLEIN-GORDON EQUATION

$$< \sum_{n=N+1}^{\infty} \frac{1}{\nu^2} \{[1-\exp(k(x-s))]\nu \exp(\nu(x-s)) + k\exp(\nu(x-s))\}$$

$$= [1-\exp(k(x-s))] \sum_{n=N+1}^{\infty} \frac{1}{\nu} \exp(\nu(x-s)) + k \sum_{n=N+1}^{\infty} \frac{1}{\nu^2} \exp(\nu(x-s))$$

$$\leq [1-\exp(k(x-s))] \sum_{n=N+1}^{\infty} \frac{1}{\nu} \exp(\nu(x-s)) + k \sum_{n=N+1}^{\infty} \frac{1}{\nu^2}$$

Thus, by recalling the value of ν as expressed in terms of n and rewriting the remainders of the above two series in an explicit form, the above inequality is rewritten as

$$|R_N(x,y;s,t)| < \frac{b}{\pi}[1-\exp(k(x-s))]\left[\sum_{n=1}^{\infty} \frac{1}{n} \exp\left(\frac{n\pi}{b}(x-s)\right)\right.$$

$$\left. - \sum_{n=1}^{N} \frac{1}{n} \exp\left(\frac{n\pi}{b}(x-s)\right)\right] + k\frac{b^2}{\pi^2}\left[\sum_{n=1}^{\infty} \frac{1}{n^2} - \sum_{n=1}^{N} \frac{1}{n^2}\right]$$

From the development in Section 5.1, one recalls the following summation formulas for the above two infinite series

$$\sum_{n=1}^{\infty} \frac{1}{n} \exp\left(\frac{n\pi}{b}(x-s)\right) = -\ln\left(1-\exp(\frac{\pi}{b}(x-s))\right)$$

and

$$\sum_{n=1}^{\infty} \frac{1}{n^2} = \frac{\pi^2}{6}$$

This finally yields

$$|R_N(x,y;s,t)| < \frac{b}{\pi}\left\{[1-\exp(k(x-s))]\left[\ln\left(1-\exp(\frac{\pi}{b}(x-s))\right)\right.\right.$$

$$\left.\left. - \sum_{n=1}^{N} \frac{1}{n} \exp\left(\frac{n\pi}{b}(x-s)\right)\right] + k\frac{b}{\pi}\left[\frac{\pi^2}{6} - \sum_{n=1}^{N} \frac{1}{n^2}\right]\right\} \quad (5.102)$$

Hence, the series of eqn (5.99) converges uniformly and its accurate evaluation can be accomplished by the direct truncation. Based on the estimate derived in eqn (5.102), one can appropriately compute the value of the truncating parameter N required for attaining a desired level of accuracy.

For the next example, we set up the mixed boundary value problem

$$u(x,0) = \frac{\partial u(x,b)}{\partial y} = 0$$

on the infinite strip Ω_s.

Proceeding in compliance with the suggested technique, the Green's function of Klein-Gordon equation for the above formulation is obtained in a quite compact form as

$$G(x,y;s,t) = \frac{1}{2\pi} \ln \frac{E_1(z-\zeta)E_2(z-\overline{\zeta})}{E_2(z-\zeta)E_1(z-\overline{\zeta})}$$

$$-\frac{1}{b}\sum_{n=1}^{\infty}\left[\frac{\exp(\nu(x-s))}{\nu} - \frac{\exp(h(x-s))}{h}\right]\sin(\nu y)\sin(\nu t) \qquad (5.103)$$

where $\nu = (2n-1)\pi/2b$ and the real-valued functions E_1 and E_2 of complex variables are defined (see Section 5.1) as

$$E_1(p) = \left|\exp\left(\frac{\pi p}{2b}\right)+1\right|, \quad E_2(p) = \left|\exp\left(\frac{\pi p}{2b}\right)-1\right|$$

Notice that the coefficient of the above series is written for $x \leq s$, whereas its expression for $x \geq s$ can be obtained by the interchanging of x with s.

The series occurring in eqn (5.103) is uniformly convergent in Ω_s. This can be shown in the same exact fashion as for the Dirichlet problem considered earlier, and the estimate of its remainder can readily be obtained.

In the next two examples, we consider different boundary value problems for Klein-Gordon equation stated on the semi-strip $\Omega_{ss} = \{(x,y): x>0, 0<y<b\}$. As the first of them, we set up the Dirichlet problem

$$u(0,y) = u(x,0) = u(x,b) = 0$$

whose Green's function, obtained by the suggested technique, is given by

$$G(x,y;s,t) = \frac{1}{2\pi} \ln \frac{E(z-\overline{\zeta})E(z+\overline{\zeta})}{E(z-\zeta)E(z+\zeta)}$$

$$-\frac{2}{b}\sum_{n=1}^{\infty}\left(\frac{\sinh(\nu x)}{\nu \exp(\nu s)} - \frac{\sinh(hx)}{h \exp(hs)}\right)\sin(\nu y)\sin(\nu t) \qquad (5.104)$$

The coefficient of the above series implies that $x \leq s$ whereas its expression for $x \geq s$ is obtained by interchanging of x with s. The series converges uniformly and the estimation of its remainder can be carried out by using the approach similar to that developed earlier in this section.

As our second example for the semi-strip, we set up the mixed boundary value problem written as

$$u(x,0) = \frac{\partial u(x,b)}{\partial y} = \frac{\partial u(0,y)}{\partial x} - \beta u(0,y) = 0, \quad \beta \geq 0$$

5.4. KLEIN-GORDON EQUATION

in which different kinds of boundary conditions are imposed on different parts of the contour of Ω_{ss}.

The Green's function for this formulation is finally obtained in the form

$$G(x,y;s,t) = \frac{1}{2\pi} \ln \frac{E_1(z-\zeta)E_2(z+\zeta)E_1(z+\overline{\zeta})E_2(z-\overline{\zeta})}{E_1(z+\zeta)E_2(z-\zeta)E_1(z-\overline{\zeta})E_2(z+\overline{\zeta})}$$

$$-\frac{2}{b}\sum_{n=1}^{\infty}\left(\frac{\cosh(\nu x)}{\nu \exp(\nu s)} - \frac{\cosh(hx)}{h \exp(hs)}\right)\sin(\nu y)\sin(\nu t)$$

$$-\frac{2\beta}{b}\sum_{n=1}^{\infty}\frac{\exp(-h(x+s))}{h(h+\beta)}\sin(\nu y)\sin(\nu t), \quad x \le s \qquad (5.105)$$

where

$$h = \sqrt{\nu^2 + k^2}, \quad \nu = \frac{\pi}{2b}(2n-1)$$

The coefficient of the first of the above two series implies that $x \le s$, whereas its expression for $x \ge s$ can be obtained from that above by the interchanging of x with s. Clearly, the coefficient of the second of those series remains invariant if x is interchanged with s.

Notice that both of the series in eqn (5.105) are uniformly convergent, and one can readily estimate the remainder of the first of those series by using the approach that has already been implemented while the series of eqn (5.99) was analyzed. To estimate the remainder $R_N(x,y;s,t)$ of the second of those series, we proceed as follows

$$|R_N(x,y;s,t)| = \left|\sum_{n=N+1}^{\infty}\frac{\exp(-h(x+s))}{h(h+\beta)}\sin(\nu y)\sin(\nu t)\right|$$

$$\le \sum_{n=N+1}^{\infty}\left|\frac{\exp(-h(x+s))}{h(h+\beta)}\right| = \sum_{n=N+1}^{\infty}\frac{\exp(-h(x+s))}{h(h+\beta)}$$

$$< \sum_{n=N+1}^{\infty}\frac{\exp(-\nu(x+s))}{\nu(\nu+\beta)}$$

The last step of this development can be clearly explained. Indeed, we just replaced h with ν. But since h is greater that ν, such a replacement simply reduces the numerator and at the same time increases the denominator of the fraction.

Recalling the earlier introduced expression for ν in terms of n

$$\nu = \frac{\pi}{2b}(2n-1)$$

we explicitly rewrite the last estimate in terms of n

$$|R_N(x,y;s,t)| < \sum_{n=N+1}^{\infty} \exp\left(-\frac{\pi(2n-1)}{2b}(x+s)\right) \frac{4b^2}{\pi^2(2n-1)[(2n-1)+\beta_0]}$$

$$= \frac{4b^2}{\pi^2} \sum_{n=N+1}^{\infty} \left[\exp\left(-\frac{\pi}{2b}(x+s)\right)\right]^{2n-1} \frac{1}{(2n-1)[(2n-1)+\beta_0]}$$

where $\beta_0 = 2b\beta/\pi$. For $\beta_0 \geq 1$, one obtains

$$|R_N(x,y;s,t)| < \frac{4b^2}{\pi^2 v} \sum_{n=N+1}^{\infty} \frac{v^{2n}}{(2n-1)2n} \qquad (5.106)$$

where, for the notational convenience, we introduce

$$v = \exp\left(-\frac{\pi}{2b}(x+s)\right)$$

To proceed further with the estimate obtained in eqn (5.106), we will derive the summation formula for the series

$$\sum_{n=1}^{\infty} \frac{v^{2n}}{(2n-1)2n}$$

whose remainder occurs in eqn (5.106). In doing so, let us take advantage of the known [29] relation

$$\sum_{n=1}^{\infty} \frac{v^{2n-1}}{2n-1} = \frac{1}{2}\ln\left(\frac{1+v}{1-v}\right) \qquad (5.107)$$

which is valid for $v^2 < 1$. Since the above series uniformly converges for $v \in (0,1)$, one is allowed to integrate it in a term-by-term fashion. This yields

$$\int_0^v \sum_{n=1}^{\infty} \frac{t^{2n-1}}{2n-1} dt = \sum_{n=1}^{\infty} \frac{1}{2n-1} \int_0^v t^{2n-1} dt = \sum_{n=1}^{\infty} \frac{v^{2n}}{(2n-1)2n}$$

On the other hand, upon integrating the right-hand side of eqn (5.107), one obtains

$$\int_0^v \frac{1}{2}\ln\left(\frac{1+t}{1-t}\right) dt = \frac{1}{2}[(1+v)\ln(1+v) + (1-v)\ln(1-v)]$$

Hence, we have finally derived the summation formula

$$\sum_{n=1}^{\infty} \frac{v^{2n}}{(2n-1)2n} = \frac{1}{2}[(1+v)\ln(1+v) + (1-v)\ln(1-v)] \qquad (5.108)$$

5.4. KLEIN-GORDON EQUATION

which is important in the fulfillment of the estimation which is under way.

We now return to the inequality of eqn (5.106) and rewrite it as

$$|R_N(x,y;s,t)| < \frac{4b^2}{\pi^2 v} \sum_{n=N+1}^{\infty} \frac{v^{2n}}{(2n-1)2n}$$

$$= \frac{4b^2}{\pi^2 v} \left(\sum_{n=1}^{\infty} \frac{v^{2n}}{(2n-1)2n} - \sum_{n=1}^{N} \frac{v^{2n}}{(2n-1)2n} \right)$$

This can be completed in view of the relation that has just been derived in eqn (5.108). That is

$$|R_N(x,y;s,t)| <$$

$$\frac{4b^2}{\pi^2 v} \left\{ \frac{1}{2}[(1+v)\ln(1+v) + (1-v)\ln(1-v)] - \sum_{n=1}^{N} \frac{v^{2n}}{(2n-1)2n} \right\}$$

where the variable v was introduced immediately after eqn (5.106).

Thus, from the estimate just obtained, it follows that the representation of the Green's function under consideration, presented in eqn (5.105), can be accurately computed by the direct truncation of the series in it.

As can easily be seen, the estimate obtained above is not uniform. Indeed, it depends on the positions of the field and source points for which the estimate is applied. This makes it possible, therefore, to use different truncations of the series derived in eqn (5.105) on different subzones of the region Ω_{ss} in order to keep a certain fixed level of the desired accuracy for the entire region.

The suggested technique provides a productive tool for constructing Green's functions for boundary value problems for Klein-Gordon equation on regions of various configurations.

We now focus on the Dirichlet problem set up on the rectangle $\Omega_r = \{(x,y): 0 < x < a, 0 < y < b\}$. Tracing out the suggested technique, one obtains the series representation of this Green's function

$$G(x,y;s,t) = \frac{2}{b} \sum_{n=1}^{\infty} g_n(x,s) \sin(\nu y) \sin(\nu t), \quad \nu = \frac{n\pi}{b} \qquad (5.109)$$

whose coefficient $g_n(x,s)$, for $x \leq s$, appears in the form

$$g_n(x,s) = \frac{1}{4h \sinh(ha)} [\exp(h(x-s-a)) - \exp(h(x+s-a))$$

$$+ \exp(-h(x-s-a)) - \exp(-h(x+s-a))]$$

where again $h = \sqrt{\nu^2 + k^2}$.

It can be shown that the above expression for $g_n(x,s)$ reduces to

$$g_n(x,s) = -\frac{\sinh(hx)\sinh(hs)}{h\exp(ha)\sinh(ha)}$$

$$+ \frac{1}{2h}[\exp(-h(x-s)) - \exp(-h(x+s))]$$

Upon adding and subtracting the function

$$\frac{1}{2\nu}[\exp(\nu(x+s)) - \exp(-\nu(x-s))]$$

to the bracket term of $g_n(x,s)$ and performing then a trivial algebra, one obtains the final representation

$$G(x,y;s,t) = \frac{1}{2\pi}\ln\frac{E(z-\overline{\zeta})E(z+\overline{\zeta})}{E(z-\zeta)E(z+\zeta)}$$

$$-\frac{2}{b}\sum_{n=1}^{\infty}\left(\frac{\sinh(\nu x)}{\nu\exp(\nu s)} - \frac{\sinh(hx)}{h\exp(hs)}\right)\sin(\nu y)\sin(\nu t)$$

$$-\frac{2}{b}\sum_{n=1}^{\infty}\frac{\sinh(hx)\sinh(hs)}{h\exp(ha)\sinh(ha)}\sin(\nu y)\sin(\nu t), \quad x \leq s \tag{5.110}$$

of the Green's function for the Dirichlet problem set up for Klein-Gordon equation on the rectangle Ω_r.

Notice that, for $x \geq s$, the first of the two series in eqn (5.110) can be obtained from that above by the interchanging of x with s, whereas the second of those series is invariant to the interchanging of x with s.

Notice also that, as a approaches infinity, the rectangle Ω_r transforms into the semi-strip Ω_{ss}. On the other hand, the representation of the Green's function for the semi-strip Ω_{ss}, that was earlier obtained in eqn (5.104), immediately follows from that of eqn (5.110) when a is taken to infinity, because the first two terms in eqn (5.110) do not depend on a, whereas the limit of the last term, as a goes to infinity, is equal to zero

$$\lim_{a\to\infty}\sum_{n=1}^{\infty}\frac{\sinh(hx)\sinh(hs)}{h\exp(ha)\sinh(ha)}\sin(\nu y)\sin(\nu t) = 0$$

Speaking of computational implementations based on the representation in eqn (5.110), it can be seen that the first series in it uniformly converges and its remainder can be estimated by using the approach similar to that developed earlier in this section (see the estimate of the series in eqn (5.99)). The second series of eqn (5.110), however, is not uniformly convergent; it

5.4. KLEIN-GORDON EQUATION

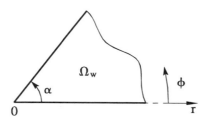

Figure 5.6   Infinite wedge

possesses the logarithmic singularity when both the observation and the source points are taken to the edge $x=a$ of the rectangle. This singularity can be successfully decomposed from the regular part by following the approach used earlier in Section 5.1, when we treated the Dirichlet problem for Laplace's equation.

The last set of examples for Klein-Gordon equation that will be presented herein, concern boundary value problems set up in polar coordinates. We first discuss the Dirichlet problem

$$u(r,0) = u(r,\alpha) = 0, \quad |u(\infty,\phi)| < \infty$$

set up over the infinite wedge $\Omega_w = \{(r,\phi): 0 < r < \infty, 0 < \phi < \alpha\}$ of angle α as shown in Figure 5.6.

Omitting the detailed description of our algorithm in this case, we present just the final series representation of the Green's function, which is found in the form

$$G(r,\phi;s,t) = \frac{2}{\alpha} \sum_{n=1}^{\infty} I_\nu(kr) K_\nu(ks) \sin(\nu\phi) \sin(\nu t)$$

where $r \leq s$ and $\nu = n\pi/\alpha$. $I_\nu(kr)$ and $K_\nu(ks)$ represent the modified Bessel functions of the first and the second kind, respectively, of order ν (see, for example, [1, 17, 63]).

Notice that for $x \geq s$, the coefficient $I_\nu(kr)K_\nu(ks)$ in $G(r,\phi;s,t)$ must be replaced with $I_\nu(ks)K_\nu(kr)$.

As a particular case from the above expansion, one obtains the Green's function for a half-plane $\Omega_{hp} = \{(r,\phi): 0 < r < \infty, 0 < \phi < \pi\}$ in the form

$$G(r,\phi;s,t) = \frac{2}{\pi} \sum_{n=1}^{\infty} I_n(kr) K_n(ks) \sin(n\phi) \sin(nt)$$

$$= \frac{1}{\pi} \sum_{n=1}^{\infty} I_n(kr) K_n(ks) [\cos(n(\phi-t)) - \cos(n(\phi+t))]$$

This series is summable (see, for example, [1, 17, 63]) providing

$$G(r,\phi;s,t) = \frac{1}{2\pi}\left[K_0(k|z-\zeta|) - K_0(k|z-\overline{\zeta}|)\right] \qquad (5.111)$$

where

$$z = r(\cos(\phi) + i\sin(\phi)), \quad \zeta = s(\cos(t) + i\sin(t))$$

Notice that this expression of the Green's function for a half-plane can be obtained from the fundamental solution $K_0(k|z-\zeta|)$ of Klein-Gordon equation by the reflection method.

For the next example, we present the Green's function

$$G(r,\phi;s,t) = \frac{1}{2\pi}\left[K_0(k|z-\zeta|) - K_0(k|z-\overline{\zeta}|)\right]$$

$$-\frac{2}{\pi}\sum_{n=1}^{\infty}\frac{I_n(kr)I_n(ks)K_n(kR)}{I_n(kR)}\sin(n\phi)\sin(nt) \qquad (5.112)$$

of the Dirichlet problem

$$u(R,\phi) = u(r,0) = u(r,\pi) = 0$$

set up on the semi-circle $\Omega_{sc} = \{(r,\phi): 0 < r < R,\ 0 < \phi < \pi\}$ of radius R.

Clearly, the Green's function for the half-plane Ω_{hp}, presented in eqn (5.111), follows from that for the semi-circle shown in eqn (5.112), as R approaches infinity. Indeed, the series part of the latter vanishes as $R \to \infty$, because the modified Bessel functions of the second kind vanish at infinity, that is

$$\lim_{R \to \infty} K_n(kR) = 0$$

As our last example in this section, we consider the following mixed boundary value problem

$$\frac{\partial u(R,\phi)}{\partial r} + \beta u(R,\phi) = u(r,0) = u(r,\pi) = 0, \quad \beta \geq 0$$

set up throughout the semi-circle Ω_{sc} of radius R. The Green's function for this problem can also be found by means of our technique. After splitting off the singular component, it is expressed as

$$G(r,\phi;s,t) = \frac{1}{2\pi}\left[K_0(k|z-\zeta|) - K_0(k|z-\overline{\zeta}|)\right]$$

$$-\frac{2}{\pi}\sum_{n=1}^{\infty}\frac{I_n(kr)I_n(ks)[K_n'(kR) + \beta K_n(kR)]}{I_n'(kR) + \beta I_n(kR)}\sin(n\phi)\sin(nt) \qquad (5.113)$$

5.5. REVIEW EXERCISES

Two other Green's functions can be obtained from this representation as particular cases. First, it reduces to the expression of eqn (5.112) (the Green's function of the Dirichlet problem for the semi-circle), when the coefficient β in the boundary condition approaches infinity. Second, as β goes to zero, the representation in eqn (5.113) reduces to the Green's function

$$G(r,\phi;s,t) = \frac{1}{2\pi}\left[K_0(k|z-\zeta|) - K_0(k|z-\overline{\zeta}|)\right]$$

$$-\frac{2}{\pi}\sum_{n=1}^{\infty}\frac{I_n(kr)I_n(ks)K'_n(kR)}{I'_n(kR)}\sin(n\phi)\sin(nt) \quad (5.114)$$

for another mixed boundary value problem

$$\frac{\partial u(R,\phi)}{\partial r} = u(r,0) = u(r,\pi) = 0$$

on the semi-circle Ω_{sc}.

A variety of boundary value problems for Laplace's and Klein-Gordon equations, for which Green's functions can be constructed with the aid of the suggested technique, is not limited to those cases considered in this chapter. Many other Green's functions are easily obtainable. The reader will find compact representations for some of them while solving the review exercises. Exercises 5.1 through 5.5 are associated with Laplace's equation, while exercises 5.6 through 5.9 relate to Klein-Gordon equation.

5.5 Review Exercises

5.1 Construct the Green's functions of Laplace's equation on an infinite strip $\Omega_s = \{(x,y): -\infty < x < \infty, 0 < y < b\}$ for the following boundary value problems:

(a) $u(x,0) = 0$, $u(x,b) = 0$;
(b) $u(x,0) = 0$, $\partial u(x,b)/\partial y = 0$.

5.2 Construct the Green's functions of Laplace's equation on a semi-strip $\Omega_{ss} = \{(x,y): 0 < x < \infty, 0 < y < b\}$ for the following boundary value problems:

(a) $u(0,y) = u(x,0) = \partial u(x,b)/\partial y = 0$;
(b) $\partial u(0,y)\partial x = u(x,0) = \partial u(x,b)/\partial y = 0$;
(c) $u(0,y) = \partial u(x,0)/\partial y = \partial u(x,b)/\partial y = 0$;

(d) $\partial u(0,y)/\partial x - \beta u(0,y) = \partial u(x,0)/\partial y = \partial u(x,b)/\partial y = 0$, $\beta \geq 0$.

5.3 Construct the Green's functions of Laplace's equation on a rectangle $\Omega_r = \{(x,y): 0 < x < a, 0 < y < b\}$ for the following boundary value problems:

(a) $u(0,y) = u(x,0) = u(x,b) = \partial u(a,y)/\partial x = 0$;

(b) $u(0,y) = u(x,0) = u(x,b) = \partial u(a,y)/\partial x + \beta u(a,y) = 0$, $\beta \geq 0$;

(c) $u(0,y) = u(x,0) = \partial u(x,b)/\partial y = \partial u(a,y)/\partial x + \beta u(a,y) = 0$, $\beta \geq 0$;

(d) $\partial u(0,y)/\partial x = u(x,0) = u(x,b) = \partial u(a,y)/\partial x + \beta u(a,y) = 0$, $\beta \geq 0$;

5.4 Construct the Green's functions of Laplace's equation on an infinite wedge $\Omega_w = \{(r,\phi): 0 < r < \infty, 0 < \phi < \alpha; \alpha \in (0, 2\pi)\}$ for the following boundary value problems:

(a) $u(r,0) = 0$, $u(r,\alpha) = 0$;

(b) $u(r,0) = 0$, $\partial u(r,\alpha)/\partial \phi = 0$,

and obtain, as particular cases (for $\alpha = \pi$), the Green's functions for the corresponding boundary value problems on a half-plane.

5.5 Construct the Green's functions of Laplace's equation on a circular sector $\Omega_{cs} = \{(r,\phi): 0 < r < R, 0 < \phi < \alpha; \alpha \in (0, 2\pi)\}$ for the following boundary value problems:

(a) $u(r,0) = u(r,\alpha) = \partial u(R,\phi)/\partial r + \beta u(R,\phi) = 0$, $\beta \geq 0$;

(b) $u(r,0) = \partial u(r,\alpha)/\partial \phi = \partial u(R,\phi)/\partial r = 0$;

(c) $u(r,0) = \partial u(r,\alpha)/\partial \phi = u(R,\phi) = 0$.

5.6 Construct the Green's functions of Klein-Gordon equation on a semi-strip $\Omega_{ss} = \{(x,y): 0 < x < \infty, 0 < y < b\}$ for the following boundary value problems:

(a) $\partial u(0,y)/\partial x = u(x,0) = u(x,b) = 0$;

(b) $\partial u(0,y)/\partial x - \beta u(0,y) = u(x,0) = u(x,b) = 0$.

5.7 Construct the Green's function of Klein-Gordon equation on an infinite wedge $\Omega_w = \{(r,\phi): 0 < r < \infty, 0 < \phi < \alpha\}$ for the boundary value problem

$$u(r,0) = 0, \quad \partial u(r,\alpha)/\partial \phi = 0$$

and obtain, as a particular case (for $\alpha = \pi$), the Green's function for the corresponding mixed boundary value problem on a half-plane.

5.5. REVIEW EXERCISES

5.8 Construct the Green's functions of Klein-Gordon equation on a semi-circle $\Omega_{sc} = \{(r, \phi)\colon 0 < r < R,\ 0 < \phi < \pi\}$ of radius R for the following boundary value problems:

(a) $u(r, 0) = \partial u(r, \pi)/\partial \phi = u(R, \phi) = 0$;

(b) $u(r, 0) = \partial u(r, \pi)/\partial \phi = \partial u(R, \phi)/\partial r = 0$;

(c) $u(r, 0) = \partial u(r, \pi)/\partial \phi = \partial u(R, \phi)/\partial r + \beta u(R, \phi) = 0,\quad \beta \geq 0$.

5.9 Construct the Green's functions of Klein-Gordon equation on a disk $\Omega_c = \{(r, \phi)\colon 0 < r < R,\ 0 \leq \phi < 2\pi\}$ for the following boundary value problems:

(a) $u(R, \phi) = 0$;

(b) $\partial u(R, \phi)/\partial r + \beta u(R, \phi) = 0,\quad \beta > 0$.

Chapter 6

Problems of Solid Mechanics

In the previous chapter, we were concerned with influence functions for such phenomena of continuum mechanics that reduce to boundary value problems for either Laplace's or Klein-Gordon equations (potential and related fields). It was shown that there exists a problem class for these equations, for which influence functions can readily be constructed with the aid of the technique proposed in [24, 42]. The material to be discussed in the present chapter is conceptually similar to that of Chapter 5. The author's technique will be implemented to the construction of influence functions and matrices for a variety of problems in mechanics of solids, which are modeled with either single partial differential equations of higher order or with systems of such equations.

So far in this book, we followed the rule to make the style of presentation as self-contained as possible, to reduce to a minimum the necessity for references to other sources. Indeed, the text of Chapter 5 contains rather detailed description of each step of the development. This makes the material accessible to the reader with the standard background in undegraduate calculus, ODEs, and PDEs.

The presentation in the present chapter will not always be that detailed, mainly because the procedures, which are utilized here for the construction of influence functions and matrices for problems of solid mechsanics, are actually fairly similar to those of Chapter 5, where the reader can find, if necessary, answers to arising questions.

In this chapter, we will consider some traditional formulations from the classical Poisson-Kirchhoff plate theory that reduces the bending of thin plates to boundary value problems formulated for the biharmonic equation. We will also extend the influence function method to the mathematical formulation based on the so-called Reissner [65] plate theory. Some of the displacement formulations for isotropic and orthotropic media from the plane problem in theory of elasticity will also be discussed. In addition, in the last section of this chapter, we will show that the elastic equilibrium of thin shells of revolution can also be considered in the scope of the suggested technique.

It is worth reminding that the influence function of a concentrated unit force for a problem in solid mechanics can always be identified with the Green's matrix of a boundary value problem for a system of differential equations that models the phenomenon under consideration.

For the purpose of the present study, the Green's matrix for a system of elliptic PDEs is defined in the following manner. Let Ω represent a simply connected region in two-dimensional Euclidean space and let Γ denote its piecewise smooth contour. Consider the boundary value problem

$$L[U(P)] = -F(P), \quad P \in \Omega \qquad (6.1)$$

$$M[U(P)] = 0, \quad P \in \Gamma \qquad (6.2)$$

where L and M represent the matrix-operators of the governing system of PDEs and of the boundary conditions, respectively. $U(P)$ and $F(P)$ are the solution vector and a vector-function of the right-hand side, respectively.

We assume that the problem posed by eqns (6.1) and (6.2) has a unique solution. That is, the corresponding homogeneous problem ($F(P) \equiv 0$) has only the trivial solution.

DEFINITION: If, for any allowable right-hand side vector $F(P)$, the solution vector $U(P)$ is expressible in the form

$$U(P) = \iint_\Omega G(P,Q) F(Q) d\Omega(Q) \qquad (6.3)$$

then the kernel-matrix $G(P,Q)$ of this representation is said to be the *influence matrix* of the homogeneous problem corresponding to that posed by eqns (6.1) and (6.2).

Notice that the term "allowable", in regard to $F(P)$, implies that the integral of $F(P)$ over Ω is bounded

$$\left| \iint_\Omega F(P) d\Omega(P) \right| < \infty$$

That is, $F(P)$ is assumed to be integrable over Ω. This reflects an obvious physical limitation on the statement in eqns (6.1) and (6.2). Namely, the amount of energy, provided with the load $F(P)$ to the solid under consideration, should be a finite quantity.

Properties of the influence matrix $G(P,Q)$ are determined individually, in every particular case, by the operators L and M of the governing system posed by eqns (6.1) and (6.2). Whereas, the structure of $G(P,Q)$ is standard for elliptic systems and is viewed as

$$G(P,Q) = S(P,Q) + R(P,Q)$$

where $S(P,Q)$ represents the *fundamental solution matrix* of the governing system (the singular component of $G(P,Q)$) and $R(P,Q)$ is the regular component of $G(P,Q)$ and represents such a solution of the homogeneous system $L[R(P,Q)] = 0$ that, for every fixed location of the source point $Q \in \Omega$, allows $G(P,Q)$ to satisfy the boundary conditions in eqn (6.2).

6.1 Poisson-Kirchhoff's plates

In this section, we will face a variety of problems on bending of thin plates. Those are treated in the scope of the classical Poisson-Kirchhoff theory and are formulated as boundary value problems for the biharmonic equation

$$D \Delta \Delta w(P) = q(P)$$

where Δ represents Laplace's operator written in terms of the coordinates of P, while $w(P)$ and $q(P)$ represent the lateral deflection of the plate's middle plane and the transverse distributed load applied to the plate, respectively. The coefficient D, which is referred to as the plate's flexural rigidity, is defined as

$$D = \frac{Eh^3}{12(1-\sigma^2)}$$

where E and σ represent the elasticity modulus and Poisson ratio, respectively, of the material of which the plate is composed, and h represents the thickness of the plate.

Notice that in this case a single equation governs the problem. Subsequently, we will be dealing here with Green's functions rather then with Green's matrices.

The fundamental solution $S(P,Q)$ of the homogeneous biharmonic equation is known (see, for example, [38, 72]) in the form

$$S(P,Q) = \frac{1}{8\pi}|P-Q|^2 \ln \frac{1}{|P-Q|} \qquad (6.4)$$

A limited number of influence functions for Poisson-Kirchhoff's plates are available in the existing literature. The most complete list of those can be found, probably, in [72]. The most productive traditional methods for obtaining influence functions for plate problems are: (i) expansion in trigonometric series (in plate and shell theory [72], it is often referred to

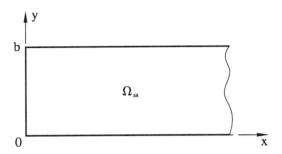

Figure 6.1 Semi-strip-shaped plate

as either Navier's (double series) or Levy's (single series) method), (ii) the reflection or image method [31, 68, 70, 72], and (iii) the complex variables method [68, 72] based on the presentation of a biharmonic function by means of two harmonic functions.

In this section, it will be shown that the version of the eigenfunction expansion method that was discussed in Chapter 5 in conjunction with problems of potential, is also efficient in obtaining influence functions for Poisson-Kirchhoff's plate problems. As the first example, we consider a semi-strip-shaped plate of a uniform thickness, which occupies the region $\Omega_{ss} = \{(x, y)\colon 0 < x < \infty, 0 < y < b\}$, with the edges $y = 0$ and $y = b$ being simply supported while the edge $x = 0$ is clamped (see Figure 6.1). This yields the following boundary value problem

$$\frac{\partial^4 w(x,y)}{\partial x^4} + 2\frac{\partial^4 w(x,y)}{\partial x^2 \partial y^2} + \frac{\partial^4 w(x,y)}{\partial y^4} = \frac{q(x,y)}{D} \equiv -f(x,y) \qquad (6.5)$$

$$w = \left.\frac{\partial^2 w}{\partial y^2}\right|_{y=0,b} = 0, \quad w = \left.\frac{\partial w}{\partial x}\right|_{x=0} = 0 \qquad (6.6)$$

written in Cartesian coordinates. The right-hand term $f(x, y)$ is introduced in eqn (6.5) for the notational convenience in the development that follows. The loading function $q(x, y)$ is assumed to be integrable over Ω_{ss}. In addition to the boundary conditions in eqn (6.6), the components of the stress-strain state of the plate are required to be bounded as x approaches infinity. This ensures a unique solvability of the original problem.

Clearly, the Green's function $G(x, y; s, t)$ of the homogeneous boundary value problem corresponding to that of eqns (6.5) and (6.6) can be interpreted as the influence function of the plate under consideration. It can be associated with the plate's deflection at the point (x, y) due to a transverse unit force concentrated at the point (s, t).

6.1. POISSON-KIRCHHOFF'S PLATES

To obtain the influence function, we seek the solution of the problem posed by eqns (6.5) and (6.6) in the form of the Fourier sine-series

$$w(x,y) = \sum_{n=1}^{\infty} w_n(x)\sin(\nu y), \quad \nu = \frac{n\pi}{b} \quad (6.7)$$

and also expend the loading function $f(x,y)$ in the sine-series

$$f(x,y) = \sum_{n=1}^{\infty} f_n(x)\sin(\nu y) \quad (6.8)$$

The above representation of $w(x,y)$ satisfies the first two boundary conditions in eqn (6.6). Upon substituting expansions from eqns (6.7) and (6.8) into the original formulation and by equating then the corresponding coefficients of the two Fourier series that appeared on the right and on the left of the equal sign, one obtains the following set $(n=1,2,3,...)$ of self-adjoint boundary value problems

$$\frac{d^4 w_n(x)}{dx^4} - 2\nu^2 \frac{d^2 w_n(x)}{dx^2} + \nu^4 w_n(x) = -f_n(x), \quad x \in (0,\infty) \quad (6.9)$$

$$w_n(0) = \frac{dw_n(0)}{dx} = 0, \quad |w_n(\infty)| < \infty, \quad \left|\frac{dw_n(\infty)}{dx}\right| < \infty \quad (6.10)$$

in the coefficients $w_n(x)$ of the series from eqn (6.7).

The Green's function $g_n(x,s)$ associated with the homogeneous $(f_n(x) \equiv 0)$ problem corresponding to that of eqns (6.9) and (6.10), can be found in Section 1.3 of Chapter 1 (see eqn (1.106)). We here rewrite it in terms of our current notations. This yields

$$g_n(x,s) = \frac{1}{4\nu^3} \begin{cases} (1+\nu(x+s)+2\nu^2 xs)e^{-\nu(x+s)} - (1-\nu(x-s))e^{\nu(x-s)} \\ (1+\nu(x+s)+2\nu^2 xs)e^{-\nu(x+s)} - (1-\nu(s-x))e^{\nu(s-x)} \end{cases} \quad (6.11)$$

where the upper and the lower branches are, as usual, defined for $x \leq s$ and $x \geq s$, respectively. The symmetry of the above representation clearly reflects the self-adjointness of the boundary value problem posed by eqns (6.9) and (6.10).

From Theorem 1.3 of Chapter 1, it follows that the solution of the problem posed by eqns (6.9) and (6.10) can be written in terms of $g_n(x,s)$ from eqn (6.11) as the improper integral

$$w_n(x) = \int_0^\infty g_n(x,s) f_n(s) ds$$

which converges, given that $f(x,y)$ is integrable over Ω_{ss} (that is, $f_n(x)$ is integrable over the interval $(0,\infty)$).

In compliance with the series expansion of eqn (6.8), the Fourier coefficients $f_n(s)$ of the right-hand term $f(x,y)$ of eqn (6.5) can be written in terms of the Fourier-Euler formula as follows

$$f_n(s) = \frac{2}{b}\int_0^b f(s,t)\sin(\nu t)dt$$

providing for $w_n(x)$ the integral representation

$$w_n(x) = \frac{2}{b}\int_0^b \int_0^\infty g_n(x,s)\sin(\nu t)f(s,t)dsdt$$

Upon substituting this into eqn (6.7) and assuming a possibility for interplacing the order of operations in it, the solution of the original boundary value problem posed by eqns (6.5) and (6.6) is finally expressed as

$$w(x,y) = \int_0^b \int_0^\infty \left(\frac{2}{b}\sum_{n=1}^\infty g_n(x,s)\sin(\nu y)\sin(\nu t)\right) f(s,t)dsdt$$

From the definition introduced in the prefacing part of this chapter, it follows that the kernel function

$$G(x,y;s,t) = \frac{2}{b}\sum_{n=1}^\infty g_n(x,s)\sin(\nu y)\sin(\nu t) \qquad (6.12)$$

of the last integral represents the influence function of a transverse concentrated unit force for the plate under consideration.

When substituting $g_n(x,s)$ from eqn (6.11) into the series of eqn (6.12), we break down the latter into two parts

$$G(x,y;s,t) = \frac{2}{b}\left[xs\sum_{n=1}^\infty \frac{e^{-\nu(x+s)}}{2\nu}\sin(\nu y)\sin(\nu t)\right.$$

$$\left. + \sum_{n=1}^\infty \tilde{g}_n(x,s)\sin(\nu y)\sin(\nu t)\right]$$

where, for $x \leq s$

$$\tilde{g}_n(x,s) = \frac{1+\nu(x+s)}{4\nu^3}e^{-\nu(x+s)} - \frac{1-\nu(x-s)}{4\nu^3}e^{\nu(x-s)}$$

The expression for $\tilde{g}_n(x,s)$ valid for $x \geq s$ can be obtained from that above by interchanging of x with s.

The rate of convergence of the first of the two series in the above representation for $G(x,y;s,t)$ is very low. This is especially true if both x and

6.1. POISSON-KIRCHHOFF'S PLATES

s are taken close to the edge $x=0$ of the plate. This makes it therefore practically difficult to accurately compute values of the function, which is approximated with that series, by its straightforward truncation. However, the series can be summed up by using the standard summation formula

$$\sum_{n=1}^{\infty} \frac{p^n}{n} \cos(n\alpha) = -\ln\sqrt{1-2p\cos\alpha+p^2} \qquad (6.13)$$

that has been widely employed in the developments of Chapter 5. This finally provides

$$G(x,y;s,t) = \frac{xs}{2\pi} \ln \frac{E(z+\zeta)}{E(z+\bar{\zeta})} + \frac{2}{b}\sum_{n=1}^{\infty} \tilde{g}_n(x,s)\sin(\nu y)\sin(\nu t) \qquad (6.14)$$

where $z=x+iy$ and $\zeta=s+it$ are the field and the source points, respectively. The real-valued function $E(p)$ of a complex variable defined as

$$E(p) = \left|1-\exp\left(\frac{\pi p}{b}\right)\right|$$

was earlier introduced in Chapter 5.

Such a summation radically increases the practicality of the representation in eqn (6.14). Indeed, the remaining series in it converges as fast as $1/n^2$ (see its coefficient $\tilde{g}_n(x,s)$ presented earlier). Hence, one can accurately compute the entire representation in eqn (6.14) by appropriately truncating its series.

Later in this section, we will discuss the convergence of the series in eqn (6.12) in more detail and analyze its differential properties which are of great importance for computational implementations.

By applying the suggested technique, one can construct influence functions for the semi-strip-shaped plate subjected to different boundary conditions. If, for example, the edges $y=0$ and $y=b$ are again simply supported, whereas the edge $x=0$ is free, then the boundary conditions imposed earlier by eqn (6.6) at $x=0$ should be replaced with

$$\left(\frac{\partial^2 w}{\partial x^2} + \sigma \frac{\partial^2 w}{\partial y^2}\right)\bigg|_{x=0} = 0$$

$$\frac{\partial}{\partial x}\left(\frac{\partial^2 w}{\partial x^2} + (2-\sigma)\frac{\partial^2 w}{\partial y^2}\right)\bigg|_{x=0} = 0$$

These conditions assign to zero the bending moment and the shear force along the plate's edge $x=0$ (see [72]).

The influence function of a concentrated unit force for such a plate can also be found in the form of the expansion from eqn (6.12) whose coefficient

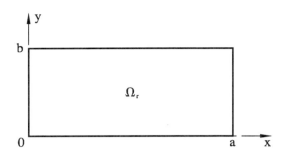

Figure 6.2 Rectangular $a \times b$ plate

$g_n(x, s)$ represents, in this case, the Green's function for the homogeneous boundary value problem

$$\frac{d^4 w_n(x)}{dx^4} - 2\nu^2 \frac{d^2 w_n(x)}{dx^2} + \nu^4 w_n(x) = 0, \quad x \in (0, \infty)$$

$$\frac{d^2 w(x)}{dx^2} - \sigma \nu^2 w_n(0) = \frac{d^3 w_n(0)}{dx^3} - (2-\sigma)\nu^2 \frac{dw_n(0)}{dx} = 0$$

$$|w_n(\infty)| < \infty, \quad \left|\frac{dw_n(\infty)}{dx}\right| < \infty$$

For $x \leq s$, $g_n(x, s)$ is found to be in the form

$$g_n(x, s) = -\frac{1 - \nu(x-s)}{4\nu^3} e^{\nu(x-s)}$$

$$- \frac{(4 + (1+\sigma)^2) + \nu(1-\sigma)^2(x+s+2\nu x s)}{4\nu^3 (1-\sigma)(3+\sigma)} e^{-\nu(x+s)}$$

Due to the self-adjointness of the boundary value problem just indicated, the expression of $g_n(x, s)$, valid for $x \geq s$, can be obtained from that above by interchanging of x with s in it.

The influence function of a concentrated unit force for the semi-strip-shaped plate with all the edges being simply supported is again given by the expansion from eqn (6.12), with $g_n(x, s)$ being, for $x \leq s$, defined as

$$g_n(x, s) = \frac{1 + \nu(x+s)}{4\nu^3} e^{-\nu(x+s)} - \frac{1 - \nu(x-s)}{4\nu^3} e^{\nu(x-s)} \qquad (6.15)$$

The suggested approach allows one to obtain influence functions of a rectangular plate for various combinations of the edge conditions imposed. Let us consider, for example, a plate occupying the region $\Omega_r = \{(x, y) \colon 0 < x < a, 0 < y < b\}$ as shown in Figure 6.2. Let the edge $x = 0$ be simply

6.1. POISSON-KIRCHHOFF'S PLATES

supported and the edge $x=a$ be clamped, while both the edges $y=0$ and $y=b$ be simply supported. That is, the boundary conditions in this case are imposed as

$$w = \frac{\partial^2 w}{\partial y^2}\bigg|_{y=0,b} = 0, \quad w = \frac{\partial^2 w}{\partial x^2}\bigg|_{x=0} = 0, \quad w = \frac{\partial w}{\partial x}\bigg|_{x=a} = 0$$

For such a plate, the influence function of a transverse concentrated unit force also reduces to the series of eqn (6.12), whose coefficient $g_n(x,s)$ represents the Green's function of the following self-adjoint boundary value problem

$$\frac{d^4 w_n(x)}{dx^4} - 2\nu^2 \frac{d^2 w_n(x)}{dx^2} + \nu^4 w_n(x) = 0, \quad x \in (0,a)$$

$$w_n(0) = \frac{d^2 w_n(0)}{dx^2} = 0, \quad w_n(a) = \frac{dw_n(a)}{dx} = 0$$

For $x \leq s$, $g_n(x,s)$ is found in the form

$$g_n(x,s) = \frac{1}{2\nu^3 \Delta^*} \{\nu x \cosh(\nu x)[2\nu(s-a)\cosh(\nu s) - \sinh(\nu(s-2a))$$

$$- \sinh(\nu s)] - \sinh(\nu x)[\nu s \cosh(\nu(s-2a)) - \sinh(\nu(s-2a))$$

$$+ \nu(s-2a)\cosh(\nu s) + (2\nu^2 a(s-a) - 1)\sinh(\nu s)]\}$$

where $\Delta^* = \sinh(2\nu a) - 2\nu a$.

Due to the self-adjointness of the last boundary value problem, the expression for $g_n(x,s)$, valid for $x \geq s$, can be written by interchanging of x with s in that above.

We now return to the convergence issue for the series in eqn (6.12). Notice that if both the observation (x,y) and the source (s,t) points belong to the interior of Ω_{ss}, then the expansion in eqn (6.12) represents a uniformly convergent series. It rapidly converges so that its first order partial derivatives can be taken on the term-by-term basis. That is, the series obtained from that of eqn (6.12) by the term-by-term partial differentiation with respect to either x or y also converges uniformly. Hence, the computational implementations based on either the series in eqn (6.12) itself or on its first order partial derivatives can be accurately carried out by an appropriate truncation of the series.

Let us recall that the influence function of a plate is viewed as a deflection at the observation point due to a transverse concentrated unit force applied at the source point. On the other hand, from the Poisson-Kirchhoff theory (see, for example, [72]), it follows that the bending moments $M_x(x,y)$ and $M_y(x,y)$ in a plate

$$M_x(x,y) = -D\left(\frac{\partial^2 w(x,y)}{\partial x^2} + \sigma\frac{\partial^2 w(x,y)}{\partial y^2}\right)$$

and

$$M_y(x,y) = -D\left(\frac{\partial^2 w(x,y)}{\partial y^2} + \sigma\frac{\partial^2 w(x,y)}{\partial x^2}\right)$$

are expressed in terms of the second order partial derivatives of the deflection function $w(x,y)$. Hence, the second order partial differentiation of the influence function is required to get the bending moments caused by concentrated forces applied to the plate.

The term-by-term second order partial differentiation of the series in eqn (6.12) yields, however, non-uniformly convergent series. These diverge logarithmically when the observation point approaches the source point. This agrees with our expectation. Indeed, for a Poisson-Kirchhoff plate, the bending moments possess logarithmic singularity at a point of application of a transverse concentrated force.

To address the last issue in more detail, let us determine the bending moments M_x and M_y in the semi-strip-shaped plate whose edges are simply supported and which is subjected to a transverse unit force $P_0 = 1$ concentrated at the point (s,t). Upon interpreting the influence function $G(x,y;s,t)$ of the plate as its deflection at the point (x,y) due to P_0, we obtain the following representations

$$M_x(x,y;s,t) = -D\left(\frac{\partial^2 G(x,y;s,t)}{\partial x^2} + \sigma\frac{\partial^2 G(x,y;s,t)}{\partial y^2}\right)$$

and

$$M_y(x,y;s,t) = -D\left(\frac{\partial^2 G(x,y;s,t)}{\partial y^2} + \sigma\frac{\partial^2 G(x,y;s,t)}{\partial x^2}\right)$$

for the bending moments caused by P_0.

After one substitutes the expression for $G(x,y;s,t)$ from eqn (6.12), with the coefficients $g_n(x,s)$ found in eqn (6.15), into the above representations, the latter convert, for $x \le s$, to

$$M_x^+(x,y;s,t) = -\frac{D}{2b}\sum_{n=1}^{\infty}\left\{\left[\frac{1+\sigma}{\nu}+(1-\sigma)(x-s)\right]e^{\nu(x-s)}\right.$$

$$\left.+\left[(1-\sigma)(x+s)-\frac{1+\sigma}{\nu}\right]e^{-\nu(x+s)}\right\}\sin(\nu y)\sin(\nu t)$$

6.1. POISSON-KIRCHHOFF'S PLATES

and

$$M_y^+(x,y;s,t) = -\frac{D}{2b}\sum_{n=1}^{\infty}\left\{\left[\frac{1+\sigma}{\nu}-(1-\sigma)(x-s)\right]e^{\nu(x-s)}\right.$$

$$\left. -\left[(1-\sigma)(x+s)+\frac{1+\sigma}{\nu}\right]e^{-\nu(x+s)}\right\}\sin(\nu y)\sin(\nu t)$$

The expressions of M_x and M_y that hold for $x \geq s$ follow from the above ones by interchanging of x with s.

The series in the above representations for bending moments can readily be summed up with the aid of the standard summation formulae. One of those formulae is that recently used when we summed up the series in eqn (6.12) for the first example in this section (see eqn (6.13)), and the other one (see, for example, [29]) is

$$\sum_{n=0}^{\infty} p^n \cos(n\alpha) = \frac{1-p\cos(\alpha)}{1-2p\cos(\alpha)+p^2} \qquad (6.16)$$

that holds for $p^2 < 1$ and $0 \leq \alpha < 2\pi$.

This finally yields for the bending moments

$$M_x^+(x,y;s,t) = -\frac{D}{4b}\left\{\frac{b(1+\sigma)}{\pi}\ln\frac{E(z-\bar{\zeta})E(z+\bar{\zeta})}{E(z-\zeta)E(z+\zeta)}\right.$$

$$-(1-\sigma)\left[(x-s)\left(\frac{R_0(z-\bar{\zeta})}{E^2(z-\bar{\zeta})}-\frac{R_0(z-\zeta)}{E^2(z-\zeta)}\right)\right.$$

$$\left.\left. -(x+s)\left(\frac{R_0(-(z+\bar{\zeta}))}{E^2(-(z+\bar{\zeta}))}-\frac{R_0(-(z+\zeta))}{E^2(-(z+\zeta))}\right)\right]\right\}$$

and

$$M_y^+(x,y;s,t) = -\frac{D}{4b}\left\{\frac{b(1+\sigma)}{\pi}\ln\frac{E(z-\bar{\zeta})E(z+\bar{\zeta})}{E(z-\zeta)E(z+\zeta)}\right.$$

$$+(1-\sigma)\left[(x-s)\left(\frac{R_0(z-\bar{\zeta})}{E^2(z-\bar{\zeta})}-\frac{R_0(z-\zeta)}{E^2(z-\zeta)}\right)\right.$$

$$\left.\left. -(x+s)\left(\frac{R_0(-(z+\bar{\zeta}))}{E^2(-(z+\bar{\zeta}))}-\frac{R_0(-(z+\zeta))}{E^2(-(z+\zeta))}\right)\right]\right\}$$

where, in addition to the recently recalled function $E(p)$, $R_0(p)$ represents a real-valued function of complex variables introduced as

$$R_0(p) = \Re\left(1-\exp\left(\frac{\pi p}{b}\right)\right)$$

with $\Re$ denoting the real part of a function.

Thus, the bending moments in the semi-strip-shaped simply supported plate, subjected to a concentrated force, are obtained in a closed form.

It is clearly seen that the above representations for the bending moments possess logarithmic singularity if the field point z approaches the source point ζ. In both $M_x(x,y;s,t)$ and $M_y(x,y;s,t)$, the singularity is caused by the logarithmic terms $\ln(E(z-\zeta))$. No other singularities are available. Indeed, the non-logarithmic terms containing $E(z-\zeta)$ in the denominators represent a removable singularity, because of the factor $(x-s)$.

We now turn the reader's attention to a validation example. That is, a clamped circular plate of radius a and of a uniform thickness h. Assume that the plate is comprised of an isotropic homogeneous material (with D representing the plate's flexural rigidity) and occupies the region $\Omega_c = \{(r,\phi): 0 < r < a, \ 0 \leq \phi < 2\pi\}$. For this plate, the influence function

$$G(z,\zeta) = \frac{1}{16\pi D}\left[\frac{1}{a^2}\left(a^2 - |z|^2\right)\left(a^2 - |\zeta|^2\right) - |z-\zeta|^2 \ln \frac{|a^2 - z\overline{\zeta}|^2}{a^2|z-\zeta|^2}\right] \quad (6.17)$$

of a transverse concentrated unit force is available in the existing literature. It can be found, for instance, in [38, 72].

Clearly, $G(z,\zeta)$ represents the deflection of the plate, occurring at the observation point $z = r(\cos(\phi) + i\sin(\phi))$, due to a transverse concentrated unit force applied at the source point $\zeta = \rho(\cos(\psi) + i\sin(\psi))$.

Let us show that the technique, described earlier in this section, provides an alternative way for deriving the representation for $G(z,\zeta)$ shown in eqn (6.17). In doing so, we consider the following boundary value problem for biharmonic equation written in polar coordinates

$$\left(\frac{1}{r}\frac{\partial}{\partial r}\left(r\frac{\partial}{\partial r}\right) + \frac{1}{r^2}\frac{\partial^2}{\partial \phi^2}\right)^2 w(r,\phi) = -f(r,\phi), \quad (r,\phi) \in \Omega_c \quad (6.18)$$

$$w(a,\phi) = 0, \quad \frac{\partial w(a,\phi)}{\partial r} = 0 \quad (6.19)$$

which models the deflection $w(r,\phi)$ caused by the transverse distributed load

$$q(r,\phi) = Df(r,\phi)$$

applied to the plate under consideration.

Based on the 2π-periodicity of the above formulation, we assume, for the deflection function $w(r,\phi)$, the following trigonometric Fourier expansion

$$w(r,\phi) = \frac{1}{2}w_0(r) + \sum_{n=1}^{\infty}(w_n^c(r)\cos(n\phi) + w_n^s(r)\sin(n\phi)) \quad (6.20)$$

6.1. POISSON-KIRCHHOFF'S PLATES

Let also the right-hand term of eqn (6.18) be expressed by the Fourier series in a general form

$$f(r,\phi) = \frac{1}{2}f_0(r) + \sum_{n=1}^{\infty}(f_n^c(r)\cos(n\phi) + f_n^s(r)\sin(n\phi)) \quad (6.21)$$

Upon substituting these representations into eqn (6.18), one obtains, for the coefficients $w_n(r)$ of the expansion in eqn (6.20), the following set ($n = 0, 1, 2, ...$) of ordinary differential equations

$$\left(\frac{d^4}{dr^4} + \frac{2}{r}\frac{d^3}{dr^3} - \frac{1+2n^2}{r^2}\frac{d^2}{dr^2} + \frac{1+2n^2}{r^3}\frac{d}{dr} + \frac{n^2(n^2-4)}{r^4}\right)w_n(r) = -f_n(r)$$

where, for notational convenience, we omit the superscripts "s" and "c" on $w_n(r)$ and $f_n(r)$, because these are not important until a certain stage in the development.

The differential operator in the above equation can readily be converted to a self-adjoint form by implementing the integrating factor r. Indeed, the differential operator

$$r\frac{d^4}{dr^4} + 2\frac{d^3}{dr^3} - \frac{1+2n^2}{r}\frac{d^2}{dr^2} + \frac{1+2n^2}{r^2}\frac{d}{dr} + \frac{n^2(n^2-4)}{r^3}$$

is self-adjoint. Hence, the Green's function of the boundary value problem

$$\left(r\frac{d^4}{dr^4} + 2\frac{d^3}{dr^3} - \frac{1+2n^2}{r}\frac{d^2}{dr^2} + \frac{1+2n^2}{r^2}\frac{d}{dr} + \frac{n^2(n^2-4)}{r^3}\right)w_n(r) = 0 \quad (6.22)$$

$$|w_n(0)| < \infty, \quad \left|\frac{d^2w_n(0)}{dr^2}\right| < \infty, \quad w_n(a) = \frac{dw_n(a)}{dr} = 0 \quad (6.23)$$

has to be symmetric.

The first two relations in eqn (6.23) are imposed to ensure the boundedness of the solution $w(r,\phi)$ of the original boundary value problem posed by eqns (6.18) and (6.19) at zero, while the third and the fourth relations in eqn (6.23) directly follow from the conditions of a clamped edge in eqn (6.19) and from the Fourier expansion shown in eqn (6.20).

Notice that it is impossible to obtain a single representation of the Green's function of the problem in eqns (6.22) and (6.23) for the entire range of the parameter n. Indeed, three individual subcases must be distinguished and considered separately.

First, for $n = 0$, the fundamental solution set of eqn (6.22) is represented with the functions

$$1, \quad \ln(r), \quad r^2, \quad \text{and} \quad r^2\ln(r)$$

and that branch of the Green's function $g_0(r,\rho)$, which is valid for $r \leq \rho$, is obtained for this case in the form

$$g_0(r,\rho) = \frac{1}{8}\left[\frac{1}{a^2}(a^2-\rho^2)(a^2+r^2) + 2(r^2+\rho^2)\ln\left(\frac{\rho}{a}\right)\right] \quad (6.24)$$

Either the procedure based on the defining properties of a Green's function or the method of variation of parameters can be used for obtaining the expressions in eqns (6.24)–(6.26).

Second, for $n = 1$, the fundamental solution set of eqn (6.22) is different from that of $n = 0$. It is represented in this case with the functions

$$r^{-1}, \quad r, \quad r^3, \quad \text{and} \quad r\ln(r)$$

and that branch of the Green's function $g_1(r,\rho)$, which is valid for $r \leq \rho$, is obtained for this case in the form

$$g_1(r,\rho) = \frac{r(\rho^2-a^2)}{16a^4\rho}\left[r^2(a^2-\rho^2) + 2a^2\rho^2\right] - \frac{1}{4}r\rho\ln\left(\frac{\rho}{a}\right) \quad (6.25)$$

Finally, for $n \geq 2$, the fundamental solution set of eqn (6.22) is represented with the functions

$$r^n, \quad r^{-n}, \quad r^{n+2}, \quad \text{and} \quad r^{-n+2}$$

and that branch of the Green's function $g_n(r,\rho)$, which is valid for $r \leq \rho$, is found in this case in the form

$$g_n(r,\rho) = -\frac{1}{8}\left\{\frac{r\rho}{n-1}\left[\left(\frac{r\rho}{a^2}\right)^{n-1} - \left(\frac{r}{\rho}\right)^{n-1}\right]\right.$$

$$\left. + \frac{r^2+\rho^2}{n}\left[\left(\frac{r}{\rho}\right)^n - \left(\frac{r\rho}{a^2}\right)^n\right] + \frac{r\rho}{n+1}\left[\left(\frac{r\rho}{a^2}\right)^{n+1} - \left(\frac{r}{\rho}\right)^{n+1}\right]\right\} \quad (6.26)$$

The boundary value problem stated by eqns (6.22) and (6.23) is in a self-adjoint form regardless of the value of n. Hence, the branches of the Green's functions $g_0(r,\rho)$, $g_1(r,\rho)$, and $g_n(r,\rho)$, which are valid for $r \geq \rho$, can be obtained from the corresponding ones presented by eqns (6.24)–(6.26) by interchanging of r with ρ.

Tracing out our procedure, the influence function $G(r,\phi;\rho,\psi)$ of the clamped circular plate is found in the form of a series

$$G(r,\phi;\rho,\psi) = \frac{1}{2\pi D}\left[g_0(r,\rho) + 2\sum_{n=1}^{\infty} g_n(r,\rho)\cos(n(\phi-\psi))\right]$$

6.1. POISSON-KIRCHHOFF'S PLATES

Let us show that the above series can be summed up. Notice that it does not matter which of the two branches of its coefficients is taken for the summation procedure – either is appropriate. In what follows, we take advantage of those branches presented in eqns (6.24)–(6.26).

Since the coefficient $g_1(r,\rho)$ of the first term in the above series is obtained in a form different from the rest of the coefficients $g_n(r,\rho)$, $n=2,3,\ldots$, we separate the entire first term $g_1(r,\rho)\cos(\phi-\psi)$ of the series and rewrite it as

$$G(r,\phi;\rho,\psi) = \frac{1}{2\pi D}\bigg[g_0(r,\rho) + 2g_1(r,\rho)\cos(\phi-\psi)$$

$$+ 2\sum_{n=2}^{\infty} g_n(r,\rho)\cos(n(\phi-\psi))\bigg] \quad (6.27)$$

Upon substituting the expressions for $g_0(r,\rho)$ and $g_1(r,\rho)$ presented in eqns (6.24) and (6.25) into this expansion, the first two of its terms are rewritten as

$$g_0(r,\rho) + 2g_1(r,\rho)\cos(\phi-\psi) =$$

$$= \frac{1}{8}\bigg[\frac{1}{a^2}(a^2-\rho^2)(a^2+r^2) + 2(r^2+\rho^2)\ln\left(\frac{\rho}{a}\right)\bigg]$$

$$-\bigg[\frac{r(a^2-\rho^2)}{8a^4\rho}\left(r^2(a^2-\rho^2)+2a^2\rho^2\right) + \frac{1}{2}r\rho\ln\left(\frac{\rho}{a}\right)\bigg]\cos(\phi-\psi)$$

Combining the logarithmic terms, one obtains

$$g_0(r,\rho) + 2g_1(r,\rho)\cos(\phi-\psi) =$$

$$= \frac{1}{8}\bigg\{\frac{1}{a^2}(a^2-\rho^2)(a^2+r^2) + 2\left[r^2-2r\rho\cos(\phi-\psi)+\rho^2\right]\ln\left(\frac{\rho}{a}\right)$$

$$-\frac{r(a^2-\rho^2)}{a^4\rho}\left[r^2(a^2-\rho^2)+2a^2\rho^2\right]\cos(\phi-\psi)\bigg\} \quad (6.28)$$

We will later recall this expression when the series in eqn (6.27) is ultimately ready to be summed up.

By substituting the expression for $g_n(r,\rho)$ from eqn (6.26) into the series part of eqn (6.27), one obtains

$$2\sum_{n=2}^{\infty} g_n(r,\rho)\cos(n(\phi-\psi)) =$$

$$-\frac{1}{4}\bigg\{r\rho\sum_{n=2}^{\infty}\frac{1}{n-1}\bigg[\left(\frac{r\rho}{a^2}\right)^{n-1} - \left(\frac{r}{\rho}\right)^{n-1}\bigg]\cos(n(\phi-\psi))$$

$$+(r^2+\rho^2)\sum_{n=2}^{\infty}\frac{1}{n}\left[\left(\frac{r}{\rho}\right)^n-\left(\frac{r\rho}{a^2}\right)^n\right]\cos(n(\phi-\psi))$$

$$+r\rho\sum_{n=2}^{\infty}\frac{1}{n+1}\left[\left(\frac{r\rho}{a^2}\right)^{n+1}-\left(\frac{r}{\rho}\right)^{n+1}\right]\cos(n(\phi-\psi))\bigg\} \quad (6.29)$$

Each of the series in this expression requires an individual treatment. To partially sum up the first of them, we change its summation variable n by making the substitution $k=n-1$. This yields

$$\sum_{n=2}^{\infty}\frac{1}{n-1}\left[\left(\frac{r\rho}{a^2}\right)^{n-1}-\left(\frac{r}{\rho}\right)^{n-1}\right]\cos(n(\phi-\psi))$$

$$=\sum_{k=1}^{\infty}\frac{1}{k}\left[\left(\frac{r\rho}{a^2}\right)^k-\left(\frac{r}{\rho}\right)^k\right]\cos((k+1)(\phi-\psi))$$

$$=\sum_{k=1}^{\infty}\frac{1}{k}\left[\left(\frac{r\rho}{a^2}\right)^k-\left(\frac{r}{\rho}\right)^k\right]$$
$$\times[\cos(k(\phi-\psi))\cos(\phi-\psi)-\sin(k(\phi-\psi))\sin(\phi-\psi)]$$

$$=\cos(\phi-\psi)\sum_{k=1}^{\infty}\frac{1}{k}\left[\left(\frac{r\rho}{a^2}\right)^k-\left(\frac{r}{\rho}\right)^k\right]\cos(k(\phi-\psi))$$

$$-\sin(\phi-\psi)\sum_{k=1}^{\infty}\frac{1}{k}\left[\left(\frac{r\rho}{a^2}\right)^k-\left(\frac{r}{\rho}\right)^k\right]\sin(k(\phi-\psi))$$

Clearly, the above cosine-series can be summed up with the aid of the summation formula from eqn (6.13), whereas the sine-series will be left in its current form. This yields, for the entire first term in eqn (6.29)

$$r\rho\sum_{n=2}^{\infty}\frac{1}{n-1}\left[\left(\frac{r\rho}{a^2}\right)^{n-1}-\left(\frac{r}{\rho}\right)^{n-1}\right]\cos(n(\phi-\psi))$$

$$=r\rho\cos(\phi-\psi)\left[\frac{1}{2}\ln\left(1-2\frac{r}{\rho}\cos(\phi-\psi)+\left(\frac{r}{\rho}\right)^2\right)\right.$$
$$\left.-\frac{1}{2}\ln\left(1-2\frac{r\rho}{a^2}\cos(\phi-\psi)+\left(\frac{r\rho}{a^2}\right)^2\right)\right]$$

$$-r\rho\sin(\phi-\psi)\sum_{k=1}^{\infty}\frac{1}{k}\left[\left(\frac{r\rho}{a^2}\right)^k-\left(\frac{r}{\rho}\right)^k\right]\sin(k(\phi-\psi)) \quad (6.30)$$

6.1. POISSON-KIRCHHOFF'S PLATES

The remaining series in eqn (6.30) can be also summed up. However, in view of the development that follows, we leave it in its current form.

To sum up the series of the second term in eqn (6.29), we rewrite it as

$$(r^2+\rho^2) \sum_{n=2}^{\infty} \frac{1}{n} \left[\left(\frac{r}{\rho}\right)^n - \left(\frac{r\rho}{a^2}\right)^n \right] \cos(n(\phi-\psi))$$

$$= (r^2+\rho^2) \sum_{n=1}^{\infty} \frac{1}{n} \left[\left(\frac{r}{\rho}\right)^n - \left(\frac{r\rho}{a^2}\right)^n \right] \cos(n(\phi-\psi))$$

$$+ (r^2+\rho^2) \left[\left(\frac{r\rho}{a^2}\right) - \left(\frac{r}{\rho}\right) \right] \cos(\phi-\psi)$$

This clearly allows us to obtain

$$(r^2+\rho^2) \sum_{n=2}^{\infty} \frac{1}{n} \left[\left(\frac{r}{\rho}\right)^n - \left(\frac{r\rho}{a^2}\right)^n \right] \cos(n(\phi-\psi))$$

$$= -(r^2+\rho^2) \left[\frac{1}{2} \ln\left(1 - 2\frac{r}{\rho}\cos(\phi-\psi) + \left(\frac{r}{\rho}\right)^2\right) \right.$$

$$\left. - \frac{1}{2} \ln\left(1 - 2\frac{r\rho}{a^2}\cos(\phi-\psi) + \left(\frac{r\rho}{a^2}\right)^2\right) \right]$$

$$+ (r^2+\rho^2) \left[\left(\frac{r\rho}{a^2}\right) - \left(\frac{r}{\rho}\right) \right] \cos(\phi-\psi) \qquad (6.31)$$

Finally, to partially sum up the series of the last term in eqn (6.29), we change its summation variable n by introducing $k=n+1$. This yields

$$r\rho \sum_{n=2}^{\infty} \frac{1}{n+1} \left[\left(\frac{r\rho}{a^2}\right)^{n+1} - \left(\frac{r}{\rho}\right)^{n+1} \right] \cos(n(\phi-\psi))$$

$$= r\rho \sum_{k=3}^{\infty} \frac{1}{k} \left[\left(\frac{r\rho}{a^2}\right)^k - \left(\frac{r}{\rho}\right)^k \right] \cos((k-1)(\phi-\psi))$$

$$= r\rho \sum_{k=1}^{\infty} \frac{1}{k} \left[\left(\frac{r\rho}{a^2}\right)^k - \left(\frac{r}{\rho}\right)^k \right] \cos((k-1)(\phi-\psi))$$

$$- r\rho \left[\left(\frac{r\rho}{a^2}\right) - \left(\frac{r}{\rho}\right) \right] - \frac{r\rho}{2} \left[\left(\frac{r\rho}{a^2}\right)^2 - \left(\frac{r}{\rho}\right)^2 \right] \cos(\phi-\psi)$$

$$= r\rho \cos(\phi-\psi) \sum_{k=1}^{\infty} \frac{1}{k} \left[\left(\frac{r\rho}{a^2}\right)^k - \left(\frac{r}{\rho}\right)^k \right] \cos(k(\phi-\psi))$$

$$+r\rho\sin(\phi-\psi)\sum_{k=1}^{\infty}\frac{1}{k}\left[\left(\frac{r\rho}{a^2}\right)^k-\left(\frac{r}{\rho}\right)^k\right]\sin(k(\phi-\psi))$$

$$-r\rho\left[\left(\frac{r\rho}{a^2}\right)-\left(\frac{r}{\rho}\right)\right]-\frac{r\rho}{2}\left[\left(\frac{r\rho}{a^2}\right)^2-\left(\frac{r}{\rho}\right)^2\right]\cos(\phi-\psi)$$

By summing up the above cosine-series and leaving the sine-series in its current form, the last term in eqn (6.29) is finally obtained as

$$r\rho\sum_{n=2}^{\infty}\frac{1}{n+1}\left[\left(\frac{r\rho}{a^2}\right)^{n+1}-\left(\frac{r}{\rho}\right)^{n+1}\right]\cos(n(\phi-\psi))$$

$$=r\rho\cos(\phi-\psi)\left[\frac{1}{2}\ln\left(1-2\frac{r}{\rho}\cos(\phi-\psi)+\left(\frac{r}{\rho}\right)^2\right)\right.$$

$$\left.-\frac{1}{2}\ln\left(1-2\frac{r\rho}{a^2}\cos(\phi-\psi)+\left(\frac{r\rho}{a^2}\right)^2\right)\right]$$

$$+r\rho\sin(\phi-\psi)\sum_{k=1}^{\infty}\frac{1}{k}\left[\left(\frac{r\rho}{a^2}\right)^k-\left(\frac{r}{\rho}\right)^k\right]\sin(k(\phi-\psi))$$

$$-r\rho\left[\left(\frac{r\rho}{a^2}\right)-\left(\frac{r}{\rho}\right)\right]-\frac{r\rho}{2}\left[\left(\frac{r\rho}{a^2}\right)^2-\left(\frac{r}{\rho}\right)^2\right]\cos(\phi-\psi) \quad (6.32)$$

If one substitutes the expressions from eqns (6.30), (6.31), and (6.32) into eqn (6.29), then the two sine-series (one from each of equations (6.30) and (6.32)) are cancelled out, all of the logarithmic terms as well as the two double-underlined terms are accordingly combined. This provides, for the series part of the influence function in eqn (6.27)

$$2\sum_{n=2}^{\infty}g_n(r,\rho)\cos(n(\phi-\psi))=$$

$$-\left[r^2-2r\rho\cos(\phi-\psi)+\rho^2\right]\ln\frac{\rho^2\left[a^4-2a^2r\rho\cos(\phi-\psi)+r^2\rho^2\right]}{a^4\left[r^2-2r\rho\cos(\phi-\psi)+\rho^2\right]}$$

$$+r\rho\left[\left(\frac{r\rho}{a^2}\right)-\left(\frac{r}{\rho}\right)\right]+\frac{r(a^2-\rho^2)}{a^4\rho}\left[r^2(a^2-\rho^2)+2a^2\rho^2\right]\cos(\phi-\psi)$$

Upon substituting this expression along with that of eqn (6.28) all into eqn (6.27), the two double-underlined terms are cancelled out, the two logarithmic and the two simply underlined terms are accordingly combined.

6.1. POISSON-KIRCHHOFF'S PLATES

This provides the final representation for the influence function $G(r,\phi;\rho,\psi)$ for the clamped circular plate of radius a in the form

$$G(r,\phi;\rho,\psi) = \frac{1}{16\pi D}\left\{\frac{1}{a^2}(a^2-\rho^2)(a^2-r^2)\right.$$

$$\left.-\left[r^2-2r\rho\cos(\phi-\psi)+\rho^2\right]\ln\frac{a^4-2a^2r\rho\cos(\phi-\psi)+r^2\rho^2}{a^2\left[r^2-2r\rho\cos(\phi-\psi)+\rho^2\right]}\right\}$$

Clearly, this can be identified with the classical representation shown in eqn (6.17). Indeed, r and ρ are the moduli of the observation z and the source ζ points, respectively, the expression of $[r^2 - 2r\rho\cos(\phi - \psi) + \rho^2]$ represents a square of the distance $|z-\zeta|$ between z and ζ, and the expression of $[a^4-2a^2r\rho\cos(\phi-\psi)+r^2\rho^2]$ represents a square of the modulus $|a^2 - z\bar{\zeta}|$.

For the next example, we consider a semi-circular plate of radius a, whose middle plane occupies the region $\Omega_{sc} = \{(r,\phi): 0 < r < a, 0 < \phi < \pi\}$, with the rectilinear edge being simply supported while the curvilinear edge is clamped.

Omitting the detailed development that can be accomplished by simply tracing out the scheme utilized for the previous problem, we present just the final expression of the influence function for this plate in the following closed form

$$G(r,\phi;\rho,\psi) =$$

$$\frac{1}{16\pi D}\left\{\left[r^2-2r\rho\cos(\phi+\psi)+\rho^2\right]\ln\frac{a^4-2a^2r\rho\cos(\phi+\psi)+r^2\rho^2}{a^2\left[r^2-2r\rho\cos(\phi+\psi)+\rho^2\right]}\right.$$

$$\left.-\left[r^2-2r\rho\cos(\phi-\psi)+\rho^2\right]\ln\frac{a^4-2a^2r\rho\cos(\phi-\psi)+r^2\rho^2}{a^2\left[r^2-2r\rho\cos(\phi-\psi)+\rho^2\right]}\right\}$$

The reader is encouraged to go through the procedure of derivation of the above representation.

The examples just completed bring a strong confidence in the suggested technique that happens to be productive in a number of situations where influence functions are either not available at all or their existing representations are not convenient enough for numerical implementations. The next example is one of those. We consider a simply supported circular plate of radius a. Although the complex variable-based method for the construction of the influence function for this problem was described in journal articles many years ago (the information concerning this issue can be found in [72]), the influence function itself is not available in the existing handbooks on the subject matter.

The boundary conditions on the simply supported edge $r = a$ are imposed in the form

$$w(a, \phi) = 0, \quad M_r(a, \phi) \equiv \left(\frac{\partial^2}{\partial r^2} + \frac{\sigma}{a} \left(\frac{\partial}{\partial r} + \frac{1}{a} \frac{\partial^2}{\partial \phi^2} \right) \right) w(a, \phi) = 0$$

prescribing the deflection and the radial bending moment to be zero.

Notice that, since the deflection $w(r, \phi)$ is identically equal to zero along the edge $r = a$ of the plate, all the derivatives of $w(r, \phi)$ with respect to the tangential variable ϕ must also be zero. Hence, the second of the above relations reduces to

$$\left(\frac{\partial^2}{\partial r^2} + \frac{\sigma}{a} \frac{\partial}{\partial r} \right) w(a, \phi) = 0$$

Following the suggested procedure, the influence function of the simply supported circular plate is also obtained in the series form of eqn (6.27), whose coefficients, for $r \leq \rho$, are found in this case as

$$g_0(r, \rho) = \frac{1}{8} \left\{ \frac{a^2 - \rho^2}{a^2} \left[(a^2 + r^2) + \frac{2}{1+\sigma}(a^2 - r^2) \right] + 2(r^2 + \rho^2) \ln\left(\frac{\rho}{a}\right) \right\}$$

$$g_1(r, \rho) = \frac{1}{8} \left\{ \frac{1+\sigma}{3+\sigma} \left[(r^2 + \rho^2) - a^2 \right] \frac{r\rho}{a^2} - \frac{r^3}{2\rho} \left(1 - \frac{1-\sigma}{3+\sigma} \frac{\rho^4}{a^4} \right) - 2r\rho \ln\left(\frac{\rho}{a}\right) \right\}$$

and

$$g_n(r, \rho) = -\frac{1}{8} \left\{ \frac{r\rho}{n-1} \left[\left(\frac{r\rho}{a^2}\right)^{n-1} - \left(\frac{r}{\rho}\right)^{n-1} \right] + \frac{r\rho}{n+1} \left[\left(\frac{r\rho}{a^2}\right)^{n+1} - \left(\frac{r}{\rho}\right)^{n+1} \right] \right.$$

$$\left. - \frac{r^2 + \rho^2}{n} \left[\left(\frac{r\rho}{a^2}\right)^n - \left(\frac{r}{\rho}\right)^n \right] + \frac{1}{n+\omega} \frac{(r^2 - a^2)(a^2 - \rho^2)}{a^2} \left(\frac{r\rho}{a^2}\right)^n \right\}$$

where the parameter ω is introduced in terms of Poisson ratio of the material as $\omega = (1+\sigma)/2$.

The series in eqn (6.27) cannot be entirely summed up in this case. This is so because of the last term in the above expression for $g_n(r, \rho)$, one containing the parameter ω. The partial summation, though, provides a quite compact representation for the influence function under consideration, that is

$$G(r, \phi; \rho, \psi) =$$

$$\frac{1}{16\pi D} \left\{ \frac{(a^2 - \rho^2)(a^2 - r^2)}{a^2} \left[\frac{3+\sigma}{1+\sigma} + 2 \sum_{n=1}^{\infty} \frac{1}{n+\omega} \left(\frac{r\rho}{a^2}\right)^n \cos(n(\phi - \psi)) \right] \right.$$

$$\left. - \left[r^2 - 2r\rho \cos(\phi - \psi) + \rho^2 \right] \ln \frac{a^4 - 2a^2 r\rho \cos(\phi - \psi) + r^2 \rho^2}{a^2 \left[r^2 - 2r\rho \cos(\phi - \psi) + \rho^2 \right]} \right\} \quad (6.33)$$

6.1. POISSON-KIRCHHOFF'S PLATES

This representation can be successfully used for computing values of the influence function $G(r,\phi;\rho,\psi)$ inside of the circle, because the series in the brackets uniformly converges if both the field (r,ϕ) and the source (ρ,ψ) points are interior for the circle, provided $r\rho/a^2 < 1$. Notice, however, that the rate of its convergence depends on the closeness of r and ρ to the plate's edge. Indeed, the convergence notably slows down if both the field and the source points approach the edge of the plate (both r and ρ approach a).

We will show now how this situation can be radically improved by splitting off the slowly convergent component of the series and summing it up. In doing so, the coefficient of the series in eqn (6.33) is presented in the form

$$\frac{1}{n+\omega} = \frac{1}{n} - \frac{\omega}{n(n+\omega)}$$

yielding

$$2\sum_{n=1}^{\infty}\frac{1}{n+\omega}\left(\frac{r\rho}{a^2}\right)^n \cos(n(\phi-\psi))$$

$$= 2\sum_{n=1}^{\infty}\left[\frac{1}{n}-\frac{\omega}{n(n+\omega)}\right]\left(\frac{r\rho}{a^2}\right)^n \cos(n(\phi-\psi))$$

$$= 2\sum_{n=1}^{\infty}\frac{1}{n}\left(\frac{r\rho}{a^2}\right)^n \cos(n(\phi-\psi))$$

$$-2\omega\sum_{n=1}^{\infty}\frac{1}{n(n+\omega)}\left(\frac{r\rho}{a^2}\right)^n \cos(n(\phi-\psi))$$

The first of these series is summable with the aid of the summation formula from eqn (6.13). This yields

$$2\sum_{n=1}^{\infty}\frac{1}{n+\omega}\left(\frac{r\rho}{a^2}\right)^n \cos(n(\phi-\psi))$$

$$= -\ln\left(1-2\frac{r\rho}{a^2}\cos(\phi-\psi)+\left(\frac{r\rho}{a^2}\right)^2\right)$$

$$-2\omega\sum_{n=1}^{\infty}\frac{1}{n(n+\omega)}\left(\frac{r\rho}{a^2}\right)^n \cos(n(\phi-\psi))$$

In view of the foregoing transformation, the expression for the influence function of the simply supported circular plate, appearing in eqn (6.33), can be rewritten in the form

$$G(r,\phi;\rho,\psi) =$$

$$\frac{1}{16\pi D}\left\{\frac{(a^2-\rho^2)(a^2-r^2)}{a^2}\left[\frac{3+\sigma}{1+\sigma}-\ln\left(1-2\frac{r\rho}{a^2}\cos(\phi-\psi)+\left(\frac{r\rho}{a^2}\right)^2\right)\right.\right.$$

$$-2\omega \sum_{n=1}^{\infty} \frac{1}{n(n+\omega)} \left(\frac{r\rho}{a^2}\right)^n \cos(n(\phi-\psi))\Bigg]$$

$$-\left[r^2 - 2r\rho\cos(\phi-\psi) + \rho^2\right] \ln \frac{a^4 - 2a^2 r\rho \cos(\phi-\psi) + r^2\rho^2}{a^2\left[r^2 - 2r\rho\cos(\phi-\psi)+\rho^2\right]}\Bigg\} \qquad (6.34)$$

Upon comparing the above representation with that of eqn (6.33), it is easily seen that the series in eqn (6.34) converges much faster than that of eqn (6.33). Indeed, the rate of its convergence is of the order of $1/n^2$, against $1/n$ for the series in eqn (6.33). The estimate of the remainder of the series from eqn (6.34) can be found in Chapter 5 (see eqn (5.26)). This ensures an accurate truncation of this series when computing.

The above expression can be further transformed in terms of the observation point $z = r(\cos(\phi) + i \sin(\phi))$ and source point $\zeta = \rho(\cos(\psi) + i\sin(\psi))$. After some elementary algebra, the logarithmic terms of eqn (6.34) are combined and rearranged. This finally yields

$$G(r, \phi; \rho, \psi) =$$

$$\frac{1}{8\pi D}\Bigg\{|z-\zeta|^2 \ln \frac{|z-\zeta|}{a} - \frac{|a^2 - z\overline{\zeta}|^2}{a^2} \ln \frac{|a^2 - z\overline{\zeta}|^2}{a^2}$$

$$+\frac{(a^2-\rho^2)(a^2-r^2)}{2a^2}\left[\frac{3+\sigma}{1+\sigma} - 2\omega \sum_{n=1}^{\infty} \frac{1}{n(n+\omega)}\left(\frac{r\rho}{a^2}\right)^n \cos(n(\phi-\psi))\right]\Bigg\}$$

Clearly, the first logarithmic term in $G(r, \phi; \rho, \psi)$ explicitly represents the singular component of the influence function.

6.2 Reissner's plates

In this section, the influence function method is extended to the mathematical formulation of thin plate problems based on the Reissner theory (see [65]), which accounts for the effect of transverse normal stress and transverse shear deformation and admits any standard physically feasible boundary conditions (clamped, simply supported, elastically supported, or free edge). The technique, described in the previous sections of this book, is here employed to analytically construct influence matrices that can be used for numerical solution of actual plate problems. A validation example is presented revealing computational properties of such an approach.

Consider a thin plate occupying a region Ω whose boundary is a piecewise smooth curve Γ. Suppose the plate is subjected to a transverse distributed load $q = q(x,y)$ per unit area. Assume, in addition, that the plate has a

6.2. REISSNER'S PLATES

uniform thickness h and is composed of a homogeneous isotropic elastic material having the modulus of elasticity E and Poisson ratio σ.

Let the equilibrium state of the plate be described in compliance with the Reissner theory [65] with the following system of equations

$$\frac{\partial^2 \beta_x}{\partial x^2} + \frac{1-\sigma}{2}\frac{\partial^2 \beta_x}{\partial y^2} + \frac{1+\sigma}{2}\frac{\partial^2 \beta_y}{\partial x \partial y} - \frac{5(1-\sigma)}{h^2}\left(\beta_x + \frac{\partial w}{\partial x}\right) = -\frac{1}{D}\frac{\partial}{\partial x}\left(\frac{\sigma h^2 q}{10(1-\sigma)}\right)$$

$$\frac{\partial^2 \beta_y}{\partial y^2} + \frac{1-\sigma}{2}\frac{\partial^2 \beta_y}{\partial x^2} + \frac{1+\sigma}{2}\frac{\partial^2 \beta_x}{\partial x \partial y} - \frac{5(1-\sigma)}{h^2}\left(\beta_y + \frac{\partial w}{\partial y}\right) = -\frac{1}{D}\frac{\partial}{\partial y}\left(\frac{\sigma h^2 q}{10(1-\sigma)}\right)$$

$$\frac{\partial^2 w}{\partial x^2} + \frac{\partial^2 w}{\partial y^2} + \frac{\partial \beta_x}{\partial x} + \frac{\partial \beta_y}{\partial y} = -\frac{12(1+\sigma)}{5Eh}q \tag{6.35}$$

in the out-of-plane deflection $w = w(x,y)$ and rotations $\beta_x = \beta_x(x,y)$ and $\beta_y = \beta_y(x,y)$ of a point (x,y) of the middle plane in the plate. The stress resultants can be written in terms of the basic unknowns of the above system in the form

$$M_x = D\left(\frac{\partial \beta_x}{\partial x} + \sigma \frac{\partial \beta_y}{\partial y}\right) + \frac{\sigma h^2 q}{10(1-\sigma)}, \quad M_y = D\left(\frac{\partial \beta_y}{\partial y} + \sigma \frac{\partial \beta_x}{\partial x}\right) + \frac{\sigma h^2 q}{10(1-\sigma)}$$

$$M_{xy} = D\frac{1-\sigma}{2}\left(\frac{\partial \beta_x}{\partial y} + \frac{\partial \beta_y}{\partial x}\right) \tag{6.36}$$

$$V_x = \frac{5Eh}{12(1+\sigma)}\left(\beta_x + \frac{\partial w}{\partial x}\right), \quad V_y = \frac{5Eh}{12(1+\sigma)}\left(\beta_y + \frac{\partial w}{\partial y}\right) \tag{6.37}$$

where M_x, M_y, and M_{xy} represent the bending and twisting moments, respectively, V_x and V_y are the shear forces, and $D = Eh^3/(12(1-\sigma^2))$ represents the flexural rigidity of the plate.

Notice that for the first time in this text, we are involved with a formulation that includes a system of partial differential equations of higher order. The total order of the system in eqn (6.35) is six. Thus, to complete the problem formulation, three linearly independent boundary conditions are to be imposed at each point on the contour Γ.

Assume the plate occupies a rectangular region $\Omega_r = \{(x,y): 0 < x < a, 0 < y < b\}$ and let the edges $y=0$ and $y=b$ be simply supported yielding the following formulation of the boundary conditions along them

$$w|_{y=0,b} = 0, \quad \beta_x|_{y=0,b} = 0, \quad M_y|_{y=0,b} = 0 \tag{6.38}$$

Any combination of the conventional boundary conditions (simply supported, clamped, elastically supported, or free edge) is allowed on the edges $x=0$ and $x=a$.

We are now in a position to construct the influence matrix of a transverse concentrated unit force applied at an arbitrary point in the plate. In doing so, we represent the loading function $q(x,y)$ by means of the Fourier series

$$q(x,y) = \sum_{n=1}^{\infty} q_n(x) \sin(\nu y), \quad \nu = \frac{n\pi}{b} \qquad (6.39)$$

The solution vector $\mathbf{U}(x,y) = (w(x,y), \beta_x(x,y), \beta_y(x,y))^T$ of the boundary value problem just formulated is expanded in the following manner

$$\mathbf{U}(x,y) = \sum_{n=0}^{\infty} Q_n(y) \mathbf{U}_n(x) \qquad (6.40)$$

where we denote

$$Q_n(y) = \begin{pmatrix} \sin(\nu y) & 0 & 0 \\ 0 & \sin(\nu y) & 0 \\ 0 & 0 & \cos(\nu y) \end{pmatrix}, \quad \mathbf{U}_n(x) = \begin{pmatrix} w_n(x) \\ \beta_{xn}(x) \\ \beta_{yn}(x) \end{pmatrix}$$

Notice that the expansion in eqn (6.40) satisfies the boundary conditions of eqn (6.38). The components of the vector $\mathbf{U}_n(x)$ are, subsequently, to satisfy the following system of ordinary differential equations

$$\frac{d^2 \beta_{xn}}{dx^2} - \frac{1-\sigma}{2} \nu^2 \beta_{xn} - \frac{1+\sigma}{2} \nu \frac{d\beta_{yn}}{dx} - \frac{5(1-\sigma)}{h^2}\left(\beta_{xn} + \frac{dw_n}{dx}\right) = -\frac{6\sigma(1+\sigma)}{5Eh} \frac{dq_n}{dx}$$

$$\frac{d^2 \beta_{yn}}{dx^2} - \frac{2\nu^2}{1-\sigma} \beta_{yn} + \frac{1+\sigma}{1-\sigma} \nu \frac{d\beta_{xn}}{dx} - \frac{10}{h^2}(\beta_{yn} + \nu w_n) = -\frac{12\sigma(1+\sigma)}{5Eh(1-\sigma)} \nu q_n$$

$$\frac{d^2 w_n}{dx^2} - \nu^2 w_n + \frac{d\beta_{xn}}{dx} - \nu \beta_{yn} = -\frac{12(1+\sigma)}{5Eh} q_n \qquad (6.41)$$

This system can be subject to any of the following combinations of boundary conditions at $x=0$ and $x=a$:

$$w_n = 0, \quad \frac{d\beta_{xn}}{dx} = 0, \quad \beta_{yn} = 0 \qquad (6.42)$$

(as associated with a simply supported edge in the original statement);

$$w_n = 0, \quad \beta_{xn} = 0, \quad \beta_{yn} = 0 \qquad (6.43)$$

6.2. REISSNER'S PLATES

(as associated with a clamped edge);

$$\beta_{xn} + \frac{dw_n}{dx} = 0, \quad \nu\beta_{xn} + \frac{d\beta_{yn}}{dx} = 0, \quad \sigma\frac{d\beta_{xn}}{dx} - \nu\beta_{yn} = 0 \qquad (6.44)$$

(as associated with a free edge).

Clearly, the system in eqn (6.41) is linear and has constant coefficients. Hence, for the corresponding homogeneous ($q_n(x) \equiv 0$) system, the fundamental solution set (which is a set of six linearly independent vector-functions) can be found with the aid of standard procedures. Omitting the elementary but quite cumbersome algebra, we present the fundamental solution set in the form

$$\mathbf{U}_n^{(1)}(x) = \begin{pmatrix} e^{\nu x} \\ -\nu e^{\nu x} \\ -\nu e^{\nu x} \end{pmatrix}, \quad \mathbf{U}_n^{(2)}(x) = \begin{pmatrix} xe^{\nu x} \\ -(\nu x+\omega+1)e^{\nu x} \\ -(\nu x+\omega)e^{\nu x} \end{pmatrix}$$

$$\mathbf{U}_n^{(3)}(x) = \begin{pmatrix} e^{-\nu x} \\ \nu e^{-\nu x} \\ -\nu e^{-\nu x} \end{pmatrix}, \quad \mathbf{U}_n^{(4)}(x) = \begin{pmatrix} xe^{-\nu x} \\ (\nu x-\omega-1)e^{-\nu x} \\ -(\nu x-\omega)e^{-\nu x} \end{pmatrix}$$

$$\mathbf{U}_n^{(5)}(x) = \begin{pmatrix} 0 \\ e^{px} \\ p_1 e^{px} \end{pmatrix}, \quad \mathbf{U}_n^{(6)}(x) = \begin{pmatrix} 0 \\ e^{-px} \\ -p_1 e^{-px} \end{pmatrix} \qquad (6.45)$$

The parameters ω, p, and p_1 in the above expressions are defined as

$$\omega = \frac{2\nu^2 h^2}{5(1-\sigma)}, \quad p = \sqrt{\nu^2 + \frac{10}{h^2}}, \quad p_1 = \frac{p}{\nu}$$

In compliance with Lagrange's method of variation of parameters, the general solution of the system in eqn (6.41) can be sought in the form

$$\mathbf{U}_n(x) = P_n(x)\mathbf{C}_n(x) \qquad (6.46)$$

where

$$P_n(x) = \left(\mathbf{U}_n^{(j)}(x)\right), \quad j = 1, \ldots, 6$$

is a rectangular matrix of order 3×6 whose columns are the vectors from eqn (6.45), while $\mathbf{C}_n(x)$ is a column-vector of order six, whose entries are undetermined functions.

Proceeding further with Lagrange's method, one obtains the system of six linear algebraic equations

$$P_n^*(x)\mathbf{C}_n'(x) = \mathbf{F}_n^*(x) \qquad (6.47)$$

in the components of the derivative $\mathbf{C}'_n(x)$ of the vector $\mathbf{C}_n(x)$.

Structure of this system can be viewed by the horizontal partitioning of its coefficient matrix $P_n^*(x)$ into two equidimensional submatrices. Its upper 3×6 submatrix represents matrix $P_n(x)$ from eqn (6.46), while its lower 3×6 submatrix represents the derivative of $P_n(x)$. Partitioning analogously the right-hand side vector $\mathbf{F}_n^*(x)$, we view its upper subvector as a three-dimensional zero-vector, while its lower subvector is viewed as the right-hand side vector

$$\mathbf{F}_n(x) = \left(-\frac{6\sigma(1+\sigma)}{5Eh}\frac{dq_n}{dx}, \ -\frac{12\sigma(1+\sigma)}{5Eh(1-\sigma)}\nu q_n, \ -\frac{12(1+\sigma)}{5Eh}q_n \right)^T$$

of the system in eqn (5.41). That is

$$P_n^*(x) = \begin{pmatrix} P_n(x) \\ P_n'(x) \end{pmatrix}, \quad \mathbf{F}_n^*(x) = \begin{pmatrix} 0 \\ \mathbf{F}_n(x) \end{pmatrix}$$

It is evident that the coefficient matrix of the system in eqn (6.47) represents the Wronskian of the fundamental solution set of the homogeneous system corresponding to that of eqn (6.41). The system in eqn (6.47) has, consequently, a unique solution that can be written in terms of the inverse of $P_n^*(x)$ as

$$\mathbf{C}'_n(x) = (P_n^*(x))^{-1}\mathbf{F}_n^*(x)$$

Recalling the vector's $\mathbf{F}_n^*(x)$ structure (its first three components are zeros), the above equation can be rewritten in terms of the vector $\mathbf{F}_n(x)$ as

$$\mathbf{C}'_n(x) = R_n(x)\mathbf{F}_n(x) \tag{6.48}$$

where $R_n(x)$ represents a matrix of order 6×3, which is a submatrix of $(P_n^*(x))^{-1}$ consisting of the latter's fourth, fifth, and sixth columns.

To obtain an explicit expression for $\mathbf{C}_n(x)$, we integrate the relation in eqn (6.48). This yields

$$\mathbf{C}_n(x) = \int_0^x R_n(s)\mathbf{F}_n(s)ds + \mathbf{D}_n$$

Upon substituting this expression for $\mathbf{C}_n(x)$ into eqn (6.46) and taking the factor $P_n(x)$ under the integral sign, one obtains

$$\mathbf{U}_n(x) = \int_0^x S_n(x,s)\mathbf{F}_n(s)ds + P_n(x)\mathbf{D}_n \tag{6.49}$$

where the kernel-matrix $S_n(x,s) = P_n(x)R_n(s)$ is a square matrix of order 3×3, whose entries $S_{ij}^n(x,s)$ are found in the form

$$S_{11}^n(x,s) = \frac{1}{\omega\nu}[(\omega+1)\sinh(\nu(x-s)) - \nu(x-s)\cosh(\nu(x-s))]$$

6.2. REISSNER'S PLATES

$$S_{12}^n(x,s) = -\frac{1}{2\nu}[(x-s)\sinh(\nu(x-s))]$$

$$S_{13}^n(x,s) = \frac{1}{\omega(p^2-\nu^2)}[\nu(x-s)\cosh(\nu(x-s)) - \sinh(\nu(x-s))]$$

$$S_{21}^n(x,s) = \frac{1}{\omega}[\nu(x-s)\sinh(\nu(x-s))]$$

$$S_{22}^n(x,s) = \frac{1}{2p\nu}[p(\omega+1)\sinh(\nu(x-s))$$
$$+p\nu(x-s)\cosh(\nu(x-s)) - \omega\nu\sinh(p(x-s))]$$

$$S_{23}^n(x,s) = \frac{\nu}{\omega(p^2-\nu^2)}[\omega\cosh(\nu(x-s))$$
$$-\nu(x-s)\sinh(\nu(x-s)) - \omega\cosh(\nu(x-s))]$$

$$S_{31}^n(x,s) = \frac{1}{\omega}[\nu(x-s)\cosh(\nu(x-s)) - \sinh(\nu(x-s))]$$

$$S_{32}^n(x,s) = \frac{1}{2\nu}[\omega\cosh(\nu(x-s))$$
$$+\nu(x-s)\sinh(\nu(x-s)) - \omega\cosh(p(x-s))]$$

$$S_{33}^n(x,s) = \frac{1}{\omega(p^2-\nu^2)}[p\omega\sinh(p(x-s))$$
$$-\nu^2(x-s)\cosh(\nu(x-s)) - \nu(\omega-1)\sinh(\nu(x-s))]$$

In the development that follows, we take advantage of a special feature of the above entries of $S_n(x,s)$. The point is that, being functions of the two variables x and s, they actually represent functions of a single variable $x-s$, that is $S_{ij}^n(x,s) = S_{ij}^n(x-s)$. Notice also that for all of the entries $S_{ij}^n(x,x) = 0$. This feature is taken into account later when the differentiation of the vector $\mathbf{U}_n(x)$ is performed while satisfying the boundary conditions.

To make the subsequent development as compact as possible, we introduce the operator form

$$B_{1n}[w_n, \beta_{xn}, \beta_{yn}] = 0, \quad B_{2n}[w_n, \beta_{xn}, \beta_{yn}] = 0$$

$$B_{3n}[w_n, \beta_{xn}, \beta_{yn}] = 0 \qquad (6.50)$$

for the boundary conditions at $x=0$ and $x=a$, imposed by eqns (6.42)–(6.44).

Satisfying the boundary conditions by the components of the vector $\mathbf{U}_n(x)$ from eqn (6.49), one obtains a system of six linear algebraic equations

$$T_n \mathbf{D}_n = \mathbf{K}_n \qquad (6.51)$$

in the components of the vector $\mathbf{D}_n$ appearing in eqn (6.49). The first, second, and third rows of the coefficient matrix T_n of the above system are defined as $B_{1n}[P_n(0)], B_{2n}[P_n(0)]$, and $B_{3n}[P_n(0)]$, respectively, while the fourth, fifth, and sixth rows are defined as $B_{1n}[P_n(a)], B_{2n}[P_n(a)]$, and $B_{3n}[P_n(a)]$, respectively. The first, second, and third entries of the right-hand side vector $\mathbf{K}_n$ are zeros, while its fourth, fifth, and sixth entries are defined as

$$\int_0^a Z_n(a,s)\mathbf{F}_n(s)ds$$

where $Z_n(x,s)$ is a square matrix of dimension 3×3, whose rows are defined as $-B_{1n}[S_n(x,s)], -B_{2n}[S_n(x,s)]$, and $-B_{3n}[S_n(x,s)]$, respectively.

In compliance with the described structure of the coefficient matrix and the right-hand side vector $\mathbf{K}_n$ of the system in eqn (6.51), its solution can be written in the form

$$\mathbf{D}_n = \int_0^a N_n(s)\mathbf{F}_n(s)ds \qquad (6.52)$$

with $N_n(s)$ being expressed in terms of the inverse T_n^{-1} of the coefficient matrix of the system in eqn (6.51) as

$$N_n(s) = T_n^{-1} \times H_n(a,s),$$

where $H_n(a,s)$ is a matrix of order 6×3, whose first three rows are zeros, while its remaining 3×3 submatrix represents the matrix $Z_n(a,s)$ recently introduced.

By substituting the expression for $\mathbf{D}_n$ derived in eqn (6.52) into eqn (6.49), one obtains

$$\mathbf{U}_n(x) = \int_0^x S_n(x,s)\mathbf{F}_n(s)ds + \int_0^a P_n(x)N_n(s)\mathbf{F}_n(s)ds$$

This can readily be put in the form of a single integral

$$\mathbf{U}_n(x) = \int_0^a g_n(x,s)\mathbf{F}_n(s)ds \qquad (6.53)$$

where $g_n(x,s)$ is defined in pieces as

$$g_n(x,s) = \begin{cases} P_n(x)N_n(s) + S_n(x,s), & x \geq s \\ P_n(x)N_n(s), & x \leq s \end{cases}$$

This represents Green's matrix of the homogeneous boundary value problem corresponding to that posed by eqns (6.41) and (6.50).

6.2. REISSNER'S PLATES

Substituting the expression for $\mathbf{U}_n(x)$ from eqn (6.53) into the expansion in eqn (6.40), and then proceeding in conformity with the relation from eqn (6.39), we finally obtain the following integral representation

$$\mathbf{U}(x,y) = \iint_{\Omega_r} \left[\frac{2}{b} \sum_{n=1}^{\infty} Q_n(y) g_n(x,s) Q_n(t) \right] \mathbf{F}(s,t) d\Omega_r(s,t) \qquad (6.54)$$

for the solution vector $\mathbf{U}(x,y)$ of the original boundary value problem for the system in eqn (6.35). Here $\mathbf{F}(s,t)$ represents a vector whose components are the right-hand terms of the system in eqn (6.35). Thus, in accordance with the definition introduced in the opening part of this chapter, the kernel-matrix

$$G(x,y;s,t) = \frac{2}{b} \sum_{n=1}^{\infty} Q_n(y) g_n(x,s) Q_n(t) \qquad (6.55)$$

of the representation in eqn (6.54) is the influence matrix of a transverse concentrated unit force applied at an arbitrary point in the rectangular $a \times b$ plate considered in the scope of the Reissner's theory.

One finds below the entries of $G(x,y;s,t)$ obtained for the plate, all edges of which are simply supported. We present the branches $G_{ij}^+ = G_{ij}^+(x,y;s,t)$ valid for $x \leq s$ in the form

$$G_{11}^+ = \frac{2}{b} \sum_{n=1}^{\infty} \left[\Phi_{11}^n(x) S_{11}^n(a,s) + \Phi_{12}^n(x) \overline{S}_{21}^n(a,s) + \Phi_{13}^n(x) S_{31}^n(a,s) \right] \sin(\nu y) \sin(\nu t),$$

$$G_{12}^+ = \frac{2}{b} \sum_{n=1}^{\infty} \left[\Phi_{11}^n(x) S_{12}^n(a,s) + \Phi_{12}^n(x) \overline{S}_{22}^n(a,s) + \Phi_{13}^n(x) S_{32}^n(a,s) \right] \sin(\nu y) \sin(\nu t),$$

$$G_{13}^+ = \frac{4}{b(1-\sigma)} \sum_{n=1}^{\infty} \left[\Phi_{11}^n(x) S_{13}^n(a,s) + \Phi_{12}^n(x) \overline{S}_{23}^n(a,s) + \Phi_{13}^n(x) S_{33}^n(a,s) \right]$$
$$\times \sin(\nu y) \cos(\nu t),$$

and

$$G_{21}^+ = \frac{2}{b} \sum_{n=1}^{\infty} \left[\Phi_{12}^n(x) S_{11}^n(a,s) + \Phi_{22}^n(x) \overline{S}_{21}^n(a,s) + \Phi_{23}^n(x) S_{31}^n(a,s) \right] \sin(\nu y) \sin(\nu t),$$

$$G_{22}^+ = \frac{2}{b} \sum_{n=1}^{\infty} \left[\Phi_{12}^n(x) S_{12}^n(a,s) + \Phi_{22}^n(x) \overline{S}_{22}^n(a,s) + \Phi_{23}^n(x) S_{32}^n(a,s) \right] \sin(\nu y) \sin(\nu t),$$

$$G_{23}^+ = \frac{4}{b(1-\sigma)} \sum_{n=1}^{\infty} \left[\Phi_{21}^n(x) S_{13}^n(a,s) + \Phi_{22}^n(x) \overline{S}_{23}^n(a,s) + \Phi_{23}^n(x) S_{33}^n(a,s) \right]$$
$$\times \sin(\nu y) \cos(\nu t),$$

and

$$G_{31}^+ = \frac{2}{b} \sum_{n=1}^{\infty} \left[\Phi_{31}^n(x) S_{11}^n(a,s) + \Phi_{32}^n(x) \overline{S}_{21}^n(a,s) + \Phi_{33}^n(x) S_{31}^n(a,s) \right] \cos(\nu y) \sin(\nu t),$$

$$G_{32}^+ = \frac{2}{b} \sum_{n=1}^{\infty} \left[\Phi_{31}^n(x) S_{12}^n(a,s) + \Phi_{32}^n(x) \overline{S}_{22}^n(a,s) + \Phi_{33}^n(x) S_{32}^n(a,s) \right] \cos(\nu y) \sin(\nu t),$$

$$G_{33}^+ = \frac{4}{b(1-\sigma)} \sum_{n=1}^{\infty} \left[\Phi_{31}^n(x) S_{13}^n(a,s) + \Phi_{32}^n(x) \overline{S}_{23}^n(a,s) + \Phi_{33}^n(x) S_{33}^n(a,s) \right]$$

$$\times \cos(\nu y) \cos(\nu t),$$

where we use notations as follows

$$\overline{S}_{21}^n(a,s) = (S_{21}^n(x,s))_{,x}\big|_{x=a}, \quad \overline{S}_{22}^n(a,s) = (S_{22}^n(x,s))_{,x}\big|_{x=a}$$

$$\overline{S}_{23}^n(a,s) = (S_{23}^n(x,s))_{,x}\big|_{x=a}$$

and

$$\Phi_{11}^n(x) = \frac{\sinh(\nu x)}{\sinh(\nu a)}, \quad \Phi_{12}^n(x) = \frac{a\cosh(\nu a)\sinh(\nu x) - x\cosh(\nu x)\sinh(\nu a)}{2\nu \sinh^2(\nu a)}$$

$$\Phi_{13}^n(x) = \frac{x\cosh(\nu x)\sinh(\nu a) - a\cosh(\nu a)\sinh(\nu x)}{2\nu \sinh^2(\nu a)}$$

$$\Phi_{21}^n(x) = \nu \left(\frac{\nu \cosh(px)}{p\sinh(pa)} - \frac{\cosh(\nu x)}{\sinh(\nu a)} \right)$$

$$\Phi_{22}^n(x) = \frac{1}{2} \left(\frac{x\sinh(\nu x)}{\sinh(\nu a)} + \frac{(\omega+1)\sinh(\nu a) - \nu a\cosh(\nu a)}{\nu \sinh^2(\nu a)(\cosh(\nu x))^{-1}} - \frac{\omega \cosh(px)}{p\sinh(pa)} \right)$$

$$\Phi_{23}^n(x) = \frac{1}{2} \left(\frac{(\omega+2)\cosh(px)}{p_1 \sinh(pa)} - \frac{\nu x\sinh(\nu x)}{\sinh(\nu a)} - \frac{(\omega+1)\sinh(\nu a) - \nu a\cosh(\nu a)}{\sinh^2(\nu a)(\cosh(\nu x))^{-1}} \right)$$

$$\Phi_{31}^n(x) = \nu \left(\frac{\sinh(px)}{\sinh(pa)} - \frac{\sinh(\nu x)}{\sinh(\nu a)} \right)$$

$$\Phi_{32}^n(x) = \frac{1}{2} \left(\frac{x\cosh(\nu x)}{\sinh(\nu a)} + \frac{\omega \sinh(\nu a) - \nu a\cosh(\nu a)}{\nu a \sinh^2(\nu a)(\sinh(\nu x))^{-1}} - \frac{\omega \sinh(px)}{\nu \sinh(pa)} \right)$$

$$\Phi_{33}^n(x) = \frac{1}{2} \left(\frac{(\omega+2)\sinh(px)}{\sinh(pa)} - \frac{\omega \sinh(\nu a) - \nu a\cosh(\nu a)}{\sinh^2(\nu a)(\sinh(\nu x))^{-1}} - \frac{x\cosh(\nu x)}{\sinh(\nu a)} \right)$$

6.2. REISSNER'S PLATES

The branches $G_{ij}^- = G_{ij}^-(x,y;s,t)$ of $G(x,y;s,t)$, valid for $x \geq s$, can readily be obtained from G_{ij}^+ in compliance with Betti's reciprocal work theorem, whose implementation to this case provides

$$G_{11}^-(x,y;s,t) = G_{11}^+(s,t;x,y), \quad \frac{1}{D}G_{12}^-(x,y;s,t) = \frac{12(1+\sigma)}{5Eh}G_{21}^+(s,t;x,y)$$

$$\frac{1}{D}G_{13}^-(x,y;s,t) = \frac{12(1+\sigma)}{5Eh}G_{31}^+(s,t;x,y), \quad G_{22}^-(x,y;s,t) = G_{22}^+(s,t;x,y)$$

and

$$G_{23}^-(x,y;s,t) = G_{32}^+(s,t;x,y), \quad G_{33}^-(x,y;s,t) = G_{33}^+(s,t;x,y)$$

In the discussion that follows, it is necessary to accurately compute the domain integrals of the entries of the influence matrix just derived as well as the domain integrals of the partial derivatives of these entries. Note that some of the entries possess singularities. That is, not all of the series from eqn (6.55) converge uniformly. Thus, the differentiation and integration mentioned above cannot be accomplished in a straightforward manner. For that reason, before proceeding any further with the development of the computational algorithm, we will clarify the integral and differential properties of the series from eqn (6.55).

In doing so, we stress that for any fixed position of the observation point (x,y) within the basic region Ω_r, the series in eqn (6.55) converges uniformly to a corresponding entry $G_{ij}(x,y;s,t)$ of the influence matrix everywhere in Ω_r except perhaps at the point $(s,t) = (x,y)$. The practical convergence takes place as long as $(s,t) \in \Omega_r^* = \Omega_r \backslash B(x,y)$, where $B(x,y)$ is an immediate vicinity of the observation point (x,y).

A similar property also remains valid for the partial derivatives of the first order of the series of eqn (6.55). They converge uniformly to the corresponding derivatives $\partial G_{ij}/\partial x$ or $\partial G_{ij}/\partial y$ as $(s,t) \in \Omega_r^*$. In other words, it is possible to differentiate the series of eqn (6.55) with respect to either x or y in a term-by-term fashion inside the region Ω_r^*.

The improper domain integrals of the form

$$\iint_{\Omega_0} G_{ij}(x,y;s,t) d\Omega_0(s,t), \quad \iint_{\Omega_0} \frac{\partial}{\partial x} G_{ij}(x,y;s,t) d\Omega_0(s,t)$$

and

$$\iint_{\Omega_0} \frac{\partial}{\partial y} G_{ij}(x,y;s,t) d\Omega_0(s,t)$$

converge absolutely over a subregion Ω_0 of Ω_r. Thus, the series of eqn (6.55) can be integrated in a term-by-term manner over a subregion of the basic region Ω_r, again resulting in a uniformly convergent series.

Based on the influence matrix recently derived, one can readily rewrite the integral representation of the solution vector $\mathbf{U}(x,y)$ from eqn (6.54) in the expanded form

$$\begin{pmatrix} w(P) \\ \beta_x(P) \\ \beta_y(P) \end{pmatrix} = \int_0^a \int_0^b \begin{pmatrix} G_{11}(P;Q) & G_{12}(P;Q) & G_{13}(P;Q) \\ G_{11}(P;Q) & G_{12}(P;Q) & G_{13}(P;Q) \\ G_{11}(P;Q) & G_{12}(P;Q) & G_{13}(P;Q) \end{pmatrix} \begin{pmatrix} F_1(Q) \\ F_2(Q) \\ F_3(Q) \end{pmatrix} dQ$$

where $F_j(Q)$, $j = 1, 2, 3$, represent the components of the loading vector of the right-hand side of the system in eqn (6.35), while P and Q are the observation point (x, y) and the source point (s, t), respectively.

To compute the approximate values $w^h(x, y)$, $\beta_x^h(x, y)$, and $\beta_y^h(x, y)$ of the components of $\mathbf{U}(x, y)$, we partition the basic region Ω_r into a set of elementary rectangles Ω_m, $m = 1, \ldots, M$ by using the scheme as shown

$$\Omega_r = \bigcup_{m=1}^{M} \Omega_m, \quad \Omega_m = \{(x,y) \in \Omega_r, \ x_m^1 \leq x \leq x_m^2, \ y_m^1 \leq y \leq y_m^2\} \quad (6.56)$$

and then apply the following cubature formulae

$$w^h(x,y) = \sum_{m=1}^{M} \left\{ \sum_{j=1}^{3} F_j(x_m, y_m) \int_{x_m^1}^{x_m^2} \int_{y_m^1}^{y_m^2} G_{1j}(x,y;s,t) dt ds \right\},$$

$$\beta_x^h(x,y) = \sum_{m=1}^{M} \left\{ \sum_{j=1}^{3} F_j(x_m, y_m) \int_{x_m^1}^{x_m^2} \int_{y_m^1}^{y_m^2} G_{2j}(x,y;s,t) dt ds \right\},$$

and

$$\beta_y^h(x,y) = \sum_{m=1}^{M} \left\{ \sum_{j=1}^{3} F_j(x_m, y_m) \int_{x_m^1}^{x_m^2} \int_{y_m^1}^{y_m^2} G_{3j}(x,y;s,t) dt ds \right\} \quad (6.57)$$

based on the mean value theorem of integration. Here

$$x_m = \frac{1}{2}(x_m^1 + x_m^2), \quad y_m = \frac{1}{2}(y_m^1 + y_m^2)$$

Clearly, approximate expressions for the first order partial derivatives of the functions from eqn (6.57) can be obtained analytically by straightforward differentiation. For the derivative $\partial w^h / \partial x$, for instance, one obtains

$$\frac{\partial w^h(x,y)}{\partial x} = \sum_{m=1}^{M} \left\{ \sum_{j=1}^{3} F_j(x_m, y_m) \int_{x_m^1}^{x_m^2} \int_{y_m^1}^{y_m^2} \frac{\partial G_{1j}(x,y;s,t)}{\partial x} dt ds \right\} \quad (6.58)$$

6.2. REISSNER'S PLATES

All other derivatives needed for the evaluation of the stress resultants from eqns (6.36) and (6.37) can be written in a similar manner.

As we have already mentioned, the integrals over the elementary rectangles Ω_m in eqns (6.57) and (6.58) converge absolutely, providing a suitable basis for approximate computation of the components of $\mathbf{U}(x,y)$ and its derivatives.

One can readily derive the error estimates for the approximations obtained in eqns (6.57) and (6.58). Indeed, let us assume that the functions $F_j(s,t)$, $j=1,2,3$ satisfy a Lipschitz condition

$$|F_j(s_2,t_2) - F_j(s_1,t_1)| \leq \alpha_m \sqrt{(s_2-s_1)^2 + (t_2-t_1)^2}$$

in each of the elementary rectangles Ω_m. Taking into account that the entries $G_{ij}(x,y;s,t)$ of the influence matrix are absolutely integrable functions of s and t over Ω_r, we obtain, for any fixed position of the observation point $(x,y) \in \Omega_r$

$$\left| w(x,y) - w^h(x,y) \right| \leq \max_m (\alpha_m D_m) \int_{x_m^1}^{x_m^2} \int_{y_m^1}^{y_m^2} \sum_{j=1}^{3} |G_{1j}(x,y;s,t)| \, dt\,ds,$$

$$\left| \beta_x(x,y) - \beta_x^h(x,y) \right| \leq \max_m (\alpha_m D_m) \int_{x_m^1}^{x_m^2} \int_{y_m^1}^{y_m^2} \sum_{j=1}^{3} |G_{2j}(x,y;s,t)| \, dt\,ds,$$

and

$$\left| \beta_y(x,y) - \beta_y^h(x,y) \right| \leq \max_m (\alpha_m D_m) \int_{x_m^1}^{x_m^2} \int_{y_m^1}^{y_m^2} \sum_{j=1}^{3} |G_{3j}(x,y;s,t)| \, dt\,ds, \quad (6.59)$$

where with D_m we denote the length of the biggest side of the elementary rectangle Ω_m.

It is evident that the error estimates for the first order partial derivatives of the components of the vector $\mathbf{U}^h(x,y)$ can be derived in a similar manner. That is, the error estimate for the stress components can also be evaluated.

From the estimates derived in eqn (6.59), it appears that the approximate solution vector $\mathbf{U}^h(x,y)$, whose components are given in eqn (6.57), converges to the true solution of the original boundary value problem for the system in eqn (6.35) as the partitioning parameter M in eqn (6.56) approaches infinity

$$\lim_{M \to \infty} \left(\mathbf{U}(x,y) - \mathbf{U}^h(x,y) \right) = 0$$

This conclusion is based on the following observation. The limit of the first factor in the right-hand side of the inequalities in eqn (6.59) equals zero, because $D_m \to 0$ as M approaches infinity (this is true at least for the uniform partitioning specified by eqn (6.56)), while the limit of the second factor (double integral) is bounded from above, because of the absolute integrability of the entries of the influence matrix over Ω_r.

Speaking of the computational procedure, the mathematical basis of which has just been described, it is worth noting that we here completely avoid numerical differentiation while computing stress components. Consequently, values of the moment and shear resultants are actually computed, within this study, as accurately as those of the deflection and rotation functions. This observation is reiterated in the validation example that is later presented.

We complete the discussion in this section by solving a validation example. Let us consider a test problem whose exact solution is known. That is, a simply supported square plate of a uniform thickness h, occupying the region $\Omega_{sq} = \{(x,y): 0 < x, y < a\}$ and subjected to a transverse load

$$q(x,y) = q_0 \sin\left(\frac{\pi x}{a}\right) \sin\left(\frac{\pi y}{a}\right).$$

The boundary conditions of a simple support can be written in terms of the deflection, rotations, and bending moments. Clearly, those vanish along the edges providing

$$w|_{x=0,a} = 0, \quad M_x|_{x=0,a} = 0, \quad \beta_y|_{x=0,a} = 0$$

$$w|_{y=0,a} = 0, \quad M_y|_{y=0,a} = 0, \quad \beta_x|_{y=0,a} = 0 \tag{6.60}$$

It can easily be verified by inspection that the deflection function $w(x,y)$ taken in the form of the following trigonometric function

$$w(x,y) = \frac{q_0}{D} \frac{(2-\sigma)\chi^2 + \delta_1}{4\delta_1 \chi^4} \sin\left(\frac{\pi x}{a}\right) \sin\left(\frac{\pi y}{a}\right)$$

along with the rotation functions $\beta_x(x,y)$ and $\beta_y(x,y)$ taken as

$$\beta_x(x,y) = \frac{q_0}{D} \frac{\delta_2 \chi^2 - 1}{4\chi^3} \cos\left(\frac{\pi x}{a}\right) \sin\left(\frac{\pi y}{a}\right)$$

and

$$\beta_y(x,y) = \frac{q_0}{D} \frac{\delta_2 \chi^2 - 1}{4\chi^3} \sin\left(\frac{\pi x}{a}\right) \cos\left(\frac{\pi y}{a}\right)$$

represent in this case the components of the true solution vector $\mathbf{U}(x,y)$ for the boundary value problem posed by eqns (6.35) and (6.60). Here we use the following notations

$$\chi = \frac{\pi}{a}, \quad \delta_1 = \frac{5(1-\sigma)}{h^2}, \quad \delta_2 = \frac{\sigma}{\delta_1}$$

6.3. ELASTIC ISOTROPIC MEDIA

Table 6.1. Relative errors of the approximate solution for the test problem posed by eqns (6.35) and (6.60)

Partitioning, $M = m_x \times m_y$	Relative error, %			
	Deflection	Rotations	Moments	Shear forces
5×5	3.16	3.36	3.42	3.46
8×8	1.24	1.26	1.34	1.39
10×10	0.78	0.80	0.82	0.85
12×12	0.65	0.66	0.68	0.71

In actually computing the components of the vector $\mathbf{U}^h(x,y)$ for this test problem, we uniformly partitioned the basic region Ω_{sq} into $M = m_x \times m_y$ elementary rectangles. The uniformly convergent series, which represent the domain integrals over the elementary rectangles Ω_m (see eqns (6.56)–(6.58)), have been truncated to attain the level of accuracy of 10^{-5}. The physical and geometrical parameters in the statement are $E = 0.21 \times 10^6 MPa$, $\sigma = 0.3$, $a = 1.0m$, and $h = 0.025m$.

In Table 6.1, some data are displayed as obtained within this study. The maximal values of relative errors are shown for the deflection, rotations, bending moments, and shear forces, versus the dimension $M = m_x \times m_y$ of the partitioning used.

Two evident observations follow from the data of Table 6.1. First, these data show a high degree of practical convergence of the computational procedure developed in this study. Indeed, for the partitioning of 5×5 (the partitioning parameter $M = 25$), the relative error slightly varies from column to column but stays within the range of about 3.5%. It then drops down to the value of about 0.7% for the partitioning of 12×12 ($M = 144$). Hence, the accuracy level for all of the components is nearly directly proportional to the partitioning parameter M.

The second observation is even more impressive. From the data presented, it is evident that for a fixed partitioning, the displacements and stresses have been computed with an equal level of accuracy. This appearance is not accidental, it is typical for computational implementations based on the influence function method, representing one of its most distinguishable and superior features.

6.3 Elastic isotropic media

We will now extend the technique developed in the preceding section to the construction of influence matrices for some problem classes occurring in the displacement formulation of the plane problem in theory of elasticity.

Consider the boundary value problem

$$\Lambda\left(\frac{\partial}{\partial x},\frac{\partial}{\partial y};\lambda,\mu\right)\mathbf{U}(x,y)=-\mathbf{F}(x,y),\quad (x,y)\in\Omega_r \qquad (6.61)$$

$$B_{1x}[\mathbf{U}(0,y)]=0,\quad B_{2x}[\mathbf{U}(a,y)]=0 \qquad (6.62)$$

$$B_{1y}[\mathbf{U}(x,0)]=0,\quad B_{2y}[\mathbf{U}(x,b)]=0 \qquad (6.63)$$

for the rectangle $\Omega_r=\{(x,y):0<x<a,0<y<b\}$ which is composed of an isotropic homogeneous elastic material whose Lame's coefficients are denoted as λ and μ. Here the vector $\mathbf{U}(x,y)$, defined as

$$\mathbf{U}(x,y)=\begin{pmatrix}u_x(x,y)\\u_y(x,y)\end{pmatrix},$$

represents the displacement vector of a point $(x,y)\in\Omega_r$, with $u_x(x,y)$ and $u_y(x,y)$ being its x component and y component, respectively. The vector $\mathbf{F}(x,y)$, defined as

$$\mathbf{F}(x,y)=\begin{pmatrix}f_x(x,y)\\f_y(x,y)\end{pmatrix},$$

represents the body force vector applied to Ω_r.

The system in eqn (6.61) is usually referred to as Lame's system of the plane problem in theory of elasticity. The entries Λ_{ij} of the matrix-operator Λ, as is known [74], are expressed as

$$\Lambda_{11}\equiv(\lambda+\mu)\frac{\partial^2}{\partial x^2}+\mu\Delta,\quad \Lambda_{12}\equiv(\lambda+\mu)\frac{\partial^2}{\partial x\partial y}$$

$$\Lambda_{21}\equiv(\lambda+\mu)\frac{\partial^2}{\partial x\partial y},\quad \Lambda_{22}\equiv(\lambda+\mu)\frac{\partial^2}{\partial y^2}+\mu\Delta$$

where Δ represents Cartesian Laplacian. The operators B_{jx} and B_{jy}, $(j=1,2)$ in eqns (6.62) and (6.63) specify the way in which the given rectangle interacts with the surrounding medium.

As is known [74], the components $\sigma_{xx}(x,y)$, $\sigma_{yy}(x,y)$, and $\sigma_{xy}(x,y)$ of the stress tensor are expressed in terms of the components $u_x(x,y)$ and $u_y(x,y)$ of the displacement vector as

$$\sigma_{xx}=(\lambda+2\mu)\frac{\partial u_x}{\partial x}+\mu\frac{\partial u_y}{\partial y},\quad \sigma_{yy}=(\lambda+2\mu)\frac{\partial u_y}{\partial y}+\mu\frac{\partial u_x}{\partial x},$$

and

$$\sigma_{xy}=\mu\left(\frac{\partial u_x}{\partial y}+\frac{\partial u_y}{\partial x}\right).$$

6.3. ELASTIC ISOTROPIC MEDIA

To obtain the influence matrix of the homogeneous ($\mathbf{F}(x,y)=\mathbf{0}$) boundary value problem corresponding to that of eqns (6.61)–(6.63), we express the displacement vector $\mathbf{U}(x,y)$ and the right-hand side vector $\mathbf{F}(x,y)$ in eqn (6.61) by the Fourier series

$$\mathbf{U}(x,y)=\sum_{n=0}^{\infty} Q_n(y)\mathbf{U}_n(x), \quad \mathbf{F}(x,y)=\sum_{n=0}^{\infty} Q_n(y)\mathbf{F}_n(x) \qquad (6.64)$$

where

$$Q_n(y)=\begin{pmatrix} \cos(\nu y) & 0 \\ 0 & \sin(\nu y) \end{pmatrix}, \quad \nu=\frac{n\pi}{b}$$

Note that the representation of the vector-function $\mathbf{U}(x,y)$ in eqn (6.64) must satisfy the boundary conditions imposed by eqn (6.63). This occurs, for example, in a case of practical importance for which the operators B_{1y} and B_{2y} are defined as

$$B_{1y} = B_{2y} \equiv \begin{pmatrix} 0 & I \\ \mu\,\partial/\partial y & \mu\,\partial/\partial x \end{pmatrix}$$

This case of boundary conditions can be interpreted as a contact with the absolutely rigid half-planes $y\leq 0$ and $y\geq b$ without friction and detachment.

Upon substituting the expressions for $\mathbf{U}(x,y)$ and $\mathbf{F}(x,y)$ from eqn (6.64) into eqns (6.61) and (6.62), one obtains a set ($n=0,1,2,...$) of systems of linear ordinary differential equations

$$\Lambda_n\left(\frac{d}{dx}; \lambda, \mu, \nu\right)\mathbf{U}_n(x)=-\mathbf{F}_n(x), \quad x\in(0,a) \qquad (6.65)$$

in the coefficients $\mathbf{U}_n(x)$ of the first of the series of eqn (6.64).

Notice that the case of $n=0$ must be considered separately from the general case of $n=1,2,3,\ldots$, because when $n=0$, the above system reduces to a single equation provided the fundamental solution sets are different for these two cases. Therefore, in what follows, we will first go through the general case and then turn to the case of $n=0$.

Clearly, $\mathbf{U}_n(x)$ should satisfy boundary conditions imposed as

$$B_{1x}^n\left(\frac{d}{dx}\right)\mathbf{U}_n(0)=0, \quad B_{2x}^n\left(\frac{d}{dx}\right)\mathbf{U}_n(a)=0 \qquad (6.66)$$

The entries Λ_{ij}^n of the matrix-operator Λ_n are expressed as

$$\Lambda_{11}^n \equiv (\lambda+2\mu)\frac{d^2}{dx^2}-\nu^2\mu, \quad \Lambda_{12}^n \equiv (\lambda+\mu)\nu\frac{d}{dx}$$

$$\Lambda_{21}^n \equiv -(\lambda+\mu)\nu\frac{d}{dx}, \quad \Lambda_{22}^n \equiv \mu\frac{d^2}{dx^2} - (\lambda+2\mu)\nu^2$$

Since the system in eqn (6.65) has constant coefficients, its fundamental solution set

$$\mathbf{U}_{n1}(x) = \begin{pmatrix} e^{\nu x} \\ -e^{\nu x} \end{pmatrix}, \quad \mathbf{U}_{n2}(x) = \begin{pmatrix} -(\lambda+\mu)\nu x e^{\nu x} \\ ((\lambda+\mu)\nu x + (\lambda+3\mu))e^{\nu x} \end{pmatrix}$$

$$\mathbf{U}_{n3}(x) = \begin{pmatrix} e^{-\nu x} \\ e^{-\nu x} \end{pmatrix}, \quad \mathbf{U}_{n4}(x) = \begin{pmatrix} (\lambda+\mu)\nu x e^{-\nu x} \\ ((\lambda+\mu)\nu x - (\lambda+3\mu))e^{-\nu x} \end{pmatrix} \quad (6.67)$$

for $n = 1, 2, 3, \ldots$ can be readily obtained by using standard techniques.

In compliance with Lagrange's method of variation of parameters, the general solution of the system in eqn (6.65) can be expressed as

$$\mathbf{U}_n(x) = P_n(x)\mathbf{C}_n(x) \quad (6.68)$$

where $P_n(x) = (\mathbf{U}_{nj}(x))$, with $j = \overline{1,4}$, represents a 2×4 matrix whose columns are the vectors from eqn (6.68), while $\mathbf{C}_n(x)$ is a vector whose entries are unknown functions. The latter can be obtained following the standard procedure of the method of variation of parameters, according to which one obtains the system of linear algebraic equations

$$P_n^*(x)\mathbf{C}_n'(x) = \mathbf{F}_n^*(x)$$

in the derivative $\mathbf{C}_n'(x)$ of the vector function $\mathbf{C}_n(x)$. The coefficient matrix $P_n^*(x)$ of the above system represents a 4×4 matrix whose structure can be described by its horizontal partitioning into two 2×4 matrices the upper of which is $P_n(x)$ as introduced in eqn (6.68) and the lower represents the derivative $P_n'(x)$ of $P_n(x)$. Structure of the right-hand side vector $\mathbf{F}_n^*(x)$ can also be described by using its partitioning in such a way that the upper two entries are zeros, while the lower sub vector is expressed in terms of the right-hand side vector $-\mathbf{F}_n(x)$ of the system in eqn (6.65) as

$$\mathbf{F}_n^*(x) = \begin{pmatrix} (\lambda+2\mu)^{-1} f_x^n(x) \\ \mu^{-1} f_y^n(x) \end{pmatrix}$$

with $f_x^n(x)$ and $f_y^n(x)$ representing the components of the right-hand side vector $-\mathbf{F}_n(x)$ of the system in eqn (6.65).

Since the determinant of the coefficient matrix of the above system represents Wronskian of a fundamental solution set of a homogeneous system of linear differential equations, the system is well-posed and its solution vector can consequently be written in terms of the inverse $(P_n^*(x))^{-1}$ of $P_n^*(x)$ as

$$\mathbf{C}_n'(x) = (P_n^*(x))^{-1} \mathbf{F}_n^*(x)$$

6.3. ELASTIC ISOTROPIC MEDIA

This relation can be rewritten in terms of the vector $\mathbf{F}_n(x)$ as

$$\mathbf{C}'_n(x) = -R_n(x)\mathbf{F}_n(x) \tag{6.69}$$

where $R_n(x)$ is a 4×2 matrix whose columns represent the third and fourth columns of $(P_n^*(x))^{-1}$ divided by $\lambda+2\mu$ and μ, respectively.

A straightforward integration of the relation in eqn (6.69) yields the following expression for the vector $\mathbf{C}_n(x)$

$$\mathbf{C}_n(x) = -\int_0^x R_n(s)\mathbf{F}_n(s)ds + \mathbf{D}_n$$

with $\mathbf{D}_n$ denoting a column-vector consisting of arbitrary constant coefficients that will be determined later upon satisfying the boundary conditions imposed by eqn (6.66).

Substituting the above expression for $\mathbf{C}_n(x)$ into eqn (6.68) and bringing the factor $P_n(x)$ under the integral sign, one obtains

$$\mathbf{U}_n(x) = \int_0^x S_n(x,s)\mathbf{F}_n(s)ds + P_n(x)\mathbf{D}_n \tag{6.70}$$

where $S_n(x,s) = P_n(x)R_n(s)$ represents a 2×2 matrix whose entries $S_{ij}^n(x,s)$, being determined for the fundamental solution set shown by eqn (6.67), are obtained as

$$S_{11}^n(x,s) = -\frac{\lambda+\mu}{2\mu}(x-s)\cosh(\nu(x-s)) + \frac{\lambda+3\mu}{2\nu\mu}\sinh((\nu(x-s)),$$

$$S_{12}^n(x.s) = -\frac{\lambda+\mu}{2(\lambda+2\mu)}(x-s)\sinh(\nu(x-s)),$$

$$S_{21}^n(x,s) = \frac{\lambda+\mu}{2\mu}(x-s)\sinh(\nu(x-s)),$$

and

$$S_{22}^n(x,s) = \frac{\lambda+\mu}{2(\lambda+2\mu)}(x-s)\cosh(\nu(x-s))$$
$$+ \frac{\lambda+3\mu}{2\nu(\lambda+2\mu)}\sinh((\nu(x-s)).$$

The components of the vector $\mathbf{D}_n$ in eqn (6.70) can be found by satisfying the boundary conditions imposed by eqn (6.66). In doing so, it is important to keep in mind that the entries $S_{ij}^n(x,s)$ of $S_n(x,s)$ hold the property $S_{ij}^n(x,x) = 0$. Hence, the differentiation of $\mathbf{U}_n(x)$, defined in the form of an integral depending on a parameter (see eqn (6.70)), can be carried out by simply taking derivatives of $S_{ij}^n(x,s)$ with respect to x under the integral sign. This yields the system of four linear algebraic equations

$$T_n\mathbf{D}_n = \mathbf{\Psi}_n \tag{6.71}$$

in the components of the vector $\mathbf{D}_n$. The first two rows of the coefficient matrix T_n of this system are obtained as

$$B_{1x}^n\left(\frac{d}{dx}\right)[P_n(0)]$$

while the third and fourth rows are obtained as

$$B_{2x}^n\left(\frac{d}{dx}\right)[P_n(a)]$$

The first two components of the right-hand side vector $\mathbf{\Psi}_n$ are zero, while its third and fourth components are given by

$$\int_0^a Z_n(a,s)\mathbf{F}_n(s)ds$$

where

$$Z_n(x,s) = -B_{2x}^n\left(\frac{d}{dx}\right)[S_n(x,s)]$$

At this stage of the development, it is clear that any set of physically feasible boundary conditions imposed on the edges $x = 0$ and $x = a$ of the rectangle Ω_r can readily be treated within the scope of the present technique. Notice that only the expressions for entries of the matrix T_n and of the vector $\mathbf{\Psi}_n$ are actually affected by the boundary operators B_{1x}^n and B_{2x}^n.

After some standard algebra, the solution vector $\mathbf{D}_n$ to the system in eqn (6.71) can be presented in the form

$$\mathbf{D}_n = \int_0^a W_n(s)\mathbf{F}_n(s)ds \qquad (6.72)$$

where $W_n(s) = T_n^{-1} Z_n^*(a,s)$, with $Z_n^*(a,s)$ representing a 4×2 matrix whose first two rows are zero, while the third and fourth rows are given by $Z_n(a,s)$.

Upon substituting the expression for $\mathbf{D}_n$ from eqn (6.72) into eqn (6.70), the vector function $\mathbf{U}_n(x)$ is found to be in the form

$$\mathbf{U}_n(x) = \int_0^x S_n(x,s)\mathbf{F}_n(s)ds + \int_0^a P_n(x)W_n(s)\mathbf{F}_n(s)ds$$

or by combining both of the above integrals, one obtains for $\mathbf{U}_n(x)$ a compact single integral representation as follows

$$\mathbf{U}_n(x) = \int_0^a g_n(x,s)\mathbf{F}_n(s)ds, \quad n=1,2,3,\ldots \qquad (6.73)$$

6.3. ELASTIC ISOTROPIC MEDIA

The kernel matrix $g_n(x,s)$ of the above expression defined in pieces as

$$g_n(x,s) = \begin{cases} S_n(x,s) + P_n(x)W_n(s) & \text{for } x \geq s \\ P_n(x)W_n(s) & \text{for } x \leq s \end{cases}$$

represents in fact the Green's matrix to the homogeneous boundary value problem corresponding to that of eqns (6.65) and (6.66).

As we have already pointed out, the case $n=0$ of the system in eqn (6.65) degenerates to a single differential equation in the x component of the vector $\mathbf{U}_0(x)$. Clearly, the Green's function $g_0(x,s)$ to the boundary value problem associated with this case can also be obtained by our technique.

We will now show how the influence matrix of the homogeneous boundary value problem corresponding to that of eqns (6.61)–(6.63) can be expressed in terms of the Green's function $g_0(x,s)$ and the Green's matrices $g_n(x,s)$ recently obtained. According to the fundamental rule for the coefficients of a Fourier series (Fourier-Euler formula [31, 68]), we write the following expression

$$\mathbf{F}_n(s) = \frac{\varepsilon_n}{b}\int_0^b Q_n(t)\mathbf{F}(s,t)dt, \quad \varepsilon_n = \begin{cases} 1, & n=0 \\ 2, & n \geq 1 \end{cases}$$

for the vector function $\mathbf{F}_n(s)$ which represents the coefficients of the second of the series from eqn (6.64) in terms of the vector function $\mathbf{F}(s,t)$.

Upon substituting the vector function $\mathbf{U}_n(x)$ from eqn (6.73) into the first of the expansions of eqn (6.64), one finally obtains the solution to the boundary value problem posed by eqns (6.61)–(6.63) as follows

$$\mathbf{U}(x,y) = \iint_{\Omega_r} \left[\frac{1}{b}\sum_{n=0}^{\infty}\varepsilon_n Q_n(y) g_n(x,s) Q_n(t)\right] \mathbf{F}(s,t) d\Omega_r(s,t)$$

Thus, in light of the definition introduced in the opening part of this chapter, it follows that the kernel matrix

$$G(x,y;s,t) = \frac{1}{b}\sum_{n=0}^{\infty}\varepsilon_n Q_n(y) g_n(x,s) Q_n(t) \qquad (6.74)$$

of the above integral representation is the influence matrix of the homogeneous boundary value problem stated by eqns (6.61)–(6.63).

One can obtain explicit expressions for the entries of the influence matrix $G(x,y;s,t)$ for a particular set of the boundary conditions imposed by eqn (6.62). We below present the entries $G_{ij}(x,y;s,t)$ of $G(x,y;s,t)$ for the case

$$B_{1x} \equiv \begin{pmatrix} I & 0 \\ \mu\partial/\partial y & \mu\partial/\partial x \end{pmatrix}, \quad B_{2x} \equiv \begin{pmatrix} I & 0 \\ 0 & I \end{pmatrix}$$

in compliance with which the x component of the displacement vector $\mathbf{U}(x,y)$ and the shear component of the stress tensor are prescribed to zero along the edge $x=0$ of Ω_r (clearly, these can be interpreted either as conditions of symmetry along the edge $x=0$ or as a contact with the absolutely rigid half-plane $x \leq 0$ without friction and detachment), while in accordance with B_{2x} the edge $x=a$ is assumed to be fixed. For $x \leq s$ one obtains

$$G_{11}(x,y;s,t) = \frac{1}{b(\lambda+2\mu)} \left\{ \frac{x(s-a)}{a} \right.$$

$$\left. + \sum_{n=1}^{\infty} \frac{1}{\Delta^*} [S_{11}^n(a,s)\Phi_1(x,a) + S_{21}^n(a,s)\Phi_2(x,a)] \cos(\nu y)\cos(\nu t) \right\},$$

$$G_{12}(x,y;s,t) = \frac{1}{b\mu} \sum_{n=1}^{\infty} \frac{1}{\Delta^*} [S_{12}^n(a,s)\Phi_1(x,a)$$

$$+ S_{22}^n(a,s)\Phi_2(x,a)] \cos(\nu y)\sin(\nu t),$$

$$G_{21}(x,y;s,t) = \frac{1}{b(\lambda+2\mu)} \sum_{n=1}^{\infty} \frac{1}{\Delta^*} [S_{11}^n(a,s)\Phi_3(x,a)$$

$$+ S_{21}^n(a,s)\Phi_4(x,a)] \sin(\nu y)\cos(\nu t),$$

and

$$G_{22}(x,y;s,t) = \frac{1}{b\mu} \sum_{n=1}^{\infty} \frac{1}{\Delta^*} [S_{12}^n(a,s)\Phi_3(x,a)$$

$$+ S_{22}^n(a,s)\Phi_4(x,a)] \sin(\nu y)\sin(\nu t),$$

where the components $S_{ij}^n(x,s)$ can be found in the relations that immediately follow eqn (6.70), while expressions for Δ^* and Φ_i are given as

$$\Delta^* = \frac{\lambda+3\mu}{\nu a(\lambda+\mu)} \cosh(\nu a)\sinh(\nu a) - 1,$$

$$\Phi_1 = \frac{2x}{a} \cosh(\nu a)\cosh(\nu x)$$

$$- 2\sinh(\nu x) \left[\sinh(\nu a) + \frac{\lambda+3\mu}{\nu a(\lambda+\mu)} \cosh(\nu a) \right],$$

$$\Phi_2 = \frac{2}{a} [x\sinh(\nu a)\cosh(\nu x) - a\cosh(\nu a)\sinh(\nu x)],$$

$$\Phi_3 = \frac{2}{a} [a\sinh(\nu a)\cosh(\nu x) - x\cosh(\nu a)\sinh(\nu x)],$$

and

$$\Phi_4 = 2\cosh(\nu x) \left[\cosh(\nu a) - \frac{\lambda+3\mu}{\nu a(\lambda+\mu)} \sinh(\nu a) \right]$$

6.3. ELASTIC ISOTROPIC MEDIA

$$-\frac{2x}{a}\sinh(\nu a)\sinh(\nu x).$$

Notice that the self-adjointness of the boundary value problem posed by eqns (6.61)–(6.63) yields symmetry for the entries of the influence matrix, that is $G(x, y; s, t) = G(s, t; x, y)$. Therefore, the expressions for $G_{ij}(x, y; s, t)$ valid for $x \geq s$ can be obtained from those above by just interchanging of x with s.

In some particular cases, expansions of the kind in eqn (6.74) can be completely summed up, providing an opportunity to express entries of the influence matrix in closed form (in terms of finite combinations of elementary functions). To illustrate this point, we consider the problem posed by eqns (6.61)–(6.63) on the semi-strip $\Omega_{ss} = \{(x, y): 0 < x < \infty, 0 < y < b\}$. That is, the parameter a in the statement for the rectangle Ω_r should be replaced with infinity. Let again, as in the previous case, the matrix operator B_{1x} of the boundary condition in eqn (6.62) be imposed as

$$B_{1x} \equiv \begin{pmatrix} I & 0 \\ \mu \partial/\partial y & \mu \partial/\partial x \end{pmatrix},$$

and assume, in addition, that the vector $\mathbf{U}(x, y)$ is bounded when x approaches infinity (this assumption replaces the second of the conditions in eqn (6.62)).

Clearly, the procedure of derivation of an influence matrix is not affected by boundary conditions until after we build the coefficient matrix T_n and the right-hand side vector $\mathbf{\Psi}_n$ for the system in eqn (6.71).

In the case under consideration, the components $D_{nj}, (j = \overline{1,4})$ of the vector $\mathbf{D}_n$ are obtained from the system in eqn (6.71) in the form

$$D_{n1} = -D_{n3} =$$

$$-\int_0^\infty \frac{e^{-\nu s}}{4\nu\mu(\lambda+2\mu)} \{[(\lambda+3\mu)+(\lambda+\mu)\nu s]F_{n1}(s)+(\lambda+\mu)\nu s F_{n2}(s)\}ds,$$

and

$$D_{n2} = -D_{n4} = -\int_0^\infty \frac{e^{-\nu s}}{4\nu\mu(\lambda+2\mu)}[F_{n1}(s)+F_{n2}(s)]ds.$$

By substituting these expressions for D_{nj} into eqn (6.70), the components $U_{n1}(x)$ and $U_{n2}(x)$ of the vector $\mathbf{U}_n(x)$ are found as

$$U_{n1}(x) = \int_0^x \omega\{[\alpha\nu(x-s)\cosh(\nu(x-s))-\beta\sinh(\nu(x-s))]F_{n1}(s)$$

$$+\alpha\nu(x-s)\sinh(\nu(x-s))F_{n2}(s)\}ds$$

$$+ \int_0^\infty \omega e^{-\nu s}\{[(\alpha\nu s + \beta)\sinh(\nu x) - \alpha\nu x \cosh(\nu x)]F_{n1}(s)$$
$$+ [\alpha\nu(s\sinh(\nu x) - x\cosh(\nu x)]F_{n2}(s)\}ds$$

and

$$U_{n2}(x) = -\int_0^x \omega\{(\alpha\nu(x-s)\sinh(\nu(x-s))F_{n1}(s)$$
$$+ [\alpha\nu(x-s)\cosh(\nu(x-s)) + \beta\sinh(\nu(x-s))]F_{n2}(s)\}ds$$
$$+ \int_0^\infty \omega e^{-\nu s}\{[\alpha\nu(x\sinh(\nu x) - s\cosh(\nu x)]F_{n1}(s)$$
$$- [(\alpha\nu s - \beta)\cosh(\nu x) - \alpha\nu x \sinh(\nu x)]F_{n2}(s)\}ds,$$

where, for the sake of compactness, we introduced parameters $\alpha = \lambda + \mu$, $\beta = \lambda + 3\mu$ and $\omega = 2[\nu(\alpha^2 - \beta^2)]^{-1}$. This yields explicit representations of the components $g_{ij}^n(x,s)$ of the kernel matrix $g_n(x,s)$ in eqn (6.73). After some elementary but quite cumbersome algebra, they are found, for $x \leq s$, in the form

$$g_{11}^n(x.s) = \frac{\omega}{2}\left\{[\beta - \alpha\nu(x-s)]e^{\nu(x-s)} - [\beta + \alpha\nu(x+s)]e^{-\nu(x+s)}\right\},$$

$$g_{12}^n(x,s) = \frac{\omega\nu\alpha}{2}\left[(x-s)e^{\nu(x-s)} + (x+s)e^{-\nu(x+s)}\right],$$

$$g_{21}^n(x,s) = \frac{\omega\nu\alpha}{2}\left[(x-s)e^{\nu(x-s)} - (x+s)e^{-\nu(x+s)}\right],$$

and

$$g_{22}^n(x.s) = \frac{\omega}{2}\left\{[\beta + \alpha\nu(x-s)]e^{\nu(x-s)} + [\beta - \alpha\nu(x+s)]e^{-\nu(x+s)}\right\}.$$

Notice that due to the self-adjointness of the problem under consideration, those expressions for $g_{ij}^n(x,s)$ which are valid for $x \geq s$ are obtainable from the corresponding above ones by the interchange of x with s.

As we already mentioned, the case of $n = 0$ is to be treated separately. In conformity with the described procedure, one obtains for $x \leq s$

$$g_{11}^0(x,s) = -\frac{2x}{\alpha+\beta}, \quad g_{12}^0 = g_{21}^0 = g_{22}^0 = 0$$

With the expressions for $g_{ij}^n(x,y)$, with $n = 0, 1, 2, ...$, just presented, the expansions in eqn (6.74) are entirely summable. This is possible with the aid of the summation formula from eqn (6.13) along with two other known (see, for example, [1, 29]) relations

$$\sum_{n=0}^\infty p^n \cos(n\gamma) = \frac{1 - p\cos(\gamma)}{1 - 2p\cos(\gamma) + p^2}$$

6.3. ELASTIC ISOTROPIC MEDIA

and
$$\sum_{n=1}^{\infty} p^n \sin(n\gamma) = \frac{p\sin(\gamma)}{1-2p\cos(\gamma)+p^2}$$

that are valid for $p^2 < 1$ and $0 \leq \gamma < 2\pi$. By performing the summation, the entries of the influence matrix of the problem under consideration for the semi-strip of width $b = \pi$ are obtained in the form

$$G_{11}(x,y;s,t) = \frac{1}{4\pi\mu(\lambda+2\mu)} \left\{ (\lambda+3\mu) \ln \frac{E(z-\zeta)E(z-\bar{\zeta})}{E(z+\zeta)E(z+\bar{\zeta})} \right.$$

$$-4\mu T(x,s) + (\lambda+\mu)(x+s)\left[\frac{P(z+\zeta)}{E^2(z+\zeta)} + \frac{P(z+\bar{\zeta})}{E^2(z+\bar{\zeta})}\right]$$

$$\left. - (\lambda+\mu)|x-s|\left[\frac{P(z-\zeta)}{E^2(z-\zeta)} + \frac{P(z-\bar{\zeta})}{E^2(z-\bar{\zeta})}\right] \right\},$$

$$G_{12}(x,y;s,t) = \frac{\lambda+\mu}{4\pi\mu(\lambda+2\mu)} \left\{ (x+s)\left[\frac{S(z+\zeta)}{E^2(z+\zeta)} - \frac{S(z+\bar{\zeta})}{E^2(z+\bar{\zeta})}\right] \right.$$

$$\left. + (x-s)\left[\frac{S(z-\zeta)}{E^2(z-\zeta)} - \frac{S(z-\bar{\zeta})}{E^2(z-\bar{\zeta})}\right] \right\},$$

$$G_{21}(x,y;s,t) = \frac{\lambda+\mu}{4\pi\mu(\lambda+2\mu)} \left\{ (x+s)\left[\frac{S(z+\zeta)}{E^2(z+\zeta)} + \frac{S(z+\bar{\zeta})}{E^2(z+\bar{\zeta})}\right] \right.$$

$$\left. - (x-s)\left[\frac{S(z-\zeta)}{E^2(z-\zeta)} - \frac{S(z-\bar{\zeta})}{E^2(z-\bar{\zeta})}\right] \right\},$$

and
$$G_{22}(x,y;s,t) = \frac{1}{4\pi\mu(\lambda+2\mu)} \left\{ (\lambda+3\mu) \ln \frac{E(z-\zeta)E(z+\bar{\zeta})}{E(z+\zeta)E(z-\bar{\zeta})} \right.$$

$$+ (\lambda+\mu)(x+s)\left[\frac{P(z+\bar{\zeta})}{E^2(z+\bar{\zeta})} - \frac{P(z+\zeta)}{E^2(z+\zeta)}\right]$$

$$\left. - (\lambda+\mu)|x-s|\left[\frac{P(z-\bar{\zeta})}{E^2(z-\bar{\zeta})} - \frac{P(z-\zeta)}{E^2(z-\zeta)}\right] \right\},$$

where, as usual, by $z = x+iy$ and $\zeta = s+it$ we denote the observation and the source point, respectively. And in addition to the real-valued function $E(p)$ of a complex variable p that was introduced in Chapter 5 (see eqn (5.6)) and used also in Section 6.1 of the present chapter, we denote

$$P(p) = \Re(e^p - 1), \quad S(p) = \Im(e^p - 1), \quad \text{and} \quad T(x,s) = \begin{cases} x, & x \leq s \\ s, & x \geq s \end{cases}$$

The suggested technique can be utilized for obtaining influence matrices to a number of plane problems of theory of elasticity stated over regions of standard configurations such as an infinite strip, semi-strip, rectangle, circle, circular sector, and so on. To illustrate this point, we consider the first interior problem

$$\Lambda\left(\frac{\partial}{\partial r},\frac{\partial}{\partial \phi};\lambda,\mu\right)\mathbf{U}(r,\phi)=0 \quad (x,y)\in\Omega_c \tag{6.75}$$

$$\mathbf{U}(a,\phi)=0 \tag{6.76}$$

for a circle $\Omega_c = \{(r,\phi): 0<r<a, 0\le \phi < 2\pi\}$ of radius a, with the displacement vector $\mathbf{U}(x,y)$ being fixed on the contour $r=a$. Lame's system in eqn (5.75) is written in polar coordinates. The entries Λ_{ij} of the matrix-operator Λ, as is known (see, for example, [74]), can be presented in the form

$$\Lambda_{11}\equiv(\lambda+\mu)\frac{\partial}{\partial r}\left(\frac{\partial}{\partial r}+\frac{1}{r}\right)-\frac{\mu}{r^2}+\mu\Delta,$$

$$\Lambda_{12}\equiv(\lambda+\mu)\frac{\partial}{\partial r}\left(\frac{1}{r}\frac{\partial}{\partial \phi}\right)-\frac{2\mu}{r^2}\frac{\partial}{\partial \phi},$$

$$\Lambda_{21}\equiv(\lambda+\mu)\frac{1}{r}\frac{\partial}{\partial \phi}\left(\frac{\partial}{\partial r}+\frac{1}{r}\right)+\frac{2\mu}{r^2}\frac{\partial}{\partial \phi},$$

and

$$\Lambda_{22}\equiv(\lambda+\mu)\frac{1}{r^2}\frac{\partial^2}{\partial \phi^2}-\frac{\mu}{r^2}+\mu\Delta,$$

where Δ represents the polar Laplacian

$$\Delta \equiv \frac{1}{r}\frac{\partial}{\partial r}\left(r\frac{\partial}{\partial r}\right)+\frac{1}{r^2}\frac{\partial^2}{\partial \phi^2}$$

Omitting details of the derivation procedure, we exhibit just final expressions of the entries $G_{ij}=G_{ij}(r,\phi;\rho,\psi)$ of the influence matrix for the boundary value problem posed by eqns (6.75) and (6.76) as follows

$$G_{11}=\frac{1}{8\pi(1-\sigma)}\left\{2(1+\sigma)r\rho\sin^2(\phi-\psi)\frac{(a^2-r^2)(a^2-\rho^2)}{(|z-\zeta||z\bar{\zeta}-a^2|)^2}\right.$$

$$+2(3-\sigma)\cos(\phi-\psi)\ln\frac{a|z-\zeta|}{|z\bar{\zeta}-a^2|^2}-\frac{1+\sigma}{3-\sigma}\frac{(a^2-r^2)(a^2-\rho^2)}{a^2|z\bar{\zeta}-a^2|^2}$$

$$\times\left[2(1-\sigma)(a^2\cos(\phi-\psi)-r\rho)-(1+\sigma)\frac{(a^4+r^2\rho^2)(a^2\cos(\phi-\psi)-2r\rho)}{|z\bar{\zeta}-a^2|^2}\right]\right\},$$

$$G_{12} = \frac{\sin(\phi-\psi)}{8\pi(1-\sigma)} \left\{ 2(1+\sigma) \frac{(a^2-r^2)[\rho^2(a^2+r^2)-(a^2+\rho^2)r\rho\cos(\phi-\psi)]}{(|z-\zeta||z\bar{\zeta}-a^2|)^2} \right.$$

$$+2(3-\sigma)\ln\frac{a|z-\zeta|}{|z\bar{\zeta}-a^2|} + \frac{1+\sigma}{3-\sigma}\frac{a^2}{|z\bar{\zeta}-a^2|^4}[(1+\sigma)(a^4-r^2\rho^2)$$

$$\left. \times(a^2-r^2+\rho(r-\rho))+2a^4(2r(r-\rho)+(1-\sigma)(a^2-r^2))]\right\},$$

$$G_{21} = \frac{\sin(\phi-\psi)}{8\pi(1-\sigma)} \left\{ 2(1+\sigma) \frac{(a^2-r^2)[\rho^2(a^2+r^2)-(a^2+\rho^2)r\rho\cos(\phi-\psi)]}{(|z-\zeta||z\bar{\zeta}-a^2|)^2} \right.$$

$$-2(3-\sigma)\ln\frac{a|z-\zeta|}{|z\bar{\zeta}-a^2|} - \frac{1+\sigma}{3-\sigma}\frac{a^2}{|z\bar{\zeta}-a^2|^4}[(1+\sigma)(a^4-r^2\rho^2)$$

$$\left. \times(a^2-r^2+\rho(r-\rho))+2(1-\sigma)a^4(2a^2+r(r-\rho))]\right\},$$

and

$$G_{22} = \frac{1}{8\pi(1-\sigma)} \left\{ 2(1+\sigma)r\rho\sin^2(\phi-\psi)\frac{(r^2-a^2)(a^2-\rho^2)}{(|z-\zeta||z\bar{\zeta}-a^2|)^2} \right.$$

$$+2(3-\sigma)\cos(\phi-\psi)\ln\frac{a|z-\zeta|}{|z\bar{\zeta}-a^2|^2} + \frac{1+\sigma}{3-\sigma}\frac{(a^2-r^2)(a^2-\rho^2)}{a^2|z\bar{\zeta}-a^2|^2}$$

$$\left. \times \left[4(a^2\cos(\phi-\psi)-r\rho)+(1+\sigma)\frac{(a^4+r^2\rho^2)(a^2\cos(\phi-\psi)-2r\rho)}{|z\bar{\zeta}-a^2|^2}\right]\right\},$$

where, as usual for problems stated in polar coordinates, by $z = r(\cos(\phi) + i\sin(\phi))$ and $\zeta = \rho(\cos(\psi) + i\sin(\psi))$ we denote the observation (r, ϕ) and the source (ρ, ψ) point, respectively.

6.4 Orthotropic media

In this section, we will demonstrate a possibility to utilize the suggested technique to a certain problem class of the plane problem of theory of elasticity for regions composed of orthotropic materials. For this purpose, the reader's attention is turned to the problem of static equilibrium of a semi-strip $\Omega_{ss} = \{(x,y): x > 0, 0 < y < b\}$ of width b composed of a homogeneous orthotropic material whose principal directions of orthotropy coincide with the coordinate axes x and y. Let $E_1, E_2; \sigma_1, \sigma_2$, and S represent the elasticity moduli in the directions of x and y; Poisson ratios, and shear modulus of the material, respectively.

We formulate the problem in displacements, with $\mathbf{U}(x,y)$ representing the displacement vector of the point $(x,y) \in \Omega_{ss}$. The system of differential equations that models the static equilibrium of the semi-strip Ω_{ss}, which undergoes the distributed body load $\mathbf{F}(x,y)$, is taken in the form

$$\Lambda\left(\frac{\partial}{\partial x}, \frac{\partial}{\partial y}; A_{11}, A_{12}, A_{22}, A_{66}\right)\mathbf{U}(x,y) = -\mathbf{F}(x,y), \quad (x,y) \in \Omega_{ss} \quad (6.77)$$

Here the entries Λ_{ij} of the matrix-operator Λ are given (see [39]) by

$$\Lambda_{11} \equiv A_{11}\frac{\partial^2}{\partial x^2} + A_{66}\frac{\partial^2}{\partial y^2}, \quad \Lambda_{12} \equiv (A_{12}+A_{66})\frac{\partial^2}{\partial x \partial y}$$

$$\Lambda_{21} \equiv (A_{12}+A_{66})\frac{\partial^2}{\partial x \partial y}, \quad \Lambda_{22} \equiv A_{66}\frac{\partial^2}{\partial x^2} + A_{22}\frac{\partial^2}{\partial y^2}$$

where the coefficients are defined in terms of the physical properties of the material as

$$A_{11} = \frac{E_1}{\omega_{12}}, \quad A_{12} = \frac{\sigma_1 E_2}{\omega_{12}}, \quad A_{22} = \frac{E_2}{\omega_{12}}, \quad A_{66} = S, \quad \omega_{12} = 1 - \sigma_1 \sigma_2$$

We follow here the conventional system of notations for orthotropic materials as that introduced in [39].

The equilibrium state of Ω_{ss} is assumed to be symmetric with respect to the y axis. This yields, for the x and y components of the displacement vector $\mathbf{U}(x,y) = (u(x,y), v(x,y))^T$, the following boundary conditions

$$u(0,y) = 0, \quad \frac{\partial v(0,y)}{\partial x} = 0 \quad (6.78)$$

along the line $x = 0$.

We also assume that the strip is contacting the absolutely rigid half-planes $y \leq 0$ and $y \geq b$ without friction and detachment. This yields the boundary conditions

$$v(x,0) = v(x,b) = 0, \quad \frac{\partial u(x,0)}{\partial y} = \frac{\partial u(x,b)}{\partial y} = 0 \quad (6.79)$$

along the boundary lines $y = 0$ and $y = b$. In addition, the components $u(x,y)$ and $v(x,y)$ of the displacement vector are assumed to be bounded as x approaches infinity.

Representing, analogously to the development in the previous section, the vectors $\mathbf{U}(x,y)$ and $\mathbf{F}(x,y)$ by series expansions written as

$$\mathbf{U}(x,y) = \sum_{n=0}^{\infty} Q_n(y)\mathbf{U}_n(x), \quad \mathbf{F}(x,y) = \sum_{n=0}^{\infty} Q_n(y)\mathbf{F}_n(x) \quad (6.80)$$

6.4. ORTHOTROPIC MEDIA

where
$$Q_n(y) = \begin{pmatrix} \cos(\nu y) & 0 \\ 0 & \sin(\nu y) \end{pmatrix}, \quad \nu = \frac{n\pi}{b}$$

one obtains, for $n = 1, 2, 3, \ldots$, the system of ordinary differential equations

$$\Lambda_n\left(\frac{d}{dx}; A_{11}, A_{12}, A_{22}, A_{66}\right) \mathbf{U}_n(x) = \mathbf{F}_n(x), \quad n = 1, 2, 3, \ldots \quad (6.81)$$

subject to boundary conditions given by

$$u_n(0) = \frac{dv_n(0)}{dx} = 0, \quad |u_n(\infty)| < \infty, \quad |v_n(\infty)| < \infty \quad (6.82)$$

where $u_n(x)$ and $v_n(x)$ represent the x and y components, respectively, of the vector $\mathbf{U}_n(x)$. The case of $n = 0$ should be considered separately. We will turn to this point later.

The entries Λ_{ij}^n of the matrix-operator Λ_n are expressed as

$$\Lambda_{11}^n \equiv A_{11}\frac{d^2}{dx^2} - A_{66}\nu^2, \quad \Lambda_{22}^n \equiv A_{66}\frac{d^2}{dx^2} - A_{22}\nu^2$$

$$\Lambda_{12}^n = -\Lambda_{21}^n \equiv (A_{12} + A_{66})\nu\frac{d}{dx}, \quad \nu = \frac{n\pi}{b}$$

The fundamental solution set of the homogeneous system corresponding to that of eqn (6.81) can be represented by the following four vectors

$$\mathbf{U}_n^*(x; \pm p), \quad \mathbf{U}_n^*(x; \pm q) \quad (6.83)$$

where the upper and lower cases are considered separately. Here

$$\mathbf{U}_n^*(x; \omega) = \left(\frac{A_{12} + A_{66}}{\nu(A_{66} - \omega^2 A_{11})} e^{\nu\omega x}, \frac{1}{\nu\omega} e^{\nu\omega x}\right)^T$$

and the values of p and q in eqn (6.83) are defined as the upper and lower cases, respectively, of the expression

$$\left(\frac{A_{11}A_{22} - A_{12}^2 - 2A_{12}A_{66} \pm B^{1/2}}{2A_{11}A_{66}}\right)^{1/2}$$

with B being defined as

$$B = (A_{12}^2 - A_{11}A_{22})\left[(A_{12}^2 - A_{11}A_{22}) + 4A_{66}(A_{12} + A_{66})\right]$$

Taking the four vectors from eqn (6.83) and following the standard procedure of the method of variation of parameters, one can obtain the general solution to the system in eqn (6.81) written as

$$\mathbf{U}_n(x) = \int_0^x S_n(x,s)\mathbf{F}_n(s)ds + P_n(x)\mathbf{C}_n \qquad (6.84)$$

The entries $S_{ij}^n(x,s)$ of the kernel matrix $S_n(x,s)$ in the above integral representation are defined as

$$S_{11}^n(x,s) = \frac{pD_q \sinh(\nu p(x-s)) - \cdots}{A_{11}R},$$

$$S_{12}^n(x,s) = \frac{\alpha[\cosh(\nu p(x-s)) + \cdots]}{A_{66}(D_p - D_q)},$$

$$S_{21}^n(x,s) = \frac{D_p D_q [\cosh(\nu p(x-s)) - \cdots]}{\alpha A_{11}R},$$

and

$$S_{22}^n(x,s) = \frac{qD_p \sinh(\nu p(x-s)) - \cdots}{A_{66}pq(D_p - D_q)},$$

where with 'dots', we denote the term that can be obtained from the preceding one by interchanging of p with q, while the parameters D_p, D_q, α, and R are defined as

$$D_p = A_{66} - A_{11}p^2, \quad D_q = A_{66} - A_{11}q^2$$
$$\alpha = A_{12} + A_{66}, \quad R = p^2 D_q - q^2 D_p.$$

The matrix $P_n(x) = (\mathbf{U}_n^*(x;\omega))$ in eqn (6.84) represents a 2×4 matrix whose columns are the vectors representing the fundamental solution set shown in eqn (6.83). $\mathbf{C}_n$ in eqn (6.84) represents a four-dimensional vector of arbitrary coefficients that can be found by satisfying the boundary conditions imposed by eqn (6.82). This yields

$$\mathbf{C}_n = \int_0^\infty W_n(s)\mathbf{F}_n(s)ds$$

Upon substituting this expression into eqn (6.84), one obtains the following integral representation

$$\mathbf{U}_n(x) = \int_0^\infty G_n(x,s)\mathbf{F}_n(s)ds \qquad (6.85)$$

for the solution vector $\mathbf{U}_n(x)$ of the problem posed by eqns (6.81) and (6.82). The kernel matrix $G_n(x,s)$ of this representation found as

$$G_n(x,s) = \begin{cases} S_n(x,s) + P_n(x)W_n(s), & x \geq s \\ P_n(x)W_n(s), & x \leq s \end{cases}$$

6.4. ORTHOTROPIC MEDIA

represents the Green's matrix to the homogeneous boundary value problem corresponding to that of eqns (6.81) and (6.82). The entries $G_{ij}^n(x,s)$ of $G_n(x,s)$ are obtained in the form

$$G_{11}^n(x,s) = -\frac{k_{11}}{\nu}\begin{cases} pD_q e^{-\nu px}\sinh(\nu ps) - \ldots, & x \geq s \\ pD_q e^{-\nu ps}\sinh(\nu px) - \ldots, & x \leq s, \end{cases}$$

$$G_{12}^n(x,s) = \frac{k_{12}}{\nu}\begin{cases} e^{-\nu px}\cosh(\nu ps) - \ldots, & x \geq s \\ -e^{-\nu ps}\sinh(\nu px) + \ldots, & x \leq s, \end{cases}$$

$$G_{21}^n(x,s) = \frac{k_{21}}{\nu}\begin{cases} e^{-\nu px}\sinh(\nu ps) - \ldots, & x \geq s \\ -e^{-\nu ps}\cosh(\nu px) + \ldots, & x \leq s, \end{cases}$$

and

$$G_{22}^n(x,s) = \frac{k_{22}}{\nu}\begin{cases} pD_q e^{-\nu px}\cosh(\nu ps) - \ldots, & x \geq s \\ pD_q e^{-\nu ps}\cosh(\nu px) - \ldots, & x \leq s, \end{cases}$$

where the coefficients k_{ij} are expressed as

$$k_{11} = (A_{11}R)^{-1}, \quad k_{12} = \alpha\left[A_{66}(D_p - D_q)\right]^{-1}$$

$$k_{21} = D_p D_q(\alpha A_{11}R)^{-1}, \quad k_{22} = [pqA_{66}(D_p - D_q)]^{-1}$$

Recall that for $n=0$ the system in eqn (6.81) reduces to a single equation and the Green's function $G_{11}^0(x,s)$ of the corresponding boundary value problem can be found as

$$G_{11}^0(x,s) = -\frac{1}{A_{11}}\begin{cases} x, & x \leq s \\ s, & x \geq s \end{cases}$$

Upon substituting expressions for the components of the vector $\mathbf{F}_n(s)$, which are obtainable by the fundamental rule for coefficients of a Fourier series, into eqn (6.85) and substituting then $\mathbf{U}_n(x)$ from eqn (6.85) into the first of the expansions of eqn (6.80), one finally finds the vector solution $\mathbf{U}(x,y)$ of the boundary value problem posed by eqns (6.77)–(6.79) in the form

$$\mathbf{U}(x,y) = \int_0^b \int_0^\infty G(x,y;s,t)\mathbf{F}(s,t)\,ds\,dt \tag{6.86}$$

Hence, in light of the definition introduced in the opening part of this chapter, it follows that the kernel matrix $G(x,y;s,t)$

$$G(x,y;s,t) = \frac{1}{b}\sum_{n=0}^\infty \varepsilon_n Q_n(y) G_n(x,s) Q_n(t), \quad \varepsilon_n = \begin{cases} 1, & n=0 \\ 2, & n \geq 1 \end{cases}$$

in eqn (6.86) represents the influence matrix of a unit concentrated body force for the homogeneous boundary value problem corresponding to that found in eqns (6.77)–(6.79). It appears that the series expansions of the entries of $G(x,y;s,t)$ are entirely summable. The summation can be carried out with the aid of the standard (see, for example, [29]) summation formula

$$\sum_{n=1}^{\infty} \frac{p^n}{n} \sin(n\alpha) = \arctan \frac{p\sin(\alpha)}{1-p\cos(\alpha)}, \quad p<1, \ 0\leq\alpha<2\pi \quad (6.87)$$

along with that from eqn (6.13). Finally, the entries $G_{ij} = G_{ij}(x,y;s,t)$ of the influence matrix for the orthotropic semi-strip Ω_{ss} of width b are obtained in the form

$$G_{11} = G_{11}^0(x,s) + \frac{k_{11}}{4\pi}\left[pD_q \ln \frac{E(x,y;-s,t;p)E(x,y;-s,-t;p)}{E(x,y;s,t;p)E(x,y;s,-t;p)} - \ldots\right],$$

$$G_{12} = \frac{k_{12}}{2\pi}\{[\arctan(M(x,y;s,t;p))-\ldots]+\delta[\arctan(M(x,y;-s,t;p))-\ldots]$$
$$+[\arctan(M(x,y;s,-t;q))-\ldots]+\delta[\arctan(M(x,y;-s,-t;q))-\ldots]\},$$

$$G_{21} = \frac{k_{21}}{2\pi}\{[\arctan(M(x,y;s,t;q))-\ldots]+\delta[\arctan(M(x,y;-s,t;p))-\ldots]$$
$$+[\arctan(M(x,y;s,-t;q))-\ldots]+\delta[\arctan(M(x,y;-s,-t;p))-\ldots]\},$$

and

$$G_{22} = \frac{k_{22}}{4\pi}\left[pD_q \ln \frac{E(x,y;s,-t;p)E(x,y;-s,-t;p)}{E(x,y;s,t;p)E(x,y;-s,t;p)} - \ldots\right],$$

where with 'dots' we again denote terms that can be obtained from the preceding ones by interchanging of p with q and vice versa. Expressions for the functions $E(x,y;s,t;p)$, $\delta=\delta(x,s)$, and $M(x,y;s,t;\omega)$ are given by

$$E(x,y;s,t;p) = \left|1 - 2\exp\left(-\frac{\pi p}{b}|x+s|\right)\cos\left(\frac{\pi}{b}(y+t)\right) + \exp\left(-\frac{2\pi p}{b}|x+s|\right)\right|,$$

$$\delta(x,s) = \begin{cases} 1, & x \geq s \\ -1, & x \leq s, \end{cases}$$

and

$$M(x,y;s,t;\omega) = \sin\left(\frac{\pi}{b}(y+t)\right)\left[\exp\left(\frac{\pi\omega}{b}|x+s|\right) - \cos\left(\frac{\pi}{b}(y+t)\right)\right]^{-1}$$

Thus, in this section it has been shown that the suggested algorithm of the construction of influence matrices can be successfully implemented to the plane problem of theory of elasticity for regions composed of orthotropic

6.5 Elastic equilibrium of thin shells

In this section, we will focus on one more problem class from solid mechanics, to which the suggested technique of the construction of influence matrices can also be successfully implemented. It will be shown that the technique can be productive for problems simulating the static equilibrium of thin elastic shells of revolution. It is worth noting that this problem class was the very first in solid mechanics to which the Green's (influence) matrix method was applied more than three decades ago (see, for example, [28]).

Consider the geometrically linear elastic equilibrium of a thin shell whose middle surface represents a surface of revolution closed in the longitudinal direction, with $x \in (0, l)$ and $\phi \in [0, 2\pi)$ representing its meridional (latitudinal) and longitudinal coordinates, respectively. The meridional cross section of the shell is depicted in Figure 6.3. A system of partial differential equations modeling the equilibrium state is written in displacements

$$\Lambda\left(\frac{\partial}{\partial x}, \frac{\partial}{\partial \phi}, x\right) \mathbf{W}(x, \phi) = \mathbf{F}(x, \phi) \tag{6.88}$$

where

$$\mathbf{W}(x, \phi) = \begin{pmatrix} u(x, \phi) \\ v(x, \phi) \\ w(x, \phi) \end{pmatrix} \quad \text{and} \quad \mathbf{F}(x, \phi) = \begin{pmatrix} X(x, \phi) \\ Y(x, \phi) \\ Z(x, \phi) \end{pmatrix}$$

are the displacement and the load vectors, respectively, with $u(x, \phi), v(x, \phi)$, and $w(x, \phi)$ representing the components of the displacement vector in the directions of x, ϕ, and the normal to the middle surface, respectively, and with $X(x, \phi), Y(x, \phi)$, and $Z(x, \phi)$ representing the components of the loading vector in the corresponding directions. The coefficients of the entries Λ_{ij} of the matrix-operator Λ

$$\Lambda\left(\frac{\partial}{\partial x}, \frac{\partial}{\partial \phi}, x\right) = \left(\Lambda_{ij}\left(\frac{\partial}{\partial x}, \frac{\partial}{\partial \phi}, x\right)\right)_{3\times 3}$$

represent functions of the meridional coordinate x. In accordance with the total order of the system in eqn (6.88), which is equal to eight, the boundary

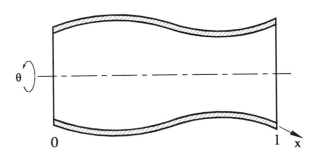

Figure 6.3 Thin elastic shell of revolution

conditions on the edges $x=0$ and $x=l$ are written in the form

$$B_0\left(\frac{\partial}{\partial x},\frac{\partial}{\partial \phi}\right)\mathbf{W}(0,\phi)=0, \quad B_l\left(\frac{\partial}{\partial x},\frac{\partial}{\partial \phi}\right)\mathbf{W}(l,\phi)=0 \qquad (6.89)$$

where B_0 and B_l represent 4×3 matrix-operators. That is, four linearly independent boundary conditions are imposed at each edge of the shell.

To construct the influence matrix of the homogeneous boundary value problem corresponding to that posed by eqns (6.88) and (6.89), we expand the vectors $\mathbf{W}(x,\phi)$ and $\mathbf{F}(x,\phi)$ in the following trigonometric series with respect to the variable ϕ

$$\mathbf{W}(x,\phi)=\sum_{n=0}^{\infty}Q_n(\phi)\mathbf{W}_n(x), \quad \mathbf{F}(x,\phi)=\sum_{n=0}^{\infty}Q_n(\phi)\mathbf{F}_n(x) \qquad (6.90)$$

where the transformation matrix $Q_n(\phi)$ is given as

$$Q_n(\phi)=\begin{pmatrix}\cos(n\phi) & 0 & 0 \\ 0 & \sin(n\phi) & 0 \\ 0 & 0 & \cos(n\phi)\end{pmatrix}$$

The expansions in eqn (6.90) presume a symmetry of the equilibrium state of the shell with respect to an axial plane.

The vectors $\mathbf{W}_n(x)$ and $\mathbf{F}_n(x)$ in eqn (6.90) are expressed as

$$\mathbf{W}_n(x)=\begin{pmatrix}u_n(x) \\ v_n(x) \\ w_n(x)\end{pmatrix}, \quad \mathbf{F}_n(x)=\begin{pmatrix}X_n(x) \\ Y_n(x) \\ Z_n(x)\end{pmatrix}$$

Upon substituting the expansions from eqn (6.90) into eqn (6.88), we obtain

$$\Lambda\left(\frac{\partial}{\partial x},\frac{\partial}{\partial \phi},x\right)\left[\sum_{n=0}^{\infty}Q_n(\phi)\mathbf{W}_n(x)\right]=\sum_{n=0}^{\infty}Q_n(\phi)\mathbf{F}_n(x)$$

6.5. ELASTIC EQUILIBRIUM OF THIN SHELLS

or

$$\sum_{n=0}^{\infty} Q_n(\phi)\left[\Lambda_n\left(\frac{d}{dx},x\right)\mathbf{W}_n(x)\right]=\sum_{n=0}^{\infty}Q_n(\phi)\mathbf{F}_n(x)$$

From this equation, it follows that the vectors $\mathbf{W}_n(x)$ should satisfy the following system of ordinary differential equations

$$\Lambda_n\left(\frac{d}{dx},x\right)\mathbf{W}_n(x)=\mathbf{F}_n(x),\quad (n=0,1,2,\ldots) \qquad (6.91)$$

Since the displacement vector $\mathbf{W}(x,\phi)$ is expressed in terms of the series in eqn (6.90), the boundary conditions from eqn (6.89) result in

$$\sum_{n=0}^{\infty}Q_n(\phi)\left[B_{0n}\left(\frac{d}{dx}\right)\mathbf{W}_n(0)\right]=0$$

$$\sum_{n=0}^{\infty}Q_n(\phi)\left[B_{ln}\left(\frac{d}{dx}\right)\mathbf{W}_n(l)\right]=0$$

or

$$B_{0n}\left(\frac{d}{dx}\right)\mathbf{W}_n(0)=0,\quad B_{ln}\left(\frac{d}{dx}\right)\mathbf{W}_n(l)=0 \qquad (6.92)$$

Thus, the original boundary value problem stated by eqns (6.88) and (6.89) has reduced to a set ($n = 0, 1, 2, \ldots$) of systems of linear ordinary differential equations (6.91) subjected to the boundary conditions in eqn (6.92). The system in eqn (6.91) has variable (generally speaking) coefficients (notice that only cylindrical and conical shells yield systems with constant coefficients) and the total order of the system is eight. As it is known [35], such systems do not allow analytical solutions. Hence, their fundamental solution sets required for the construction of Green's matrices can be computed by using numerical methods. We below describe one of the possible approaches to such a computation.

In doing so, the system in eqn (6.91) is converted to the normal form

$$\frac{dy_i(x)}{dx}=\sum_{j=1}^{8}\alpha_{ij}(x)y_j(x)+f_i(x),\quad (i=\overline{1,8}) \qquad (6.93)$$

where the unknown functions $y_i(x)$ are defined in terms of the components of the vector $\mathbf{W}_n(x)$ as

$$y_1(x)=u_n(x),\quad y_2(x)=\frac{du_n(x)}{dx}$$

$$y_3(x)=v_n(x),\quad y_4(x)=\frac{dv_n(x)}{dx}$$

$$y_5(x) = w_n(x), \quad y_6(x) = \frac{dw_n(x)}{dx}$$

$$y_7(x) = \frac{d^2 w_n(x)}{dx^2}, \quad y_8(x) = \frac{d^3 w_n(x)}{dx^3} \qquad (6.94)$$

The coefficients $\alpha_{ij}(x)$ of the system in eqn (6.93) and its right-hand side $f_i(x)$ are defined by the operator Λ_n and the right-hand side vector $\mathbf{F}_n(x)$, respectively, of the system in eqn (6.91).

To show how the boundary conditions in eqn (6.92) can be expressed in terms of the newly introduced functions $y_i(x)$, we consider a particular case of the edge conditions. Let, for example, the edges $x = 0$ and $x = l$ of the shell be clamped and simply supported, respectively. This yields, for the matrices B_{0n} and B_{ln}

$$B_{0n}\left(\frac{d}{dx}\right) \equiv \begin{pmatrix} I & 0 & 0 \\ 0 & I & 0 \\ 0 & 0 & I \\ 0 & 0 & d/dx \end{pmatrix}, \quad B_{ln}\left(\frac{d}{dx}\right) \equiv \begin{pmatrix} I & 0 & 0 \\ 0 & I & 0 \\ 0 & 0 & I \\ 0 & 0 & d^2/dx^2 \end{pmatrix}$$

Thus, in this case, in light of the relations from eqn (6.94), the boundary conditions stated by eqn (6.92) can be reformulated in terms of $y_i(x)$ as

$$y_i(0) = 0, \quad \text{for } i = 1, 3, 5, 6$$
$$y_i(l) = 0, \quad \text{for } i = 1, 3, 5, 7 \qquad (6.95)$$

To obtain the Green's matrix for the homogeneous system

$$\frac{dy_i(x)}{dx} = \sum_{j=1}^{8} \alpha_{ij}(x) y_j(x), \quad (i = \overline{1,8}) \qquad (6.96)$$

associated with that found in eqn (6.93) and subjected to the boundary conditions in eqn (6.95), one is required to have the fundamental solution set for the above system. That is the set $(j = 1, 2, \ldots, 8)$ of its eight linearly independent vector-solutions $\{y_{ij}(x)\}$. These can be computed with the aid of standard procedures for numerical solution of Cauchy problems for systems of ordinary differential equations. Notice that for cylindrical and conical shells the fundamental solution sets can be obtained analytically, since the corresponding systems of differential equations have constant coefficients.

In compliance with Lagrange's method of variation of parameters, the general solution of the system in eqn (6.93) can be written in terms of the fundamental solution set $y_{ij}(x)$ as

$$y_i(x) = \sum_{j=1}^{8} y_{ij}(x) C_j(x), \quad (i = \overline{1,8}) \qquad (6.97)$$

6.5. ELASTIC EQUILIBRIUM OF THIN SHELLS

where $C_j(x)$ represent unknown functions whose derivatives $C'_j(x)$ have to satisfy the following system of linear algebraic equations

$$\sum_{j=1}^{8} y_{ij}(x)C'_j(x) = f_i(x), \quad (i = \overline{1,8}) \tag{6.98}$$

Recall that $f_i(x)$ are the right-hand terms of the system in eqn (6.93).

We now write the solution of the system from eqn (6.98) in the form

$$C'_i(x) = \sum_{j=1}^{8} y_{ij}^{-1}(x)f_j(x), \quad (i = \overline{1,8})$$

with $y_{ij}^{-1}(x)$ representing entries of the inverse of the coefficient matrix of the system in eqn (6.98).

Upon integrating the above expressions for $C'_i(x)$, one obtains the following integral representations for $C_i(x)$

$$C_i(x) = \int_0^x \sum_{j=1}^{8} y_{ij}^{-1}(s)f_j(s)ds + D_i, \quad (i = \overline{1,8})$$

where D_i are arbitrary coefficients. This allows the general solution of the system in eqn (6.93) to be written in the form

$$y_i(x) = \int_0^x \sum_{j=1}^{8} T_{ij}(x,s)f_j(s)ds + \sum_{j=1}^{8} y_{ij}(x)D_j \tag{6.99}$$

where

$$T_{ij}(x,s) = \sum_{k=1}^{8} y_{ik}(x)y_{kj}^{-1}(s)$$

By virtue of the boundary conditions imposed by eqn (6.95), one obtains, for the coefficients D_i, the following system of linear algebraic equations

$$\sum_{j=1}^{8} r_{ij}D_j = S_i, \quad (i=\overline{1,8}) \tag{6.100}$$

The entries r_{ij} of the first four rows of the 8×8 coefficient matrix R of this system are defined as

$$r_{ij} = y_{mj}(0), \quad (i=\overline{1,4},\ m=1,3,5,6)$$

The entries of the last four rows of that matrix are defined as

$$r_{ij} = y_{mj}(l), \quad (i=\overline{5,8},\ m=1,3,5,7)$$

The components S_i of the right-hand side vector of the system in eqn (6.100) are expressed as

$$S_i = \int_0^l \sum_{j=1}^8 T_{ij}^*(l,s) f_j(s) ds, \quad (i=\overline{1,8}) \qquad (6.101)$$

with $T_{ij}^*(l,s)$ being defined by

$$T_{ij}^*(x,s) = \begin{cases} 0, & (i=\overline{1,4}) \\ -T_{ij}(l,s), & (i=\overline{5,8}) \end{cases}$$

where $T_{ij}(l,s)$ represent the boundary values of the functions $T_{ij}(x,s)$ defined earlier through $y_{ik}(x)$ and $y_{kj}^{-1}(s)$ in the equation that immediately follows that of (6.99).

Thus, the solution of the system in eqn (6.100) can be found in terms of the inverse of its coefficient matrix R as

$$D_i = \sum_{j=1}^8 r_{ij}^{-1} S_j, \quad (i=\overline{1,8})$$

where r_{ij}^{-1} represents the entries of the inverse R^{-1} of R.

Substituting the components S_j from eqn (6.101) into the above relation, one obtains for D_i

$$D_i = \int_0^l \sum_{j=1}^8 P_{ij}(l,s) f_j(s) ds,$$

where

$$P_{ij}(l,s) = \sum_{k=1}^8 r_{ik}^{-1} T_{kj}^*(l,s).$$

Substitution of the above expression for D_i into eqn (6.99) yields, for the general solution of the system from eqn (6.93)

$$y_i(x) = \int_0^x \sum_{j=1}^8 T_{ij}(x,s) f_j(s) ds + \int_0^l \sum_{j=1}^8 H_{ij}(x,s) f_j(s) ds, \qquad (6.102)$$

where

$$H_{ij}(x,s) = \sum_{j=1}^8 y_{ik}(x) P_{kj}(l,s).$$

Finally, the expression for $y_i(x)$ in eqn (6.102) can be written in a compact integral form

$$y_i(x) = \int_0^l \sum_{j=1}^8 G_{ij}(x,s) f_j(s) ds, \quad i=\overline{1,8} \qquad (6.103)$$

6.5. ELASTIC EQUILIBRIUM OF THIN SHELLS

in which

$$G_{ij}(x,s) = \begin{cases} H_{ij}(x,s), & x \leq s \\ H_{ij}(x,s) + T_{ij}(x,s), & x \geq s \end{cases} \quad (6.104)$$

Thus, the solution of the boundary value problem posed by eqns (6.93) and (6.95) is finally found in the form of the definite integral in eqn (6.103). Subsequently, in view of the relations from eqn (6.94), the vector

$$\mathbf{W}_n(x) = \begin{pmatrix} u_n(x) \\ v_n(x) \\ w_n(x) \end{pmatrix}$$

that represents solution to the problem stated by eqns (6.91) and (6.92), can be written as

$$\mathbf{W}_n(x) = \int_0^l G^n(x,s) \mathbf{F}_n(s) ds$$

or, in an extended form

$$\begin{pmatrix} u_n(x) \\ v_n(x) \\ w_n(x) \end{pmatrix} = \int_0^l \begin{pmatrix} G_{12}(x,s) & G_{14}(x,s) & G_{18}(x,s) \\ G_{32}(x,s) & G_{34}(x,s) & G_{38}(x,s) \\ G_{52}(x,s) & G_{54}(x,s) & G_{58}(x,s) \end{pmatrix} \begin{pmatrix} X_n(s) \\ Y_n(s) \\ Z_n(s) \end{pmatrix} ds \quad (6.105)$$

The entries $G_{ij}(x,s)$ (with $i=1,3,5$ and $j=2,4,8$) of the kernel-matrix $G^n(x,s)$ of the integral representation in eqn (6.105) can be found among the other entries of the 8×8 kernel-matrix from eqn (6.103). They are shown in eqn (6.104).

Upon substituting expressions for the components of $\mathbf{W}_n(x)$ from eqn (6.105) into the first of the expansions of eqn (6.90) and then applying the Fourier-Euler formula

$$\mathbf{F}_n(s) = \frac{\varepsilon_n}{2\pi} \int_0^{2\pi} Q_n(\psi) \mathbf{F}(s,\psi) d\psi, \quad \varepsilon_n = \begin{cases} 1, & n=0 \\ 2, & n>0 \end{cases}$$

for the coefficients of the second of the expansions from eqn (6.90), one obtains the following integral representation

$$\mathbf{W}(x,\phi) = \int_0^l \int_0^{2\pi} G(x,\phi;s,\psi) \mathbf{F}(s,\psi) ds d\psi \quad (6.106)$$

of the solution to the original boundary value problem posed by eqns (6.88) and (6.89).

Hence, in light of the definition introduced in the prefatory part of this chapter, it follows that the kernel-matrix

$$G(x,\phi;s,\psi) = \sum_{n=0}^{\infty} \frac{\varepsilon_n}{2\pi} Q_n(\phi) G^n(x,s) Q_n(\psi)$$

of the integral in eqn (6.106) represents the influence matrix of the homogeneous problem corresponding to that posed by eqns (6.88) and (6.89).

Thus, what has just been presented is the algorithm for constructing influence matrices for problems modeling static equilibrium of thin elastic shells of revolution. Recall that the variability of the coefficients of the original equation (6.88) is taken care of by numerically solving a set of linearly independent Cauchy problems for the system in eqn (6.96). In regard to the numerical solution of such problems, it worth noting that, due to the *stiffness* of the original equation (6.88), which is caused by its small leading coefficients, effectiveness of the numerical solution significantly depends on the length of the shell under consideration. For 'long' shells, special care is required to obtain accurate results.

6.6 *Review Exercises*

6.1 Obtain the influence function of a transverse concentrated unit force for the infinite strip-shaped Poisson-Kirchhoff plate occupying the region $\Omega_s = \{(x, y): -\infty < x < \infty, 0 < y < b\}$, with simply supported edges.

6.2 Obtain the influence functions for the rectangular Poisson-Kirchhoff plate occupying the region $\Omega_r = \{(x, y): 0 < x < a, 0 < y < b\}$, with the following edge conditions:

(a) $w(0, y) = \partial^2 w(0, y)/\partial x^2 = 0$, $w(a, y) = \partial^2 w(a, y)/\partial x^2 = 0$
$w(x, 0) = \partial^2 w(x, 0)/\partial y^2 = 0$, $w(x, b) = \partial^2 w(x, b)/\partial y^2 = 0$
(all the edges are simply supported);

(b) $w(0, y) = \partial^2 w(0, y)/\partial x^2 = 0$, $\partial^2 w(a, y)/\partial x^2 = \partial^3 w(a, y)/\partial x^3 = 0$
$w(x, 0) = \partial^2 w(x, 0)/\partial y^2 = 0$, $w(x, b) = \partial^2 w(x, b)/\partial y^2 = 0$
(three edges are simply supported, while one is free);

(c) $w(0, y) = \partial w(0, y)/\partial x = 0$, $w(a, y) = \partial w(a, y)/\partial x = 0$
$w(x, 0) = \partial^2 w(x, 0)/\partial y^2 = 0$, $w(x, b) = \partial^2 w(x, b)/\partial y^2 = 0$
(two opposite edges are simply supported, while the other two are clamped);

(d) $w(0, y) = \partial^2 w(0, y)/\partial x^2 = 0$, $w(a, y) = \partial w(a, y)/\partial x = 0$
$w(x, 0) = \partial^2 w(x, 0)/\partial y^2 = 0$, $w(x, b) = \partial^2 w(x, b)/\partial y^2 = 0$
(three edges are simply supported, while one is clamped).

6.3 Obtain the influence function of a transverse concentrated unit force for the circular Poisson-Kirchhoff plate whose edge $r = a$ is elastically supported.

6.6. REVIEW EXERCISES

6.4 Obtain the influence matrix of a transverse concentrated unit force for the infinite strip-shaped Reissner plate occupying the region $\Omega_s = \{(x,y): -\infty < x < \infty, 0 < y < b\}$, with simply supported edges.

6.5 Obtain the influence matrix of a concentrated unit body force for an elastic medium occupying an infinite strip-shaped region $\Omega_s = \{(x,y): -\infty < x < \infty, 0 < y < b\}$ and subjected to the boundary conditions:

$$|u(-\infty, y)| < \infty, \quad |v(-\infty, y)| < \infty, \quad |u(\infty, y)| < \infty, \quad |v(\infty, y)| < \infty$$

$$v(x, 0) = 0, \quad \frac{\partial u(x, 0)}{\partial y} = 0, \quad v(x, b) = 0, \quad \frac{\partial u(x, b)}{\partial y} = 0$$

where $u(x,y)$ and $v(x,y)$ represent the x and y components, respectively, of the displacement vector at the point $(x,y) \in \Omega_s$.

6.6 Obtain the influence matrix of a concentrated unit body force for an elastic medium occupying a semi-strip shaped region $\Omega_{ss} = \{(x,y): 0 < x < \infty, 0 < y < b\}$ and subjected to the boundary conditions:

$$u(0, y) = v(0, y) = 0, \quad |u(\infty, y)| < \infty, \quad |v(\infty, y)| < \infty$$

$$v(x, 0) = 0, \quad \frac{\partial u(x, 0)}{\partial y} = 0, \quad v(x, b) = 0, \quad \frac{\partial u(x, b)}{\partial y} = 0$$

where $u(x,y)$ and $v(x,y)$ represent the x and y components, respectively, of the displacement vector at the point $(x,y) \in \Omega_{ss}$.

6.7 Obtain the influence matrix of a concentrated unit body force for an elastic medium occupying a rectangular region $\Omega_r = \{(x,y): 0 < x < a, 0 < y < b\}$ and subjected to the boundary conditions:

$$u(0, y) = v(0, y) = 0, \quad u(a, y) = v(a, y) = 0$$

$$v(x, 0) = 0, \quad \frac{\partial u(x, 0)}{\partial y} = 0, \quad v(x, b) = 0, \quad \frac{\partial u(x, b)}{\partial y} = 0$$

where $u(x,y)$ and $v(x,y)$ represent the x and y components, respectively, of the displacement vector at the point $(x,y) \in \Omega_r$.

6.8 Obtain the influence matrix of a concentrated unit body force for an elastic medium occupying a circular region of radius a that contacts a surrounding absolutely rigid plane without friction and detaching.

Chapter 7

Compound Media

In the two preceding chapters, we focused on the developing of algorithms for obtaining influence functions and matrices for problems of continuum mechanics formulated in regions occupied with materials whose physical properties either do not vary from point to point in the region or vary continuously. This implies that coefficients of governing differential equations represent continuous functions of spatial variables. In the present chapter, we will extend the notion of an influence function (matrix) to problems of continuum mechanics stated in compound media whose properties discontinuously vary within regions under consideration. Such an extension is possible using the material of Chapter 4.

We first consider fields of potential occurring in compound media. In extending the concept of an influence function to compound media, we consider the region $\Omega = \bigcup \Omega_i$, $(i=1,\ldots,m)$ of a compound structure (see Figure 7.1) belonging to two-dimensional Euclidean space. Let the constants λ_i specify physical properties (conductivities) of the homogeneous materials of which the fragments Ω_i are composed. Let also each of the functions $u_i(P)$ be defined in the corresponding fragment Ω_i of Ω and let each of them satisfy Poisson's equation

$$\Delta u_i(P) = -f_i(P), \quad P \in \Omega_i, \quad (i=\overline{1,m}) \tag{7.1}$$

subject to the boundary conditions

$$u_i(P) = 0, \quad P \in \Gamma_0, \quad (i=\overline{1,m}) \tag{7.2}$$

imposed along the outer contour Γ_0 of Ω and to the conditions of ideal contact

$$u_k(P) = u_{k+1}(P), \quad p \in \Gamma_k, \quad (k=\overline{1,m-1}) \tag{7.3}$$

$$\lambda_k \frac{\partial u_k(P)}{\partial n_k} = \lambda_{k+1} \frac{\partial u_{k+1}(P)}{\partial n_k}, \quad P \in \Gamma_k, \quad (k=\overline{1,m-1}) \tag{7.4}$$

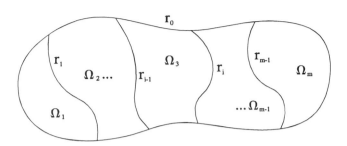

Figure 7.1 Compound region

formulated along the interfaces Γ_k. Here Δ represents the two-dimensional Laplace's operator and n_k indicate normal directions to Γ_k.

Assume that the boundary value problem posed by eqns (7.1)–(7.4) has a unique solution. That is, the corresponding homogeneous ($f_i(P) \equiv 0$) problem is assumed to have only the trivial solution such that all $u_i(P) = 0$. For the sake of compactness in the development that follows, we introduce two vector-functions defined over Ω as follows

$$\mathbf{U}(P) = (U_i(P))_{i=\overline{1,m}}, \quad \mathbf{F}(P) = (F_i(P))_{i=\overline{1,m}}$$

with their components $U_i(P)$ and $F_i(P)$ defined in pieces as

$$U_i(P) = \begin{cases} u_i(P), & P \in \Omega_i \\ 0, & P \in \Omega \setminus \Omega_i, \end{cases} \quad F_i(P) = \begin{cases} f_i(P), & P \in \Omega_i \\ 0, & P \in \Omega \setminus \Omega_i \end{cases} \quad (7.5)$$

We are now in a position to extend the concept of an influence function to compound media. In doing so, we introduce the following definition.

Definition: If for any allowable vector-function $\mathbf{F}(P)$, the vector-function $\mathbf{U}(P)$ is expressible in terms of the integral

$$\mathbf{U}(P) = \iint_\Omega G(P, Q) \mathbf{F}(Q) d\Omega(Q) \quad (7.6)$$

then let us call the kernel-matrix $G(P, Q)$ in this integral representation the *influence matrix* (*matrix of Green's type*) of the homogeneous problem corresponding to that posed by eqns (7.1)–(7.4).

The term 'allowable' with respect to $\mathbf{F}(P)$ implies that this vector-function is integrable over Ω. As it is commonly accepted in influence function related discussions, P and Q are referred to as the *field* (*observation*) and the *source* point, respectively.

7.1. POTENTIAL FIELDS IN COMPOUND REGIONS

The influence matrix $G(P,Q)$ represents an $m \times m$ matrix, with the entries $G_{ij}(P,Q)$, being defined for $P \in \Omega_i$ and $Q \in \Omega_j$, and meeting the following properties:

1. For $i \neq j$, $G_{ij}(P,Q)$ are harmonic functions in Ω_i, that is:

$$\Delta G_{ij}(P,Q) = 0, \quad \text{for} \quad P \in \Omega_i \text{ and } Q \notin \Omega_i;$$

2. For $i = j$, $G_{ii}(P,Q)$ are harmonic everywhere in Ω_i except for $P = Q$;

3. For $P = Q$, $G_{ii}(P,Q)$ possess logarithmic singularity of the type:

$$\frac{1}{2\pi} \ln \frac{1}{|P-Q|};$$

4. $G_{ij}(P,Q)$ satisfy all of the boundary and contact conditions imposed by eqn (7.2)–(7.4), in which they are involved.

In the discussion that follows, we develop an approach based on a version of Lagrange's method of variation of parameters to show how influence matrices for piecewise homogeneous media can practically be constructed for various situations in continuum mechanics. A number of particular examples from the potential theory and theory of elasticity are later considered.

7.1 Potential fields in compound regions

To proceed with a detailed description of the proposed algorithm for obtaining influence matrices for compound regions, we consider an infinite strip $\Omega_s = \{(x,y) \colon -\infty < x < \infty, 0 < y < \pi\}$ of width π (see Figure 7.2), comprised of two semi-strips: $\Omega_1 = \{(x,y) \colon -\infty < x < 0, 0 < y < \pi\}$ and $\Omega_2 = \{(x,y) \colon 0 < x < \infty, 0 < y < \pi\}$ occupied with materials whose physical

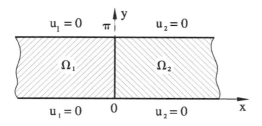

Figure 7.2 Compound strip

properties (conductivities) are specified with constants λ_1 and λ_2, respectively.

Let $u_1(x,y)$ and $u_2(x,y)$ be functions defined in Ω_1 and Ω_2, respectively, that satisfy Poisson's equations

$$\frac{\partial^2 u_1(x,y)}{\partial x^2}+\frac{\partial^2 u_1(x,y)}{\partial y^2}=-f_1(x,y), \quad (x,y)\in\Omega_1 \tag{7.7}$$

and

$$\frac{\partial^2 u_2(x,y)}{\partial x^2}+\frac{\partial^2 u_2(x,y)}{\partial y^2}=-f_2(x,y), \quad (x,y)\in\Omega_2 \tag{7.8}$$

subject to the boundary conditions

$$u_1(x,0)=u_2(x,0)=u_1(x,\pi)=u_2(x,\pi)=0 \tag{7.9}$$

imposed along the edges $y=0$ and $y=\pi$ of Ω_s and to the contact conditions

$$|u_1(-\infty,y)|<\infty, \quad |u_2(\infty,y)|<\infty \tag{7.10}$$

$$u_1(0,y)=u_2(0,y), \quad \frac{\partial u_1(0,y)}{\partial x}=\lambda\frac{\partial u_2(0,y)}{\partial x} \tag{7.11}$$

formulated along the interface line $x=0$. Here the parameter λ in the second condition of eqn (7.11) represents the ratio of the coefficients λ_2 and λ_1, that is $\lambda=\lambda_2/\lambda_1$.

Expand the unknown functions $u_i(x,y)$ and the right-hand terms $f_i(x,y)$, $(i=1,2)$ from the above formulation in Fourier series as

$$u_i(x,y)=\sum_{n=1}^{\infty}u_{i,n}(x)\sin(ny), \quad f_i(x,y)=\sum_{n=1}^{\infty}f_{i,n}(x)\sin(ny) \tag{7.12}$$

Upon substituting these expansions into the original formulation, one obtains the following set ($n=1,2,3,\ldots$) of three-point posed boundary value problems

$$\frac{d^2u_{i,n}(x)}{dx^2}-n^2u_{i,n}(x)=-f_{i,n}(x), \quad (i=1,2) \tag{7.13}$$

$$|u_{1,n}(-\infty)|<\infty, \quad |u_{2,n}(\infty)|<\infty \tag{7.14}$$

$$u_{1,n}(0)=u_{2,n}(0), \quad \frac{du_{1,n}(0)}{dx}=\lambda\frac{du_{2,n}(0)}{dx} \tag{7.15}$$

7.1. POTENTIAL FIELDS IN COMPOUND REGIONS

in the coefficients $u_{1,n}(x)$ and $u_{2,n}(x)$ of the first of the expansions in eqn (7.12), whose domains of definition are $(-\infty, 0)$ and $(0, \infty)$, respectively.

Clearly, the fundamental solution set for the homogeneous equation associated with (7.13) can be composed, for example, of the following two functions

$$e^{nx} \quad \text{and} \quad e^{-nx}$$

Hence, following the standard procedure of Lagrange's method of variation of parameters, we can seek the solution to the boundary value problem posed by eqns (7.13)–(7.15) in the form

$$u_{i,n}(x) = C_{i,n}(x) e^{nx} + D_{i,n}(x) e^{-nx}, \quad (i=1,2) \tag{7.16}$$

In compliance with this method, one obtains the following well-posed system of linear algebraic equations

$$\begin{pmatrix} e^{nx} & e^{-nx} \\ ne^{nx} & -ne^{-nx} \end{pmatrix} \begin{pmatrix} C'_{i,n}(x) \\ D'_{i,n}(x) \end{pmatrix} = \begin{pmatrix} 0 \\ -f_{i,n}(x) \end{pmatrix}, \quad (i=1,2) \tag{7.17}$$

in the derivatives of the variable coefficients $C_{i,n}(x)$ and $D_{i,n}(x)$ from eqn (7.16). This system yields

$$C'_{i,n}(x) = -\frac{1}{2n} e^{-nx} f_{i,n}(x), \quad D'_{i,n}(x) = \frac{1}{2n} e^{nx} f_{i,n}(x), \quad (i=1,2)$$

A straightforward integration of the above expressions provides

$$C_{1,n}(x) = -\frac{1}{2n} \int_{-\infty}^{x} e^{-ns} f_{1,n}(s) ds + \gamma_1,$$

$$D_{1,n}(x) = \frac{1}{2n} \int_{-\infty}^{x} e^{ns} f_{1,n}(s) ds + \delta_1,$$

$$C_{2,n}(x) = -\frac{1}{2n} \int_{0}^{x} e^{-ns} f_{2,n}(s) ds + \gamma_2,$$

and

$$D_{2,n}(x) = \frac{1}{2n} \int_{0}^{x} e^{ns} f_{2,n}(s) ds + \delta_2.$$

Note that the integration for $C_{1,n}(x)$ and $D_{1,n}(x)$ is carried out from negative infinity, whereas to obtain $C_{2,n}(x)$ and $D_{2,n}(x)$, we integrate from zero. The point is that these are the left-end points of the domains for $u_{1,n}(x)$ and $u_{2,n}(x)$, respectively.

Substituting the expressions just found into eqn (7.16) and properly rearranging the integral terms, one obtains

$$u_{1,n}(x) = \frac{1}{2n} \int_{-\infty}^{x} \left[e^{-n(x-s)} - e^{n(x-s)} \right] f_{1,n}(s) ds + \gamma_1 e^{nx} + \delta_1 e^{-nx} \tag{7.18}$$

and

$$u_{2,n}(x) = \frac{1}{2n}\int_0^x \left[e^{-n(x-s)} - e^{n(x-s)}\right] f_{2,n}(s)ds + \gamma_2 e^{nx} + \delta_2 e^{-nx} \qquad (7.19)$$

To obtain values of the constant coefficients γ_1, γ_2, δ_1, and δ_2, we take advantage of the boundary and contact conditions imposed by eqns (7.14) and (7.15). Clearly, the first condition in eqn (7.14) requires $\delta_1 = 0$. To satisfy the second condition in eqn (7.14), we rewrite the representation for $u_{2,n}(x)$ from eqn (7.19) in the form

$$u_{2,n}(x) = \left[\delta_2 + \frac{1}{2n}\int_0^x e^{ns} f_{2,n}(s)ds\right] e^{-nx}$$

$$+ \left[\gamma_2 - \frac{1}{2n}\int_0^x e^{-ns} f_{2,n}(s)ds\right] e^{nx}$$

from which it follows that to ensure the boundedness of $u_{2,n}(x)$ as x is taken to infinity, the coefficient of the positive exponential function e^{nx} in the above representation for $u_{2,n}(x)$ has to be zero. This yields

$$\gamma_2 = \frac{1}{2n}\int_0^\infty e^{-ns} f_{2,n}(s)ds$$

Upon satisfying the contact conditions in eqn (7.15), we derive the following well-posed system of linear algebraic equations

$$\gamma_1 - \delta_2 = \frac{1}{2n}\int_{-\infty}^0 \left(e^{-ns} - e^{ns}\right) f_{1,n}(s)ds + \frac{1}{2n}\int_0^\infty e^{-ns} f_{2,n}(s)ds,$$

$$\gamma_1 + \lambda\delta_2 = \frac{1}{2n}\int_{-\infty}^0 \left(e^{-ns} + e^{ns}\right) f_{1,n}(s)ds + \frac{\lambda}{2n}\int_0^\infty e^{-ns} f_{2,n}(s)ds$$

in γ_1 and δ_2. From this system, it follows

$$\gamma_1 = \frac{1}{2n}\int_{-\infty}^0 \left(e^{-ns} - \frac{\lambda-1}{\lambda+1}e^{ns}\right) f_{1,n}(s)ds + \frac{1}{2n}\int_0^\infty \frac{2\lambda}{\lambda+1} e^{-ns} f_{2,n}(s)ds$$

and

$$\delta_2 = \frac{1}{2n}\int_{-\infty}^0 \frac{2}{\lambda+1} e^{ns} f_{1,n}(s)ds + \frac{1}{2n}\int_0^\infty \frac{\lambda-1}{\lambda+1} e^{-ns} f_{2,n}(s)ds$$

Substituting the expressions for γ_1, γ_2, δ_1, and δ_2 just found into eqns (7.18) and (7.19), one obtains the solution to the boundary value problem posed by eqns (7.13)–(7.15) in the form

$$u_{1,n}(x) = \frac{1}{2n}\int_{-\infty}^x \left[e^{-n(x-s)} - e^{n(x-s)}\right] f_{1,n}(s)ds$$

7.1. POTENTIAL FIELDS IN COMPOUND REGIONS

$$+\frac{1}{2n}\int_{-\infty}^{0}\left[e^{n(x-s)}-\frac{\lambda-1}{\lambda+1}e^{n(x+s)}\right]f_{1,n}(s)ds$$

$$+\frac{1}{2n}\int_{0}^{\infty}\frac{2\lambda}{\lambda+1}e^{n(x-s)}f_{2,n}(s)ds$$

and

$$u_{2,n}(x)=\frac{1}{2n}\int_{0}^{x}\left[e^{-n(x-s)}-e^{n(x-s)}\right]f_{2,n}(s)ds$$

$$+\frac{1}{2n}\int_{-\infty}^{0}\frac{2}{\lambda+1}e^{-n(x+s)}f_{1,n}(s)ds$$

$$+\frac{1}{2n}\int_{0}^{\infty}\left[e^{n(x-s)}+\frac{\lambda-1}{\lambda+1}e^{-n(x+s)}\right]f_{2,n}(s)ds.$$

These expressions for $u_{1,n}(x)$ and $u_{2,n}(x)$ can be rewritten in a more compact form as

$$u_{1,n}(x)=\int_{-\infty}^{0}g_{11}^{n}(x,s)f_{1,n}(s)ds+\int_{0}^{\infty}g_{12}^{n}(x,s)f_{2,n}(s)ds, \quad x\in(-\infty,0] \quad (7.20)$$

and

$$u_{2,n}(x)=\int_{-\infty}^{0}g_{21}^{n}(x,s)f_{1,n}(s)ds+\int_{0}^{\infty}g_{22}^{n}(x,s)f_{2,n}(s)ds, \quad x\in[0,\infty) \quad (7.21)$$

The kernel-functions $g_{ij}^{n}(x,s)$ in these integral representations can be written as

$$g_{11}^{n}(x,s)=\frac{1}{2n(\lambda+1)}\begin{cases}(\lambda+1)e^{n(x-s)}-(\lambda-1)e^{n(x+s)}, & -\infty<x\le s\le 0\\(\lambda+1)e^{n(s-x)}-(\lambda-1)e^{n(x+s)}, & -\infty<s\le x\le 0,\end{cases}$$

$$g_{12}^{n}(x,s)=\frac{\lambda}{n(\lambda+1)}e^{n(x-s)}, \quad -\infty<x\le 0\le s<\infty,$$

$$g_{21}^{n}(x,s)=\frac{1}{n(\lambda+1)}e^{n(s-x)}, \quad -\infty<s\le 0\le x<\infty,$$

and

$$g_{22}^{n}(x,s)=\frac{1}{2n(\lambda+1)}\begin{cases}(\lambda+1)e^{n(x-s)}+(\lambda-1)e^{-n(x+s)}, & 0\le x\le s<\infty\\(\lambda+1)e^{n(s-x)}+(\lambda-1)e^{-n(x+s)}, & 0\le s\le x<\infty.\end{cases}$$

Recall the second Fourier series from eqn (7.12). By means of the fundamental rule for Fourier coefficients (Fourier-Euler formula), the coefficients $f_{i,n}(x)$ of that series can be written as

$$f_{i,n}(s) = \frac{2}{\pi} \int_0^\pi f_i(s,t) \sin(nt) dt, \quad (i=1,2), \quad n=1,2,3,\ldots$$

Upon substituting the above expressions for $f_{1,n}(s)$ and $f_{2,n}(s)$ into eqns (7.20) and (7.21) and substituting thereupon the expressions for $u_{1,n}(x)$ and $u_{2,n}(x)$ into the first series in eqn (7.12), one finally obtains the solution of the boundary value problem posed by eqns (7.7)–(7.11) in the form

$$u_1(x,y) = \int_0^\pi \int_{-\infty}^0 \left(\frac{2}{\pi} \sum_{n=1}^\infty g_{11}^n(x,s) \sin(ny) \sin(nt) \right) f_1(s,t) ds dt$$

$$+ \int_0^\pi \int_0^\infty \left(\frac{2}{\pi} \sum_{n=1}^\infty g_{12}^n(x,s) \sin(ny) \sin(nt) \right) f_2(s,t) ds dt$$

and

$$u_2(x,y) = \int_0^\pi \int_{-\infty}^0 \left(\frac{2}{\pi} \sum_{n=1}^\infty g_{21}^n(x,s) \sin(ny) \sin(nt) \right) f_1(s,t) ds dt$$

$$+ \int_0^\pi \int_0^\infty \left(\frac{2}{\pi} \sum_{n=1}^\infty g_{22}^n(x,s) \sin(ny) \sin(nt) \right) f_2(s,t) ds dt;$$

Thus, in light of the definition introduced in the opening part of this chapter, the series

$$G_{ij}(x,y;s,t) = \frac{2}{\pi} \sum_{n=1}^\infty g_{ij}^n(x,s) \sin(ny) \sin(nt), \quad (i,j=1,2) \quad (7.22)$$

represents the entries of the matrix of Green's type for the homogeneous boundary value problem corresponding to that of eqns (7.7)–(7.11). That is, we ultimately derived the influence matrix of a concentrated unit source for the compound strip Ω_s.

The series expansions for $G_{ij}(x,y;s,t)$ in eqn (7.22) can readily be summed up with the aid of the standard summation formula

$$\sum_{n=1}^\infty \frac{p^n}{n} \cos(n\alpha) = -\ln\sqrt{1 - 2p\cos\alpha + p^2}$$

that has repeatedly been used in Chapters 5 and 6. After the summation is completed, one obtains

$$G_{11}(z,\zeta) = \frac{1}{2\pi} \left[\frac{\lambda-1}{\lambda+1} \ln \frac{|1-e^{z+\bar{\zeta}}|}{|1-e^{z+\zeta}|} + \ln \frac{|1-e^{z-\bar{\zeta}}|}{|1-e^{z-\zeta}|} \right],$$

7.1. POTENTIAL FIELDS IN COMPOUND REGIONS

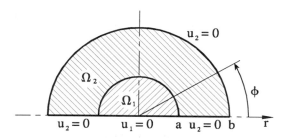

Figure 7.3 Compound semi-circle

$$G_{12}(z,\zeta) = \frac{\lambda}{\pi(\lambda+1)} \ln \frac{|1-e^{z-\bar{\zeta}}|}{|1-e^{z-\zeta}|},$$

$$G_{21}(z,\zeta) = \frac{1}{\pi(\lambda+1)} \ln \frac{|1-e^{z-\bar{\zeta}}|}{|1-e^{z-\zeta}|},$$

and

$$G_{22}(z,\zeta) = \frac{1}{2\pi} \left[\frac{1-\lambda}{\lambda+1} \ln \frac{|1-e^{z+\bar{\zeta}}|}{|1-e^{z+\zeta}|} + \ln \frac{|1-e^{z-\bar{\zeta}}|}{|1-e^{z-\zeta}|} \right],$$

where $z = x+iy$ and $\zeta = s+it$ denote the field and the source point, respectively, and the bar on ζ denotes a conjugate value.

Clearly, the above representations of the entries of the influence matrix for the compound strip are real-valued functions of the complex variables z and ζ which are introduced here just for the sake of compactness.

As it can readily be seen, if the parameter λ is equal to one (that is, the material occupying the entire strip is homogeneous), then the above expressions for the entries of the influence matrix reduce to the well-known (see, for example, [63, 68, 70]) closed form

$$G(z,\zeta) = \frac{1}{2\pi} \ln \frac{|1-e^{z-\bar{\zeta}}|}{|1-e^{z-\zeta}|}$$

of the Green's function of the Dirichlet problem for Laplace's equation set up on the infinite strip $\Omega_s = \{(x,y): -\infty < x < \infty, 0 < y < \pi\}$.

For the next example, we formulate a boundary value problem in polar coordinates. Let the semi-circle $\Omega_{sc} = \{(r,\phi): 0 < r < b, 0 < \phi < \pi\}$ consist of the two fragments $\Omega_1 = \{(r,\phi): 0 < r < a, 0 < \phi < \pi\}$ and $\Omega_2 = \{(r,\phi): a < r < b, 0 < \phi < \pi\}$ as shown in Figure 7.3. Throughout Ω_{sc}, consider Laplace's equations

$$\Delta u_i(r,\phi) = 0, \quad (r,\phi) \in \Omega_i, \quad (i=1,2) \tag{7.23}$$

subject to the following boundary and contact conditions

$$|u_1(0,\phi)|<\infty, \quad u_2(b,\phi)=0 \tag{7.24}$$

$$u_1(a,\phi)=u_2(a,\phi), \quad \lambda_1\frac{\partial u_1(a,\phi)}{\partial r}=\lambda_2\frac{\partial u_2(a,\phi)}{\partial r} \tag{7.25}$$

$$u_1(r,0)=u_2(r,0)=u_1(r,\pi)=u_2(r,\pi)=0 \tag{7.26}$$

where

$$\Delta = \frac{1}{r}\frac{\partial}{\partial r}\left(r\frac{\partial}{\partial r}\right)+\frac{1}{r^2}\frac{\partial^2}{\partial \phi^2}$$

represents the polar Laplacian, while λ_1 and λ_2 represent the conductivities of the materials of which the regions Ω_1 and Ω_2, respectively, are composed.

Following the procedure whose detailed description can be found in the development for the problem posed by eqns (7.7)–(7.11), we expand $u_1(r,\phi)$ and $u_2(r,\phi)$ in Fourier series

$$u_i(r,\phi)=\sum_{n=1}^{\infty} u_{i,n}(r)\sin(n\phi), \quad (i=1,2) \tag{7.27}$$

This reduces the original formulation to a set $(n=1,2,3,\dots)$ of the three-point posed boundary value problems

$$\frac{1}{r}\frac{d}{dr}\left(r\frac{du_{1,n}(r)}{dr}\right)-\frac{n^2}{r^2}u_{1,n}(r)=0, \quad r\in(0,a) \tag{7.28}$$

$$\frac{1}{r}\frac{d}{dr}\left(r\frac{du_{2,n}(r)}{dr}\right)-\frac{n^2}{r^2}u_{2,n}(r)=0, \quad r\in(a,b) \tag{7.29}$$

$$|u_{1,n}(0)|<\infty, \quad u_{2,n}(b)=0 \tag{7.30}$$

$$u_{1,n}(a)=u_{2,n}(a), \quad \lambda_1\frac{du_{1,n}(a)}{dr}=\lambda_2\frac{du_{2,n}(a)}{dr} \tag{7.31}$$

for ordinary differential equations in the coefficients $u_{i,n}(r)$ of the series introduced in eqn (7.27).

The matrix of Green's type to the above formulation can be derived by using the procedure described earlier in this section. We leave the derivation

7.1. POTENTIAL FIELDS IN COMPOUND REGIONS

for the review exercises. The entries $g_{ij}^n(r,\rho)$ of that matrix are found in the form

$$g_{11}^n(r,\rho) = \frac{1}{2n\Delta^*}\left[(1+\lambda)\left(\left(\frac{r\rho}{b^2}\right)^n - \left(\frac{r}{\rho}\right)^n\right)\right.$$

$$\left. + (1-\lambda)\left(\left(\frac{ra^2}{\rho b^2}\right)^n - \left(\frac{r\rho}{a^2}\right)^n\right)\right], \quad 0 \le r \le \rho \le a$$

$$g_{12}^n(r,\rho) = \frac{\lambda}{n\Delta^*}\left[\left(\frac{r\rho}{b^2}\right)^n - \left(\frac{r}{\rho}\right)^n\right], \quad 0 \le r \le a \le \rho \le b$$

$$g_{21}^n(r,\rho) = \frac{1}{n\Delta^*}\left[\left(\frac{r\rho}{b^2}\right)^n - \left(\frac{\rho}{r}\right)^n\right], \quad 0 \le \rho \le a \le r \le b$$

and

$$g_{22}^n(r,\rho) = \frac{1}{2n\Delta^*}\left[(1+\lambda)\left(\left(\frac{r\rho}{b^2}\right)^n - \left(\frac{r}{\rho}\right)^n\right)\right.$$

$$\left. + (1-\lambda)\left(\left(\frac{a^2}{r\rho}\right)^n - \left(\frac{\rho a^2}{rb^2}\right)^n\right)\right], \quad a \le r \le \rho \le b$$

with

$$\Delta^* = [(1-\lambda)(a/b)^{2n} - (1+\lambda)]$$

and λ denoting again the ratio $\lambda = \lambda_2/\lambda_1$ of the conductivities of the materials of which Ω_2 and Ω_1 are composed.

The entries $G_{ij}(r,\phi;\rho,\psi)$ of the influence matrix to the original boundary value problem posed by eqns (7.23)–(7.26) are ultimately found in terms of the entries $g_{ij}^n(r,\rho)$ of the matrix of Green's type to the three-point posed problem stated by eqns (7.28)–(7.31) as

$$G_{ij}(r,\phi;\rho,\psi) = \frac{2}{\pi}\sum_{n=1}^{\infty} g_{ij}^n(r,\rho)\sin(n\phi)\sin(n\psi), \quad (i,j=1,2) \qquad (7.32)$$

Note that the series representing the peripheral entries $G_{12}(r,\phi;\rho,\psi)$ and $G_{21}(r,\phi;\rho,\psi)$ converge uniformly. The logarithmic singularities contained in the principal diagonal entries $G_{11}(r,\phi;\rho,\psi)$ and $G_{22}(r,\phi;\rho,\psi)$ of the influence matrix can be split off to finally provide

$$G_{11}(r,\phi;\rho,\psi) = \frac{1}{2\pi}\left[\frac{1-\lambda}{1+\lambda}\ln\frac{|a^2-z\overline{\zeta}|}{|a^2-z\zeta|} + \ln\frac{|z-\overline{\zeta}|}{|z-\zeta|}\right]$$

$$+\frac{4\lambda}{\pi(1+\lambda)}\sum_{n=1}^{\infty}\frac{1}{n\Delta^*}\left(\frac{r\rho}{b^2}\right)^n \sin(n\phi)\sin(n\psi)$$

and

$$G_{22}(r,\phi;\rho,\psi) = \frac{1}{2\pi}\left[\frac{1-\lambda}{1+\lambda}\ln\frac{|a^2-\overline{z\zeta}|}{|a^2-z\zeta|} + \ln\frac{|z-\overline{\zeta}||b^2-z\overline{\zeta}|}{|z-\zeta||b^2-z\zeta|}\right]$$

$$+ \frac{1-\lambda}{\pi}\sum_{n=1}^{\infty}\frac{1}{n\Delta^*}\left(\frac{a}{b}\right)^{2n}\left[\left(\frac{r\rho}{b^2}\right)^n + \frac{1-\lambda}{1+\lambda}\left(\frac{a^2}{r\rho}\right)^n - \frac{r^{2n}+\rho^{2n}}{r\rho}\right]\sin(n\phi)\sin(n\psi)$$

with z and ζ representing the field and source points, respectively, defined as

$$z = r(\cos(\phi) + i\sin(\phi)), \quad \zeta = \rho(\cos(\psi) + i\sin(\psi))$$

Since the trigonometric series appearing in the above expressions for G_{11} and G_{22} converge uniformly, values of all the entries of the influence matrix can be practically computed by truncating the series. That is, the representations just derived are suitable for direct utilization in computational algorithms.

It can be shown that if $\lambda = 1$ (that is, the material occupying Ω_{sc} is homogeneous), the above expressions for G_{ij} reduce to the well-known closed form

$$G(r,\phi;\rho,\psi) = \frac{1}{2\pi}\ln\frac{|z-\overline{\zeta}||b^2-z\overline{\zeta}|}{|z-\zeta||b^2-z\zeta|}$$

of the Green's function of the Dirichlet problem for Laplace's equation set up on the semi-circle of radius b (see eqn (5.60) in Chapter 5).

7.2 Potential fields on plates and shells

This section concerns the construction of influence matrices of a concentrated unit source for the equation of potential stated on various joint thin-walled structures consisting of shell and plate elements. As the first example, we consider two surfaces joined together to form a cylindrical shell closed at one edge with a spherical cap. The axial cross-section of the structure is depicted in Figure 7.4.

Define the region $\Omega = \Omega_1 \cup \Omega_2$ consisting of the hemisphere $\Omega_1 = \{(\phi,\theta): 0 < \phi < \pi/2, 0 \le \theta < 2\pi\}$ and the cylinder $\Omega_2 = \{(x,y): 0 < x < a, 0 \le y < 2\pi\}$, both of unit radius. Let Ω_1 and Ω_2 be occupied with materials whose conductivities are λ_1 and λ_2, respectively.

We analyze the potential field $u_1(\phi,\theta)$ as defined throughout Ω_1 satisfying Poisson's equation

$$\Delta u_1(\phi,\theta) = -f_1(\phi,\theta), \quad (\phi,\theta) \in \Omega_1 \qquad (7.33)$$

7.2. POTENTIAL FIELDS ON PLATES AND SHELLS

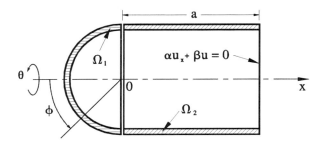

Figure 7.4 Cylindrical shell with a spherical cap

where Δ represents Laplace's operator

$$\Delta = \frac{1}{\sin(\phi)} \frac{\partial}{\partial \phi} \left(\sin(\phi) \frac{\partial}{\partial \phi} \right) + \frac{1}{\sin^2(\phi)} \frac{\partial^2}{\partial \theta^2}$$

written in spherical coordinates on the unit sphere. The potential field $u_2(x,y)$ throughout Ω_2 is also defined by Poisson's equation

$$\frac{\partial^2 u_2(x,y)}{\partial x^2} + \frac{\partial^2 u_2(x,y)}{\partial y^2} = -f_2(x,y), \quad (x,y) \in \Omega_2 \qquad (7.34)$$

where Laplacian is, however, written in Cartesian form.

The unknown functions $u_1(\phi, \theta)$ and $u_2(x, y)$ are subject to the boundary and contact conditions imposed as

$$|u_1(0,\theta)| < \infty, \quad \alpha \frac{\partial u_2(a,y)}{\partial x} + \beta u_2(a,y) = 0 \qquad (7.35)$$

$$u_1(\pi/2, \theta) = u_2(0, y), \quad \lambda_1 \frac{\partial u_1(\pi/2, \theta)}{\partial \phi} = \lambda_2 \frac{\partial u_2(0, y)}{\partial x} \qquad (7.36)$$

$$u_1(\phi, 0) = u_1(\phi, 2\pi), \quad \frac{\partial u_1(\phi, 0)}{\partial \theta} = \frac{\partial u_1(\phi, 2\pi)}{\partial \theta} \qquad (7.37)$$

$$u_2(x, 0) = u_2(x, 2\pi), \quad \frac{\partial u_2(x, 0)}{\partial y} = \frac{\partial u_2(x, 2\pi)}{\partial y} \qquad (7.38)$$

Note that eqns (7.37) and (7.38) reflect closedness of the fragment Ω_1 in the direction of θ and of the fragment Ω_2 in the direction of y. In other words, the relations in these equations specify a periodic nature of the functions $u_1(\phi,\theta)$ and $u_2(x,y)$ with respect to θ and y, respectively.

Keeping in mind the way in which the influence matrix is defined in Section 7.1 (see eqn (7.6)), we intend to obtain the solution to the boundary value problem stated by eqns (7.33)–(7.38) as expressed in the following fashion

$$u_1(\phi,\theta) = \iint_{\Omega_1} G_{11}(\phi,\theta;\psi,\tau) f_1(\psi,\tau) d\Omega_1(\psi,\tau)$$
$$+ \iint_{\Omega_2} G_{12}(\phi,\theta;s,t) f_2(s,t) d\Omega_2(s,t)$$

and

$$u_2(x,y) = \iint_{\Omega_1} G_{21}(x,y;\psi,\tau) f_1(\psi,\tau) d\Omega_1(\psi,\tau)$$
$$+ \iint_{\Omega_2} G_{22}(x,y;s,t) f_2(s,t) d\Omega_2(s,t)$$

with G_{ij}, $(i=1,2)$ representing the entries of the influence matrix.

In view of the 2π-periodicity of $u_1(x,y)$ and $f_1(x,y)$ with respect to y, and of $u_2(\phi,\theta)$ and $f_2(\phi,\theta)$ with respect to θ (see eqns (7.37) and (7.38)), we assume that there exist their Fourier expansions as follows

$$u_1(\phi,\theta) = \frac{1}{2} u_{10}(\phi) + \sum_{n=1}^{\infty} u_{1n}^c(\phi) \cos(n\theta) + u_{1n}^s(\phi) \sin(n\theta),$$

$$f_1(\phi,\theta) = \frac{1}{2} f_{10}(\phi) + \sum_{n=1}^{\infty} f_{1n}^c(\phi) \cos(n\theta) + f_{1n}^s(\phi) \sin(n\theta) \qquad (7.39)$$

and

$$u_2(x,y) = \frac{1}{2} u_{20}(x) + \sum_{n=1}^{\infty} u_{2n}^c(x) \cos(ny) + u_{2n}^s(x) \sin(ny),$$

$$f_2(x,y) = \frac{1}{2} f_{20}(x) + \sum_{n=1}^{\infty} f_{2n}^c(x) \cos(ny) + f_{2n}^s(x) \sin(ny) \qquad (7.40)$$

Substituting these expansions into eqns (7.33)–(7.38), one obtains the following set $(n=0,1,2,\dots)$ of three-point posed boundary value problems for ordinary differential equations, occurring in the coefficients of the Fourier series for $u_1(\phi,\theta)$ and $u_2(x,y)$. For the case of $n=0$, it follows that

$$\frac{1}{\sin(\phi)} \frac{d}{d\phi}\left(\sin(\phi)\frac{du_{10}(\phi)}{d\phi}\right) = -f_{10}(\phi), \quad 0<\phi<\pi/2 \qquad (7.41)$$

$$\frac{d^2 u_{20}(x)}{dx^2} = -f_{20}(x), \quad 0<x<a \qquad (7.42)$$

7.2. POTENTIAL FIELDS ON PLATES AND SHELLS

$$|u_{10}(0)| < \infty, \quad \alpha\frac{du_{20}(a)}{dx} + \beta u_{20}(a) = 0 \tag{7.43}$$

$$u_{10}(\pi/2) = u_{20}(0), \quad \frac{du_{10}(\pi/2)}{d\phi} = \lambda\frac{du_{20}(0)}{dx} \tag{7.44}$$

while for the case of $n = 1, 2, 3, \ldots$, one obtains

$$\frac{1}{\sin(\phi)}\frac{d}{d\phi}\left(\sin(\phi)\frac{du_{1n}(\phi)}{d\phi}\right) - \frac{n^2}{\sin^2(\phi)}u_{1n}(\phi)$$

$$= -f_{1n}(\phi), \quad 0 < \phi < \pi/2 \tag{7.45}$$

$$\frac{d^2 u_{2n}(x)}{dx^2} - n^2 u_{2n}(x) = -f_{2n}(x), \quad 0 < x < a \tag{7.46}$$

$$|u_{1n}(0)| < \infty, \quad \alpha\frac{du_{2n}(a)}{dx} + \beta u_{2n}(a) = 0 \tag{7.47}$$

$$u_{1n}(\pi/2) = u_{2n}(0), \quad \frac{du_{1n}(\pi/2)}{d\phi} = \lambda\frac{du_{2n}(0)}{dx} \tag{7.48}$$

In both of the above formulations, the parameter λ is again defined in terms of the conductivities of the materials involved as $\lambda = \lambda_2/\lambda_1$. Note that in formulating the problem in eqns (7.45)–(7.48), we omit the superscripts 'c' and 's' on $u_{1n}(\phi)$ and $u_{2n}(x)$ until some later stages in the development, in which they need to be distinguished.

We first consider the case of $n = 0$. To compute the matrix of Green's type to the homogeneous boundary value problem corresponding to that of eqns (7.41)–(7.44), we recall from Chapter 5 that the fundamental solution set to the homogeneous ($f_{10}(\phi) = 0$) equation corresponding to eqn (7.41) can be composed of the functions

$$u_{10}^{(1)}(\phi) = 1 \quad \text{and} \quad u_{10}^{(2)}(\phi) = \ln(\tan(\phi/2)),$$

while the functions

$$u_{20}^{(1)}(x) = 1 \quad \text{and} \quad u_{20}^{(2)}(x) = x$$

can be used as the fundamental solution set to the homogeneous ($f_{20}(x) = 0$) equation corresponding to eqn (7.42).

Tracing out the standard procedure of Lagrange's method of variation of parameters, we seek the complementary solutions to equation (7.41) as

$$u_{10}(\phi) = c_1(\phi) + c_2(\phi)\ln\left(\tan(\frac{\phi}{2})\right),$$

while for equation (7.42), we have

$$u_{20}(x) = d_1(x) + d_2(x)x$$

Proceeding further with Lagrange's method, we obtain, for $u_{10}(\phi)$

$$u_{10}(\phi) = -\int_0^\phi \ln\left(\frac{\tan(\phi/2)}{\tan(\psi/2)}\right) f_{10}(\psi)d\psi + C_1 + C_2\ln\left(\tan(\frac{\phi}{2})\right) \qquad (7.49)$$

and for $u_{20}(x)$ we derive

$$u_{20}(x) = -\int_0^x (x-s)f_{20}(s)ds + D_1 + D_2 x \qquad (7.50)$$

where C_1, C_2, D_1, and D_2 represent arbitrary constants that are obtained by satisfying the boundary and contact conditions imposed by eqns (7.43) and (7.44). The boundedness of $u_{10}(\phi)$ as $\phi \to 0$ (see the first relation in eqn (7.43)) yields $C_2 = 0$. From the rest of the conditions in eqns (7.43) and (7.44), it follows that

$$D_1 = \int_0^{\pi/2} \frac{\beta a + \alpha}{\lambda\beta} f_{10}(\psi)\sin(\psi)d\psi$$

$$+ \int_0^a \frac{1}{\beta}[\beta(a-s)+\alpha]f_{20}(s)ds,$$

$$D_2 = -\int_0^{\pi/2} \frac{\sin(\psi)}{\lambda} f_{10}(\psi)d\psi,$$

and

$$C_1 = \int_0^{\pi/2}\left[\frac{\beta a + \alpha}{\lambda\beta} - \ln\left(\tan(\frac{\psi}{2})\right)\right] f_{10}(\psi)\sin(\psi)d\psi$$

$$+ \int_0^a \frac{1}{\beta}[\beta(a-s)+\alpha]f_{20}(s)ds.$$

Substituting the above values of the coefficients C_1, C_2, D_1, and D_2 into eqns (7.49) and (7.50) and properly regrouping their terms, one obtains

$$u_{10}(\phi) = \int_0^{\pi/2} g_{11}^0(\phi,\psi)f_{10}(\psi)\sin(\psi)d\psi + \int_0^a g_{12}^0(\phi,s)f_{20}(s)ds$$

7.2. POTENTIAL FIELDS ON PLATES AND SHELLS

and
$$u_{20}(x) = \int_0^{\pi/2} g_{21}^0(x,\psi) f_{10}(\psi) \sin(\psi) d\psi + \int_0^a g_{22}^0(x,s) f_{20}(s) ds$$

with the kernel-functions g_{ij}^0 defined as

$$g_{11}^0(\phi,\psi) = \frac{1}{\beta\lambda} \begin{cases} (\beta a + \alpha) - \beta\lambda \ln(\tan(\psi/2)), & 0 \leq \phi \leq \psi \leq \pi/2 \\ (\beta a + \alpha) - \beta\lambda \ln(\tan(\phi/2)), & 0 \leq \psi \leq \phi \leq \pi/2, \end{cases}$$

$$g_{12}^0(\phi,s) = \frac{1}{\beta}[\beta(a-s) + \alpha], \quad 0 \leq \phi \leq \frac{\pi}{2},\ 0 \leq s \leq a,$$

$$g_{21}^0(x,\psi) = \frac{1}{\lambda\beta}[\beta(a-x) + \alpha], \quad 0 \leq x \leq a,\ 0 \leq \psi \leq \frac{\pi}{2},$$

and
$$g_{22}^0(x,s) = \frac{1}{\beta} \begin{cases} \beta(a-s) + \alpha, & 0 \leq x \leq s \leq a \\ \beta(a-x) + \alpha, & 0 \leq s \leq x \leq a. \end{cases}$$

Thus, we have constructed the matrix of Green's type for the homogeneous boundary value problem corresponding to that of eqns (7.41)–(7.44).

We now turn to the three-point posed boundary value problem stated by eqns (7.45)–(7.48). To proceed with the constructing procedure for its matrix of Green's type, recall from Chapter 5 that the fundamental solution set for the homogeneous equation associated with eqn (7.45) can be comprised of the functions

$$u_{1n}^{(1)}(\phi) = \tan^n(\phi/2) \quad \text{and} \quad u_{1n}^{(2)}(\phi) = \cot^n(\phi/2),$$

while such a set for the homogeneous equation corresponding to eqn (7.46)

$$u_{2n}^{(1)}(x) = e^{nx} \quad \text{and} \quad u_{2n}^{(2)}(x) = e^{-nx}$$

has already been recalled earlier in Section 7.1.

The complementary solutions to equations (7.45) and (7.46) can also be derived by Lagrange's procedure. In doing so, one obtains $u_{1n}(\phi)$ and $u_{2n}(x)$ in the form

$$u_{1n}(\phi) = \int_0^\phi \frac{1}{2n} \left[\tan^n\left(\frac{\psi}{2}\right) \cot^n\left(\frac{\phi}{2}\right) - \tan^n\left(\frac{\phi}{2}\right) \cot^n\left(\frac{\psi}{2}\right) \right] f_{1n}(\psi) \sin(\psi) d\psi$$

$$+ C_1 \tan^n\left(\frac{\phi}{2}\right) + C_2 \cot^n\left(\frac{\phi}{2}\right) \tag{7.51}$$

and

$$u_{2n}(x) = -\int_0^x \frac{1}{n}\sinh(n(x-s))f_{2n}(s)ds + D_1 e^{nx} + D_2 e^{-nx} \qquad (7.52)$$

By the first condition in eqn (7.46), it is necessary that $C_2 = 0$. Applying the rest of the boundary and contact conditions from eqns (7.47) and (7.48) one derives a well-posed system of linear algebraic equations in C_1, D_1, and D_2. The determinant Δ^* of the coefficient matrix of that system is

$$\Delta^* = [(1-\lambda)(\beta - \alpha n)e^{-na} + (1+\lambda)(\beta + \alpha n)e^{na}]$$

and the solution to the system is found as

$$D_1 = \frac{\lambda(\alpha n - \beta)e^{-na}}{n\Delta^*}\int_0^{\pi/2}\tan^n\left(\frac{\psi}{2}\right)f_{1n}(\psi)\sin(\psi)d\psi$$

$$+\frac{1+\lambda}{n\Delta^*}\int_0^a [\alpha n \cosh(n(s-a)) - \beta \sinh(n(s-a))]f_{2n}(s)ds,$$

$$D_2 = \frac{\lambda(\alpha n + \beta)e^{na}}{n\Delta^*}\int_0^{\pi/2}\tan^n\left(\frac{\psi}{2}\right)f_{1n}(\psi)\sin(\psi)d\psi$$

$$+\frac{1-\lambda}{n\Delta^*}\int_0^a [\alpha n \cosh(n(s-a)) - \beta \sinh(n(s-a))]f_{2n}(s)ds,$$

and

$$C_1 = \int_0^{\pi/2}\left\{\frac{\lambda[(\beta+\alpha n)e^{na} - (\beta-\alpha n)e^{-na}]}{n\Delta^*}\tan^n\left(\frac{\psi}{2}\right)\right.$$

$$\left. -\frac{1}{2n}\left[\tan^n\left(\frac{\psi}{2}\right) - \cot^n\left(\frac{\psi}{2}\right)\right]\right\}f_{1n}(\psi)\sin(\psi)d\psi$$

$$+\frac{2}{n\Delta^*}\int_0^a [\alpha n \cosh(n(s-a)) - \beta \sinh(n(s-a))]f_{2n}(s)ds.$$

Upon substituting the values of C_1, C_2, D_1, and D_2 just found into eqns (7.51) and (7.52) and properly rearranging their terms, one obtains for $u_{1n}(\phi)$ and $u_{2n}(x)$

$$u_{1n}(\phi) = \int_0^{\pi/2} g_{11}^n(\phi,\psi)f_{1n}(\psi)\sin(\psi)d\psi + \int_0^a g_{12}^n(\phi,s)f_{2n}(s)ds \qquad (7.53)$$

and

$$u_{2n}(x) = \int_0^{\pi/2} g_{21}^n(x,\psi)f_{1n}(\psi)\sin(\psi)d\psi + \int_0^a g_{22}^n(x,s)f_{2n}(s)ds \qquad (7.54)$$

7.2. POTENTIAL FIELDS ON PLATES AND SHELLS

with the kernel-functions g_{ij}^n defined as

$$g_{11}^n(\phi,\psi) = \frac{1}{2n\Delta^*} \begin{cases} (\beta+\alpha n)e^{na}\tan^n(\psi/2)[(1+\lambda)\cot^n(\phi/2) \\ -(1-\lambda)\tan^n(\phi/2)]+(\beta-\alpha n)e^{-na}\tan^n(\psi/2) \\ \times[(1-\lambda)\cot^n(\phi/2)-(1+\lambda)\tan^n(\phi/2)], & \psi \leq \phi \\ \\ (\beta+\alpha n)e^{na}\tan^n(\phi/2)[(1+\lambda)\cot^n(\psi/2) \\ -(1-\lambda)\tan^n(\psi/2)]+(\beta-\alpha n)e^{-na}\tan^n(\phi/2) \\ \times(1-\lambda)\cot^n(\psi/2)]-(1+\lambda)\tan^n(\psi/2), & \phi \leq \psi \end{cases}$$

$$g_{12}^n(\phi,s) = \frac{2}{n\Delta^*}[\alpha n\cosh(n(a-s))-\beta\sinh(n(a-s))]$$
$$\times \tan^n(\phi/2), \quad 0\leq\phi\leq\pi/2,\ 0\leq s\leq a,$$

$$g_{21}^n(x,\psi) = \frac{2\lambda}{n\Delta^*}[\alpha n\cosh(n(a-x))-\beta\sinh(n(a-x))]$$
$$\times \tan^n(\psi/2), \quad 0\leq\psi\leq\pi/2,\ 0\leq x\leq a,$$

and

$$g_{22}^n(x,s) = \frac{1}{n\Delta^*} \begin{cases} [\alpha n\cosh(n(x-a))-\beta\sinh(n(x-a))] \\ \times[(1+\lambda)e^{ns}+(1-\lambda)e^{-ns}], & 0\leq s\leq x\leq a \\ \\ [\alpha n\cosh(n(s-a))-\beta\sinh(n(s-a))] \\ \times[(1+\lambda)e^{nx}+(1-\lambda)e^{-nx}], & 0\leq x\leq s\leq a. \end{cases}$$

According to the fundamental rule for Fourier coefficients (Fourier-Euler formulae), the coefficients $f_{1n}^c(\psi)$, $f_{1n}^s(\psi)$, $f_{2n}^c(s)$, and $f_{2n}^s(s)$ of the series in eqns (7.39) and (7.40) can be written as

$$f_{1n}^c(\psi) = \frac{1}{\pi}\int_0^{2\pi} f_1(\psi,\tau)\cos(n\tau)d\tau, \quad n=0,1,2,\ldots,$$

$$f_{1n}^s(\psi) = \frac{1}{\pi}\int_0^{2\pi} f_1(\psi,\tau)\sin(n\tau)d\tau, \quad n=1,2,3,\ldots,$$

$$f_{2n}^c(s) = \frac{1}{\pi}\int_0^{2\pi} f_2(s,t)\cos(nt)dt, \quad n=0,1,2,\ldots,$$

and

$$f_{2n}^s(s) = \frac{1}{\pi}\int_0^{2\pi} f_2(s,t)\sin(nt)dt, \quad n=1,2,3,\ldots$$

Substituting the above expressions for $f_{1n}^c(\psi)$ and $f_{2n}^c(s)$ into eqns (7.53) and (7.54), we obtain

$$u_{1n}^c(\phi) = \int_0^{\pi/2}\int_0^{2\pi}\frac{1}{\pi}g_{11}^n(\phi,\psi)\cos(n\tau)f_1(\psi,\tau)\sin(\psi)d\tau d\psi$$

$$+ \int_0^a \int_0^{2\pi} \frac{1}{\pi} g_{12}^n(\phi, s) \cos(nt) f_2(s,t) dt ds, \quad n=0,1,2,\ldots$$

and

$$u_{2n}^c(x) = \int_0^{\pi/2} \int_0^{2\pi} \frac{1}{\pi} g_{21}^n(x, \psi) \cos(n\tau) f_1(\psi, \tau) \sin(\psi) d\tau d\psi$$

$$+ \int_0^a \int_0^{2\pi} \frac{1}{\pi} g_{22}^n(x, s) \cos(nt) f_2(s,t) dt ds, \quad n=0,1,2,\ldots.$$

While substituting the expressions for $f_{1n}^s(\psi)$ and $f_{2n}^s(s)$ into eqns (7.53) and (7.54), we have

$$u_{1n}^s(\phi) = \int_0^{\pi/2} \int_0^{2\pi} \frac{1}{\pi} g_{11}^n(\phi, \psi) \sin(n\tau) f_1(\psi, \tau) \sin(\psi) d\tau d\psi$$

$$+ \int_0^a \int_0^{2\pi} \frac{1}{\pi} g_{12}^n(\phi, s) \sin(nt) f_2(s,t) dt ds, \quad n=1,2,3,\ldots$$

and

$$u_{2n}^s(x) = \int_0^{\pi/2} \int_0^{2\pi} \frac{1}{\pi} g_{21}^n(x, \psi) \sin(n\tau) f_1(\psi, \tau) \sin(\psi) d\tau d\psi$$

$$+ \int_0^a \int_0^{2\pi} \frac{1}{\pi} g_{22}^n(x, s) \sin(nt) f_2(s,t) dt ds, \quad n=1,2,3,\ldots.$$

Substitute the expressions for $u_{1n}^c(\phi)$ and $u_{1n}^s(\phi)$ into the expression for $u_1(\phi, \theta)$ in eqn (7.39) and interchange the summation and integration in it. This yields

$$u_1(\phi, \theta) = \int_0^{\pi/2} \int_0^{2\pi} G_{11}(\phi, \theta; \psi, \tau) f_1(\psi, \tau) \sin(\psi) d\tau d\psi$$

$$+ \int_0^a \int_0^{2\pi} G_{12}(\phi, \theta; s, t) f_2(s,t) dt ds, \quad (\phi, \theta) \in \Omega_1 \qquad (7.55)$$

where

$$G_{11}(\phi, \theta; \psi, \tau) = \frac{1}{2\pi} g_{11}^0(\phi, \psi) + \frac{1}{\pi} \sum_{n=1}^{\infty} g_{11}^n(\phi, \psi) \cos(n(\theta - \tau))$$

Recalling the expressions derived earlier for $g_{11}^0(\phi, \psi)$ and $g_{11}^n(\phi, \psi)$, we write explicitly, for $\phi \leq \psi$

$$G_{11}(\phi, \theta; \psi, \tau) = \frac{1}{2\pi} \left[\frac{\lambda}{\beta}(\alpha + \beta a) - \ln\left(\tan(\frac{\psi}{2})\right) \right]$$

$$+ \frac{1}{\pi} \sum_{n=1}^{\infty} \frac{1}{2n\Delta^*} \left\{ (\beta + \alpha n) e^{na} \tan^n(\frac{\phi}{2}) \left[(1+\lambda) \cot^n(\frac{\psi}{2}) \right. \right.$$

7.2. POTENTIAL FIELDS ON PLATES AND SHELLS

$$-(1-\lambda)\tan^n(\frac{\psi}{2})\Big] + (\beta-\alpha n)e^{-na}\tan^n(\frac{\phi}{2})$$

$$\times\left[(1-\lambda)\cot^n(\frac{\psi}{2})-(1+\lambda)\tan^n(\frac{\psi}{2})\right]\Big\}\cos(n(\theta-\tau)) \quad (7.56)$$

Note that to compute values of G_{11} for $\phi \geq \psi$ one can use the above expression where, however, ϕ must be interchanged with ψ.

An explicit expression for $G_{12}(\phi,\theta;s,t)$ is defined accordingly

$$G_{12}(\phi,\theta;s,t) = \frac{1}{2\pi\beta}[\beta(a-s)+\alpha]$$

$$+\sum_{n=1}^{\infty}\frac{1}{2\pi n\Delta^*}\left[(\beta+\alpha n)e^{n(a-s)}-(\beta-\alpha n)e^{n(s-a)}\right]\tan^n(\frac{\phi}{2})\cos(n(\theta-t))$$

By substituting the expressions for $u_{2n}^c(x)$ and $u_{2n}^s(x)$ into the expression for $u_2(x,y)$ from eqn (7.40) and interchanging again the summation and integration in it, we obtain

$$u_2(x,y) = \int_0^{\pi/2}\int_0^{2\pi}G_{21}(x,y;\psi,\tau)f_1(\psi,\tau)\sin(\psi)d\tau d\psi$$

$$+\int_0^a\int_0^{2\pi}G_{22}(x,y;s,t)f_2(s,t)dtds, \quad (x,y)\in\Omega_2 \quad (7.57)$$

where

$$G_{21}(x,y;\psi,\tau) = \frac{\lambda}{2\pi\beta}[\beta(a-x)+\alpha]$$

$$+\sum_{n=1}^{\infty}\frac{\lambda}{\pi n\Delta^*}\left[(\beta+\alpha n)e^{n(a-x)}-(\beta-\alpha n)e^{n(x-a)}\right]\tan^n(\frac{\psi}{2})\cos(n(y-\tau))$$

and for G_{22} when $x \leq s$, we have

$$G_{22}(x,y;s,t) = \frac{1}{2\pi\beta}[\beta(a-s)+\alpha]$$

$$+\sum_{n=1}^{\infty}\frac{1}{2\pi n\Delta^*}\left[(1-\lambda)e^{-nx}+(1+\lambda)e^{nx}\right]$$

$$\times\left[(\beta+\alpha n)e^{n(a-s)}-(\beta-\alpha n)e^{n(s-a)}\right]\cos(n(y-t)).$$

To compute values of G_{22} for $x \geq s$ one can use the above expression where x must, however, be interchanged with s.

Notice that the series involved in G_{12} and G_{21} converge uniformly and are thus already expressed in a form appropriate for numerical calculation. However, the representations for G_{11} and G_{22} are entirely inappropriate for computational purposes as the series in them converge non-uniformly due to the singularities available. The logarithmic singularity in G_{22}, for example, is found in the term

$$\sum_{n=1}^{\infty} \frac{(1+\lambda)(\alpha n+\beta)e^{n(x-s)}e^{2na}}{n[(1-\lambda)(\beta-\alpha n)+(1+\lambda)(\beta+\alpha n)e^{2na}]} \cos(n(y-t))$$

whose convergence rate is, indeed, of an order of $1/n$ which, as we know from Chapter 5, indicates the logarithmic singularity. To split off this singularity, we add and subtract to the summand of the above series the term

$$\frac{e^{n(x-s)}}{n} \cos(n(y-t))$$

This ultimately yields for G_{22} the following representation

$$G_{22}(x,y;s,t) = \frac{1}{2\pi\beta}[\beta(a-s)+\alpha]$$

$$+ \sum_{n=1}^{\infty} \frac{1}{2\pi n \Delta^*} \left\{ \left[2(1-\lambda)\cosh(n(x-s)) + (1+\lambda)e^{n(x+s)}\right] \right.$$

$$\left. \times (\alpha n - \beta)e^{-na} + (1-\lambda)(\beta+\alpha n)e^{n(a-x-s)} \right\} \cos(n(y-t))$$

$$- \frac{1}{2\pi} \ln \sqrt{1 - 2e^{x-s}\cos(y-t) + e^{2(x-s)}} \qquad (7.58)$$

whose singular component is written explicitly while the regular component is expressed with a uniformly convergent series.

The singularity available in the representations of G_{11} in eqn (7.56) can be split off in a similar manner.

Thus, all entries of the influence matrix of a unit point source for a cylindrical shell with a spherical cap are ultimately found in a form consisting of elementary functions and uniformly convergent Fourier series. That is, they are appropriate for numerical calculation.

It is interesting to focus on one particular case of the boundary value problem stated by eqns (7.33)–(7.38), for which the series representing the entries G_{ij} of its influence matrix can be completely summed up. That is the case where both fragments Ω_1 and Ω_2 are occupied with the same material (that is, $\lambda_1 = \lambda_2$ and, subsequently, $\lambda = 1$) and, in addition, the second boundary condition of eqn (7.35) reduces to the Dirichlet type ($\alpha = 0$

7.2. POTENTIAL FIELDS ON PLATES AND SHELLS

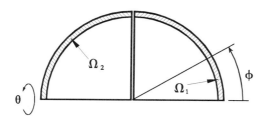

Figure 7.5 Compound hemispherical shell

and $\beta = 1$). We leave for the review exercises the verification of the fact that entries of the influence matrix in this case can be found as

$$G_{11}(\phi,\theta;\psi,\tau) = \frac{1}{2\pi}\ln\sqrt{\frac{e^{2a}-2\Phi\Psi\cos(\theta-\tau)+e^{-2a}\Phi^2\Psi^2}{\Phi^2-2\Phi\Psi\cos(\theta-\tau)+\Psi^2}},$$

$$G_{12}(\phi,\theta;s,t) = \frac{1}{2\pi}\ln\sqrt{\frac{e^{2a}-2e^s\Phi\cos(\theta-t)+e^{2(s-a)}\Phi^2}{e^{2s}-2e^s\Phi\cos(\theta-t)+\Phi^2}},$$

$$G_{21}(x,y;\psi,\tau) = \frac{1}{2\pi}\ln\sqrt{\frac{e^{2a}-2e^x\Psi\cos(y-\tau)+e^{2(x-a)}\Psi^2}{e^{2x}-2e^x\Psi\cos(y-\tau)+\Psi^2}},$$

and

$$G_{22}(x,y;s,t) = \frac{1}{2\pi}\ln\sqrt{\frac{e^{2(a-x)}-2e^{(s-x)}\cos(y-t)+e^{2(s-a)}}{1-2e^{(s-x)}\cos(y-t)+e^{2(s-x)}}}$$

where

$$\Phi = \tan(\phi/2), \quad \Psi = \tan(\psi/2).$$

For the next example, we consider a thin hemispherical shell whose middle surface is a hemisphere $\Omega = \{(\phi,\theta): 0<\phi<\pi,\ 0<\theta<\pi\}$ of unit radius, consisting of two congruent fragments: $\Omega_1 = \{(\phi,\theta): 0<\phi<\pi/2,\ 0<\theta<\pi\}$ and $\Omega_2 = \{(\phi,\theta): \pi/2<\phi<\pi,\ 0<\theta<\pi\}$. Let Ω_1 and Ω_2 be occupied with different materials whose conductivities are λ_1 and λ_2, respectively. The axial cross-section of the structure is depicted in Figure 7.5.

Throughout the entire region Ω, we consider the field of potential described by Poisson's equations

$$\Delta u_i(\phi,\theta) = -f_i(\phi,\theta), \quad (\phi,\theta) \in \Omega_i, \quad (i=1,2) \qquad (7.59)$$

in each of the fragments Ω_1 and Ω_2. Here Δ represents Laplace's operator

$$\Delta = \frac{1}{\sin(\phi)}\frac{\partial}{\partial\phi}\left(\sin(\phi)\frac{\partial}{\partial\phi}\right) + \frac{1}{\sin^2(\phi)}\frac{\partial^2}{\partial\theta^2}$$

written in terms of spherical coordinates on a unit sphere.

Equations (7.59) are subject to the boundary conditions

$$u_1(\phi,0)=u_1(\phi,\pi)=0, \quad u_2(\phi,0)=u_2(\phi,\pi)=0 \qquad (7.60)$$

and to the conditions of ideal contact imposed along the interface of Ω_1 and Ω_2 as

$$u_1(\pi/2,\theta)=u_2(\pi/2,\theta), \quad \lambda_1\frac{\partial u_1(\pi/2,\theta)}{\partial \phi}=\lambda_2\frac{\partial u_2(\pi/2,\theta)}{\partial \phi} \qquad (7.61)$$

Expanding the functions $u_i(\phi,\theta)$ and the right-hand terms $f_i(\phi,\theta)$, $(i=1,2)$ from the above formulation in Fourier series

$$u_i(\phi,\theta)=\sum_{n=1}^{\infty}u_{in}(\phi)\sin(n\theta), \quad f_i(\phi,\theta)=\sum_{n=1}^{\infty}f_{in}(\phi)\sin(n\theta) \qquad (7.62)$$

one obtains the following set $(n=1,2,3,\dots)$ of three-point posed boundary value problems

$$\frac{1}{\sin(\phi)}\frac{d}{d\phi}\left(\sin(\phi)\frac{du_{in}(\phi)}{d\phi}\right)-\frac{n^2}{\sin^2(\phi)}u_{in}(\phi)=-f_{in}(\phi), \quad (i=1,2) \quad (7.63)$$

$$|u_{1n}(0)|<\infty, \quad |u_{2n}(\pi)|<\infty \qquad (7.64)$$

$$u_{1n}(\pi/2)=u_{2n}(\pi/2), \quad \frac{du_{1n}(\pi/2)}{d\phi}=\lambda\frac{du_{2n}(\pi/2)}{d\phi} \qquad (7.65)$$

in the coefficients $u_{1n}(\phi)$ and $u_{2n}(\phi)$ of the first of the series expansions from eqn (7.62).

As we have already recalled in the preceding example, the fundamental solution set of the homogeneous equation corresponding to eqn (7.63) can be represented with the functions $\tan^n(\phi/2)$ and $\cot^n(\phi/2)$. Based on this set and applying Lagrange's procedure, one obtains the complementary solutions to equations (7.63) as

$$u_{1n}(\phi)=\frac{1}{2n}\int_0^{\phi}\left[\tan^n\left(\frac{\psi}{2}\right)\cot^n\left(\frac{\phi}{2}\right)-\tan^n\left(\frac{\phi}{2}\right)\cot^n\left(\frac{\psi}{2}\right)\right]f_{1n}(\psi)\sin(\psi)d\psi$$

$$+C_{11}\tan^n\left(\frac{\phi}{2}\right)+C_{12}\cot^n\left(\frac{\phi}{2}\right) \qquad (7.66)$$

7.2. POTENTIAL FIELDS ON PLATES AND SHELLS

and

$$u_{2n}(\phi) = \frac{1}{2n}\int_{\pi/2}^{\phi}\left[\tan^n\left(\frac{\psi}{2}\right)\cot^n\left(\frac{\phi}{2}\right) - \tan^n\left(\frac{\phi}{2}\right)\cot^n\left(\frac{\psi}{2}\right)\right]f_{2n}(\psi)\sin(\psi)d\psi$$

$$+ C_{21}\tan^n\left(\frac{\phi}{2}\right) + C_{22}\cot^n\left(\frac{\phi}{2}\right) \qquad (7.67)$$

where C_{11}, C_{12}, C_{21}, and C_{22} represent arbitrary coefficients that can be obtained when the above expressions for $u_{1n}(\phi)$ and $u_{2n}(\phi)$ are used to satisfy the boundary and contact conditions from eqns (7.64) and (7.65). In doing so, one obtains

$$C_{12} = 0, \quad C_{21} = \frac{1}{2n}\int_{\pi/2}^{\pi}\cot^n\left(\frac{\psi}{2}\right)f_{2n}(\psi)\sin(\psi)d\psi,$$

$$C_{11} = \frac{1}{2n}\int_0^{\pi/2}\left[\cot^n\left(\frac{\psi}{2}\right) + \frac{1-\lambda}{1+\lambda}\tan^n\left(\frac{\psi}{2}\right)\right]f_{1n}(\psi)\sin(\psi)d\psi$$

$$+ \frac{\lambda}{n(1+\lambda)}\int_{\pi/2}^{\pi}\cot^n\left(\frac{\psi}{2}\right)f_{2n}(\psi)\sin(\psi)d\psi,$$

and

$$C_{22} = \frac{1}{n(1+\lambda)}\int_0^{\pi/2}\tan^n\left(\frac{\psi}{2}\right)f_{1n}(\psi)\sin(\psi)d\psi$$

$$- \frac{1-\lambda}{2n(1+\lambda)}\int_{\pi/2}^{\pi}\cot^n\left(\frac{\psi}{2}\right)f_{2n}(\psi)\sin(\psi)d\psi.$$

By substituting the expressions for C_{11} and C_{12} just found into eqn (7.66), we obtain

$$u_{1n}(\phi) = \frac{1}{2n}\int_0^{\phi}\left[\tan^n\left(\frac{\psi}{2}\right)\cot^n\left(\frac{\phi}{2}\right) - \tan^n\left(\frac{\phi}{2}\right)\cot^n\left(\frac{\psi}{2}\right)\right]f_{1n}(\psi)\sin(\psi)d\psi$$

$$+ \frac{1}{2n}\int_0^{\pi/2}\left[\cot^n\left(\frac{\psi}{2}\right) + \frac{1-\lambda}{1+\lambda}\tan^n\left(\frac{\psi}{2}\right)\right]\tan^n\left(\frac{\phi}{2}\right)f_{1n}(\psi)\sin(\psi)d\psi$$

$$+ \frac{\lambda}{n(1+\lambda)}\int_{\pi/2}^{\pi}\cot^n\left(\frac{\psi}{2}\right)\tan^n\left(\frac{\phi}{2}\right)f_{2n}(\psi)\sin(\psi)d\psi.$$

From this representation for $u_{1n}(\phi)$, it follows that the entries of the first row of the matrix of Green's type to the boundary value problem posed by eqns (7.63)–(7.65) are

$$g_{11}^n(\phi,\psi) = \frac{1}{2n}\left[\cot^n\left(\frac{\psi}{2}\right) + \frac{1-\lambda}{1+\lambda}\tan^n\left(\frac{\psi}{2}\right)\right]\tan^n\left(\frac{\phi}{2}\right), \quad 0 \leq \phi \leq \psi \leq \frac{\pi}{2}$$

and
$$g_{12}^n(\phi,\psi) = \frac{\lambda}{n(1+\lambda)} \cot^n\left(\frac{\psi}{2}\right) \tan^n\left(\frac{\phi}{2}\right), \quad 0 \le \phi \le \frac{\pi}{2} \le \psi \le \pi.$$

Substituting the recently found expressions for C_{21} and C_{22} into eqn (7.67), we have

$$u_{2n}(\phi) = \frac{1}{2n} \int_{\pi/2}^{\phi} \left[\tan^n\left(\frac{\psi}{2}\right) \cot^n\left(\frac{\phi}{2}\right) - \tan^n\left(\frac{\phi}{2}\right) \cot^n\left(\frac{\psi}{2}\right) \right] f_{1n}(\psi) \sin(\psi) d\psi$$

$$+ \frac{1}{2n} \int_{\pi/2}^{\pi} \left[\tan^n\left(\frac{\phi}{2}\right) - \frac{1-\lambda}{1+\lambda} \cot^n\left(\frac{\phi}{2}\right) \right] \cot^n\left(\frac{\psi}{2}\right) f_{1n}(\psi) \sin(\psi) d\psi$$

$$+ \frac{1}{n(1+\lambda)} \int_0^{\pi} \tan^n\left(\frac{\psi}{2}\right) \cot^n\left(\frac{\phi}{2}\right) f_{2n}(\psi) \sin(\psi) d\psi$$

from which it immediately follows that the entries of the second row of the matrix of Green's type to the boundary value problem posed by eqns (7.63)–(7.65) are expressed as

$$g_{21}^n(\phi,\psi) = \frac{1}{n(1+\lambda)} \tan^n\left(\frac{\psi}{2}\right) \cot^n\left(\frac{\phi}{2}\right), \quad 0 \le \psi \le \frac{\pi}{2} \le \phi \le \pi$$

and
$$g_{22}^n(\phi,\psi) = \frac{1}{2n} \left[\tan^n\left(\frac{\phi}{2}\right) - \frac{1-\lambda}{1+\lambda} \cot^n\left(\frac{\phi}{2}\right) \right] \cot^n\left(\frac{\psi}{2}\right), \quad \frac{\pi}{2} \le \phi \le \psi \le \pi.$$

In light of the trigonometric expansions from eqn (7.62), the entries of the influence matrix of a concentrated unit source in the hemispherical compound shell occupying Ω can be expressed in terms of the entries $g_{ij}(\phi,\psi)$ of the matrix of Green's type just found as

$$G_{ij}(\phi,\theta;\psi,\tau) = \frac{2}{\pi} \sum_{n=1}^{\infty} g_{ij}^n(\phi,\psi) \sin(n\theta) \sin(n\tau), \quad i,j = 1,2 \qquad (7.68)$$

These series are completely summable so that one finally obtains the following expressions

$$G_{11}(\phi,\theta;\psi,\tau) = \frac{1}{4\pi} \left[\Gamma_0(\Phi,\theta;\Psi,\tau) + \frac{1-\lambda}{1+\lambda} \Gamma_1(\Phi,\theta;\Psi,\tau) \right],$$

$$G_{12}(\phi,\theta;\psi,\tau) = \frac{\lambda}{2\pi(1+\lambda)} \Gamma_0(\Phi,\theta;\Psi,\tau),$$

$$G_{21}(\phi,\theta;\psi,\tau) = \frac{1}{2\pi(1+\lambda)} \Gamma_0(\Phi,\theta;\Psi,\tau),$$

7.2. POTENTIAL FIELDS ON PLATES AND SHELLS

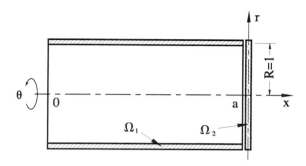

Figure 7.6 Cylindrical shell with a circular plate

and
$$G_{22}(\phi,\theta;\psi,\tau) = \frac{1}{4\pi}\left[\Gamma_0(\Phi,\theta;\Psi,\tau) - \frac{1-\lambda}{1+\lambda}\Gamma_1(\Phi,\theta;\Psi,\tau)\right],$$

for the entries of the influence matrix for the hemispherical compound shell consisting of two congruent fragments composed of different materials. Here Γ_0 and Γ_1 are introduced as

$$\Gamma_0(\Phi,\theta;\Psi,\tau) = \ln\frac{\Phi^2 - 2\Phi\Psi\cos(\theta+\tau) + \Psi^2}{\Phi^2 - 2\Phi\Psi\cos(\theta-\tau) + \Psi^2}$$

and
$$\Gamma_1(\Phi,\theta;\Psi,\tau) = \ln\frac{1 - 2\Phi\Psi\cos(\theta+\tau) + \Phi^2\Psi^2}{1 - 2\Phi\Psi\cos(\theta-\tau) + \Phi^2\Psi^2}$$

with Φ and Ψ having been introduced earlier in this section as

$$\Phi = \tan(\phi/2), \quad \Psi = \tan(\psi/2).$$

We leave the summation that leads to the above finite representations for the review exercises.

For the next example, we consider a thin cylindrical shell of unit radius and of length a closed at one edge with a circular plate as shown in Figure 7.6. Let λ_1 and λ_2 represent conductivities of the materials of which the shell and the plate are composed.

The field of potential in such a thin-walled structure can be modeled with Poisson's equations

$$\frac{\partial^2 u_1(x,y)}{\partial x^2} + \frac{\partial^2 u_1(x,y)}{\partial y^2} = -f_1(x,y), \quad (x,y) \in \Omega_1 \qquad (7.69)$$

where $\Omega_1 = \{(x, y): 0 < x < a,\ 0 \leq y < 2\pi\}$ and

$$\frac{1}{r}\frac{\partial}{\partial r}\left(r\frac{\partial u_2(r,\phi)}{\partial r}\right) + \frac{1}{r^2}\frac{\partial^2 u_2(r,\phi)}{\partial \theta^2} = -f_2(r,\theta), \quad (r,\theta) \in \Omega_2 \qquad (7.70)$$

where $\Omega_2 = \{(r,\theta): 0 < r < 1,\ 0 \leq \theta < 2\pi\}$. Equations (7.69) and (7.70) are subject to the boundary and contact conditions as shown

$$u_1(a, y) = 0, \quad |u_2(0,\theta)| < \infty \qquad (7.71)$$

$$u_1(0, y) = u_2(1, \theta), \quad \lambda_1 \frac{\partial u_1(0, y)}{\partial x} = \lambda_2 \frac{\partial u_2(1, \theta)}{\partial r} \qquad (7.72)$$

Following the technique developed earlier in the section, the solution to the above boundary value problem is found in the form

$$u_1(x, y) = \int_0^{2\pi}\!\!\int_0^a \frac{1}{\pi}\left[\frac{1}{2}g_{11}^0(x,s) + \sum_{n=1}^{\infty} g_{11}^n(x,s)\cos(n(y-t))\right] f_1(s,t)\,ds\,dt$$

$$+ \int_0^{2\pi}\!\!\int_0^1 \frac{1}{\pi}\left[\frac{1}{2}g_{12}^0(x,\rho) + \sum_{n=1}^{\infty} g_{12}^n(x,\rho)\cos(n(y-\tau))\right] f_2(\rho,\tau)\rho\,d\rho\,d\tau \qquad (7.73)$$

and

$$u_2(r, \theta) = \int_0^{2\pi}\!\!\int_0^a \frac{1}{\pi}\left[\frac{1}{2}g_{21}^0(r,s) + \sum_{n=1}^{\infty} g_{21}^n(r,s)\cos(n(\theta-t))\right] f_1(s,t)\,ds\,dt$$

$$+ \int_0^{2\pi}\!\!\int_0^1 \frac{1}{\pi}\left[\frac{1}{2}g_{22}^0(r,\rho) + \sum_{n=1}^{\infty} g_{22}^n(r,\rho)\cos(n(\theta-\tau))\right] f_2(\rho,\tau)\rho\,d\rho\,d\tau \qquad (7.74)$$

where the components g_{ij}^0 in the brackets are defined as

$$g_{11}^0(x,s) = \begin{cases} a-s, & x \leq s \\ a-x, & s \leq x, \end{cases} \qquad g_{12}^0(x,\rho) = \lambda(a-x),$$

$$g_{21}^0(r,s) = a-s, \qquad g_{22}^0(r,\rho) = \begin{cases} \lambda a - \ln(\rho), & r \leq \rho \\ \lambda a - \ln(r), & \rho \leq r, \end{cases}$$

while the components g_{ij}^n are found as

$$g_{11}^n(x,s) = \frac{1}{n\Delta^*}\begin{cases} \sinh(n(a-s))[(1+\lambda)e^{nx} + (1-\lambda)e^{-nx}], & x \leq s \\ \sinh(n(a-x))[(1+\lambda)e^{ns} + (1-\lambda)e^{-ns}], & s \leq x \end{cases}$$

7.2. POTENTIAL FIELDS ON PLATES AND SHELLS

$$g_{12}^n(x,\rho) = \frac{2\lambda}{n\Delta^*}\rho^n \sinh(n(a-x)), \quad g_{21}^n(r,s) = \frac{2\lambda}{n\Delta^*} r^n \sinh(n(a-s))$$

and

$$g_{22}^n(r,\rho) = \frac{1}{2n}\left\{\left(\frac{r}{\rho}\right)^n - \frac{r\rho}{\Delta^*}\left[(1-\lambda)e^{na} + (1+\lambda)e^{-na}\right]\right\}, \quad r \leq \rho$$

where $\Delta^* = (1-\lambda)e^{-na} + (1+\lambda)e^{na}$.

Thus, the kernel-functions

$$G_{11}(x,y;s,t) = \frac{1}{\pi}\left[\frac{1}{2}g_{11}^0(x,s) + \sum_{n=1}^{\infty} g_{11}^n(x,s)\cos(n(y-t))\right],$$

$$G_{12}(x,y;\rho,\tau) = \frac{1}{\pi}\left[\frac{1}{2}g_{12}^0(x,\rho) + \sum_{n=1}^{\infty} g_{12}^n(x,\rho)\cos(n(y-\tau))\right],$$

$$G_{21}(r,\theta;s,t) = \frac{1}{\pi}\left[\frac{1}{2}g_{21}^0(r,s) + \sum_{n=1}^{\infty} g_{21}^n(r,s)\cos(n(\theta-t))\right],$$

and

$$G_{22}(r,\theta;\rho,\tau) = \frac{1}{\pi}\left[\frac{1}{2}g_{22}^0(r,\rho) + \sum_{n=1}^{\infty} g_{22}^n(r,\rho)\cos(n(\theta-\tau))\right],$$

of the integrals in eqns (7.73) and (7.74) represent the entries of the influence matrix of a concentrated unit source for the structure under consideration.

The above series does not allow the complete summation because Δ^* is not a monomial type expression. If, however, both the shell and the plate are composed of the same material (that is, $\lambda = 1$ and expression for Δ^* reduces to $2e^{na}$), then the series representing the entries G_{ij} of the influence matrix are summable and we finally obtain

$$G_{11}(x,y;s,t) = \frac{1}{2\pi}\ln\sqrt{\frac{e^{2(a-s)} - 2e^{x-s}\cos(y-t) + e^{2(x-a)}}{1 - 2e^{x-s}\cos(y-t) + e^{2(x-s)}}},$$

$$G_{12}(x,y;\rho,\tau) = \frac{1}{2\pi}\ln\sqrt{\frac{e^{2a} - 2\rho e^x \cos(y-\tau) + \rho^2 e^{2(x-a)}}{e^{2x} - 2\rho e^x \cos(y-\tau) + \rho^2}},$$

$$G_{21}(r,\theta;s,t) = \frac{1}{2\pi}\ln\sqrt{\frac{e^{2a} - 2re^s \cos(\theta-t) + r^2 e^{2(s-a)}}{e^{2s} - 2re^s \cos(\theta-t) + r^2}},$$

and

$$G_{22}(r,\theta;\rho,\tau) = \frac{1}{2\pi}\ln\sqrt{\frac{e^{2a} - 2r\rho\cos(\theta-\tau) + r^2\rho^2 e^{-2a}}{\rho^2 - 2r\rho\cos(\theta-\tau) + r^2}}.$$

Summation of the above series can be accomplished in a standard way and we leave it for the review exercises.

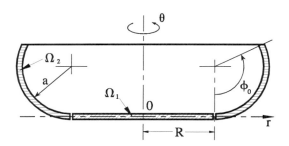

Figure 7.7 Toroidal shell with a circular plate

For the next example we consider a compound thin-walled structure $\Omega = \Omega_1 \cup \Omega_2$ consisting of a circular plate of radius R that occupies the region $\Omega_1 = \{(r, \theta): 0 < r < R, 0 \le \theta < 2\pi\}$ joined to a toroidal shell, with a circular meridional cross-section, occupying the region $\Omega_2 = \{(\phi, \theta): 0 < \phi < \phi_0, 0 \le \theta < 2\pi\}$, where $0 < \phi_0 < \pi$. Assume that the meridional cross-section of the shell represents a unit circle and the plate's radius $R > 1$ (see Figure 7.7). Assume also that both sections of Ω are composed of the same homogeneous material.

The field of potential in such a thin-walled structure can be modeled with Poisson's equations

$$\frac{1}{r}\frac{\partial}{\partial r}\left(r\frac{\partial u_1(r,\theta)}{\partial r}\right) + \frac{1}{r^2}\frac{\partial^2 u_1(r,\theta)}{\partial \theta^2} = -f_1(r,\theta), \quad (r,\theta) \in \Omega_1 \qquad (7.75)$$

and

$$\frac{1}{D}\frac{\partial}{\partial \phi}\left(D\frac{\partial u_2(\phi,\theta)}{\partial \phi}\right) + \frac{1}{D^2}\frac{\partial^2 u_2(\phi,\theta)}{\partial \theta^2} = -f_2(\phi,\theta), \quad (\phi,\theta) \in \Omega_2 \qquad (7.76)$$

where D represents a function of ϕ, defined as $D(\phi) = R + \sin(\phi)$.

Equations (7.75) and (7.76) are subject to the boundary and contact conditions imposed as

$$|u_1(0,\theta)| < \infty, \quad u_2(\phi_0, \theta) = 0 \qquad (7.77)$$

$$u_1(R,\theta) = u_2(0,\theta), \quad \frac{\partial u_1(R,\theta)}{\partial r} = \frac{\partial u_2(0,\theta)}{\partial \phi} \qquad (7.78)$$

In view of the 2π-periodicity of the above formulation with respect to the angular variable θ, we expand the solutions and the right-hand sides of equations (7.75) and (7.76) in Fourier series

$$u_1(r,\theta) = \frac{1}{2}u_{10}(r) + \sum_{n=1}^{\infty} u_{1n}^c(r)\cos(n\theta) + u_{1n}^s(r)\sin(n\theta),$$

7.2. POTENTIAL FIELDS ON PLATES AND SHELLS

$$f_1(r,\theta) = \frac{1}{2}f_{10}(r) + \sum_{n=1}^{\infty} f_{1n}^c(r)\cos(n\theta) + f_{1n}^s(r)\sin(n\theta) \qquad (7.79)$$

and

$$u_2(\phi,\theta) = \frac{1}{2}u_{20}(\phi) + \sum_{n=1}^{\infty} u_{2n}^c(\phi)\cos(n\theta) + u_{2n}^s(\phi)\sin(n\theta),$$

$$f_2(\phi,\theta) = \frac{1}{2}f_{20}(\phi) + \sum_{n=1}^{\infty} f_{2n}^c(\phi)\cos(n\theta) + f_{2n}^s(\phi)\sin(n\theta) \qquad (7.80)$$

This results in the set $(n = 0, 1, 2, \ldots)$ of three-point posed boundary value problems in the coefficients $u_{1n}(r)$ and $u_{2n}(\phi)$ of the above expansions. For the case of $n=0$, we arrive at the following formulation

$$\frac{1}{r}\frac{d}{dr}\left(r\frac{du_{10}(r)}{dr}\right) = -f_{10}(r), \quad 0 < r < R$$

$$\frac{1}{D(\phi)}\frac{d}{d\phi}\left(D(\phi)\frac{du_{20}(\phi)}{d\phi}\right) = -f_{20}(\phi), \quad 0 < \phi < \phi_0$$

$$|u_{10}(0)| < \infty, \quad u_{20}(\phi_0) = 0$$

$$u_{10}(R) = u_{20}(0), \quad \frac{du_{10}(R)}{dr} = \frac{du_{20}(0)}{d\phi},$$

while for the case of $n = 1, 2, 3, \ldots$, we have

$$\frac{1}{r}\frac{d}{dr}\left(r\frac{du_{1n}(r)}{dr}\right) - \frac{n^2}{r^2}u_{1n}(r) = -f_{1n}(r), \quad 0 < r < R$$

$$\frac{1}{D(\phi)}\frac{d}{d\phi}\left(D(\phi)\frac{du_{2n}(\phi)}{d\phi}\right) - \frac{n^2}{D^2(\phi)}u_{2n}(\phi) = -f_{2n}(\phi), \quad 0 < \phi < \phi_0$$

$$|u_{1n}(0)| < \infty, \quad u_{2n}(\phi_0) = 0$$

$$u_{1n}(R) = u_{2n}(0), \quad \frac{du_{1n}(R)}{dr} = \frac{du_{2n}(0)}{d\phi},$$

As we have repeatedly recalled in the preceding sections, the fundamental solution set of the equation

$$\frac{1}{r}\frac{d}{dr}\left(r\frac{du_{10}(r)}{dr}\right) = 0$$

can be composed of the functions

$$u_{10}^{(1)}(r) = 1, \quad \text{and} \quad u_{10}^{(2)}(r) = \ln(r)$$

To obtain the fundamental solution set for the equation

$$\frac{1}{D(\phi)} \frac{d}{d\phi}\left(D(\phi) \frac{du_{20}(\phi)}{d\phi}\right) = 0,$$

one of its particular solutions can clearly be taken as $u_{20}^{(1)}(\phi) = 1$, while its second particular solution $u_{20}^{(2)}$ that is linearly independent of $u_{20}^{(1)}$ can be found by a straightforward integration in the form

$$u_{20}^{(2)}(\phi) = \frac{2}{\sqrt{R^2-1}} \arctan \frac{1 + R\tan(\phi/2)}{\sqrt{R^2-1}} \qquad (7.81)$$

Based on the fundamental solution sets just recalled, the entries g_{ij}^0 of the matrix of Green's type to the homogeneous boundary value problem corresponding to that appearing earlier in the case of $n=0$ are found as

$$g_{11}^0(r,\rho) = \begin{cases} T(\phi_0, 0) - \ln(\rho/R), & r \leq \rho \\ T(\phi_0, 0) - \ln(r/R), & r \geq \rho \end{cases} \qquad g_{12}^0(r,\psi) = T(\phi_0, \psi),$$

$$g_{21}^0(\phi, \rho) = T(\phi_0, \phi), \qquad g_{22}^0(\phi, \psi) = \begin{cases} T(\phi_0, \psi), & \phi \leq \psi \\ T(\phi_0, \phi), & \phi \geq \psi \end{cases}$$

where $T(\phi, \psi)$ is defined in terms of the particular solution $u_{20}^{(2)}(\phi)$ introduced in eqn (7.81) as

$$T(\phi, \psi) = u_{20}^{(2)}(\phi) - u_{20}^{(2)}(\psi)$$

which after some elementary algebra can be written in the form

$$T(\phi, \psi) = \frac{2}{\sqrt{R^2-1}} \arctan\left(\frac{\sqrt{R^2-1}\sin((\phi-\psi)/2)}{R\cos((\phi-\psi)/2) + \sin((\phi-\psi)/2)}\right).$$

To construct the matrix of Green's type to the homogeneous boundary value problem corresponding to that appearing earlier for the case of $n = 1, 2, 3, \ldots$, we recall the fundamental solution set

$$u_{1n}^{(1)}(r) = r^n, \quad \text{and} \quad u_{1n}^{(2)}(r) = r^{-n} \qquad (7.82)$$

to the equation

$$\frac{1}{r}\frac{d}{dr}\left(r\frac{du_{1n}(r)}{dr}\right) - \frac{n^2}{r^2} u_{1n}(r) = 0$$

7.2. POTENTIAL FIELDS ON PLATES AND SHELLS

The fundamental solution set for the equation

$$\frac{1}{D(\phi)}\frac{d}{d\phi}\left(D(\phi)\frac{du_{2n}(\phi)}{d\phi}\right) - \frac{n^2}{D^2(\phi)}u_{2n}(\phi) = 0,$$

can be found [35] as a combination of the functions

$$S_1(\phi) = \exp\left(\frac{2n}{\sqrt{R^2-1}}\arctan\frac{1+R\tan(\phi/2)}{\sqrt{R^2-1}}\right) \qquad (7.83)$$

and

$$S_2(\phi) = \exp\left(-\frac{2n}{\sqrt{R^2-1}}\arctan\frac{1+R\tan(\phi/2)}{\sqrt{R^2-1}}\right) \qquad (7.84)$$

Based on the fundamental solution sets shown in eqns (7.82)–(7.84), in compliance with our standard procedure, the entries g_{ij}^n of the matrix of Green's type to the homogeneous problem corresponding to that appearing earlier in the case of $n = 1, 2, 3, \ldots$ are found as

$$g_{11}^n(r,\rho) = \frac{1}{2n}\left[\left(\frac{r}{\rho}\right)^n - \frac{S_1(0)S_2(\phi_0)}{S_2(0)S_1(\phi_0)}\left(\frac{r\rho}{R^2}\right)^n\right], \qquad r \leq \rho$$

while for $g_{11}^n(r,\rho)$ valid $r \geq \rho$, one interplaces r with ρ,

$$g_{12}^n(r,\psi) = \frac{1}{2n}\frac{S_1(0)}{S_1(\phi_0)}\left(\frac{r}{R}\right)^n[S_1(\phi_0)S_2(\psi) - S_2(\phi_0)S_1(\psi)],$$

$$g_{21}^n(\phi,\rho) = \frac{1}{2n}[S_2(0)S_1(\phi_0)]^{-1}\left(\frac{\rho}{R}\right)^n[S_1(\phi_0)S_2(\phi) - S_2(\phi_0)S_1(\phi)],$$

and

$$g_{22}^n(\phi,\psi) = \frac{1}{2n}\frac{S_1(\phi)}{S_1(\phi_0)}[S_1(\phi_0)S_2(\psi) - S_2(\phi_0)S_1(\psi)], \qquad \phi \leq \psi$$

while for $\phi \geq \psi$, one interplaces ϕ with ψ.

Taking into account the Fourier expansions from eqns (7.79) and (7.80), the entries G_{ij} of the influence matrix to the homogeneous boundary value problem corresponding to that posed by eqns (7.75)–(7.78) are obtained in the form of the following series

$$G_{11}(r,\theta;\rho,\tau) = \frac{1}{\pi}\left[\frac{1}{2}g_{11}^0(r,\rho) + \sum_{n=1}^{\infty}g_{11}^n(r,\rho)\cos(n(\theta-\tau))\right],$$

$$G_{12}(r,\theta;\psi,\tau) = \frac{1}{\pi}\left[\frac{1}{2}g_{12}^0(r,\psi) + \sum_{n=1}^{\infty}g_{12}^n(r,\psi)\cos(n(\theta-\tau))\right],$$

$$G_{21}(\phi,\theta;\rho,\tau)=\frac{1}{\pi}\left[\frac{1}{2}g_{21}^0(\phi,\rho)+\sum_{n=1}^{\infty}g_{21}^n(\phi,\rho)\cos(n(\theta-\tau))\right],$$

and

$$G_{22}(\phi,\theta;\psi,\tau)=\frac{1}{\pi}\left[\frac{1}{2}g_{22}^0(\phi,\psi)+\sum_{n=1}^{\infty}g_{22}^n(\phi,\psi)\cos(n(\theta-\tau))\right],$$

In this case the series are completely summable. Indeed, accomplishing some elementary but quite cumbersome algebra, one arrives at the following expressions

$$G_{11}(r,\theta;\rho,\tau)=\frac{1}{2\pi}\left[T(\phi_0,0)+\ln\sqrt{\frac{R^4-2\alpha R^2 r\rho\cos(\theta-\tau)+(\alpha r\rho)^2}{R^2(r^2-2r\rho\cos(\theta-\tau)+\rho^2)}}\right],$$

$$G_{12}(r,\theta;\psi,\tau)=\frac{1}{2\pi}\left[T(\phi_0,\psi)+\ln\sqrt{\frac{R^2-2\gamma Rr\cos(\theta-\tau)+(\gamma r)^2}{R^2-2\beta Rr\cos(\theta-\tau)+(\beta r)^2}}\right],$$

$$G_{21}(\phi,\theta;\rho,\tau)=\frac{1}{2\pi}\left[T(\phi_0,\phi)+\ln\sqrt{\frac{R^2-2\eta R\rho\cos(\theta-\tau)+(\eta\rho)^2}{R^2-2\delta R\rho\cos(\theta-\tau)+(\delta\rho)^2}}\right],$$

and

$$G_{22}(\phi,\theta;\psi,\tau)=\frac{1}{2\pi}\left[T(\phi_0,\psi)+\ln\sqrt{\frac{1-2\omega\cos(\theta-\tau)+\omega^2}{1-2\sigma\cos(\theta-\tau)+\sigma^2}}\right].$$

for the entries of the influence function for the field of potential in the thin-walled structure consisting of the circular plate and toroidal shell.

The parameter α in the above representations is introduced as

$$\alpha=\left[\frac{S_1(0)S_2(\phi_0)}{S_2(0)S_1(\phi_0)}\right]^{\frac{1}{n}}=\exp\left(\frac{4}{\sqrt{R^2-1}}\arctan\left(-\frac{\sqrt{R^2-1}\tan(\phi_0/2)}{R+\tan(\phi_0/2)}\right)\right).$$

The rest of the parameters are likewise introduced as

$$\beta=[S_1(0)S_2(\psi)]^{\frac{1}{n}},\quad \gamma=\left[\frac{S_1(0)S_2(\phi_0)S_1(\psi)}{S_1(\phi_0)}\right]^{\frac{1}{n}},$$

$$\delta=\left[\frac{S_2(\phi)}{S_2(0)}\right]^{\frac{1}{n}},\quad \eta=\left[\frac{S_1(\phi)S_2(\phi_0)}{S_1(\phi_0)S_2(0)}\right]^{\frac{1}{n}},$$

and

$$\sigma=[S_1(\phi)S_2(\psi)]^{\frac{1}{n}},\quad \omega=\left[\frac{S_1(\phi)S_2(\phi_0)S_1(\psi)}{S_1(\phi_0)}\right]^{\frac{1}{n}}.$$

7.2. POTENTIAL FIELDS ON PLATES AND SHELLS

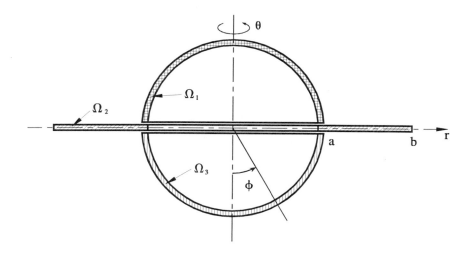

Figure 7.8 'Saturn' type thin-walled construction

The last example in this section involves an assembly of three thin-walled fragments depicted in Figure 7.8. Consider a thin hemispherical shell $\Omega_1 = \{(a,\varphi,\vartheta): 0 < \varphi < \pi/2,\, 0 \leq \vartheta < 2\pi\}$ of radius a, joined to a thin annular plate $\Omega_2 = \{(r,\vartheta): a < r < b,\, 0 \leq \vartheta < 2\pi\}$, and to another thin hemispherical shell $\Omega_3 = \{(a,\varphi,\vartheta): \pi/2 < \varphi < \pi,\, 0 \leq \vartheta < 2\pi\}$ to form a 'Saturn' type construction. Let each of the fragments be composed of a homogeneous conductive material. The thickness of the construction is assumed to be negligibly small compared to the radius a of the hemispheres and to the width $b-a$ of the annular plate.

Clearly, the two-dimensional field of potential in this construction can be determined by the following system of equations

$$\frac{1}{\sin(\varphi)}\frac{\partial}{\partial \varphi}\left(\sin(\varphi)\frac{\partial v_1(\varphi,\vartheta)}{\partial \varphi}\right) + \frac{1}{\sin^2(\varphi)}\frac{\partial^2 v_1(\varphi,\vartheta)}{\partial \vartheta^2} = 0, \quad (\varphi,\vartheta) \in \Omega_1 \quad (7.85)$$

$$\frac{1}{r}\frac{\partial}{\partial r}\left(r\frac{\partial v_2(r,\vartheta)}{\partial r}\right) + \frac{1}{r^2}\frac{\partial^2 v_2(r,\vartheta)}{\partial \vartheta^2} = 0, \quad (r,\vartheta) \in \Omega_2 \quad (7.86)$$

$$\frac{1}{\sin(\varphi)}\frac{\partial}{\partial \varphi}\left(\sin(\varphi)\frac{\partial v_3(\varphi,\vartheta)}{\partial \varphi}\right) + \frac{1}{\sin^2(\varphi)}\frac{\partial^2 v_3(\varphi,\vartheta)}{\partial \vartheta^2} = 0, \quad (\varphi,\vartheta) \in \Omega_3 \quad (7.87)$$

subjected to the boundary and contact conditions written as

$$|v_1(0,\vartheta)| < \infty, \quad v_2(b,\vartheta) = 0, \quad |v_3(\pi,\vartheta)| < \infty \quad (7.88)$$

$$v_1(\pi/2,\vartheta)=v_2(a,\vartheta)=v_3(\pi/2,\vartheta) \tag{7.89}$$

$$\lambda_1\frac{\partial v_1(\pi/2,\vartheta)}{\partial\varphi}-\lambda_2\frac{\partial v_2(a,\vartheta)}{\partial r}-\lambda_3\frac{\partial v_3(\pi/2,\vartheta)}{\partial\varphi}=0 \tag{7.90}$$

where $\lambda_i,(i=1,2,3)$ represent the conductivities of the materials of which the fragments Ω_i are composed.

The conditions of boundedness, subjected at the poles $\varphi=0$ and $\varphi=\pi$ of the hemispheres Ω_1 and Ω_3, respectively, reflect the singularity of the coefficients of eqns (7.85) and (7.87) at these points.

Notably, the functions $v_1(\varphi,\vartheta),v_2(r,\vartheta)$, and $v_3(\varphi,\vartheta)$ are 2π-periodic with respect to the variable ϑ. This allows the influence function of a point source for the entire construction (which can be identified with the matrix of Green's type $G(x,y;\xi,\eta)=(G_{ij}(x,y;\xi,\eta)),(i,j=1,2,3)$ for the boundary value problem in eqns (7.85)–(7.90)) to be presented by means of the trigonometric series

$$G_{ij}(x,y;\xi,\eta)=\frac{1}{\pi}\left[\frac{1}{2}g_{ij}^0(x,\xi)+\sum_{n=1}^\infty g_{ij}^n(x,\xi)\cos(n(y-\eta))\right] \tag{7.91}$$

where x,y and ξ,η represent common notations for the coordinates of the observation and the source points, respectively.

The coefficients $g_{ij}^n(x,\xi)$ of the above expansion represent the entries of the matrices of Green's type of the four-point posed boundary value problems for the set $(n=0,1,2,\ldots)$ of systems of ordinary differential equations

$$\frac{1}{\sin(\varphi)}\frac{d}{d\varphi}\left(\sin(\varphi)\frac{dv_{1n}(\varphi)}{d\varphi}\right)-\frac{n^2}{\sin^2(\varphi)}v_{1n}(\varphi)=0,\quad 0<\varphi<\frac{\pi}{2} \tag{7.92}$$

$$\frac{1}{r}\frac{d}{dr}\left(r\frac{dv_{2n}(r)}{dr}\right)-\frac{n^2}{r^2}v_{2n}(r)=0,\quad a<r<b \tag{7.93}$$

$$\frac{1}{\sin(\varphi)}\frac{d}{d\varphi}\left(\sin(\varphi)\frac{dv_{3n}(\varphi)}{d\varphi}\right)-\frac{n^2}{\sin^2(\varphi)}v_{3n}(\varphi)=0,\quad \frac{\pi}{2}<\varphi<\pi \tag{7.94}$$

stated on the graph shown in Figure 7.9. Recall that the background of such a graph-based statement was earlier developed in Section 4.3.

The system in eqns (7.92)–(7.94) is subject to the boundary conditions

$$|v_{1n}(0)|<\infty,\quad v_{2n}(b)=0,\quad |v_{3n}(\pi)|<\infty \tag{7.95}$$

7.2. POTENTIAL FIELDS ON PLATES AND SHELLS 353

at the end-points E_1, E_2, and E_3 of the graph and to the contact conditions

$$v_{1n}(\pi/2) = v_{2n}(a) = v_{3n}(\pi/2) \tag{7.96}$$

$$\lambda_1 \frac{dv_{1n}(\pi/2)}{d\varphi} - \lambda_2 \frac{dv_{2n}(a)}{dr} - \lambda_3 \frac{dv_{3n}(\pi/2)}{d\varphi} = 0 \tag{7.97}$$

imposed at the graph's vertex V_1.

This statement results from the procedure of separation of variables as being applied to the original problem in eqns (7.85)–(7.90).

For the sake of simplicity in the development that follows, we assume that each fragment Ω_i of the construction is composed of the same material, that is, $\lambda_1 = \lambda_2 = \lambda_3$. In addition, we assume unit radius $a = 1$ of the hemispherical shells occupying Ω_1 and Ω_3.

The matrices of Green's type for the boundary value problems in eqns (7.92)–(7.97) can be constructed, for example, by means of the method of variation of parameters. Notably, the case $n = 0$ requires an individual treatment, since the fundamental solution set for this case is different of that for $n \neq 0$. Upon applying our procedure, the entries $g_{ij}^0(x,\xi)$ of the matrix of Green's type for $n=0$ are found in the form

$$g_{11}^0(\varphi,\psi) = \begin{cases} \ln(b\cot(\psi/2)), & \varphi \leq \psi \\ \ln(b\cot(\varphi/2)), & \psi \leq \varphi, \end{cases}$$

$$g_{12}^0(\phi,\rho) = \ln(b/\rho), \quad g_{13}^0(\varphi,\psi) = \ln(b),$$

$$g_{21}^0(r,\psi) = \ln(b/r), \quad g_{22}^0(r,\rho) = \begin{cases} \ln(b/\rho), & r \leq \rho \\ \ln(b/r), & \rho \leq r, \end{cases}$$

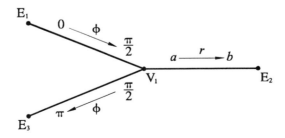

Figure 7.9 The graph hosting the problem

$$g_{23}^0(r,\psi)=\ln(b/r), \quad g_{31}^0(\varphi,\psi)=\ln(b),$$

and

$$g_{32}^0(\varphi,\rho)=\ln(b/\rho), \quad g_{33}^0(\varphi,\psi)=\begin{cases} \ln(b\tan(\varphi/2)), & \varphi\leq\psi \\ \ln(b\tan(\psi/2)), & \psi\leq\varphi. \end{cases}$$

For $n=1,2,3,\ldots$, the entries of the first row of the matrix of Green's type of the problem posed by eqns (7.92)–(7.97) are obtained as follows

$$g_{11}^n(\varphi,\psi)=\frac{1}{2n\Delta^*}\begin{cases} \tan^n(\varphi/2)[\Delta^*\cot^n(\psi/2)-(b^n+b^{-n})\tan^n(\psi/2)], & \varphi\leq\psi \\ \tan^n(\psi/2)[\Delta^*\cot^n(\varphi/2)-(b^n+b^{-n})\tan^n(\varphi/2)], & \psi\leq\varphi, \end{cases}$$

$$g_{12}^n(\varphi,\rho)=\frac{1}{n\Delta^*}\left[\left(\frac{b}{\rho}\right)^n-\left(\frac{\rho}{b}\right)^n\right]\tan^n\left(\frac{\varphi}{2}\right),$$

and

$$g_{13}^n(\varphi,\psi)=\frac{b^n-b^{-n}}{n\Delta}\tan^n\left(\frac{\varphi}{2}\right)\cot^n\left(\frac{\psi}{2}\right).$$

For the entries of the second row we have

$$g_{21}^n(r,\psi)=\frac{1}{n\Delta^*}\left[\left(\frac{b}{r}\right)^n-\left(\frac{r}{b}\right)^n\right]\tan^n\left(\frac{\psi}{2}\right),$$

$$g_{22}^n(r,\rho)=\frac{1}{2n\Delta^*}\begin{cases} (3r^n-r^{-n})[(b/\rho)^n-(\rho/b)^n], & r\leq\rho \\ (3\rho^n-\rho^{-n})[(b/r)^n-(r/b)^n], & \rho\leq r, \end{cases}$$

and

$$g_{23}^n(r,\psi)=\frac{1}{n\Delta^*}\left[\left(\frac{b}{r}\right)^n-\left(\frac{r}{b}\right)^n\right]\cot^n\left(\frac{\psi}{2}\right).$$

Finally, the entries of the third row are expressed as

$$g_{31}^n(\varphi,\psi)=\frac{b^n-b^{-n}}{n\Delta^*}\cot^n\left(\frac{\varphi}{2}\right)\tan^n\left(\frac{\psi}{2}\right),$$

$$g_{32}^n(\varphi,\rho)=\frac{1}{n\Delta^*}\left[\left(\frac{b}{\rho}\right)^n-\left(\frac{\rho}{b}\right)^n\right]\cot^n\left(\frac{\varphi}{2}\right),$$

and

$$g_{33}^n(\varphi,\psi)=\frac{1}{2n\Delta^*}\begin{cases} \cot^n(\psi/2)[\Delta^*\tan^n(\varphi/2)-(b^n+b^{-n})\cot^n(\varphi/2)], & \varphi\leq\psi \\ \cot^n(\varphi/2)[\Delta^*\tan^n(\psi/2)-(b^n+b^{-n})\cot^n(\psi/2)], & \psi\leq\varphi \end{cases}$$

where $\Delta^* = 3b^n - b^{-n}$.

Partial summation of the series shown in eqn (7.91) with the coefficients $g_{ij}^n(x,s)$ just presented can be accomplished in compliance with our standard method. This allows the singular components of the matrix of Green's type $G(x,y;\xi,\eta)$ to be expressed in terms of elementary functions, while its regular components are expressed in a form of uniformly convergent series. One finds below the entries G_{i1} of the first column of this matrix

$$G_{11}(\varphi,\vartheta;\psi,\tau) = -\frac{1}{12\pi}\left\{2\sum_{n=1}^{\infty}\frac{b^n+b^{-n}}{nb^{2n}\Delta^*}\Phi^n\Psi^n\cos(n(\vartheta-\tau))\right.$$

$$\left. - \ln\frac{b^2(1-2\Phi\Psi\Theta+\Phi^2\Psi^2)(b^4-2b^2\Phi\Psi\Theta+\Phi^2\Psi^2)}{(\Phi^2-2\Phi\Psi\Theta+\Psi^2)^3}\right\},$$

$$G_{21}(r,\vartheta;\psi,\tau) = \frac{1}{6\pi}\left\{2\sum_{n=1}^{\infty}\frac{1}{nb^{2n}\Delta^*}\left[\left(\frac{b}{r}\right)^n - \left(\frac{r}{b}\right)^n\right]\Psi^n\cos(n(\vartheta-\tau))\right.$$

$$\left. + \ln\frac{b^4-2rb^2\Psi\Theta+r^2\Psi^2}{br(r^2-2r\Psi\Theta+\Psi^2)}\right\},$$

and

$$G_{31}(\varphi,\vartheta;\psi,\tau) = \frac{1}{6\pi}\left[2\sum_{n=1}^{\infty}\frac{b^n-b^{-n}}{nb^{2n}\Delta}\Phi^{-n}\Psi^n\cos(n(\vartheta-\tau))\right.$$

$$\left. + \ln\frac{\Phi^2 b^4 - 2b^2\Phi\Psi\Theta+\Psi^2}{b(\Phi^2-2\Phi\Psi\Theta+\Psi^2)}\right]$$

where $\Phi = \tan(\varphi/2)$, $\Psi = \tan(\psi/2)$, and $\Theta = \cos(\vartheta-\tau)$.

These entries represent the response of the entire 'Saturn' type thin-walled structure to a unit source acting in an arbitrary source point $(\psi,\tau) \in \Omega_1$. Likewise one can derive expressions for the rest of the entries of the influence matrix.

7.3 Plane problem for elastic media

In this section we will explore the possibility to extend the influence matrix formalism to problems of theory of elasticity. We will define the notion of an influence matrix for the plane problem of theory of elasticity formulated in a compound medium and will develop the algorithm for obtaining such matrices for a 'sandwich' type regions.

Consider the region $\Omega = \bigcup \Omega_k$, $(k = 1, \ldots, m)$ of a compound structure (see Figure 7.1) belonging to two-dimensional Euclidean space. Let the constants λ_k and μ_k be Lame's coefficients of the homogeneous isotropic material of which the fragment Ω_k is composed. Let also $\mathbf{U}_k(x,y)$, being defined on the fragment Ω_k, represent the displacement vector of the point $(x,y) \in \Omega_k$. For the vectors $\mathbf{U}_k(x,y)$, $(k = \overline{1,m})$, formulate the boundary value problem

$$\Lambda\left(\frac{\partial}{\partial x}, \frac{\partial}{\partial y}; \lambda_k, \mu_k\right) \mathbf{U}_k(x,y) = -\mathbf{F}_k(x,y), \quad (x,y) \in \Omega_k, \quad (k=\overline{1,m}) \quad (7.98)$$

$$B_0[\mathbf{U}_k(x,y)] = 0, \quad (x,y) \in \Gamma_0 \quad (7.99)$$

$$B_k[\mathbf{U}_k(x,y), \mathbf{U}_{k+1}(x,y)] = 0, \quad (x,y) \in \Gamma_k, \quad (k=\overline{1,m-1}) \quad (7.100)$$

Here Λ represents Lame's operator of the displacement formulation for the plane problem and $\mathbf{F}_k(x,y)$ is a vector-function of the distributed body force applied to Ω_k.

The system in eqn (7.98) is usually referred to as Lame's system of the plane problem of theory of elasticity. As we have already recalled in Chapter 6, the entries Λ_{ij} of the matrix-operator Λ are expressed as

$$\Lambda_{11} \equiv (\lambda_k + \mu_k)\frac{\partial^2}{\partial x^2} + \mu_k \Delta, \quad \Lambda_{12} \equiv (\lambda_k + \mu_k)\frac{\partial^2}{\partial x \partial y}$$

$$\Lambda_{21} \equiv (\lambda_k + \mu_k)\frac{\partial^2}{\partial x \partial y}, \quad \Lambda_{22} \equiv (\lambda_k + \mu_k)\frac{\partial^2}{\partial y^2} + \mu_k \Delta$$

where Δ represents Cartesian Laplacian. The matrix-operators B_0 in eqn (7.99) and B_k in eqn (7.100) specify the boundary and contact conditions imposed along the outer contour Γ_0 of Ω and the interfaces Γ_k, respectively.

Suppose the boundary value problem posed by eqns (7.98)–(7.100) has a unique solution. That is, the corresponding homogeneous ($\mathbf{F}_k(x,y) = \mathbf{0}$) problem has only the trivial solution such that all $\mathbf{U}_k(x,y) = \mathbf{0}$. For the sake of compactness in what follows, we introduce two vector-functions defined over the entire Ω in the fashion

$$\mathbf{V}(x,y) = (\mathbf{V}_r(x,y))_{r=\overline{1,m}}, \quad \mathbf{H}(x,y) = (\mathbf{H}_r(x,y))_{r=\overline{1,m}}$$

with the components $\mathbf{V}_r(x,y)$ and $\mathbf{H}_r(x,y)$ defined in pieces as

$$\mathbf{V}_r(x,y) = \begin{cases} \mathbf{U}_r(x,y), & (x,y) \in \Omega_r \\ \mathbf{0}, & (x,y) \in \Omega \setminus \Omega_r, \end{cases} \quad (7.101)$$

and

$$\mathbf{H}_r(x,y) = \begin{cases} \mathbf{F}_r(x,y), & (x,y) \in \Omega_r \\ \mathbf{0}, & (x,y) \in \Omega \setminus \Omega_r \end{cases} \quad (7.102)$$

7.3. PLANE PROBLEM FOR ELASTIC MEDIA

We now introduce the notion of an influence matrix for the homogeneous boundary value problem corresponding to that of eqns (7.98)–(7.100).

Definition: If for any allowable vector-function $\mathbf{H}(x,y)$, the vector-function $\mathbf{V}(x,y)$ is expressible in terms of the integral

$$\mathbf{V}(x,y) = \iint_\Omega G(x,y;s,t)\mathbf{H}(s,t)\,d\Omega(s,t) \qquad (7.103)$$

then let us call the kernel-matrix $G(x,y;s,t)$ of this integral representation the *influence matrix* of the homogeneous problem corresponding to that of eqns (7.98)–(7.100).

The term 'allowable' with respect to $\mathbf{H}(x,y)$ implies that this vector-function is integrable over Ω. As it is commonly accepted in influence-function-related discussions, (x,y) and (s,t) will be referred to as the *field (observation)* and the *source* point, respectively.

The influence matrix $G(x,y;s,t)$ represents an $m \times m$ matrix, with the entries $G_{ij}(x,y;s,t)$, being defined for $(x,y)\in\Omega_i$ and $(s,t)\in\Omega_j$ and meeting the following properties:

1. For $i\neq j$, $G_{ij}(x,y;s,t)$ satisfy the homogeneous system corresponding to that of eqn (7.98) in Ω_i;

2. For $i=j$, $G_{ii}(x,y;s,t)$ satisfy the homogeneous system corresponding to that of eqn (7.98) everywhere in Ω_i except for $(x,y)=(s,t)$;

3. For $(x,y)=(s,t)$, $G_{ii}(x,y;s,t)$ possess the singularity associated with the known [37] fundamental solution (Somigliana tensor) of Lame's system;

4. $G_{ij}(x,y;s,t)$ satisfy all of the boundary and contact conditions imposed by eqn (7.99) and (7.100), in which they are involved.

To describe the algorithm for constructing influence matrices for compound elastic regions, we consider a compound rectangle $\Omega = \{(x,y): 0 < x < a_2,\ 0 < y < b\}$ that consists of two other rectangles: $\Omega_1 = \{(x,y): 0 < x < a_1,\ 0 < y < b\}$ and $\Omega_2 = \{(x,y): a_1 < x < a_2,\ 0 < y < b\}$ as shown in Figure 7.10. Let Ω_1 and Ω_2 be occupied with materials whose elastic properties are specified with Lame's coefficients λ_1, μ_1 and λ_2, μ_2, respectively.

Consider elastic equilibrium of the compound medium occupying the region Ω. In doing so, formulate the boundary value problem

$$\Lambda\left(\frac{\partial}{\partial x},\frac{\partial}{\partial y};\lambda_k,\mu_k\right)\mathbf{U}_k(x,y) = -\mathbf{F}_k(x,y),\quad (x,y)\in\Omega_k,\ (k=1,2) \quad (7.104)$$

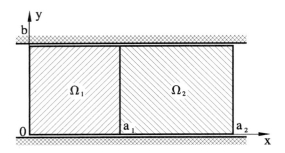

Figure 7.10 Compound elastic rectangle

$$B_{1x}[\mathbf{U}_1(0,y)]=0, \quad B_{2x}[\mathbf{U}_2(a_2,y)]=0 \qquad (7.105)$$

$$B_{1y}[\mathbf{U}_k(x,0)]=0, \quad B_{2y}[\mathbf{U}_k(x,b)]=0, \quad (k=1,2) \qquad (7.106)$$

$$B_{12}[\mathbf{U}_1(a_1,y), \mathbf{U}_2(a_1,y)]=0 \qquad (7.107)$$

Assume that the above problem has a unique solution. To trace the outline of the proposed algorithm for obtaining that solution, we do not need to specify the boundary conditions imposed by eqn (7.105). The point is that any standard combination of the displacements and stresses prescribed on the edges $x=0$ and $x=a_2$, that allows a unique solution to the problem, is feasible for our algorithm.

For the boundary and contact conditions imposed by eqns (7.106) and (7.107) to be specified, recall from Section 6.3 the following expressions

$$\sigma_{xx}=(\lambda+2\mu)\frac{\partial u_x}{\partial x}+\mu\frac{\partial u_y}{\partial y}, \quad \sigma_{yy}=(\lambda+2\mu)\frac{\partial u_y}{\partial y}+\mu\frac{\partial u_x}{\partial x},$$

and

$$\sigma_{xy}=\sigma_{yx}=\mu\left(\frac{\partial u_x}{\partial y}+\frac{\partial u_y}{\partial x}\right)$$

for the components of the stress tensor written in terms of the components $u_x(x,y)$ and $u_y(x,y)$ of the displacement vector $\mathbf{U}(x,y)$.

As far as the boundary conditions imposed by eqn (7.106) are concerned, we assume that the elastic medium occupying Ω is in a frictionless contact without detachment from the absolutely rigid half–planes $y\leq 0$ and $y\geq b$ (see Figure 7.10). That is, the y components $u_{y,1}(x,y)$ and $u_{y,2}(x,y)$ of the displacement vectors

$$\mathbf{U}_1(x,y)=\begin{pmatrix} u_{x,1}(x,y) \\ u_{y,1}(x,y) \end{pmatrix} \quad \text{and} \quad \mathbf{U}_2(x,y)=\begin{pmatrix} u_{x,2}(x,y) \\ u_{y,2}(x,y) \end{pmatrix}$$

7.3. PLANE PROBLEM FOR ELASTIC MEDIA

as well as the shear components $\sigma_{yx,1}(x,y)$ and $\sigma_{yx,2}(x,y)$ of the stress tensors equal zero along the edges $y=0$ and $y=b$ of both the fragments Ω_1 and Ω_2. Hence, the boundary conditions imposed by eqn (7.106) can be specified as

$$u_{y,k}(x,0)=0, \quad k=1,2$$

and

$$\sigma_{yx,k}(x,0) \equiv \mu_k \left(\frac{\partial u_{x,k}(x,0)}{\partial y} + \frac{\partial u_{y,k}(x,0)}{\partial x} \right) = 0, \quad k=1,2$$

Clearly, since the components $u_{y,k}(x,y)$ identically equal zero along the edge $y=0$ (the first of these relations), the partial derivative of these components with respect to x must also equal zero on $y=0$. Therefore, the above conditions can be subsequently rewritten in a more compact form

$$u_{y,k}(x,0)=0, \quad \frac{\partial u_{x,k}(x,0)}{\partial y}=0, \quad k=1,2 \tag{7.108}$$

For the edge $y=b$, we analogously have

$$u_{y,k}(x,b)=0, \quad \frac{\partial u_{x,k}(x,b)}{\partial y}=0, \quad k=1,2 \tag{7.109}$$

Speaking of the contact conditions imposed by eqn (7.107), note that two types of a contact along the interface $x=a_1$ of Ω_1 and Ω_2 are feasible within this algorithm. The first is a perfect mechanical contact in which the normal σ_{xx} and shear σ_{xy} stresses, as well as both the normal u_x and tangential u_y components of the displacement vectors are continuous through $x=a_1$. That is, the contact conditions of eqn (7.107) can be specified in this case as

$$(\lambda_1+2\mu_1)\frac{\partial u_{x,1}(a_1,y)}{\partial x} + \lambda_1 \frac{\partial u_{y,1}(a_1,y)}{\partial y}$$
$$= (\lambda_2+2\mu_2)\frac{\partial u_{x,2}(a_1,y)}{\partial x} + \lambda_2 \frac{\partial u_{y,2}(a_1,y)}{\partial y},$$

$$\mu_1 \left(\frac{\partial u_{x,1}(a_1,y)}{\partial y} + \frac{\partial u_{y,1}(a_1,y)}{\partial x} \right) = \mu_2 \left(\frac{\partial u_{x,2}(a_1,y)}{\partial y} + \frac{\partial u_{y,2}(a_1,y)}{\partial x} \right),$$

$$u_{x,1}(a_1,y)=u_{x,2}(a_1,y), \quad u_{y,1}(a_1,y)=u_{y,2}(a_1,y). \tag{7.110}$$

In the second type of the contact that is feasible within our development, the fragments Ω_1 and Ω_2 are not in a perfect mechanical contact, but instead, we presume a frictionless contact without detachment between the media occupying Ω_1 and Ω_2. That is, the continuity of the normal stresses

σ_{xx} is required, along with the continuity of the x components of the displacement vectors, and the absence of the shear stresses σ_{xy} on the interface $x=a_1$. This yields

$$(\lambda_1+2\mu_1)\frac{\partial u_{x,1}(a_1,y)}{\partial x}+\lambda_1\frac{\partial u_{y,1}(a_1,y)}{\partial y}$$

$$=(\lambda_2+2\mu_2)\frac{\partial u_{x,2}(a_1,y)}{\partial x}+\lambda_2\frac{\partial u_{y,2}(a_1,y)}{\partial y},$$

$$\frac{\partial u_{x,1}(a_1,y)}{\partial y}+\frac{\partial u_{y,1}(a_1,y)}{\partial x}=0, \quad \frac{\partial u_{x,2}(a_1,y)}{\partial y}+\frac{\partial u_{y,2}(a_1,y)}{\partial x}=0,$$

and

$$u_{x,1}(a_1,y)=u_{x,2}(a_1,y)$$

In the development that follows, we take advantage of the first of the two types of contact conditions imposed with the four relations in eqn (7.110).

To obtain the influence matrix of the homogeneous boundary value problem corresponding to that posed by eqns (7.104), (7.105), and (7.108)–(7.110), we expand the displacement vectors $\mathbf{U}_k(x,y)$ and the right-hand side vectors $\mathbf{F}_k(x,y)$, $(k=1,2)$ of eqn (7.104) in the Fourier series

$$\mathbf{U}_k(x,y)=\sum_{n=0}^{\infty}Q_n(y)\mathbf{U}_k^n(x),\quad \mathbf{F}_k(x,y)=\sum_{n=0}^{\infty}Q_n(y)\mathbf{F}_k^n(x) \qquad (7.111)$$

where the matrix of transform is given as

$$Q_n(y)=\begin{pmatrix}\cos(\nu y) & 0 \\ 0 & \sin(\nu y)\end{pmatrix},\quad \nu=\frac{n\pi}{b}$$

Note that the representations of the displacement vectors $\mathbf{U}_k(x,y)$ in eqn (7.111) satisfy the boundary conditions imposed on the edges $y=0$ and $y=b$ by eqn (7.108) and (7.109).

Upon substituting the expressions for $\mathbf{U}_k(x,y)$ and $\mathbf{F}_k(x,y)$ from eqn (7.111) into eqns (7.104), (7.105), and (7.110), one obtains a set ($n=0,1,2,\ldots$) of systems of linear ordinary differential equations

$$\Lambda_k^n\left(\frac{d}{dx};\lambda_k,\mu_k,\nu\right)\mathbf{U}_k^n(x)=-\mathbf{F}_k^n(x),\quad k=1,2 \qquad (7.112)$$

in the coefficients

$$\mathbf{U}_k^n(x)=\begin{pmatrix}u_{x,k}^n(x)\\u_{y,k}^n(x)\end{pmatrix},\quad k=1,2$$

of the first of the series appearing in eqn (7.111).

7.3. PLANE PROBLEM FOR ELASTIC MEDIA

As we have already pointed out in Chapter 6, the case of $n = 0$ must be considered separately from the general case of $n = 1, 2, 3, \ldots$, because when $n = 0$, the fundamental solution set of the system in eqn (7.112) is different of that of the general case. Therefore, in what follows, we will first go through the general case and then turn to the case of $n=0$.

The entries $\Lambda^n_{ij,k}$ of the matrix-operator Λ^n_k occurring in eqn (7.112) are expressed as

$$\Lambda^n_{11,k} \equiv (\lambda_k + 2\mu_k)\frac{d^2}{dx^2} - \nu^2\mu_k, \quad \Lambda^n_{12,k} \equiv (\lambda_k + \mu_k)\nu\frac{d}{dx},$$

$$\Lambda^n_{21,k} \equiv -(\lambda_k + \mu_k)\nu\frac{d}{dx}, \quad \Lambda^n_{22,k} \equiv \mu_k\frac{d^2}{dx^2} - (\lambda_k + 2\mu_k)\nu^2.$$

Clearly, the vector-functions $\mathbf{U}^n_k(x)$ should satisfy boundary conditions at $x = 0$ and $x = a_2$ imposed as

$$B^n_{1x}\left(\frac{d}{dx}\right)\mathbf{U}^n_1(0) = 0, \quad B^n_{2x}\left(\frac{d}{dx}\right)\mathbf{U}^n_2(a_2) = 0 \qquad (7.113)$$

as well as the contact conditions at $x = a_1$ in the form

$$(\lambda_1 + 2\mu_1)\frac{du^n_{x,1}(a_1)}{dx} + \lambda_1\nu u^n_{y,1}(a_1)$$

$$= (\lambda_2 + 2\mu_2)\frac{du^n_{x,2}(a_1)}{dx} + \lambda_2\nu u^n_{y,2}(a_1),$$

$$\mu_1\left(\nu u^n_{x,1}(a_1) - \frac{du^n_{y,1}(a_1)}{dx}\right) = \mu_2\left(\nu u^n_{x,2}(a_1) - \frac{du^n_{y,2}(a_1)}{dx}\right),$$

$$u^n_{x,1}(a_1) = u^n_{x,2}(a_1), \quad u^n_{y,1}(a_1) = u^n_{y,2}(a_1) \qquad (7.114)$$

Lagrange's method of variation of parameters will be used to obtain the solution of the three-point posed boundary value problem occurring in eqns (7.112)–(7.114). With this in mind, we recall from Chapter 6 (see eqn (6.67)) the fundamental solution set of the homogeneous system corresponding to that of eqn (7.112)

$$\mathbf{U}^n_{1,k}(x) = \begin{pmatrix} e^{\nu x} \\ -e^{\nu x} \end{pmatrix}, \quad \mathbf{U}^n_{2,k}(x) = \begin{pmatrix} e^{-\nu x} \\ e^{-\nu x} \end{pmatrix},$$

$$\mathbf{U}^n_{3,k}(x) = \begin{pmatrix} -(\lambda_k + \mu_k)\nu x e^{\nu x} \\ ((\lambda_k + \mu_k)\nu x + (\lambda_k + 3\mu_k))e^{\nu x} \end{pmatrix},$$

$$\mathbf{U}^n_{4,k}(x) = \begin{pmatrix} (\lambda_k + \mu_k)\nu x e^{-\nu x} \\ ((\lambda_k + \mu_k)\nu x - (\lambda_k + 3\mu_k))e^{-\nu x} \end{pmatrix} \qquad (7.115)$$

In compliance with Lagrange's method, the general solution of the system in eqn (7.112) can be written as a linear combination of the components of the fundamental solution set from eqn (7.115). That is

$$\mathbf{U}_k^n(x) = P_k^n(x)\mathbf{C}_k^n(x), \quad k=1,2 \tag{7.116}$$

where $P_k^n(x) = \left(\mathbf{U}_{j,k}^n(x)\right)$, with $j = \overline{1,4}$, represents a 2×4 matrix whose columns are the vectors from eqn (7.115), while $\mathbf{C}_k^n(x)$ is a vector whose entries are arbitrary functions.

Following the standard procedure of the method of variation of parameters, one obtains the system of linear algebraic equations

$$\widehat{P}_k^n(x)\mathbf{C}_k^{n\prime}(x) = \widehat{\mathbf{F}}_k^n(x), \quad k=1,2 \tag{7.117}$$

in the derivative $\mathbf{C}_k^{n\prime}(x)$ of the vector function $\mathbf{C}_k^n(x)$. The coefficient matrix $\widehat{P}_k^n(x)$ of the above system represents a 4×4 matrix whose structure can be described by its horizontal partitioning into two 2×4 matrices, the upper of which is $P_k^n(x)$ as introduced in eqn (7.116), while the lower represents the derivative $P_k^{n\prime}(x)$ of $P_k^n(x)$. Structure of the right-hand side vector $\widehat{\mathbf{F}}_k^n(x)$ can be described in a similar manner by the horizontal partitioning into two two-dimensional vectors in such a way that its upper subvector is a zero vector, while its lower subvector represents the right-hand side vector $-\mathbf{F}_k^n(x)$ of the system in eqn (7.112). That is

$$\widehat{P}_k^n(x) = \begin{pmatrix} P_k^n(x) \\ P_k^{n\prime}(x) \end{pmatrix}, \quad \widehat{\mathbf{F}}_k^n(x) = \begin{pmatrix} 0 \\ -\mathbf{F}_k^n(x) \end{pmatrix}.$$

Since the determinant of the coefficient matrix of the system in eqn (7.117) represents Wronskian of a fundamental solution set of a homogeneous system of linear differential equations, the system appeared in eqn (7.117) is well-posed and its solution vector can consequently be written by means of the inverse $(\widehat{P}_k^n(x))^{-1}$ of the coefficient matrix $\widehat{P}_k^n(x)$ as

$$\mathbf{C}_k^{n\prime}(x) = \left(\widehat{P}_k^n(x)\right)^{-1}\widehat{\mathbf{F}}_k^n(x), \quad k=1,2$$

This can be rewritten in terms of the vector $\mathbf{F}_k^n(x)$ as

$$\mathbf{C}_k^{n\prime}(x) = -R_k^n(x)\mathbf{F}_k^n(x), \quad k=1,2 \tag{7.118}$$

where $R_k^n(x)$ is a 4×2 matrix whose columns represent the third and fourth columns of $(\widehat{P}_k^n(x))^{-1}$.

A straightforward integration of the relations in eqn (7.118) yields the following expressions for the vectors $\mathbf{C}_1^n(x)$ and $\mathbf{C}_2^n(x)$

$$\mathbf{C}_1^n(x) = -\int_0^x R_1^n(s)\mathbf{F}_1^n(s)ds + \mathbf{D}_1^n$$

7.3. PLANE PROBLEM FOR ELASTIC MEDIA

and
$$\mathbf{C}_2^n(x) = -\int_{a_1}^{x} R_2^n(s)\mathbf{F}_2^n(s)ds + \mathbf{D}_2^n$$

with $\mathbf{D}_1^n$ and $\mathbf{D}_2^n$ denoting column-vectors consisting of arbitrary constant coefficients that will be determined later upon satisfying the boundary and contact conditions imposed by eqn (7.113) and (7.114). It worth noting that the lower limits of the above integrals for $\mathbf{C}_1^n(x)$ and $\mathbf{C}_2^n(x)$ are different, because they represent the left-end points of the domains of definition for $\mathbf{F}_1^n(x)$ and $\mathbf{F}_2^n(x)$, which are 0 and a_1, respectively.

Substituting the above expressions for $\mathbf{C}_1^n(x)$ and $\mathbf{C}_2^n(x)$ into eqn (7.116) and bringing the factors $P_1^n(x)$ and $P_2^n(x)$ under the integral signs, one obtains

$$\mathbf{U}_1^n(x) = \int_0^x S_1^n(x,s)\mathbf{F}_1^n(s)ds + P_1^n(x)\mathbf{D}_1^n \qquad (7.119)$$

and

$$\mathbf{U}_2^n(x) = \int_{a_1}^x S_2^n(x,s)\mathbf{F}_2^n(s)ds + P_2^n(x)\mathbf{D}_2^n \qquad (7.120)$$

where $S_k^n(x,s) = P_k^n(x)R_k^n(s)$, with $k = 1,2$, represent 2×2 matrices whose entries $S_{ij,k}^n(x,s)$, being determined for the fundamental solution set presented by eqn (7.115), are obtained as

$$S_{11,k}^n(x,s) = -\frac{\lambda_k + \mu_k}{2\mu_k}(x-s)\cosh(\nu(x-s))$$

$$+ \frac{\lambda_k + 3\mu_k}{2\nu\mu_k}\sinh((\nu(x-s)),$$

$$S_{12,k}^n(x.s) = -\frac{\lambda_k + \mu_k}{2(\lambda_k + 2\mu_k)}(x-s)\sinh(\nu(x-s)),$$

$$S_{21,k}^n(x,s) = \frac{\lambda_k + \mu_k}{2\mu_k}(x-s)\sinh(\nu(x-s)),$$

and

$$S_{22,k}^n(x,s) = \frac{\lambda_k + \mu_k}{2(\lambda_k + 2\mu_k)}(x-s)\cosh(\nu(x-s))$$

$$+ \frac{\lambda_k + 3\mu_k}{2\nu(\lambda_k + 2\mu_k)}\sinh((\nu(x-s))$$

The components of the vectors $\mathbf{D}_k^n$ in eqns (7.119) and (7.120) can be found by satisfying the boundary and contact conditions imposed by eqns (7.113) and (7.114). In doing so, it is important to keep in mind that the entries $S_{ij,k}^n(x,s)$ of $S_k^n(x,s)$ hold the property $S_{ij,k}^n(x,x) = 0$. Hence, the differentiation of the vectors $\mathbf{U}_k^n(x)$, defined by eqns (7.119) and (7.120) in the form of integrals depending on a parameter, can be carried out by simply taking derivatives of $S_{ij,k}^n(x,s)$ with respect to x under the integral signs. This yields the well-posed system of linear algebraic equations of the eighth order

$$T_n \mathbf{D}_n = \mathbf{K}_n \qquad (7.121)$$

in the components of the vector $\mathbf{D}_n$ for which the vectors $\mathbf{D}_1^n$ and $\mathbf{D}_2^n$ serve as the upper and lower subvectors, respectively.

The well-posedness of the above system follows from the uniqueness of the solution of the three-point posed boundary value problem occurring in eqns (7.112)–(7.114), which, in turn, follows from the uniqueness of the solution of the original problem posed by eqns (7.104)–(7.107).

To describe a compound structure of the coefficient matrix T_n in the system in eqn (7.121), we take advantage of its partitioning into eight rectangular 2×4 submatrices according to the scheme

$$T_n = \left(T_n^{kl} \right)_{4,2}$$

The submatrices T_n^{11} and T_n^{22} of such a partitioning are determined as

$$T_n^{11} = B_{1x}^n[P_1^n(0)] \quad \text{and} \quad T_n^{22} = B_{2x}^n[P_2^n(a_2)],$$

while the submatrices T_n^{12} and T_n^{21} represent zero-matrices.

In order to describe the lower half of the coefficient matrix T_n, let us turn to the contact conditions written earlier in eqn (7.114). Associated with those conditions, we define the following four matrix-operators

$$B_c^{11}\left(\frac{d}{dx}\right) = \begin{pmatrix} (\lambda_1+2\mu_1)d/dx & \lambda_1 \nu \\ \mu_1 \nu & -\mu_1 d/dx \end{pmatrix},$$

$$B_c^{12}\left(\frac{d}{dx}\right) = \begin{pmatrix} -(\lambda_2+2\mu_2)d/dx & -\lambda_2 \nu \\ -\mu_2 \nu & \mu_2 d/dx \end{pmatrix},$$

$$B_c^{21}\left(\frac{d}{dx}\right) = \begin{pmatrix} I & 0 \\ 0 & I \end{pmatrix}, \quad B_c^{22}\left(\frac{d}{dx}\right) = \begin{pmatrix} -I & 0 \\ 0 & -I \end{pmatrix}.$$

where with I we denote a unit operator transforming an operand into itself.

In terms of the matrix-operators just introduced, the rest of the submatrices of the coefficient matrix T_n can be written as

$$T_n^{31} = B_c^{11}\left(\frac{d}{dx}\right)[P_1^n(a_1)], \quad T_n^{32} = B_c^{12}\left(\frac{d}{dx}\right)[P_2^n(a_1)],$$

7.3. PLANE PROBLEM FOR ELASTIC MEDIA

and
$$T_n^{41} = B_c^{21}\left(\frac{d}{dx}\right)[P_1^n(a_1)], \quad T_n^{42} = B_c^{22}\left(\frac{d}{dx}\right)[P_2^n(a_1)].$$

We now turn to the right-hand side vector $\mathbf{K}_n$ of the system in eqn (7.121). The first two components K_1^n and K_2^n of that vector are zeros, while its components K_3^n and K_4^n are defined as

$$\begin{pmatrix} K_3^n \\ K_4^n \end{pmatrix} = -\int_{a_1}^{a_2} B_{2x}^n[S_2^n(a_2,s)]\mathbf{F}_2^n(s)ds.$$

The components K_5^n and K_6^n are given by

$$\begin{pmatrix} K_5^n \\ K_6^n \end{pmatrix} = -\int_0^{a_1} B_c^{11}[S_1^n(a_1,s)]\mathbf{F}_1^n(s)ds.$$

Finally, for the last two components K_7^n and K_8^n, we have

$$\begin{pmatrix} K_7^n \\ K_8^n \end{pmatrix} = -\int_{a_1}^{a_2} B_c^{21}[S_2^n(a_2,s)]\mathbf{F}_2^n(s)ds.$$

Based on these representations of the components of $\mathbf{K}_n$ and taking into account that the vector $\mathbf{F}_1^n$ is defined over the interval $(0,a_1)$, whereas the vector $\mathbf{F}_2^n$ is defined over the interval (a_1,a_2), we express all the eight components of $\mathbf{K}_n$ in the form of a single integral

$$\mathbf{K}_n = \int_0^{a_2} Z_n(s)\mathbf{\Phi}_n(s)ds,$$

where $\mathbf{\Phi}_n(s)$ represents a four-dimensional vector for which $\mathbf{F}_1^n(s)$ and $\mathbf{F}_2^n(s)$ serve as the upper and lower subvectors, respectively. That is

$$\mathbf{\Phi_n}(s) = \begin{pmatrix} \mathbf{F}_1^n(s) \\ \mathbf{F}_2^n(s) \end{pmatrix}$$

And $Z_n(s)$ represents an 8×4 matrix whose structure, for the sake of convenience of the development that follows, is described by its partitioning into four 4×2 submatrices according to the scheme

$$Z_n(s) = \left(Z_n^{kl}(s)\right)_{2,2}$$

where the submatrices $Z_n^{11}(s)$ and $Z_n^{22}(s)$ are zero-matrices, while the structure of the two other submatrices $Z_n^{12}(s)$ and $Z_n^{21}(s)$ can easier be viewed by their partitioning into two 2×2 submatrices. Namely, the upper submatrix of $Z_n^{12}(s)$ represents a zero-matrix and its lower submatrix is defined as $-B_{2x}^n[S_2^n(a_2,s)]$. The upper submatrix of $Z_n^{21}(s)$ is defined as $-B_c^{11}[S_1^n(a_1,s)]$, while its lower submatrix is $-B_c^{21}[S_1^n(a_1,s)]$.

Let us return to the system of linear algebraic equations occurring in eqn (7.121) and let $\tilde{T}_n$ denote the inverse T_n^{-1} of the coefficient matrix of that system. If so, then its solution vector $\mathbf{D}_n$ can be expressed as

$$\mathbf{D}_n = \tilde{T}_n \mathbf{K}_n$$

Recalling the expression from eqn (7.122) and taking $\tilde{T}_n$ under the integral sign, the above representation can be rewritten as

$$\mathbf{D}_n = \int_0^{a_2} \tilde{T}_n Z_n(s) \mathbf{\Phi}_n(s) ds, \tag{7.122}$$

To come forth with our procedure, we partition $\tilde{T}_n$ (which is an 8×8 matrix) into four submatrices each of the dimension 4×4 as shown

$$\tilde{T}_n = \left(\tilde{T}_n^{kl}\right)_{2,2}$$

This allows one to rewrite the relation from eqn (7.122) as two separate relations for the vectors $\mathbf{D}_1^n$ and $\mathbf{D}_2^n$ as

$$\mathbf{D}_1^n = \int_0^{a_1} W_n^{11}(s) \mathbf{F}_1^n(s) ds + \int_{a_1}^{a_2} W_n^{12}(s) \mathbf{F}_2^n(s) ds$$

$$\mathbf{D}_2^n = \int_0^{a_1} W_n^{21}(s) \mathbf{F}_1^n(s) ds + \int_{a_1}^{a_2} W_n^{22}(s) \mathbf{F}_2^n(s) ds$$

where the entries $W_n^{kl}(s)$ of the matrix $W_n(s)$ are defined in terms of the submatrices of $\tilde{T}_n$ and $Z_n(s)$ as

$$W_n^{kl}(s) = \tilde{T}_n^{k1} Z_n^{1l}(s) + \tilde{T}_n^{k2} Z_n^{2l}(s), \quad k, l = 1, 2$$

Substituting the expressions for $\mathbf{D}_1^n$ and $\mathbf{D}_2^n$ just found into eqns (7.119) and (7.120), we obtain

$$\mathbf{U}_1^n(x) = \int_0^x S_1^n(x,s) \mathbf{F}_1^n(s) ds + P_1^n(x) \bigg(\int_0^{a_1} W_n^{11}(s) \mathbf{F}_1^n(s) ds$$

$$+ \int_{a_1}^{a_2} W_n^{12}(s) \mathbf{F}_2^n(s) ds \bigg), \quad x \in [0, a_1] \tag{7.123}$$

and

$$\mathbf{U}_2^n(x) = \int_{a_1}^x S_2^n(x,s) \mathbf{F}_2^n(s) ds + P_2^n(x) \bigg(\int_0^{a_1} W_n^{21}(s) \mathbf{F}_1^n(s) ds$$

$$+ \int_{a_1}^{a_2} W_n^{22}(s) \mathbf{F}_2^n(s) ds \bigg), \quad x \in [a_1, a_2] \tag{7.124}$$

7.3. PLANE PROBLEM FOR ELASTIC MEDIA

Introduce the vector-function
$$\mathbf{U}_n^*(x) = \begin{pmatrix} \mathbf{U}_{n,1}^*(x) \\ \mathbf{U}_{n,2}^*(x) \end{pmatrix},$$

whose components represent vectors defined in terms of the vectors $\mathbf{U}_1^n(x)$ and $\mathbf{U}_2^n(x)$ as

$$\mathbf{U}_{n,1}^*(x) = \begin{cases} \mathbf{U}_1^n(x), & x \in [0, a_1] \\ 0, & x \in [a_1, a_2], \end{cases} \qquad \mathbf{U}_{n,2}^*(x) = \begin{cases} 0, & x \in [0, a_1] \\ \mathbf{U}_2^n(x), & x \in [a_1, a_2]. \end{cases}$$

Analogously, introduce the vector-function
$$\mathbf{F}_n^*(x) = \begin{pmatrix} \mathbf{F}_{n,1}^*(x) \\ \mathbf{F}_{n,2}^*(x) \end{pmatrix}$$

whose components are defined in terms of the vectors $\mathbf{F}_1^n(x)$ and $\mathbf{F}_2^n(x)$ as

$$\mathbf{F}_{n,1}^*(x) = \begin{cases} \mathbf{F}_1^n(x), & x \in [0, a_1] \\ 0, & x \in [a_1, a_2], \end{cases} \qquad \mathbf{F}_{n,2}^*(x) = \begin{cases} 0, & x \in [0, a_1] \\ \mathbf{F}_2^n(x), & x \in [a_1, a_2]. \end{cases}$$

In terms of the vectors $\mathbf{U}_n^*(x)$ and $\mathbf{F}_n^*(x)$ just introduced, the relations occurred in eqns (7.123) and (7.124) can now be rewritten in the form of a single integral representation

$$\mathbf{U}_n^*(x) = \int_0^{a_2} G_n(x,s) \mathbf{F}_n^*(s) ds \qquad (7.125)$$

The dimension of the kernel-matrix $G_n(x,s)$ of this representation is 4×4 and its 2×2 submatrices $G_n^{kl}(x,s)$ are defined as

$$G_n^{11}(x,s) = \begin{cases} S_1^n(x,s) + P_1^n(x) W_n^{11}(s), & \text{for } 0 \le s \le x \le a_1 \\ P_1^n(x) W_n^{11}(s), & \text{for } 0 \le x \le s \le a_1, \end{cases}$$

$$G_n^{12}(x,s) = P_1^n(x) W_n^{12}(s), \quad \text{for } 0 \le x \le a_1 \le s \le a_2,$$

$$G_n^{21}(x,s) = P_2^n(x) W_n^{21}(s), \quad \text{for } 0 \le s \le a_1 \le x \le a_2,$$

and

$$G_n^{22}(x,s) = \begin{cases} S_2^n(x,s) + P_2^n(x) W_n^{22}(s), & \text{for } a_1 \le s \le x \le a_2 \\ P_2^n(x) W_n^{22}(s), & \text{for } a_1 \le x \le s \le a_2. \end{cases}$$

Clearly, $G_n(x, s)$ represents the matrix of Green's type of the homogeneous boundary value problem corresponding to that posed by eqns (7.112)–(7.114).

To complete the description of the construction procedure, we now return to the particular case of the three-point posed boundary value problem in eqns (7.112)–(7.114) that occurs when $n=0$. That is, the system of differential equations

$$(\lambda_1+2\mu_1)\frac{d^2 u_{x,1}^0(x)}{dx^2} = f_{x,1}(x), \quad x \in (0, a_1)$$

and

$$(\lambda_2+2\mu_2)\frac{d^2 u_{x,2}^0(x)}{dx^2} = f_{x,2}(x), \quad x \in (a_1, a_2)$$

in the functions $u_{x,1}^0(x)$ and $u_{x,2}^0(x)$, subjected to the boundary conditions

$$B_{1x}^0\left(\frac{d}{dx}\right)u_{x,1}^0(0)=0, \quad B_{2x}^0\left(\frac{d}{dx}\right)u_{x,2}^0(a_2)=0$$

imposed at the end-points $x=0$ and $x=a_2$, along with the contact conditions

$$(\lambda_1+2\mu_1)\frac{du_{x,1}^0(a_1)}{dx} = (\lambda_2+2\mu_2)\frac{du_{x,2}^0(a_1)}{dx}$$

and

$$u_{x,1}^0(a_1) = u_{x,2}^0(a_1)$$

that are imposed at the point $x=a_1$.

The matrix of Green's type $G_0(x, s)$ to the above formulation can readily be fixed and finally, the influence matrix $G(x, y; s, t)$ of the homogeneous boundary value problem corresponding to that posed by eqns (7.104)–(7.107) for the compound region Ω depicted in Figure 7.10 is obtained as

$$G(x, y; s, t) = \frac{1}{b}\sum_{n=0}^{\infty} \varepsilon_n Q_n^*(y) G_n(x, s) Q_n^*(t)$$

where ε_n and $Q_n^*(y)$ are given as

$$\varepsilon_n = \begin{cases} 1, & n=0 \\ 2, & n\geq 1, \end{cases} \quad Q_n^*(y) = \begin{pmatrix} \cos(ny) & 0 & 0 & 0 \\ 0 & \sin(ny) & 0 & 0 \\ 0 & 0 & \cos(ny) & 0 \\ 0 & 0 & 0 & \sin(ny) \end{pmatrix}.$$

7.4 Review Exercises

7.1 Construct the influence matrix of a point source for the field of potential in an infinite strip $\Omega_s = \{(x,y): -\infty < x < \infty, 0 < y < \pi\}$ of width π (see Figure 7.2), comprised of two semi-strips: $\Omega_1 = \{(x,y): -\infty < x < 0, 0 < y < \pi\}$ and $\Omega_2 = \{(x,y): 0 < x < \infty, 0 < y < \pi\}$ occupied with materials whose conductivities are specified with constants λ_1 and λ_2, respectively. Let the potential functions $u_1(x,y)$ and $u_2(x,y)$ be subject to the boundary conditions

$$u_1(x,0) = u_2(x,0) = \frac{\partial u_1(x,\pi)}{\partial y} = \frac{\partial u_2(x,\pi)}{\partial y} = 0.$$

7.2 Construct the influence matrices of a point source for the field of potential in a compound rectangle $\Omega = \{(x,y): -a < x < a, 0 < y < b\}$ that consists of two other rectangles: $\Omega_1 = \{(x,y): -a < x < 0, 0 < y < b\}$ and $\Omega_2 = \{(x,y): 0 < x < a, 0 < y < b\}$ occupied with materials whose conductivities are specified with constants λ_1 and λ_2, respectively, and subjected to the following sets of boundary conditions:

(a) $u_1(-a,y) = u_1(x,0) = u_1(x,b) = u_2(a,y) = u_2(x,0) = u_2(x,b) = 0$;

(b) $u_1(-a,y) = u_1(x,0) = u_1(x,b) = 0$,
$\partial u_2(a,y)/\partial x = u_2(x,0) = u_2(x,b) = 0$.

7.3 Derive the entries $g_{ij}^n(r,\rho)$ of the matrix of Green's type for the boundary value problem posed by eqns (7.28)–(7.31) in Section 7.1 and check your results against those presented later in that section.

7.4 Construct the influence matrix of a point source for the field of potential in a compound semi-circle $\Omega_{sc} = \{(r,\phi): 0 < r < b, 0 < \phi < \pi\}$ consisting of the two fragments $\Omega_1 = \{(r,\phi): 0 < r < a, 0 < \phi < \pi\}$ and $\Omega_2 = \{(r,\phi): a < r < b, 0 < \phi < \pi\}$ as shown in Figure 7.3. Assume that Ω_1 and Ω_2 are occupied with materials whose conductivities are specified with constants λ_1 and λ_2, respectively, and the boundary conditions are imposed as

$$\frac{\partial u_2(b,\phi)}{\partial r} = 0, \quad u_1(r,0) = u_2(r,0) = u_1(r,\pi) = u_2(r,\pi) = 0.$$

7.5 Accomplish the summation of the series occurring in eqn (7.68) of Section 7.2 that represent the entries of the influence matrix of a unit point source for the field of potential in a compound hemispherical shell. Check your results against those presented later in that section.

7.6 Construct the influence matrix of a point source for the field of potential in a compound thin-walled structure $\Omega = \Omega_1 \bigcup \Omega_2$ consisting of a hemispherical shell $\Omega_1 = \{(\phi,\theta) : 0 < \phi < \pi/2,\ 0 \le \theta < 2\pi\}$ of radius R_1, joined to a spherical belt $\Omega_2 = \{(\phi,\theta) : \phi_0 < \phi < \pi/2,\ 0 \le \theta < 2\pi\}$ of radius R_2 that is greater than R_1. The conductivities of the materials of which the fragments are composed are λ_1 and λ_2, respectively. Assume the boundary condition $u_2(\phi_0, \theta) = 0$ on the free edge of the structure. For the sake of simplicity, assume that the following relation $R_1/R_2 = \lambda_2/\lambda_1$ takes place between the radii of Ω_1 and Ω_2 and the conductivities of the materials.

7.7 Construct the influence matrix of a point source for the field of potential in a compound thin-walled structure consisting of a hemispherical shell of unit radius $\Omega_1 = \{(\phi,\theta) : 0 < \phi < \pi/2,\ 0 \le \theta < 2\pi\}$, composed of the material whose conductivity is λ_1 and an annular plate $\Omega_2 = \{(r,\theta) : 1 < r < b,\ 0 < \theta < 2\pi\}$ the conductivity of the material of which is λ_2. Assume the boundary condition $u_2(b,\theta) = 0$ on the edge $r = b$ of the plate.

7.8 Perform the summation of the series that express the kernel-functions G_{ij} of the integral representations shown in eqns (7.55) and (7.57) of Section 7.2, if the parameters α, β, and λ are fixed as: $\alpha = 0$, $\beta = 1$, $\lambda = 1$. Check your results against those shown later in that section.

7.9 Construct the influence matrix of a transverse unit concentrated force for a thin elastic Poisson-Kirchhoff's plate occupying an infinite strip $\Omega_s = \{(x,y): -\infty < x < \infty,\ 0 < y < \pi\}$ of width π (see Figure 7.2). Assume that the edges $y = 0$ and $y = \pi$ are simply supported and Ω_s is comprised of two semi-strips: $\Omega_1 = \{(x,y): -\infty < x < 0,\ 0 < y < \pi\}$ and $\Omega_2 = \{(x,y): 0 < x < \infty,\ 0 < y < \pi\}$. Assume also that both the fragments have the same thickness h, but the elastic moduli E_1 and E_2 and Poisson ratios σ_1 and σ_2 of the materials of which the fragments are composed are different.

7.10 Construct the influence matrix of a unit concentrated body force for elastic medium occupying an infinite strip $\Omega_s = \{(x,y): -\infty < x < \infty,\ 0 < y < \pi\}$ of width π (see Figure 7.2), comprised of two semi-strips: $\Omega_1 = \{(x,y): -\infty < x < 0,\ 0 < y < \pi\}$ and $\Omega_2 = \{(x,y): 0 < x < \infty,\ 0 < y < \pi\}$ composed with different materials whose elastic properties are specified with Lame's coefficients λ_1, μ_1 and λ_2, μ_2, respectively. Assume a perfect contact (displacements and stresses are continuous) along the interface line $x = 0$.

Chapter 8

Heat Equation

So far in this text, we have been involved with time-independent problems of applied mechanics that reduce to elliptic partial differential equations. The influence function approach that was developed earlier in the text is also productive in such areas of mechanics that deal with the two other basic types (parabolic and hyperbolic) of PDEs.

The discussion in this chapter focuses on the classical heat equation which is of the parabolic type. Explicit, easily computable expressions of influence functions for some initial–boundary value problems formulated for this equation can also be obtained by utilizing the ideas that were employed in the preceding sections of the text for steady state formulations in mechanics.

The influence function approach is traditional for texts dealing with heat conduction problems. A number of influence (Green's) functions for the heat equation are available in the existing literature (e.g. [7, 14]). There are also many texts in the engineering-oriented literature (e.g. [30, 31, 70]) that provide a reasonably rigorous background to the influence-functions-based methods for all the basic types of equations of mathematical physics.

Designing this chapter, we did not intend to compete with standard texts in the field. It would be extremely hard to accomplish such a feat. Our task in this chapter is much more modest. We simply plan to show that the techniques developed earlier in this text for constructing influence functions for elliptic equations in mechanics can also be extended to at least some particular situations occurring with the heat equation.

A very brief discussion will be presented here on the potential of our approach in the area of heat conduction. We intend to show that such an approach allows one to obtain some alternative representations of influence functions for the heat equation, which are more economical computationally compared to those already available in the literature.

As a supplementary tool for our development in this chapter, some leading principles of the integral Laplace transform are recalled in Section 8.1, where we review only those topics from the subject that are essential for the present discussion. Sections 8.2 and 8.3 describe methods that can be used for obtaining compact representations of influence functions and matrices for the heat equation in one and two spatial variables.

8.1 Laplace transform

In this section, we will recall some basics of one of the classical integral transforms that is broadly used in applied mathematical physics. The Laplace transform will be discussed here and utilized later in the following sections for constructing influence (Green's) functions and matrices of some problems for the heat equation. Since topics related to the Laplace transform have traditionally been included in the standard content of the undergraduate course of ODEs (which is among of the prerequisites for this text), we assume that the reader has at least a superficial background in this area.

The purpose in this section is not to give a detailed treatment of the Laplace transform method. This is simply impossible to accomplish within the scope of a single section. Our main objective here is more limited. In reviewing the subject, we restrict ourselves to the elementary aspects of the Laplace transform that are required to proceed with the discussion in this chapter. For a more detailed analysis of the subject, the reader is referred to specialized sources.

Definition: Let $f(t)$ be a function of a real variable defined for $t \geq 0$. Then the function $F(p)$ defined as

$$F(p) = \int_0^\infty \exp(-pt) f(t) dt \tag{8.1}$$

is said to be the *Laplace transform* of $f(t)$ provided the foregoing improper integral converges.

In the sources related to the Laplace transform method, $f(t)$ is often referred to as the *original*.

We will be using in this text the short-handed notation

$$\mathbf{L}\{f(t)\} = F(p)$$

for the relation in eqn (8.1) between the original $f(t)$ and its Laplace transform $F(p)$.

The existence of the Laplace transform $F(p)$ (that is, the convergence of the improper integral in eqn (8.1)) is guaranteed if the original function $f(t)$ is piecewise continuous on the interval $[0, \infty)$ and $f(t)$ is of exponential

8.1. LAPLACE TRANSFORM

order. We say that a function $f(t)$ is of exponential order if there exist numbers σ, $M > 0$, and $t_0 > 0$ such that

$$|f(t)| \leq M \exp(\sigma t)$$

for $t > t_0$. These conditions are sufficient for the Laplace transform to exist for $p > \sigma$.

The argument p of the Laplace transform $F(p)$ in eqn (8.1) represents a complex variable and the original function $f(t)$ can be expressed in terms of $F(p)$ by the following inverse transform

$$f(t) = \lim_{R \to \infty} \int_{\sigma - iR}^{\sigma + iR} F(p) \exp(pt) dp \qquad (8.2)$$

The short-handed expression for this relation in our text is

$$\mathbf{L}^{-1}\{F(p)\} = f(t).$$

The common scheme of using the Laplace transform method in applications implies: (i) taking the direct Laplace transform (converting $f(t)$ onto $F(p)$ by eqn (8.1)) of the original problem (which is somewhat difficult to tackle directly), (ii) solving the problem occurring as a result of stage (i) (these are, as a rule, easier to solve), and (iii) taking the inverse Laplace transform by using eqn (8.2).

One can readily obtain Laplace transforms for many particular types of originals by a straightforward integration in eqn (8.1). Indeed, it is not difficult to integrate the product of the exponential function with the most of elementary and even special functions. As a result, the reader can find rather extensive tables of the $f(t) \Leftrightarrow F(p)$ relationship in the existing texts and handbooks (see, for example, [1, 29]) devoted to the Laplace transform and its applications in engineering and science.

As far as the inverse Laplace transform is concerned, note that the direct use of the relation in eqn (8.2) can often be avoided, because given a transform $F(p)$, it is often possible to find the corresponding original in the existing tables. However, in practice it might not be always the case. If so, then integration in the complex plane in compliance with eqn (8.2) becomes the only way of computing originals. If the analytical integration in eqn (8.2) fails, then numerical methods can be used.

The properties of the Laplace transform and some particular $f(t) \Leftrightarrow F(p)$ relations that are required in this chapter, are listed below.

(A) Linearity: The Laplace transform of a linear combination of functions equals a linear combination of the Laplace transforms of each of the functions. That is

$$\mathbf{L}\left\{\sum_{i=1}^{n} C_i f_i(t)\right\} = \int_0^\infty \exp(-pt) \sum_{i=1}^{n} C_i f_i(t) dt$$

$$= \sum_{i=1}^{n} C_i \int_0^\infty \exp(-pt) f_i(t) dt.$$

To justify this property, one simply recalls the linearity property of a definite (improper) integral.

(B) Differentiation of the original: This property states that

$$\mathbf{L}\left\{\frac{df(t)}{dt}\right\} = \int_0^\infty \exp(-pt)\left(\frac{df(t)}{dt}\right) dt = pF(p) - f(0). \qquad (8.3)$$

This property can be justified by taking the integral

$$\int_0^\infty \exp(-pt)\left(\frac{df(t)}{dt}\right) dt$$

by parts. In doing so, we partition the integrand as

$$\exp(-pt) = u, \quad du = -p\exp(-pt), \quad \frac{df(t)}{dt} dt = dv, \quad v = f(t)$$

This yields

$$\int_0^\infty \exp(-pt)\left(\frac{df(t)}{dt}\right) dt = \exp(-pt) f(t) \Big|_0^\infty$$
$$+ p \int_0^\infty \exp(-pt) f(t) dt = pF(p) - f(0).$$

finally justifying this property.

(C) Integration of the original: If the functions $f(t)$ and $\varphi(t)$ hold the sufficient existence conditions for the Laplace transform and if $F(p)$ and $\Phi(p)$ represent the Laplace transforms of $f(t)$ and $\varphi(t)$, respectively, then

$$\mathbf{L}\left\{\int_0^t f(\tau)\varphi(t-\tau) d\tau\right\}$$
$$= \int_0^\infty \exp(-pt)\left[\int_0^t f(\tau)\varphi(t-\tau) d\tau\right] dt = F(p)\Phi(p).$$

The above relation is often referred to as the **Convolution Theorem** for the Laplace transform. To proceed with its proof, let

$$F(p) = \int_0^\infty \exp(-p\tau) f(\tau) d\tau \quad \text{and} \quad \Phi(p) = \int_0^\infty \exp(-p\xi)\varphi(\xi) d\xi$$

represent the Laplace transforms of $f(t)$ and $\varphi(t)$, respectively.

Multiplying $F(p)$ and $\Phi(p)$, we have

$$F(p)\Phi(p) = \int_0^\infty \exp(-p\tau) f(\tau) d\tau \int_0^\infty \exp(-p\xi) f(\xi) d\xi$$

8.1. LAPLACE TRANSFORM

$$= \int_0^\infty \int_0^\infty \exp(-p(\tau+\xi))f(\tau)\varphi(\xi)d\tau d\xi$$

$$= \int_0^\infty f(\tau)d\tau \int_0^\infty \exp(-p(\tau+\xi))\varphi(\xi)d\xi.$$

Let us fix τ and introduce a new variable $t=\tau+\xi$, this yields $t-\tau=\xi$ and $dt=d\xi$ so that the foregoing expression for $F(p)\Phi(p)$ can be rewritten in terms of τ and t as

$$F(p)\Phi(p) = \int_0^\infty f(\tau)d\tau \int_\tau^\infty \exp(-pt)\varphi(t-\tau)dt.$$

Since both $f(t)$ and $\varphi(t)$ satisfy the sufficient conditions of existence for the Laplace transform, the order of integration in the above expression is not important. Based on this, we interchange that order

$$F(p)\Phi(p) = \int_0^\infty \exp(-pt)dt \int_0^t f(\tau)\varphi(t-\tau)d\tau.$$

This can finally be rewritten as

$$\int_0^\infty \exp(-pt) \left[\int_0^t f(\tau)\varphi(t-\tau)d\tau \right] dt$$

$$= \mathbf{L}\left\{ \int_0^t f(\tau)\varphi(t-\tau)d\tau \right\} = F(p)\Phi(p). \tag{8.4}$$

Thus, we have completed the proof of the convolution theorem. To formulate the integration property for the original, that is, to express the Laplace transform of the integral

$$\int_0^t f(\tau)d\tau$$

in terms of the Laplace transform $F(p)$ of $f(t)$, we first write the transform of the function $\varphi(t)=1$. Eqn (8.1) immediately provides in this elementary case $\Phi(p)=1/p$. Subsequently, when $\varphi(t)=1$, the relation derived in eqn (8.4) implies

$$\mathbf{L}\left\{ \int_0^t f(\tau)d\tau \right\} = \int_0^\infty \exp(-pt) \left[\int_0^t f(\tau)d\tau \right] dt = \frac{F(p)}{p}. \tag{8.5}$$

(D) Translation Theorem: If $F(p)$ represents the Laplace transform of $f(t)$ and if β is any real number, then

$$\mathbf{L}\{\exp(\beta t)f(t)\} = \int_0^\infty \exp(-pt)\left[\exp(\beta t)f(t)\right]dt = F(p-\beta).$$

The proof of this relation immediately follows from eqn (8.1).

In addition to the properties listed, while using the Laplace transform method in order to derive influence functions in our further discussion, we will refer to the following relations:

$$\mathbf{L}^{-1}\left\{\frac{1}{\sqrt{p}}\exp(-a\sqrt{p})\right\} = \frac{1}{\sqrt{\pi t}}\exp\left(-\frac{a^2}{4t}\right), \quad a \geq 0 \tag{8.6}$$

and

$$\mathbf{L}^{-1}\left\{\frac{\exp(-a\sqrt{p})}{\sqrt{p}(\sqrt{p}+b)}\right\} = \exp(b(at+b))\operatorname{erfc}\left(b\sqrt{t}+\frac{a}{2\sqrt{t}}\right), \quad a \geq 0, \tag{8.7}$$

where $\operatorname{erfc}(\xi)$ represents a special function which is referred to as the *complementary error function* [3, 17, 23, 29, 31, 68]. It is defined as

$$\operatorname{erfc}(\xi) = \frac{2}{\sqrt{\pi}}\int_\xi^\infty \exp(-\eta^2)d\eta.$$

One more relation will be of help in the next section, that is

$$\mathbf{L}^{-1}\{K_0(a\sqrt{p})\} = \frac{1}{2t}\exp\left(-\frac{a^2}{4t}\right), \tag{8.8}$$

where K_0 represents the modified Bessel (or Macdonald) function of the second kind of order zero, which we have already referred to in Chapter 5 while discussing the Klein-Gordon equation.

The relations in eqns (8.6)–(8.8) can be found in the standard tables (see, for example, [1, 29]).

8.2 Influence functions

To define the notion of an influence function of the heat equation, we consider heat conduction in a simply connected region Ω, belonging to two-dimensional Euclidean space, composed of a homogeneous isotropic material whose constant thermal diffusivity is denoted with κ, with an initial temperature $f(P)$ throughout. Assume that the boundary $\Gamma = \bigcup_{i=1}^m \Gamma_i$ of Ω represents a piecewise smooth contour and that some homogeneous steady-state boundary conditions are imposed on Γ at all time $t > 0$. Assume also that no heat is generated within Ω. Such a statement leads us to the heat equation

$$\frac{\partial u(P,t)}{\partial t} = \kappa \Delta u(P,t), \quad P \in \Omega, \quad t > 0 \tag{8.9}$$

8.2. INFLUENCE FUNCTIONS

where Δ represents Laplace's operator with respect to the coordinates of P. According to the statement, equation (8.9) should be subjected to the initial condition

$$u(P,0) = f(P), \quad P \in \Omega \qquad (8.10)$$

and to the boundary conditions

$$\alpha_i(P)\frac{\partial u(P,t)}{\partial n_i} + \beta_i(P)u(P,t) = 0, \quad i = \overline{1,m}, \quad P \in \Gamma_i \qquad (8.11)$$

where $\alpha_i(P)$ and $\beta_i(P)$ represent given functions defined on Γ in such a way that at least one of them is nonzero for every piece Γ_i of Γ, and n_i represents a normal direction to Γ_i at the point P.

Assume that the initial–boundary value problem posed by eqns (8.9)–(8.11) has a unique solution.

Definition: If the solution of the initial–boundary value problem posed by eqns (8.9)–(8.11) is expressible in the integral form

$$u(P,t) = \iint_\Omega g(P,t;Q)f(Q)dQ, \quad x \in \Omega, \quad t > 0 \qquad (8.12)$$

then the kernel function $g(P,t;Q)$ of this representation is often called the *influence (Green's) function of an instant point source* for the problem posed with eqns (8.9)–(8.11).

The influence function reveals the temperature response at the observation point P at time t due to an instant concentrated (point) heat source released at the point Q at time $t = 0$. In addition, it is understood that $g(P,t;Q) = 0$ for $t < 0$.

The reader is advised to take a look at the above definition from the viewpoint of the primary goal of this study, which is to develop effective procedures for the construction of influence functions. This makes our objective not to just solve the problem in eqns (8.9)–(8.11), but rather to find its solution in a form of the integral representation shown in eqn (8.12). Indeed, such a form of the solution provides an explicit expression of the influence function.

With this in mind, we will here describe some standard methods that are helpful in constructing influence functions.

EXAMPLE 1 Obtain the influence function for the entire space (that is usually referred to as the *fundamental solution*) of the heat equation in a single spatial variable.

The way in which the fundamental solution will be here obtained gives an idea about one of the approaches that will later be used to construct influence functions in more complicated cases.

Consider heat conduction in an infinite thin rod composed of a homogeneous material with constant thermal diffusivity κ, with an initial temperature $f(x)$ and whose temperature at positive and negative infinities is supposed to be finite for all time $t>0$. We also assume that the flow of heat within the rod goes only in the axial x direction, the lateral surface of the rod is insulated, and no heat is generated within the rod. Such a statement provides a one-dimensional heat equation

$$\frac{\partial u(x,t)}{\partial t} = \kappa \frac{\partial^2 u(x,t)}{\partial x^2}, \quad x \in (-\infty, \infty), \quad t>0 \tag{8.13}$$

subject to the initial

$$u(x,0) = f(x) \tag{8.14}$$

and the 'boundary' conditions

$$|u(-\infty, t)| < \infty, \quad |u(\infty, t)| < \infty \tag{8.15}$$

Take the Laplace transform of the function $u(x,t)$ with respect to the time variable t

$$U(x,p) = \int_0^\infty \exp(-pt) u(x,t) dt$$

Based on properties **(A)** and **(B)** of the Laplace transform (see Section 8.1), one obtains the following self-adjoint boundary value problem

$$\frac{d^2 U(x,p)}{dx^2} - \frac{p}{\kappa} U(x,p) = -\frac{1}{\kappa} f(x) \tag{8.16}$$

$$|U(-\infty, p)| < \infty, \quad |U(\infty, p)| < \infty \tag{8.17}$$

in the transform $U(x,p)$ of $u(x,t)$.

Upon using the method of variation of parameters, we can construct the Green's function for the above formulation. In doing so, the complementary solution to eqn (8.16) is found in the form

$$U(x,p) = \int_{-\infty}^x \frac{1}{2\kappa\sqrt{q}} [\exp(\sqrt{q}(s-x)) - \exp(\sqrt{q}(x-s))] f(s) ds$$

$$+ D_1 \exp(\sqrt{q}x) + D_2 \exp(-\sqrt{q}x)$$

where, for the sake of compactness, we introduced $q = p/\kappa$.

To satisfy the boundary conditions from eqn (8.17), we regroup the terms in the above representation for $U(x,p)$ as

$$U(x,p) = \left(D_1 - \frac{1}{2\kappa\sqrt{q}} \int_{-\infty}^x \exp(-\sqrt{q}s) f(s) ds \right) \exp(\sqrt{q}x)$$

8.2. INFLUENCE FUNCTIONS

$$+ \left(D_2 + \frac{1}{2\kappa\sqrt{q}} \int_{-\infty}^{x} \exp(\sqrt{q}s)f(s)ds\right) \exp(-\sqrt{q}x) \qquad (8.18)$$

The first of the conditions from eqn (8.17), being satisfied by this expression for $U(x,p)$, requires the coefficient of $\exp(-\sqrt{q}x)$ to be zero. This subsequently provides $D_2 = 0$, while the second of the conditions from eqn (8.17) requires the coefficient of $\exp(\sqrt{q}x)$ to be zero. This yields

$$D_1 = \frac{1}{2\kappa\sqrt{q}} \int_{-\infty}^{\infty} \exp(-\sqrt{q}s)f(s)ds$$

By substituting the values of D_1 and D_2 just found into eqn (8.18) and rearranging the like integral terms in it, one obtains

$$U(x,p) = \frac{1}{2\kappa\sqrt{q}} \left\{ \int_{-\infty}^{\infty} \exp(\sqrt{q}(x-s))f(s)ds \right.$$

$$\left. + \int_{-\infty}^{x} [\exp(\sqrt{q}(s-x)) - \exp(\sqrt{q}(x-s))] f(s)ds \right\}$$

This representation, as we usually suggest within the scope of the method of variation of parameters, can be rewritten in the form of a single integral

$$U(x,p) = \int_{-\infty}^{\infty} G(x,p;s)f(s)ds \qquad (8.19)$$

whose kernel function is defined in pieces as

$$G(x,p;s) = \frac{1}{2\kappa\sqrt{q}} \begin{cases} \exp(\sqrt{q}(x-s)), & -\infty < x \le s < \infty \\ \exp(\sqrt{q}(s-x)), & \infty > x \ge s > -\infty \end{cases}$$

We, however, can write a single-line expression for $G(x,p;s)$ by using the notion of absolute value. That is

$$G(x,p;s) = \frac{1}{2\kappa\sqrt{q}} \exp(-\sqrt{q}|x-s|)$$

In accordance with Theorem 1.3 of Chapter 1, this represents the Green's function of the homogeneous boundary value problem corresponding to that posed by eqns (8.16) and (8.17). Recalling the relation $q = p/\kappa$ introduced earlier, we rewrite $G(x,p;s)$ as

$$G(x,p;s) = \frac{1}{2\sqrt{\kappa p}} \exp\left(-\frac{|x-s|}{\sqrt{\kappa}}\sqrt{p}\right) \qquad (8.20)$$

Clearly, the solution $u(x,t)$ of the initial-boundary value problem posed by eqns (8.13)–(8.15) can be obtained as the inverse Laplace transform of $U(x,p)$ from eqn (8.19). That is

$$u(x,t) = \mathbf{L}^{-1}\left\{\int_{-\infty}^{\infty} G(x,p;s)f(s)ds\right\}$$

$$= \int_{-\infty}^{\infty} \mathbf{L}^{-1}\{G(x,p;s)\}\, f(s)ds. \qquad (8.21)$$

From the definition introduced earlier in this section, it follows that the kernel-function

$$g(x,t;s) = \mathbf{L}^{-1}\{G(x,p;s)\}$$

of the integral in eqn (8.21) represents the influence function of the problem posed by eqns (8.13)–(8.15). Hence, $g(x,t;s)$ can practically be found as the inverse Laplace transform of the Green's function $G(x,p;s)$ of the homogeneous boundary value problem corresponding to that of eqns (8.16) and (8.17). In accordance with the relation from eqn (8.6), the inverse transform of $G(x,p;s)$ shown in eqn (8.20) is found as

$$g(x,t;s) = \frac{1}{2\sqrt{\kappa\pi t}} \exp\left(-\frac{(x-s)^2}{4\kappa t}\right) \qquad (8.22)$$

It is clearly seen that eqn (8.22) presents the classical fundamental solution of the one-dimensional heat equation.

EXAMPLE 2 Derive the influence function of an instant point heat source for a semi-infinite rod whose end-point $x=0$ is held at temperature zero at all time $t>0$.

Let all the assumptions made in the preceding example hold, yielding in this case the following initial-boundary value problem

$$\frac{\partial u(x,t)}{\partial t} = \kappa \frac{\partial^2 u(x,t)}{\partial x^2}, \quad x\in(0,\infty), \quad t>0 \qquad (8.23)$$

$$u(x,0) = f(x) \qquad (8.24)$$

$$u(0,t) = 0, \quad |u(\infty,t)| < \infty \qquad (8.25)$$

This formulation implies that the Laplace transform

$$U(x,p) = \int_0^\infty \exp(-pt)u(x,t)dt$$

8.2. INFLUENCE FUNCTIONS

of its solution $u(x,t)$ represents the solution of the self-adjoint boundary value problem written as

$$\frac{d^2 U(x,p)}{dx^2} - \frac{p}{\kappa} U(x,p) = -\frac{1}{\kappa} f(x) \tag{8.26}$$

$$U(0,p) = 0, \quad |U(\infty,p)| < \infty. \tag{8.27}$$

It has been shown in the preceding example that, in compliance with the method of variation of parameters, the complementary solution of eqn (8.26) can be written as

$$U(x,p) = \frac{1}{2\sqrt{\kappa p}} \int_0^x \left[\exp\left(\frac{s-x}{\sqrt{\kappa}}\sqrt{p}\right) - \exp\left(\frac{x-s}{\sqrt{\kappa}}\sqrt{p}\right) \right] f(s) ds$$

$$+ D_1 \exp\left(\sqrt{\frac{p}{\kappa}} x\right) + D_2 \exp\left(-\sqrt{\frac{p}{\kappa}} x\right). \tag{8.28}$$

The arbitrary constants D_1 and D_2 can be found upon satisfying the boundary conditions imposed by eqn (8.27). When $x=0$, the integral term in eqn (8.28) vanishes, and the first of those conditions implies that

$$D_1 + D_2 = 0,$$

while the second condition provides

$$D_1 = \frac{1}{2\sqrt{\kappa p}} \int_0^\infty \exp\left(-\sqrt{\frac{p}{\kappa}} s\right) f(s) ds$$

yielding

$$D_2 = -\frac{1}{2\sqrt{\kappa p}} \int_0^\infty \exp\left(-\sqrt{\frac{p}{\kappa}} s\right) f(s) ds$$

Substituting these values of D_1 and D_2 into eqn (8.28) and again using the notion of absolute value, one obtains the following compact expression for $U(x,p)$

$$U(x,p) = \frac{1}{2\sqrt{\kappa p}} \int_0^\infty \left[\exp\left(-\frac{|x-s|}{\sqrt{\kappa}}\sqrt{p}\right) - \exp\left(-\frac{x+s}{\sqrt{\kappa}}\sqrt{p}\right) \right] f(s) ds.$$

Thus, in accordance with Theorem 1.3 of Chapter 1, the kernel of the above integral

$$G(x,p;s) = \frac{1}{2\sqrt{\kappa p}} \left[\exp\left(-\frac{|x-s|}{\sqrt{\kappa}}\sqrt{p}\right) - \exp\left(-\frac{x+s}{\sqrt{\kappa}}\sqrt{p}\right) \right] \tag{8.29}$$

represents the Green's function of the homogeneous boundary value problem corresponding to that of eqns (8.26) and (8.27). Subsequently, the inverse Laplace transform of $G(x, p; s)$

$$g(x, t; s) = \mathbf{L}^{-1}\{G(x, p; s)\},$$

which can also be found with the aid of the relation from eqn (8.6) as

$$g(x, t; s) = \frac{1}{2\sqrt{\pi \kappa t}} \left[\exp\left(-\frac{(x-s)^2}{4\kappa t}\right) - \exp\left(-\frac{(x+s)^2}{4\kappa t}\right) \right],$$

represents the influence function of the initial-boundary value problem posed by eqns (8.23)–(8.25).

EXAMPLE 3 Derive the influence function of an instant point heat source for a semi-infinite rod, if the convectional (Newtonean) cooling by a surrounding medium of temperature zero takes place at the end-point $x = 0$ at all time $t > 0$. That is, the boundary conditions

$$\frac{\partial u(0, t)}{\partial x} - \beta u(0, t) = 0, \quad |u(\infty, t)| < \infty, \quad \beta \geq 0 \qquad (8.30)$$

are imposed at all time $t > 0$, with β representing the so-called *heat transfer coefficient* of the cooling.

Analogously to the previous examples, we assume that the rod's lateral surface is insulated, no heat is generated within the rod, and the initial temperature is $f(x)$, so that eqns (8.23) and (8.24) also hold in this case.

The initial-boundary value problem posed by eqns (8.23), (8.24), and (8.30) implies that the Laplace transform $U(x, p)$ of its solution $u(x, t)$ represents the solution of the self-adjoint boundary value problem written as

$$\frac{d^2 U(x, p)}{dx^2} - \frac{p}{\kappa} U(x, p) = -\frac{1}{\kappa} f(x) \qquad (8.31)$$

$$\frac{dU(0, p)}{dx} - \beta U(0, p) = 0, \quad |U(\infty, p)| < \infty \qquad (8.32)$$

As we have already found, the complementary solution of eqn (8.31) is given by eqn (8.28). The second of the boundary conditions in eqn (8.32), as we learned from the previous example, yields

$$D_1 = \frac{1}{2\kappa \sqrt{q}} \int_0^\infty \exp(-\sqrt{q}s) f(s) ds$$

8.2. INFLUENCE FUNCTIONS

To proceed further with the first of the boundary conditions in eqn (8.32), we differentiate $U(x,p)$ from eqn (8.28)

$$\frac{dU(x,p)}{dx} = -\int_0^x \frac{1}{2\kappa}[\exp(\sqrt{q}(s-x))+\exp(\sqrt{q}(x-s))]f(s)ds$$

$$+ D_1\sqrt{q}\exp(\sqrt{q}x) - D_2\sqrt{q}\exp(-\sqrt{q}x)$$

where again $q=p/\kappa$.

Hence, the first of the boundary conditions in eqn (8.32) implies that

$$D_1(\sqrt{q}-\beta) - D_2(\sqrt{q}+\beta) = 0,$$

So, taking into account the value of D_1 that is already available, one obtains for D_2

$$D_2 = \frac{\sqrt{q}-\beta}{2\kappa\sqrt{q}(\sqrt{q}+\beta)}\int_0^\infty \exp(-\sqrt{q}s)f(s)ds$$

Upon substituting the values of D_1 and D_2 just found into eqn (8.28), we arrive at

$$U(x,p) = \frac{1}{2\kappa\sqrt{q}}\left\{\int_0^x [\exp(\sqrt{q}(s-x))-\exp(\sqrt{q}(x-s))]f(s)ds\right.$$

$$\left.+ \int_0^\infty \left[\exp(\sqrt{q}(x-s))+\frac{\sqrt{q}-\beta}{\sqrt{q}+\beta}\exp(-\sqrt{q}(x+s))\right]f(s)ds\right\}$$

This representation can be rewritten as a single integral

$$U(x,p) = \frac{1}{2\kappa\sqrt{q}}\int_0^\infty \left[\exp(-\sqrt{q}|x-s|)\right.$$

$$\left.+ \frac{\sqrt{q}-\beta}{\sqrt{q}+\beta}\exp(-\sqrt{q}(x+s))\right]f(s)ds$$

whose kernel function

$$G(x,p;s) = \frac{1}{2\kappa\sqrt{q}}\left[\exp(-\sqrt{q}|x-s|)+\frac{\sqrt{q}-\beta}{\sqrt{q}+\beta}\exp(-\sqrt{q}(x+s))\right]$$

$$= \frac{1}{2\sqrt{\kappa p}}\left[\exp\left(-\frac{|x-s|}{\sqrt{\kappa}}\sqrt{p}\right)+\frac{\sqrt{p}-\sqrt{\kappa}\beta}{\sqrt{p}+\sqrt{\kappa}\beta}\exp\left(-\frac{x+s}{\sqrt{\kappa}}\sqrt{p}\right)\right] \quad (8.33)$$

represents the Green's function of the homogeneous boundary value problem corresponding to that of eqns (8.31) and (8.32).

To obtain the inverse Laplace transform $\mathbf{L}^{-1}\{G(x,p;s)\}$, which, as we know, represents the influence function $g(x,t;s)$ of the initial-boundary value problem posed by eqns (8.23), (8.24), and (8.30), we rewrite $G(x,p;s)$ from eqn (8.33) in the following equivalent form

$$G(x,p;s) = \frac{1}{2\sqrt{\kappa p}}\left[\exp\left(-\frac{|x-s|}{\sqrt{\kappa}}\sqrt{p}\right) + \exp\left(-\frac{x+s}{\sqrt{\kappa}}\sqrt{p}\right)\right.$$

$$\left. - \frac{2\sqrt{\kappa}\beta}{\sqrt{p}+\sqrt{\kappa}\beta}\exp\left(-\frac{x+s}{\sqrt{\kappa}}\sqrt{p}\right)\right]$$

The inverse Laplace transform of $G(x,p;s)$ can be found with the aid of the relations form eqns (8.6) and (8.7). That is, the influence function $g(x,t;s)$ is ultimately found as

$$g(x,t;s) = \frac{1}{2\sqrt{\pi\kappa t}}\left[\exp\left(-\frac{(x-s)^2}{4\kappa t}\right) + \exp\left(-\frac{(x+s)^2}{4\kappa t}\right)\right]$$

$$- \beta\exp(\beta(x+s))\exp(\beta^2\kappa t)\operatorname{erfc}\left(\beta\sqrt{\kappa t}+\frac{x+s}{2\sqrt{\kappa t}}\right) \quad (8.34)$$

It is clearly seen that if the heat transfer coefficient β approaches zero, then the boundary condition in eqn (8.30) converts into the condition of an insulated end

$$\frac{\partial u(0,t)}{\partial x} = 0$$

On the other hand, if $\beta=0$, then eqn (8.34) transforms into

$$g(x,t;s) = \frac{1}{2\sqrt{\pi\kappa t}}\left[\exp\left(-\frac{(x-s)^2}{4\kappa t}\right) + \exp\left(-\frac{(x+s)^2}{4\kappa t}\right)\right]$$

That is, the above expression represents the influence function for a semi-infinite rod with an insulated end.

Thus, from the examples that we went through, it follows that the approach described in this section for the construction of influence functions for the heat equation is a classical one. We, however, propose a slightly different interpretation. Namely, we suggest to view the procedure as consisting of the following three stages:

(i) upon using the Laplace transform, an original initial-boundary value problems is mapped into a boundary value problem for either ordinary or elliptic partial differential equation;

(ii) by the method of variation of parameters, the Green's function is constructed of the boundary value problem occurring in point (i);

8.2. INFLUENCE FUNCTIONS

(iii) upon using the inverse Laplace transform to the Green's function obtained in point (ii), we finally derive the desired influence function.

In the next example, along with the Laplace transform method, we discuss an alternative approach that is also helpful in some situations for the construction of influence functions for the heat equation.

EXAMPLE 4 Consider heat conduction in a thin rod of length a composed of a homogeneous material, with an initial temperature $f(x)$ throughout and whose end-points $x=0$ and $x=a$ are held at temperature zero for all time $t>0$. Let all the assumptions made in the preceding examples also hold in this case, so that the heat equation

$$\frac{\partial u(x,t)}{\partial t} = \kappa \frac{\partial^2 u(x,t)}{\partial x^2}, \quad x \in (0,a), \quad t>0 \qquad (8.35)$$

is subject to the initial

$$u(x,0) = f(x) \qquad (8.36)$$

and to the boundary conditions

$$u(0,t)=0, \quad u(a,t)=0 \qquad (8.37)$$

The Laplace transform method can also be applied to this case. It reduces the above formulation to the self-adjoint boundary value problem

$$\frac{d^2 U(x,p)}{dx^2} - \frac{p}{\kappa} U(x,p) = -\frac{1}{\kappa} f(x)$$

$$U(0,p)=0, \quad U(a,p)=0$$

in the Laplace transform $U(x,p)$ of $u(x,t)$. The branch of the Green's function of this problem, which is valid for $x \le s$ is obtained as

$$G(x,p;s) = \frac{\exp(\sqrt{q}x) - \exp(-\sqrt{q}x)}{2\sqrt{\kappa p}\left[\exp(\sqrt{q}a) - \exp(-\sqrt{q}a)\right]}$$

$$\times \left[\exp(\sqrt{q}(a-s)) - \exp(\sqrt{q}(s-a))\right] \qquad (8.38)$$

where $q = p/\kappa$. Due to the self-adjointness, the branch of $G(x,p;s)$ valid for $x \ge s$ can be obtained from that in eqn (8.38) by interchanging of x with s.

The inverse Laplace transform of $G(x,p;s)$ represents the influence function of the problem posed by eqns (8.35)–(8.37). To ease the process of taking the inverse transform, we rewrite the factor

$$\frac{1}{\exp(\sqrt{q}a) - \exp(-\sqrt{q}a)} = \frac{\exp(-\sqrt{q}a)}{1 - \exp(-2\sqrt{q}a)}$$

as a sum of the geometric progression. This yields

$$\frac{1}{\exp(\sqrt{q}a)-\exp(-\sqrt{q}a)} = \exp(-\sqrt{q}a)\sum_{n=0}^{\infty}\exp(-2\sqrt{q}na)$$

Based on this, the final form of $G(x,p;s)$ is obtained as

$$G(x,p;s) = \frac{1}{2\sqrt{\kappa p}}\sum_{n=0}^{\infty}\left[\exp\left(-\sqrt{\frac{p}{\kappa}}(s-x+2na)\right) - \exp\left(-\sqrt{\frac{p}{\kappa}}(s+x+2na)\right)\right.$$
$$\left. - \exp\left(-\sqrt{\frac{p}{\kappa}}(2(n+1)a-x-s)\right) + \exp\left(-\sqrt{\frac{p}{\kappa}}(2(n+1)a+x-s)\right)\right]$$

By virtue of the relation from eqn (8.6), the influence function $g(x,t;s)$ of the initial-boundary value problem posed by eqns (8.35)–(8.37) is found in the form

$$g(x,t;s) = \mathbf{L}^{-1}\{G(x,p;s)\}$$

$$= \frac{1}{2\sqrt{\kappa\pi t}}\sum_{n=0}^{\infty}\left[\exp\left(-\frac{(s-x+2na)^2}{4\kappa t}\right) - \exp\left(-\frac{(s+x+2na)^2}{4\kappa t}\right)\right.$$
$$\left. - \exp\left(-\frac{(s+x-2(n+1)a)^2}{4\kappa t}\right) + \exp\left(-\frac{(s-x-2(n+1)a)^2}{4\kappa t}\right)\right]$$

Rearranging the summation in this expression, one rewrites it in the following symmetric form

$$g(x,t;s) = \frac{1}{2\sqrt{\kappa\pi t}}\sum_{j=-\infty}^{\infty}\left[\exp\left(-\frac{(x-s+2ja)^2}{4\kappa t}\right) - \exp\left(-\frac{(x+s+2ja)^2}{4\kappa t}\right)\right]$$

which is usually [7, 14] presented in the literature.

In what follows, we discuss another (different from the Laplace transform) approach to the construction of influence functions for the heat equation. In some cases, it allows one to derive a closed form of an influence function. The technique rooted in the classical method of eigenfunction expansion [63] will be described here.

Let us again consider the initial-boundary value problem posed by eqns (8.35)–(8.37). To solve it, we expand $u(x,t)$ and $f(x)$ in Fourier series with respect to the spatial variable x

$$u(x,t) = \sum_{n=1}^{\infty} u_n(t)\sin(\nu x), \quad f(x) = \sum_{n=1}^{\infty} f_n \sin(\nu x), \tag{8.39}$$

where, for notational convenience, we introduce $\nu = n\pi/a$.

8.2. INFLUENCE FUNCTIONS

Upon substituting these expansions into eqns (3.35)–(8.37), one arrives at the following initial value problem

$$\frac{du_n(t)}{dt} + \kappa \nu^2 u_n(t) = 0$$

$$u_n(0) = f_n$$

for the coefficients of the first series in eqn (8.39). The solution of this problem is readily found as

$$u_n(t) = f_n \exp(-\kappa \nu^2 t)$$

Expressing the values of f_n by means of the fundamental rule for Fourier coefficients (Fourier-Euler formula), we rewrite $u_n(t)$ as

$$u_n(t) = \left(\frac{2}{a} \int_0^a f(s) \sin(\nu s) ds\right) \exp(-\kappa \nu^2 t)$$

By substituting the above expression into the first of the series in eqn (8.39) and interchanging the order of the summation and integration in it, we obtain the solution of the problem in eqns (8.35)–(8.37) in the form

$$u(x,t) = \int_0^a \frac{2}{a} \sum_{n=1}^\infty \exp(-\kappa \nu^2 t) \sin(\nu x) \sin(\nu s) f(s) ds \qquad (8.40)$$

From the definition (see eqn (8.12)), it follows that, since the solution of the initial-boundary value problem posed by eqns (8.35)–(8.37) is found as an integral over the segment $[0,a]$, the kernel function

$$g(x,t;s) = \frac{2}{a} \sum_{n=1}^\infty \exp(-\kappa \nu^2 t) \sin(\nu x) \sin(\nu s) \qquad (8.41)$$

of the integral represents the influence function that we are looking for.

Probably the reader realizes that until this moment, we have simply traced out the standard algorithm of the eigenvalue expansion method. Indeed, eqn (8.41) exhibits the classical representation of the influence function under consideration, which is available in nearly every text on the subject. Note that usually, while discussing the expansion in eqn (8.41), authors outline its high convergence rate for 'large' values of time variable t and low convergence rate for 'small' values of t.

In what follows, however, the convergence rate of the series in eqn (8.41) can be ignored, at least formally. The point is that the sum of that series is expressible in terms of a special function which can be tabulated so as to make the convergence issue not a matter of the present discussion, but rather a matter of computer software development.

Before going any further with the expansion of the influence function appearing in eqn (8.41), we recall from [1, 29, 68] the so-called *Jacobi Theta function of the third kind* ϑ_3 whose series representation is

$$\vartheta_3(\alpha,\beta) = 1 + 2\sum_{n=1}^{\infty} \exp(-\pi^2\beta n^2)\cos(2n\pi\alpha)$$

To express $g(x,t;s)$ from eqn (8.41) in terms of the Theta function ϑ_3, we rewrite it in the form

$$g(x,t;s) = \frac{1}{a}\sum_{n=1}^{\infty} \exp(-\kappa\nu^2 t)[\cos(\nu(x-s)) - \cos(\nu(x+s))]$$

$$= \frac{1}{a}\left[\sum_{n=1}^{\infty} \exp(-\kappa\nu^2 t)\cos(\nu(x-s)) - \sum_{n=1}^{\infty} \exp(-\kappa\nu^2 t)\cos(\nu(x+s))\right]$$

from which it clearly follows that the above series converge providing

$$g(x,t;s) = \frac{1}{2a}\left[\vartheta_3\left(\frac{x-s}{2a},\frac{\kappa t}{a^2}\right) - 1\right] - \frac{1}{2a}\left[\vartheta_3\left(\frac{x+s}{2a},\frac{\kappa t}{a^2}\right) - 1\right]$$

$$= \frac{1}{2a}\left[\vartheta_3\left(\frac{x-s}{2a},\frac{\kappa t}{a^2}\right) - \vartheta_3\left(\frac{x+s}{2a},\frac{\kappa t}{a^2}\right)\right] \qquad (8.42)$$

Thus, eqn (8.42) presents an alternative form of the influence function of the instant point heat source for the rod of length a with the end-points being held at temperature zero.

In the next example, we will show that the approach based on the combination of the Laplace transform and the eigenfunction expansion method is efficient in constructing influence functions of a point source for some two-dimensional heat conduction problems.

EXAMPLE 5 Let us consider the heat conduction in a thin semi-strip-shaped plate occupying the region Ω_{ss} (see Figure 5.1). The plate is composed with an isotropic homogeneous material with constant thermal diffusivity κ, with an initial temperature $f(x,y)$ throughout and whose contour is held at temperature zero for all time $t>0$. We also assume that the flow of heat within the plate goes only in the x and y directions and lateral surfaces of the plate are insulated, and that no heat is generated within the plate. Such a statement provides a two-dimensional initial-boundary value problem written as

$$\frac{\partial u(x,y,t)}{\partial t} = \kappa\left(\frac{\partial^2 u(x,y,t)}{\partial x^2} + \frac{\partial^2 u(x,y,t)}{\partial y^2}\right) \quad (x,y)\in\Omega_{ss},\ t>0 \qquad (8.43)$$

8.2. INFLUENCE FUNCTIONS

$$u(x, y, 0) = f(x, y) \tag{8.44}$$

$$u(x, 0, t) = u(x, b, t) = 0, \quad u(0, y, t) = 0, \quad |u(\infty, y, t)| < \infty \tag{8.45}$$

To solve this problem, we take the Laplace transform of the function $u(x, y, t)$ with respect to the time variable t

$$U(x, y, p) = \int_0^\infty \exp(-pt) u(x, y, t) dt$$

Based on properties (**A**) and (**B**) of the Laplace transform (see Section 8.1), one obtains a self-adjoint boundary value problem for the following Klein-Gordon equation

$$\frac{\partial^2 U(x, y, p)}{\partial x^2} + \frac{\partial^2 U(x, y, p)}{\partial y^2} - \frac{p}{\kappa} U(x, y, p) = -\frac{1}{\kappa} f(x, y) \tag{8.46}$$

subject to the boundary conditions

$$U(x, 0, t) = U(x, b, t) = 0, \quad U(0, y, t) = 0, \quad |U(\infty, y, t)| < \infty \tag{8.47}$$

for the transform $U(x, y, p)$ of $u(x, y, t)$.

To obtain the solution of the boundary value problem posed by eqns (8.46) and (8.47), we expand $U(x, y, t)$ and $f(x, y)$ in the following Fourier series with respect to the variable y

$$U(x, y, p) = \sum_{n=1}^\infty U_n(x, p) \sin(\nu y), \quad f(x, y) = \sum_{n=1}^\infty f_n(x) \sin(\nu y), \tag{8.48}$$

where, for notational convenience, we denote $\nu = n\pi/b$.

This yields the boundary value problem

$$\frac{d^2 U_n(x, p)}{dx^2} - \left(\nu^2 + \frac{p}{\kappa}\right) U_n(x, p) = -\frac{1}{\kappa} f_n(x) \tag{8.49}$$

$$U_n(0, p) = 0, \quad |U_n(\infty, p)| < \infty. \tag{8.50}$$

for the coefficient $U_n(x, p)$ of the first series of eqn (8.48).

The problem of this type has been shown earlier in this section (see *EXAMPLE 2*). Based on eqn (8.29), the Green's function of the homogeneous boundary value problem corresponding to that of eqns (8.49) and (8.50) can be written as

$$G_n(x, p; \xi)$$

$$= \frac{\sqrt{\kappa}}{2\sqrt{p+\omega}}\left[\exp\left(-\frac{\sqrt{p+\omega}}{\sqrt{\kappa}}|x-\xi|\right) - \exp\left(-\frac{\sqrt{p+\omega}}{\sqrt{\kappa}}(x+\xi)\right)\right] \quad (8.51)$$

where we introduced $\omega = \nu^2 \kappa$.

Hence, in light of Theorem 1.3 of Chapter 1, the solution of the problem posed by eqns (8.49) and (8.50) is found in terms of $G_n(x, p; \xi)$ as the following integral

$$U_n(x, p) = \frac{1}{\kappa}\int_0^\infty G_n(x, p; \xi) f_n(\xi) d\xi$$

By expressing $f_n(\xi)$ in terms of the function $f(x, y)$ in accordance with Fourier-Euler formula

$$f_n(\xi) = \frac{2}{b}\int_0^b f(\xi, \eta) \sin(\nu\eta) d\eta,$$

substituting the expression for $U_n(x, p)$ just found into the first of the series in eqn (8.48), and interchanging the order of the summation and integration, we obtain the following expression for the solution of the boundary value problem posed by eqns (8.46) and (8.47)

$$U(x, y, p) = \int_0^b \int_0^\infty \left\{\frac{2}{b\kappa} \sum_{n=1}^\infty G_n(x, p; \xi) \sin(\nu y) \sin(\nu\eta)\right\} f(\xi, \eta) d\xi d\eta$$

Since $U(x, y, p)$ represents the Laplace transform of the solution $u(x, y, t)$ of the initial-boundary value problem posed by eqns (8.43)–(8.45), $u(x, y, t)$ itself can be obtained as the inverse of $U(x, y, p)$. That is

$$u(x, y, t) = \mathbf{L}^{-1}\{U(x, y, p)\}$$

$$= \mathbf{L}^{-1}\left\{\int_0^b \int_0^\infty \frac{2}{b\kappa} \sum_{n=1}^\infty G_n(x, p; \xi) \sin(\nu y) \sin(\nu\eta) f(\xi, \eta) d\xi d\eta\right\}$$

$$= \int_0^b \int_0^\infty \frac{2}{b\kappa} \sum_{n=1}^\infty \mathbf{L}^{-1}\{G_n(x, p; \xi)\} \sin(\nu y) \sin(\nu\eta) f(\xi, \eta) d\xi d\eta$$

where, according to the definition presented earlier, the kernel function $G(x, y, t; \xi, \eta)$ introduced as

$$G(x, y, t; \xi, \eta) = \frac{2}{b\kappa} \sum_{n=1}^\infty \mathbf{L}^{-1}\{G_n(x, p; \xi)\} \sin(\nu y) \sin(\nu\eta) \quad (8.52)$$

represents the influence function of an instant concentrated heat source released at the point $(\xi, \eta) \in \Omega_{ss}$ at time $t = 0$.

8.2. INFLUENCE FUNCTIONS

To obtain the inverse transform $\mathbf{L}^{-1}\{G_n(x,p;\xi)\}$, we recall the expression from eqn (8.51) and apply the relation from eqn (8.6) along with Translation theorem from Section 8.1. This yields

$$\mathbf{L}^{-1}\{G_n(x,p;\xi)\}$$

$$= \mathbf{L}^{-1}\left\{\frac{\sqrt{\kappa}}{2\sqrt{p+\omega}}\left[\exp\left(-\frac{\sqrt{p+\omega}}{\sqrt{\kappa}}|x-\xi|\right)-\exp\left(-\frac{\sqrt{p+\omega}}{\sqrt{\kappa}}(x+\xi)\right)\right]\right\}$$

$$= \frac{\sqrt{\kappa}}{2\sqrt{\pi t}}\left[\exp\left(-\frac{(x-\xi)^2}{4\kappa t}\right)-\exp\left(-\frac{(x+\xi)^2}{4\kappa t}\right)\right]\exp(-\omega t)$$

Upon substituting the above expression into eqn (8.52) and performing some elementary algebra, one obtains

$$G(x,y,t;\xi,\eta)$$

$$= \frac{1}{2b\sqrt{\kappa\pi t}}\sum_{n=1}^{\infty}\left[\exp\left(-\frac{(x-\xi)^2}{4\kappa t}\right)-\exp\left(-\frac{(x+\xi)^2}{4\kappa t}\right)\right]$$

$$\times \exp\left(-n^2\pi^2\frac{\kappa t}{b^2}\right)\left[\cos\left(n\frac{\pi}{b}(y-\eta)\right)-\cos\left(n\frac{\pi}{b}(y+\eta)\right)\right].$$

Regrouping terms in the above expression provides

$$G(x,y,t;\xi,\eta)$$

$$= \frac{1}{2b\sqrt{\kappa\pi t}}\left[\exp\left(-\frac{(x-\xi)^2}{4\kappa t}\right)-\exp\left(-\frac{(x+\xi)^2}{4\kappa t}\right)\right]$$

$$\times \left\{\sum_{n=1}^{\infty}\exp\left(-n^2\pi^2\frac{\kappa t}{b^2}\right)\cos\left(2n\pi\frac{(y-\eta)}{2b}\right)\right.$$

$$\left.-\sum_{n=1}^{\infty}\exp\left(-n^2\pi^2\frac{\kappa t}{b^2}\right)\cos\left(2n\pi\frac{(y+\eta)}{2b}\right)\right\}$$

Recalling from the previous example the series expansion

$$\vartheta_3(\alpha,\beta)=1+2\sum_{n=1}^{\infty}\exp(-\pi^2\beta n^2)\cos(2n\pi\alpha)$$

of the Jacobi Theta function of the third kind, we finally obtain the following closed form of the influence function of an instant point source for the initial-boundary value problem posed by eqns (8.43)–(8.45)

$$G(x,y,t;\xi,\eta)$$

$$= \frac{1}{4b\sqrt{\kappa\pi t}}\left[\exp\left(-\frac{(x-\xi)^2}{4\kappa t}\right) - \exp\left(-\frac{(x+\xi)^2}{4\kappa t}\right)\right]$$

$$\times \left[\vartheta_3\left(\frac{y-\eta}{2b}, \frac{\kappa t}{b^2}\right) - \vartheta_3\left(\frac{y+\eta}{2b}, \frac{\kappa t}{b^2}\right)\right] \tag{8.53}$$

For the next example, we formulate one more two-dimensional heat conduction problem whose influence function can readily be obtained by our approach.

EXAMPLE 6 Let us consider the heat conduction in a half-plane-shaped plate occupying a region $\Omega = \{(r,\phi) : 0 < r < \infty, 0 < \phi < \pi\}$, composed with the material whose thermal diffusivity is κ, if the initial temperature is given as $f(r,\phi)$, the lateral surfaces are insulated, and the edge ($\phi=0$ and $\phi=\pi$) is held at temperature zero for all time $t>0$.

The above formulation results in the following initial-boundary value problem

$$\frac{\partial u(r,\phi,t)}{\partial t} = \kappa \Delta u(r,\phi,t), \quad (r,\phi) \in \Omega, \quad t>0 \tag{8.54}$$

$$u(r,\phi,0) = f(r,\pi) \tag{8.55}$$

$$u(r,0,t) = u(r,\pi,t) = 0 \tag{8.56}$$

where Δ represents the polar Laplace operator. That is

$$\Delta = \frac{1}{r}\frac{\partial}{\partial r}\left(r\frac{\partial}{\partial r}\right) + \frac{1}{r^2}\frac{\partial^2}{\partial \phi^2}$$

In addition, we require for $u(r,\phi,t)$ to be bounded when r approaches zero and infinity.

Upon taking the Laplace transform of $u(r,\phi,t)$ with respect to t

$$U(r,\phi,p) = \mathbf{L}\{u(r,\phi,t)\}$$

the problem posed by eqns (8.54)–(8.56) maps in the boundary value problem for Klein-Gordon equation

$$\Delta U(r,\phi,p) - \frac{p}{\kappa} U(r,\phi,p) = -\frac{1}{\kappa} f(r,\phi)$$

$$U(r,0,p) = U(r,\pi,p) = 0$$

8.3. INFLUENCE MATRICES

The Green's function of the corresponding homogeneous problem can be found in Chapter 5 (see eqn (5.111)). In terms of our current notations, it is written in the form

$$G(r,\phi,p;\rho,\psi) = \frac{1}{2\pi}\left[K_0\left(\sqrt{\frac{p}{\kappa}}|z-\zeta|\right) - K_0\left(\sqrt{\frac{p}{\kappa}}|z-\overline{\zeta}|\right)\right]$$

where K_0 represents the modified Bessel function of the second kind of order zero, with $z = r(\cos(\phi)+i\sin(\phi))$ and $\zeta = \rho(\cos(\psi)+i\sin(\psi))$ representing the observation and source point, respectively.

Thus, the influence function $g(r,\phi,t;\rho,\psi)$ of the initial-boundary value problem posed by eqns (8.54)–(8.56) can be obtained as the inverse Laplace transform of $G(r,\phi,p;\rho,\psi)$. This transform can be written with the aid of the relation from eqn (8.8) of Section 8.1 as

$$g(r,\phi,t;\rho,\psi) = \frac{1}{4\sqrt{\pi t}}\left[\exp\left(-\frac{|z-\zeta|^2}{4\kappa t}\right) - \exp\left(-\frac{|z-\overline{\zeta}|^2}{4\kappa t}\right)\right]$$

This is a well-known expression of the influence function under consideration. This representation is more often obtained by the reflection method from the fundamental solution

$$\frac{1}{4\sqrt{\pi t}}\exp\left(-\frac{|z-\zeta|^2}{4\kappa t}\right)$$

of the heat equation in two dimensions.

We have considered a limited number of illustrative examples in this section. In the exercises related to this section, the reader is encouraged to derive some more influence functions for the one- and two-dimensional heat equation.

8.3 Influence matrices

In this section, we will focus the reader's attention on the constructing procedure for obtaining influence matrices of an instant point heat source for the heat conduction phenomenon taking place in media whose conductive properties discontinuously vary with the spatial coordinates within the region to be considered.

As an illustrative example, we obtain the influence matrix for an assembly of thin semi-infinite rods, shown in Figure 8.1. Each rod is composed of a homogeneous conductive material whose heat conductivity and thermal diffusivity are h_i and κ_i, respectively, with initial temperatures $f_i(x)$. The ideal contact is assumed at $x=0$. That is, both the temperature and heat flux are assumed to be continuous at $x=0$.

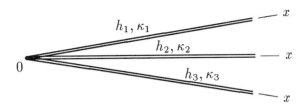

Figure 8.1 An assembly of semi-infinite rods

It is also assumed that the flows of heat within the rods go only in the x directions, that is, lateral surfaces of the rods are insulated. Such a statement is associated with the set ($i = 1, 2, 3$) of heat equations

$$\frac{\partial u_i(x,t)}{\partial t} = \kappa_i \frac{\partial^2 u_i(x,t)}{\partial x^2}, \quad x \in (0, \infty), \quad t > 0 \tag{8.57}$$

subject to the initial conditions

$$u_i(x, 0) = f_i(x), \quad i = 1, 2, 3 \tag{8.58}$$

and put into the system format by the contact conditions

$$u_1(0, t) = u_2(0, t) = u_3(0, t) \tag{8.59}$$

and

$$h_1 \frac{\partial u_1(0,t)}{\partial x} + h_2 \frac{\partial u_2(0,t)}{\partial x} + h_3 \frac{\partial u_3(0,t)}{\partial x} = 0 \tag{8.60}$$

In addition, we assume that the temperature is finite at infinity for all time $t > 0$

$$|u_i(\infty, t)| < \infty \tag{8.61}$$

Take the Laplace transform of the functions $u_i(x, t)$ with respect to the time variable t

$$U_i(x, p) = \int_0^\infty \exp(-pt) u_i(x, t) dt$$

By virtue of properties (**A**) and (**B**) of the Laplace transform (see Section 8.1), the above formulation transforms into the following multi-point posed boundary value problem in $U_i(x, p)$

$$\frac{d^2 U_i(x,p)}{dx^2} - \frac{p}{\kappa_i} U_i(x, p) = -\frac{1}{\kappa_i} f_i(x) \tag{8.62}$$

8.3. INFLUENCE MATRICES

$$U_1(0,p) = U_2(0,p) = U_3(0,p) \tag{8.63}$$

$$h_1 \frac{dU_1(0,p)}{dx} + h_2 \frac{dU_2(0,p)}{dx} + h_3 \frac{dU_3(0,p)}{dx} = 0 \tag{8.64}$$

$$|U_i(\infty, p)| < \infty \tag{8.65}$$

With the aid of the method of variation of parameters, the complementary solution of equation (8.62) can be found (see eqn (8.28)) in the form

$$U_i(x,p) = \frac{1}{2\sqrt{\kappa_i p}} \int_0^x \left[\exp\left(\frac{\xi - x}{\sqrt{\kappa_i}}\sqrt{p}\right) - \exp\left(\frac{\xi - s}{\sqrt{\kappa_i}}\sqrt{p}\right) \right] f_i(\xi) d\xi$$

$$+ M_i \exp\left(\sqrt{\frac{p}{\kappa_i}} x\right) + N_i \exp\left(-\sqrt{\frac{p}{\kappa_i}} x\right), \quad i = 1, 2, 3 \tag{8.66}$$

The arbitrary coefficients M_i and N_i can be obtained from the boundary and contact conditions imposed by eqns (8.63)–(8.65). That is, from the boundedness conditions in eqn (8.65), the values of M_i are obtained as

$$M_i = \frac{1}{2\sqrt{\kappa_i p}} \int_0^\infty \exp\left(-\sqrt{\frac{p}{\kappa_i}} \xi\right) f_i(\xi) d\xi, \quad i = 1, 2, 3$$

Based on these expressions for M_i, the values of N_i can be found from the contact conditions. That is, the continuity conditions formulated in eqn (8.63) for the temperature at $x = 0$ result in

$$N_1 - N_2 = \frac{1}{2\sqrt{p}} \int_0^\infty \left[\exp\left(-\sqrt{\frac{p}{\kappa_2}}\xi\right) \frac{f_2(\xi)}{\sqrt{\kappa_2}} - \exp\left(-\sqrt{\frac{p}{\kappa_1}}\xi\right) \frac{f_1(\xi)}{\sqrt{\kappa_1}} \right] d\xi \tag{8.67}$$

and

$$N_2 - N_3 = \frac{1}{2\sqrt{p}} \int_0^\infty \left[\exp\left(-\sqrt{\frac{p}{\kappa_3}}\xi\right) \frac{f_3(\xi)}{\sqrt{\kappa_3}} - \exp\left(-\sqrt{\frac{p}{\kappa_2}}\xi\right) \frac{f_2(\xi)}{\sqrt{\kappa_2}} \right] d\xi \tag{8.68}$$

The continuity condition formulated in eqn (8.64) for the heat flux at $x = 0$ provides

$$\frac{h_1}{\sqrt{\kappa_1}} N_1 + \frac{h_2}{\sqrt{\kappa_2}} N_2 + \frac{h_3}{\sqrt{\kappa_3}} N_3 = \frac{1}{2\sqrt{p}} \int_0^\infty \left[\frac{h_1}{\kappa_1} \exp\left(-\sqrt{\frac{p}{\kappa_1}}\xi\right) f_1(\xi) \right.$$

$$\left. + \frac{h_2}{\kappa_2} \exp\left(-\sqrt{\frac{p}{\kappa_2}}\xi\right) f_2(\xi) + \frac{h_3}{\kappa_3} \exp\left(-\sqrt{\frac{p}{\kappa_3}}\xi\right) f_3(\xi) \right] d\xi \tag{8.69}$$

The relations in eqns (8.67)–(8.69) form a well-posed system of linear algebraic equations in N_i. Upon its solution, one obtains

$$N_1 = \frac{1}{2H\sqrt{p}} \int_0^\infty \left[\left(\frac{h_1}{\sqrt{\kappa_1}} - \frac{h_2}{\sqrt{\kappa_2}} - \frac{h_3}{\sqrt{\kappa_3}} \right) \exp\left(-\sqrt{\frac{p}{\kappa_1}}\xi\right) \frac{f_1(\xi)}{\sqrt{\kappa_1}} \right.$$

$$\left. + \frac{2h_2}{\kappa_2} \exp\left(-\sqrt{\frac{p}{\kappa_2}}\xi\right) f_2(\xi) + \frac{2h_3}{\kappa_3} \exp\left(-\sqrt{\frac{p}{\kappa_3}}\xi\right) f_3(\xi) \right] d\xi,$$

$$N_2 = \frac{1}{2H\sqrt{p}} \int_0^\infty \left[\left(\frac{h_2}{\sqrt{\kappa_2}} - \frac{h_1}{\sqrt{\kappa_1}} - \frac{h_3}{\sqrt{\kappa_3}} \right) \exp\left(-\sqrt{\frac{p}{\kappa_2}}\xi\right) \frac{f_2(\xi)}{\sqrt{\kappa_2}} \right.$$

$$\left. + \frac{2h_1}{\kappa_1} \exp\left(-\sqrt{\frac{p}{\kappa_1}}\xi\right) f_1(\xi) + \frac{2h_3}{\kappa_3} \exp\left(-\sqrt{\frac{p}{\kappa_3}}\xi\right) f_3(\xi) \right] d\xi,$$

and

$$N_3 = \frac{1}{2H\sqrt{p}} \int_0^\infty \left[\left(\frac{h_3}{\sqrt{\kappa_3}} - \frac{h_2}{\sqrt{\kappa_2}} - \frac{h_1}{\sqrt{\kappa_1}} \right) \exp\left(-\sqrt{\frac{p}{\kappa_3}}\xi\right) \frac{f_3(\xi)}{\sqrt{\kappa_3}} \right.$$

$$\left. + \frac{2h_1}{\kappa_1} \exp\left(-\sqrt{\frac{p}{\kappa_1}}\xi\right) f_1(\xi) + \frac{2h_2}{\kappa_2} \exp\left(-\sqrt{\frac{p}{\kappa_2}}\xi\right) f_2(\xi) \right] d\xi,$$

where

$$H = \frac{h_1}{\sqrt{\kappa_1}} + \frac{h_2}{\sqrt{\kappa_2}} + \frac{h_3}{\sqrt{\kappa_3}}$$

As can readily be seen, once the coefficients M_i and N_i, $(i = 1, 2, 3)$ are found, eqn (8.66) provides an explicit form for the solution functions $U_i(x,p)$ to the boundary value problem posed by eqns (8.62)–(8.65) in terms of the right-hand side functions $f_i(x)$ of eqn (8.62). For the sake of compactness in the development that follows, we introduce the column-vectors $\mathbf{U}(x,p) = (U_1(x,p), U_2(x,p), U_3(x,p))^T$ and $\mathbf{f}(x) = (f_1(x), f_2(x), f_3(x))^T$.

Upon substituting the values of M_i and N_i just found into eqn (8.66) and performing some algebra, we obtain the vector $\mathbf{U}(x,p)$ expressed in terms of the vector $\mathbf{f}(x)$ as

$$\mathbf{U}(x,p) = \int_0^\infty G(x,p;\xi)\mathbf{f}(\xi)d\xi \qquad (8.70)$$

Clearly, the kernel-matrix $G(x,p;\xi) = (G_{ij}(x,p;\xi))_{3\times 3}$ of this integral represents the matrix of Green's type of the homogeneous boundary value problem corresponding to that posed by eqns (8.62)–(8.65). In what follows, we exhibit the entries $G_{ij}(x,p;\xi)$ of $G(x,p;\xi)$ in a row-wise manner. The entries of its first row are found as

$$G_{11}(x,p;\xi) = \frac{1}{2H\sqrt{\kappa_1 p}} \left[H \exp\left(-|x-\xi|\sqrt{\frac{p}{\kappa_1}}\right) \right.$$

8.3. INFLUENCE MATRICES

$$+ \left(\frac{h_1}{\sqrt{\kappa_1}} - \frac{h_2}{\sqrt{\kappa_2}} - \frac{h_3}{\sqrt{\kappa_3}}\right) \exp\left(-(x+\xi)\sqrt{\frac{p}{\kappa_1}}\right)\Bigg],$$

$$G_{12}(x,p;\xi) = \frac{h_2}{H\kappa_2\sqrt{p}} \exp\left(-(\frac{x}{\sqrt{\kappa_1}} + \frac{\xi}{\sqrt{\kappa_2}})\sqrt{p}\right),$$

and

$$G_{13}(x,p;\xi) = \frac{h_3}{H\kappa_3\sqrt{p}} \exp\left(-(\frac{x}{\sqrt{\kappa_1}} + \frac{\xi}{\sqrt{\kappa_3}})\sqrt{p}\right).$$

For the entries $G_{2j}(x,p;\xi)$ of the second row of $G(x,p;\xi)$ we have

$$G_{21}(x,p;\xi) = \frac{h_1}{H\kappa_1\sqrt{p}} \exp\left(-(\frac{x}{\sqrt{\kappa_2}} + \frac{\xi}{\sqrt{\kappa_1}})\sqrt{p}\right),$$

$$G_{22}(x,p;\xi) = \frac{1}{2H\sqrt{\kappa_2 p}} \left[H \exp\left(-|x-\xi|\sqrt{\frac{p}{\kappa_2}}\right)\right.$$

$$+ \left(\frac{h_2}{\sqrt{\kappa_2}} - \frac{h_1}{\sqrt{\kappa_1}} - \frac{h_3}{\sqrt{\kappa_3}}\right) \exp\left(-(x+\xi)\sqrt{\frac{p}{\kappa_2}}\right)\Bigg],$$

and

$$G_{23}(x,p;\xi) = \frac{h_3}{H\kappa_3\sqrt{p}} \exp\left(-(\frac{x}{\sqrt{\kappa_2}} + \frac{\xi}{\sqrt{\kappa_3}})\sqrt{p}\right).$$

Finally, the entries $G_{3j}(x,p;\xi)$ of the third row of $G(x,p;\xi)$ are found as

$$G_{31}(x,p;\xi) = \frac{h_1}{H\kappa_1\sqrt{p}} \exp\left(-(\frac{x}{\sqrt{\kappa_3}} + \frac{\xi}{\sqrt{\kappa_1}})\sqrt{p}\right),$$

$$G_{32}(x,p;\xi) = \frac{h_2}{H\kappa_2\sqrt{p}} \exp\left(-(\frac{x}{\sqrt{\kappa_3}} + \frac{\xi}{\sqrt{\kappa_2}})\sqrt{p}\right),$$

and

$$G_{33}(x,p;\xi) = \frac{1}{2H\sqrt{\kappa_3 p}} \left[H \exp\left(-|x-\xi|\sqrt{\frac{p}{\kappa_3}}\right)\right.$$

$$+ \left(\frac{h_3}{\sqrt{\kappa_3}} - \frac{h_1}{\sqrt{\kappa_1}} - \frac{h_2}{\sqrt{\kappa_2}}\right) \exp\left(-(x+\xi)\sqrt{\frac{p}{\kappa_3}}\right)\Bigg].$$

From the relation in eqn (8.70), it can be concluded that the solution vector $\mathbf{u}(x,t) = (u_1(x,t), u_2(x,t), u_3(x,t))$ of the original boundary value problem posed by eqns (8.57)–(8.61) can be found as the inverse Laplace transform of the vector $\mathbf{U}(x,p;\xi)$

$$\mathbf{u}(x,t) = \mathbf{L}^{-1}\{\mathbf{U}(x,p)\}$$

$$= \int_0^\infty \mathbf{L}^{-1}\{G(x,p;\xi)\}\mathbf{f}(\xi)d\xi = \int_0^\infty g(x,t;\xi)\mathbf{f}(\xi)d\xi$$

Hence, the inverse Laplace transforms of $G(x,p;\xi)$ represents the influence matrix $g(x,t;\xi)$ of an instant point heat source for the assembly of rods shown in Figure 8.1. The entries $g_{ij}(x,t;\xi)$ of $g(x,t;\xi)$ can be obtained from the corresponding entries $G_{ij}(x,p;\xi)$ of $G(x,p;\xi)$ by the relation from eqn (8.6). In doing so, the entries of the first row of $g(x,t;\xi)$ are found as

$$g_{11}(x,t;\xi) = \frac{1}{2H\sqrt{\pi\kappa_1 t}}\left[H\exp\left(-\frac{(x-\xi)^2}{4\kappa_1 t}\right)\right.$$
$$\left. + \left(\frac{h_1}{\sqrt{\kappa_1}} - \frac{h_2}{\sqrt{\kappa_2}} - \frac{h_3}{\sqrt{\kappa_3}}\right)\exp\left(-\frac{(x+\xi)^2}{4\kappa_1 t}\right)\right],$$

$$g_{12}(x,t;\xi) = \frac{h_2}{H\kappa_2\sqrt{\pi t}}\exp\left(-\frac{1}{4t}(\frac{x}{\sqrt{\kappa_1}}+\frac{\xi}{\sqrt{\kappa_2}})^2\right),$$

and

$$g_{13}(x,t;\xi) = \frac{h_3}{H\kappa_3\sqrt{\pi t}}\exp\left(-\frac{1}{4t}(\frac{x}{\sqrt{\kappa_1}}+\frac{\xi}{\sqrt{\kappa_3}})^2\right).$$

while for the entries of the second row of $g(x,t;\xi)$, we obtain

$$g_{21}(x,t;\xi) = \frac{h_1}{H\kappa_1\sqrt{\pi t}}\exp\left(-\frac{1}{4t}(\frac{x}{\sqrt{\kappa_2}}+\frac{\xi}{\sqrt{\kappa_1}})^2\right),$$

$$g_{22}(x,t;\xi) = \frac{1}{2H\sqrt{\pi\kappa_2 t}}\left[H\exp\left(-\frac{(x-\xi)^2}{4\kappa_2 t}\right)\right.$$
$$\left. + \left(\frac{h_2}{\sqrt{\kappa_2}} - \frac{h_1}{\sqrt{\kappa_1}} - \frac{h_3}{\sqrt{\kappa_3}}\right)\exp\left(-\frac{(x+\xi)^2}{4\kappa_2 t}\right)\right],$$

and

$$g_{23}(x,t;\xi) = \frac{h_3}{H\kappa_3\sqrt{\pi t}}\exp\left(-\frac{1}{4t}(\frac{x}{\sqrt{\kappa_2}}+\frac{\xi}{\sqrt{\kappa_3}})^2\right).$$

Finally, the entries of the third row of $g(x,t;\xi)$ are found as

$$g_{31}(x,t;\xi) = \frac{h_1}{H\kappa_1\sqrt{\pi t}}\exp\left(-\frac{1}{4t}(\frac{x}{\sqrt{\kappa_3}}+\frac{\xi}{\sqrt{\kappa_1}})^2\right),$$

$$g_{32}(x,t;\xi) = \frac{h_2}{H\kappa_2\sqrt{\pi t}}\exp\left(-\frac{1}{4t}(\frac{x}{\sqrt{\kappa_3}}+\frac{\xi}{\sqrt{\kappa_2}})^2\right),$$

and

$$g_{33}(x,t;\xi) = \frac{1}{2H\sqrt{\pi\kappa_3 t}}\left[H\exp\left(-\frac{(x-\xi)^2}{4\kappa_3 t}\right)\right.$$

$$+ \left(\frac{h_3}{\sqrt{\kappa_3}} - \frac{h_1}{\sqrt{\kappa_1}} - \frac{h_2}{\sqrt{\kappa_2}} \right) \exp\left(-\frac{(x+\xi)^2}{4\kappa_3 t} \right) \Bigg].$$

In this brief section, we have considered a single illustrative example just to demonstrate a possibility of constructing influence matrices of the heat equation for media whose conductive properties vary discontinuously with spatial variables within a region. Many other formulations of this kind can be treated in a similar manner (exercise 8.5, for instance, deals with a compound plate).

8.4 *Review Exercises*

8.1 Construct the influence function of an instant point source for the heat equation in a semi-infinite thin rod whose end-point is insulated.

8.2 Construct the influence function of an instant point source for the heat equation in a thin finite rod of length a, if its left-hand end-point $x=0$ is held at temperature zero for all time $t>0$ while the right-hand end-point is insulated.

8.3 Construct the influence function of an instant point source for the two-dimensional heat equation in an infinite strip-shaped plate occupying the region $\Omega = \{(x,y): -\infty < x < \infty, \ 0 < y < b\}$, if the lateral faces of the plate are insulated and the edges $y=0$ and $y=b$ are held at temperature zero for all time $t>0$.

8.4 Construct the influence function of an instant point source for the two-dimensional heat equation in a semi-strip-shaped plate occupying the region $\Omega = \{(x,y): 0 < x < \infty, \ 0 < y < b\}$, if the lateral faces of the plate are insulated, and the edge $x=0$ is also insulated, while the edges $y=0$ and $y=b$ are held at temperature zero.

8.5 Construct the influence matrix of an instant point source for the two-dimensional heat equation in a compound infinite strip-shaped plate occupying the region $\Omega = \{(x,y): -\infty < x < \infty, \ 0 < y < b\}$ consisting of two fragments $\Omega_1 = \{(x,y): -\infty < x < 0, \ 0 < y < b\}$ and $\Omega_2 = \{(x,y): 0 < x < \infty, \ 0 < y < b\}$ composed of the hypothetical materials with different heat conductivities h_1 and h_2 and equal thermal diffusivities $\kappa_1 = \kappa_2 = \kappa$. Assume that the faces of the plate are insulated and the edges $y=0$ and $y=b$ are held at temperature zero.

Appendix A
Catalogue of Green's Functions

The notion of a Green's (influence) function or matrix of linear boundary value problems for differential equations represents the key issue of the present study. A number of influence functions and matrices have been constructed and a wide spectrum of problem classes from applied mechanics have been analyzed here by means of this method. However, areas of applied mechanics, where influence functions and matrices can effectively be used, are not limited to those studied in the present text. There are many other areas of engineering and science where Green's-function-related methods can be productively used.

Keeping also in mind that the reader would benefit from having a list of all of the boundary value problems whose Green's functions or matrices are available in the text, we have created this appendix, where one will find a catalogue of those problems (of a total number of about one hundred fifty, with more than a half of them related to PDEs) whose Green's functions or matrices are either derived in this book or just presented here.

Notice that in Parts II through VII that are related to PDEs, for the sake of compactness, the subscriptive notations are used for partial derivatives. That is, we follow a conventional scheme where $u_{xy}(x,y)$, for example, denotes the mixed second partial derivative of $u(x,y)$ with respect to x and y.

Part I of this section is devoted to ordinary differential equations and systems, Part II to equations of potential written in different coordinate systems, Part III to Klein-Gordon equation, Part IV to the biharmonic equation, Part V to Lame's system of the plane problem of theory of elasticity, Part VI to problems of applied mechanics formulated over compound regions, and Part VII to the heat equation.

I   Ordinary Differential Equations and Systems:

1. $y''(x)=0, \quad x\in(0,a); \quad y(0)=0, \; y(a)=0$ $\boxed{\text{p. 13}}$

2. $y''(x)=0, \quad x\in(0,a); \quad y'(0)=0, \; y'(a)+my(a)=0$ $\boxed{\text{p. 14}}$

3. $y''(x)=0, \quad x\in(0,a); \quad y'(0)=0, \; y(a)=0$ $\boxed{\text{p. 14}}$

4. $y''(x)=0$, $x\in(0,a)$; $y(0)=0$, $y'(a)=0$ — p. 415

5. $y''(x)=0$, $x\in(0,a)$; $y(0)=0$, $y'(a)+hy(a)=0$ — p. 415

6. $y''(x)=0$, $x\in(0,a)$; $y'(0)-h_1 y(0)=0$, $y'(a)+h_2 y(a)=0$ — p. 415

7. $y''(x)-k^2 y(x)=0$, $x\in(0,\infty)$; $y(0)=0$, $|y(\infty)|<\infty$ — p. 16

8. $y''(x)-k^2 y(x)=0$, $x\in(0,\infty)$; $y'(0)-hy(0)=0$, $|y(\infty)|<\infty$ — p. 40

9. $y''(x)-k^2 y(x)=0$, $x\in(0,a)$; $y(0)=y(a)$, $y'(0)=y'(a)$ — p. 17

10. $y''(x)-k^2 y(x)=0$, $x\in(0,a)$; $y'(0)=0$, $y(a)=0$ — p. 39

11. $y''(x)-k^2 y(x)=0$, $x\in(0,a)$; $y'(0)-hy(0)=0$, $y(a)=0$ — p. 417

12. $y''(x)+k^2 y(x)=0$, $x\in(0,a)$; $y'(0)=0$, $y'(a)=0$ — p. 38

13. $y''(x)+k^2 y(x)=0$, $x\in(0,a)$; $y(0)=0$, $y(a)=0$ — p. 416

14. $y''(x)+k^2 y(x)=0$, $x\in(0,1)$; $y(0)=0$, $y'(1)=0$ — p. 417

15. $y''(x)+y'(x)-2y(x)=0$, $x\in(0,\infty)$; $y(0)=0$, $|y(\infty)|<\infty$ — p. 23

16. $y''(x)+3y'(x)-10y(x)=0$, $x\in(0,1)$; $y(0)=0$, $|y(\infty)|<\infty$ — p. 416

17. $((mx+p)y'(x))'=0$, $x\in(0,a)$; $y'(0)=0$, $y(a)=0$ — p. 19

18. $((mx+p)y'(x))'=0$, $x\in(0,a)$; $y(0)=0$, $y(a)=0$ — p. 415

19. $(xy'(x))'=0$, $x\in(0,a)$; $|y(0)|<\infty$, $y'(a)+hy(a)=0$ — p. 20

20. $(xy'(x))'=0$, $x\in(0,a)$; $|y(0)|<\infty$, $y(a)=0$ — pp. 20, 417

21. $(e^{\beta x} y'(x))'=0$, $x\in(0,a)$; $y(0)=0$, $y(a)=0$ — p. 415

22. $(e^{\beta x} y'(x))'=0$, $x\in(0,a)$; $y(0)=0$, $y'(a)=0$ — p. 416

23. $y^{IV}(x)=0$, $x\in(0,1)$;
    $y(0)=y'(0)=0$, $y(1)=y''(1)=0$ — pp. 21-22

24. $y^{IV}(x)=0$, $x\in(0,1)$;
    $y(0)=y'(0)=0$, $y''(1)=y'''(1)=0$ — p. 416

25. $y^{IV}(x)=0$, $x\in(0,1)$;
    $y(0)=y'(0)=0$, $y(1)=y'(1)=0$ — p. 417

APPENDIX A: CATALOGUE OF GREEN'S FUNCTIONS

26. $y^{IV}(x) = 0, \; x \in (0,1);$
$y(0) = y''(0) = 0, \; y(1) = y''(1) = 0$ — p. 417

27. $y^{IV}(x) = 0, \; x \in (0,a);$
$y(0) = y'(0) = 0, \; y(a) = y'(a) = 0$ — p. 57

28. $y^{IV}(x) = 0, \; x \in (0,a);$
$y''(0) = EIy'''(0) + k_0 y(0) = 0, \; y(a) = y'(a) = 0$ — p. 60

29. $y^{IV}(x) = 0, \; x \in (0,a);$
$y''(0) = y'''(0) = 0, \; y(a) = y'(a) = 0$ — p. 61

30. $y^{IV}(x) = 0, \; x \in (0,a);$
$y(0) = y'(0) = 0, \; y''(a) = y'''(a) = 0$ — p. 63

31. $y^{IV}(x) = 0, \; x \in (0,a);$
$y(0) = y''(0) = 0, \; y(a) = y''(a) = 0$ — p. 74

32. $y^{IV}(x) = 0, \; x \in (0,a);$
$y(0) = y''(0) = 0, \; y(a) = y'(a) = 0$ — p. 61

33. $y^{IV}(x) = 0, \; x \in (0,a);$
$y''(0) = EIy'''(0) + k_0 y(0) = 0,$
$y''(a) = EIy'''(a) - k_a y(a) = 0$ — p. 419

34. $y^{IV}(x) = 0, \; x \in (0,a);$
$y(0) = y''(0) = 0, \; y''(a) = EIy'''(a) - k_a y(a) = 0$ — p. 419

35. $y^{IV}(x) = 0, \; x \in (0,a);$
$y(0) = y''(0) = 0, \; y'(a) = y'''(a) = 0$ — p. 420

36. $y^{IV}(x) = 0, \; x \in (0,a);$
$y(0) = y'(0) = 0, \; y'(a) = y'''(a) = 0$ — p. 420

37. $y^{IV}(x) - 2k^2 y''(x) + k^4 y(x) = 0, \; x \in (0,\infty);$
$y(0) = y'(0) = 0, \; |y(\infty)| < \infty, \; |y'(\infty)| < \infty$ — p. 43

38. $y^{IV}(x) + 4k^4 y(x) = 0, \; x \in (-\infty, \infty);$
$|y(-\infty)| < \infty, \; |y'(-\infty)| < \infty,$
$|y(\infty)| < \infty, \; |y'(\infty)| < \infty$ — p. 84

39. $y^{IV}(x) + 4k^4 y(x) = 0, \; x \in (0, \infty);$
$y(0) = y'(0) = 0, \; |y(\infty)| < \infty, \; |y'(\infty)| < \infty$ — p. 89

40. $y^{IV}(x) + 4k^4 y(x) = 0, \; x \in (0, \infty);$
$y''(0) = y'''(0) = 0, \; |y(\infty)| < \infty, \; |y'(\infty)| < \infty$ — p. 91

41. $y^{IV}(x) + 4k^4 y(x) = 0$, $x \in (0, \infty)$;
$y(0) = y''(0) = 0$, $|y(\infty)| < \infty$, $|y'(\infty)| < \infty$
$\boxed{\text{p. 425}}$

42. $y^{IV}(x) - q^2 y''(x) = 0$, $x \in (0, a)$;
$y(0) = y''(0) = 0$, $y(a) = y''(a) = 0$
$\boxed{\text{p. 123}}$

43. $y^{IV}(x) - q^2 y''(x) = 0$, $x \in (0, a)$;
$y(0) = y''(0) = 0$, $y'(a) = y'''(a) = 0$
$\boxed{\text{p. 126}}$

44. $y^{IV}(x) + q^2 y''(x) = 0$, $x \in (0, a)$;
$y(0) = y''(0) = 0$, $y(a) = y''(a) = 0$
$\boxed{\text{p. 124}}$

45. $y^{IV}(x) + q^2 y''(x) = 0$, $x \in (0, a)$;
$y(0) = y''(0) = 0$, $y'(a) = y'''(a) = 0$
$\boxed{\text{p. 126}}$

46. $y^{IV}(x) + N/EI_0 y''(x) - y(x) = 0$, $x \in (0, a)$;
$y(0) = y''(0) = 0$, $y(a) = y''(a) = 0$
$\boxed{\text{p. 136}}$

47. $((mx+b)y''(x))'' = 0$, $x \in (0, a)$;
$y(0) = y'(0) = 0$, $y''(a) = y'''(a) = 0$
$\boxed{\text{p. 97}}$

48. $(mxy''(x))'' = 0$, $x \in (0, a)$;
$y(a) = y'(a) = 0$, $y''(0) = y'''(0) = 0$
$\boxed{\text{p. 426}}$

49. $(me^{bx} y''(x))'' = 0$, $x \in (0, a)$;
$y(0) = y''(0) = 0$, $y(a) = y''(a) = 0$
$\boxed{\text{p. 99}}$

50. $(me^{bx} y''(x))'' = 0$, $x \in (0, a)$;
$y(0) = y'(0) = 0$, $y''(a) = y'''(a) = 0$
$\boxed{\text{p. 427}}$

51. $y_1''(x) = 0$, $x \in (-1, 0)$;
$y_2''(x) = 0$, $x \in (0, 1)$;
$y_1(-1) = 0$, $y_2(1) = 0$,
$y_1(0) = y_2(0)$, $y_1'(0) = \lambda y_2'(0)$
$\boxed{\text{pp. 153-154}}$

52. $(xy_1'(x))' - x^{-1} y_1(x) = 0$, $x \in (0, a)$;
$(xy_2'(x))' - x^{-1} y_2(x) = 0$, $x \in (a, \infty)$;
$|y_1(0)| < \infty$, $|y_2(\infty)| < \infty$,
$y_1(a) = y_2(a)$, $y_1'(a) = \lambda y_2'(a)$
$\boxed{\text{p. 157}}$

53. $y_i''(x) = 0$, $x \in (0, 1)$, $i = 1, 2, 3$;
$y_1(0) = y_2(0) = y_3(0)$, $h_1 y_1'(0) + h_2 y_2'(0) + h_3 y_3'(0) = 0$,
$y_1(1) = y_2(1) = y_3(1) = 0$
$\boxed{\text{p. 159}}$

54. $y_1''(x) = 0$, $x \in (-a, 0)$;
$y_2''(x) - k^2 y_2(x) = 0$, $x \in (0, \infty)$;
$y_1(-a) = 0$, $|y_2(\infty)| < \infty$,
$y_1(0) = y_2(0)$, $y_1'(0) = \lambda y_2'(0)$
$\boxed{\text{p. 427}}$

APPENDIX A: CATALOGUE OF GREEN'S FUNCTIONS 405

55. $y_1''(x) - k^2 y_1(x) = 0, \quad x \in (-a, 0);$
$y_2''(x) - k^2 y_2(x) = 0, \quad x \in (0, \infty);$
$y_1(-a) = 0, \quad |y_2(\infty)| < \infty,$
$y_1(0) = y_2(0), \quad y_1'(0) = \lambda y_2'(0)$ $\boxed{\text{p. 428}}$

56. $y_1''(x) + k^2 y_1(x) = 0, \quad x \in (-a, 0);$
$y_2''(x) + k^2 y_2(x) = 0, \quad x \in (0, a);$
$y_1(-a) = 0, \quad y_2(a) = 0,$
$y_1(0) = y_2(0), \quad y_1'(0) = \lambda y_2'(0)$ $\boxed{\text{p. 428}}$

57. $y_i''(x) - k^2 y_i(x) = 0, \quad x \in (0, \infty), \quad i = 1, 2, 3;$
$y_1(0) = y_2(0) = y_3(0),$
$h_1 y_1'(0) + h_2 y_2'(0) + h_3 y_3'(0) = 0, \quad |y_i(\infty)| < \infty$ $\boxed{\text{p. 428}}$

58. $w_1^{IV}(x) = 0, \quad x \in (0, b);$
$w_2^{IV}(x) = 0, \quad x \in (b, a);$
$w_1(0) = w_1'(0) = 0, \quad w_2''(a) = w_2'''(a) = 0,$
$w_1(b) = w_2(b) = 0, \quad w_1'(b) = w_2'(b), \quad EI_1 w_1''(b) = EI_2 w_2''(b)$ $\boxed{\text{p. 165}}$

59. $w_1^{IV}(x) = 0, \quad x \in (-a, 0);$
$w_2^{IV}(x) = 0, \quad x \in (0, a);$
$w_1(-a) = w_1''(-a) = 0, \quad w_2(a) = w_2''(a) = 0,$
$w_1(0) = w_2(0) = 0, \quad w_1'(0) = w_2'(0), \quad EI_1 w_1''(0) = EI_2 w_2''(0)$ $\boxed{\text{pp. 167-168}}$

60. $w_1^{IV}(x) = 0, \quad x \in (-a, 0);$
$w_2^{IV}(x) = 0, \quad x \in (0, a);$
$w_1(-a) = w_1'(-a) = 0, \quad w_2''(a) = w_2'''(a) = 0,$
$w_1(0) = w_2(0), \quad w_1'(0) = w_2'(0), \quad EI_1 w_1''(0) = EI_2 w_2''(0)$
$EI_1 w_1'''(0) - k w_1(0) = EI_2 w_2'''(0) + k w_2(0)$ $\boxed{\text{pp. 170-171}}$

61. $w_1^{IV}(x) = 0, \quad x \in (-a, 0);$
$w_2^{IV}(x) = 0, \quad x \in (0, a);$
$w_1(-a) = w_1'(-a) = 0, \quad w_2(a) = w_2'(a) = 0,$
$w_1(0) = w_2(0) = 0, \quad w_1'(0) = w_2'(0), \quad EI_1 w_1''(0) = EI_2 w_2''(0)$ $\boxed{\text{p. 430}}$

62. $w_1^{IV}(x) = 0, \quad x \in (-a, 0);$
$w_2^{IV}(x) = 0, \quad x \in (0, a);$
$w_1(-a) = w_1'(-a) = 0, \quad w_2(a) = w_2''(a) = 0,$
$w_1(0) = w_2(0) = 0, \quad w_1'(0) = w_2'(0), \quad EI_1 w_1''(0) = EI_2 w_2''(0)$ $\boxed{\text{p. 431}}$

63. $w_1^{IV}(x) = 0, \quad x \in (0, a);$
$w_2^{IV}(x) = 0, \quad x \in (a, 2a);$
$w_3^{IV}(x) = 0, \quad x \in (2a, 3a);$
$w_1(0) = w_1'(0) = 0, \quad w_3''(3a) = w_3'''(3a) = 0,$
$w_1(a) = w_2(a) = 0, \quad w_1'(a) = w_2'(a), \quad EI_1 w_1''(a) = EI_2 w_2''(a)$
$w_2(2a) = w_3(2a) = 0, \quad w_2'(2a) = w_3'(2a),$
$EI_2 w_2''(2a) = EI_3 w_3''(2a)$ $\boxed{\text{pp. 172-174}}$

64. $w_1^{IV}(x)=0,\ x\in(0,a);$
$w_2^{IV}(x)=0,\ x\in(a,2a);$
$w_3^{IV}(x)=0,\ x\in(2a,3a);$
$w_1(0)=w_1'(0)=0,\ w_3''(3a)=w_3'''(3a)=0,$
$w_1(a)=w_2(a)=0,\ w_1'(a)=w_2'(a),\ w_1''(a)=w_2''(a)$
$w_2(2a)=w_3(2a),\ w_2'(2a)=w_3'(2a),$
$EI_1 w_2''(2a)=EI_2 w_3''(2a),\ EI_1 w_2'''(2a)=EI_2 w_3'''(2a)$
$\boxed{\text{p. 433}}$

65. $w_1^{IV}(x)=0,\ x\in(0,a);$
$w_2^{IV}(x)=0,\ x\in(a,2a);$
$w_3^{IV}(x)=0,\ x\in(2a,3a);$
$w_1''(0)=w_1'''(0)=0,\ w_3''(3a)=w_3'''(3a)=0,$
$w_1(a)=w_2(a)=0,\ w_1'(a)=w_2'(a),\ w_1''(a)=w_2''(a)$
$w_2(2a)=w_3(2a)=0,\ w_2'(2a)=w_3'(2a),\ w_2''(2a)=w_3''(2a)$
$\boxed{\text{pp. 433-434}}$

66. $u_i''(x)=0,\ x\in(0,l_i),\ (i=\overline{1,4});$
$u_1(l_1)=u_2(l_2)=u_3(l_3),$
$p_1 u_1'(l_1)+p_2 u_2'(l_2)+p_3 u_3'(l_3)=0,$
$u_3(0)=u_4(l_4),\ p_3 u_3'(0)=p_4 u_4'(l_4),$
$u_1(0)=u_2(0)=u_4(0)=0$
$\boxed{\text{p. 187}}$

II Equation of Potential:

1. Half-plane $\Omega_{hp}=\{(r,\phi):0<r<\infty,\ 0<\phi<\pi\}$:

1a. $|u(0,\phi)|<\infty,\ u(r,0)=0,\ u(r,\pi)=0$ $\boxed{\text{p. 435}}$

1b. $|u(0,\phi)|<\infty,\ u(r,0)=0,\ u_\phi(r,\pi)=0$ $\boxed{\text{p. 435}}$

2. Infinite strip $\Omega_s=\{(x,y):-\infty<x<\infty,\ 0<y<b\}$:

2a. $u(x,0)=0,\ u(x,b)=0,\ |u(\pm\infty,y)|<\infty$ $\boxed{\text{p. 434}}$

2b. $u(x,0)=0,\ u_y(x,b)=0,\ |u(\pm\infty,y)|<\infty$ $\boxed{\text{p. 434}}$

3. Semi-strip $\Omega_{ss}=\{(x,y):0<x<\infty,\ 0<y<b\}$:

3a. $u(0,y)=0,\ u(x,0)=0,\ u(x,b)=0,\ |u(\infty,y)|<\infty$ $\boxed{\text{pp. 195, 197}}$

3b. $u(0,y)=0,\ u(x,0)=0,\ u_y(x,b)=0,\ |u(\infty,y)|<\infty$ $\boxed{\text{p. 434}}$

3c. $u(0,y)=0,\ u_y(x,0)=0,\ u_y(x,b)=0,\ |u(\infty,y)|<\infty$ $\boxed{\text{p. 434}}$

3d. $u_x(0,y)=0,\ u(x,0)=0,\ u(x,b)=0,\ |u(\infty,y)|<\infty$ $\boxed{\text{p. 205}}$

… APPENDIX A: CATALOGUE OF GREEN'S FUNCTIONS

3e. $u_x(0,y)=0,\ u(x,0)=0,\ u_y(x,b)=0,\ |u(\infty,y)|<\infty$ — p. 434

3f. $|u(\infty,y)|<\infty,\ u_x(0,y)-\beta u(0,y)=0,\ \beta\geq 0,$
$u(x,0)=0,\ u(x,b)=0$ — p. 205

3g. $|u(\infty,y)|<\infty,\ u_x(0,y)-\beta u(0,y)=0,\ \beta\geq 0,$
$u(x,0)=0,\ u_y(x,b)=0$ — p. 209

3h. $|u(\infty,y)|<\infty,\ u_x(0,y)-\beta u(0,y)=0,\ \beta\geq 0,$
$u_y(x,0)=0,\ u_y(x,b)=0$ — p. 434

4. Rectangle $\Omega_r=\{(x,y):0<x<a,\ 0<y<b\}$:

4a. $u(0,y)=0,\ u(a,y)=0,\ u(x,0)=0,\ u(x,b)=0$ — pp. 209, 211, 214

4b. $u(0,y)=0,\ u_x(a,y)=0,\ u(x,0)=0,\ u(x,b)=0$ — p. 435

4c. $u(0,y)=0,\ u_x(a,y)+\beta u(a,y)=0,\ \beta\geq 0,$
$u(x,0)=0,\ u_y(x,b)=0$ — p. 435

4d. $u(0,y)=0,\ u_x(a,y)+\beta u(a,y)=0,\ \beta\geq 0,$
$u(x,0)=0,\ u(x,b)=0$ — pp. 215, 218

4e. $u_x(0,y)=0,\ u_x(a,y)+\beta u(a,y)=0,\ \beta\geq 0,$
$u(x,0)=0,\ u(x,b)=0$ — p. 435

5. Circle $\Omega_c=\{(r,\phi):0<r<R,\ 0\leq\phi<2\pi\}$:

5a. $u(R,\phi)=0$ — p. 224

5b. $u_r(R,\phi)+\beta u(R,\phi)=0,\ \beta>0$ — p. 229

6. Infinite wedge $\Omega_w=\{(r,\phi):0<r<\infty,\ 0<\phi<\alpha\}$:

6a. $|u(0,\phi)|<\infty,\ u(r,0)=0,\ u(r,\alpha)=0$ — p. 435

6b. $|u(0,\phi)|<\infty,\ u(r,0)=0,\ u_\phi(r,\alpha)=0$ — p. 435

7. Circular sector $\Omega_{cs}=\{(r,\phi):0<r<R,\ 0<\phi<\alpha\}$:

7a. $|u(0,\phi)|<\infty,\ u(R,\phi)=0,\ u(r,0)=0,\ u(r,\alpha)=0$ — p. 227

7b. $|u(0,\phi)|<\infty,\ u(R,\phi)=0,\ u(r,0)=0,\ u_\phi(r,\alpha)=0$ — p. 436

7c. $|u(0,\phi)|<\infty,\ u_r(R,\phi)=0,\ u(r,0)=0,\ u_\phi(r,\alpha)=0$ — p. 436

7d. $|u(0,\phi)|<\infty$, $u_r(R,\phi)+\beta u(R,\phi)=0$, $\beta>0$,
$u(r,0)=0$, $u(r,\alpha)=0$ p. 436

7e. $|u(0,\phi)|<\infty$, $u_r(R,\phi)+\beta u(R,\phi)=0$, $\beta>0$,
$u(r,0)=0$, $u_\phi(r,\alpha)=0$ p. 230

8. Spherical Triangle $\Omega_{st}=\{(\phi,\theta): 0<\phi<\alpha,\ 0<\theta<\beta\}$:

$|u(0,\theta)|<\infty$, $u_\phi(\alpha,\theta)=0$, $u(\phi,0)=0$, $u(\phi,\beta)=0$ p. 234

9. Spherical Segment $\Omega_{sg}=\{(\phi,\theta): 0<\phi<\alpha,\ 0\leq\theta<2\pi\}$:

$u(\alpha,\theta)=0$ p. 235

10. Toroidal Sector $\Omega_{ts}=\{(\phi,\theta): 0\leq\phi<2\pi,\ 0<\theta<\beta\}$:

$u(\phi,0)=0$, $u(\phi,\beta)=0$ p. 237

III Klein-Gordon Equation:

1. Half-plane $\Omega_{hp}=\{(r,\phi): 0<r<\infty,\ 0<\phi<\pi\}$:

1a. $|u(0,\phi)|<\infty$, $u(r,0)=0$, $u(r,\pi)=0$ pp. 249-250

1b. $|u(0,\phi)|<\infty$, $u(r,0)=0$, $u_\phi(r,\pi)=0$ p. 437

2. Infinite strip $\Omega_s=\{(x,y): -\infty<x<\infty,\ 0<y<b\}$:

2a. $u(x,0)=0$, $u(x,b)=0$, $|u(\pm\infty,y)|<\infty$ p. 241

2b. $u(x,0)=0$, $u_y(x,b)=0$, $|u(\pm\infty,y)|<\infty$ p. 244

3. Semi-strip $\Omega_{ss}=\{(x,y): 0<x<\infty,\ 0<y<b\}$:

3a. $u(0,y)=0$, $|u(\infty,y)|<\infty$, $u(x,0)=0$, $u(x,b)=0$ p. 244

3b. $u_x(0,y)=0$, $|u(\infty,y)|<\infty$, $u(x,0)=0$, $u(x,b)=0$ p. 436

3c. $|u(\infty,y)|<\infty$, $u_x(0,y)-\beta u(0,y)=0$, $\beta\geq 0$,
$u(x,0)=0$, $u(x,b)=0$ p. 436

3d. $|u(\infty,y)|<\infty$, $u_x(0,y)-\beta u(0,y)=0$, $\beta\geq 0$,
$u(x,0)=0$, $u_y(x,b)=0$ p. 245

APPENDIX A: CATALOGUE OF GREEN'S FUNCTIONS

4. Rectangle $\Omega_r = \{(x,y): 0 < x < a,\ 0 < y < b\}$:

$u(0,y) = 0,\ u(a,y) = 0,\ u(x,0) = 0,\ u(x,b) = 0$ $\boxed{\text{p. 248}}$

5. Circle $\Omega_c = \{(r,\phi): 0 < r < R,\ 0 \le \phi < 2\pi\}$:

5a. $u(R,\phi) = 0$ $\boxed{\text{p. 438}}$

5b. $u_r(R,\phi) + \beta u(R,\phi) = 0,\ \beta > 0$ $\boxed{\text{p. 438}}$

6. Infinite wedge $\Omega_w = \{(r,\phi): 0 < r < \infty,\ 0 < \phi < \alpha\}$:

6a. $|u(0,\phi)| < \infty,\ u(r,0) = 0,\ u(r,\alpha) = 0$ $\boxed{\text{p. 249}}$

6b. $|u(0,\phi)| < \infty,\ u(r,0) = 0,\ u_\phi(r,\alpha) = 0$ $\boxed{\text{p. 437}}$

7. Semi-circle $\Omega_{sc} = \{(r,\phi): 0 < r < R,\ 0 < \phi < \pi\}$:

7a. $|u(0,\phi)| < \infty,\ u(R,\phi) = 0,\ u(r,0) = 0,\ u(r,\pi) = 0$ $\boxed{\text{pp. 250-251}}$

7b. $|u(0,\phi)| < \infty,\ u_r(R,\phi) = 0,\ u(r,0) = 0,\ u(r,\pi) = 0$ $\boxed{\text{p. 251}}$

7c. $|u(0,\phi)| < \infty,\ u_r(R,\phi) + \beta u(R,\phi) = 0,\ \beta > 0,$
$u(r,0) = 0,\ u(r,\pi) = 0$ $\boxed{\text{p. 250}}$

7d. $|u(0,\phi)| < \infty,\ u(R,\phi) = 0,\ u(r,0) = 0,\ u_\phi(r,\pi) = 0$ $\boxed{\text{p. 437}}$

7e. $|u(0,\phi)| < \infty,\ u_r(R,\phi) = 0,\ u(r,0) = 0,\ u_\phi(r,\pi) = 0$ $\boxed{\text{p. 437}}$

7f. $|u(0,\phi)| < \infty,\ u_r(R,\phi) + \beta u(R,\phi) = 0,\ \beta > 0,$
$u(r,0) = 0,\ u_\phi(r,\pi) = 0$ $\boxed{\text{p. 438}}$

IV Biharmonic Equation:

1. Infinite strip $\Omega_s = \{(x,y): -\infty < x < \infty,\ 0 < y < b\}$:

$w(x,0) = 0,\ w_{yy}(x,0) = 0,\ w(x,b) = 0,\ w_{yy}(x,b) = 0$ $\boxed{\text{p. 438}}$

2. Semi-strip $\Omega_{ss} = \{(x,y): 0 < x < \infty,\ 0 < y < b\}$:

2a. $w(0,y) = 0,\ w_x(0,y) = 0,\ |w(\infty,y)| < \infty,\ |w_x(\infty,y)| < \infty,$
$w(x,0) = 0,\ w_{yy}(x,0) = 0,\ w(x,b) = 0,\ w_{yy}(x,b) = 0$ $\boxed{\text{p. 261}}$

2b. $w(0,y) = 0,\ w_{xx}(0,y) = 0,\ |w(\infty,y)| < \infty,\ |w_x(\infty,y)| < \infty,$
$w(x,0) = 0,\ w_{yy}(x,0) = 0,\ w(x,b) = 0,\ w_{yy}(x,b) = 0$ $\boxed{\text{pp. 260, 262}}$

2c. $w_{xx}(0,y) + \sigma w_{yy}(0,y) = 0$, $(w_{xx}(0,y) + (2-\sigma)w_{yy}(0,y))_x = 0$
$w(x,0) = 0$, $w_{yy}(x,0) = 0$, $w(x,b) = 0$, $w_{yy}(x,b) = 0$
$|w(\infty,y)| < \infty$, $|w_x(\infty,y)| < \infty$ pp. 260, 262

3. Rectangle $\Omega_r = \{(x,y): 0 < x < a, \ 0 < y < b\}$:

3a. $w(0,y) = 0$, $w_{xx}(0,y) = 0$, $w(a,y) = 0$, $w_{xx}(a,y) = 0$,
$w(x,0) = 0$, $w_{yy}(x,0) = 0$, $w(x,b) = 0$, $w_{yy}(x,b) = 0$ p. 438

3b. $w(0,y) = 0$, $w_{xx}(0,y) = 0$, $w(a,y) = 0$, $w_x(a,y) = 0$,
$w(x,0) = 0$, $w_{yy}(x,0) = 0$, $w(x,b) = 0$, $w_{yy}(x,b) = 0$ pp. 260, 263

3c. $w(0,y) = 0$, $w_x(0,y) = 0$, $w(a,y) = 0$, $w_x(a,y) = 0$,
$w(x,0) = 0$, $w_{yy}(x,0) = 0$, $w(x,b) = 0$, $w_{yy}(x,b) = 0$ p. 439

3d. $w(0,y) = 0$, $w_x(0,y) = 0$, $w_{xx}(a,y) = 0$, $w_{xxx}(a,y) = 0$,
$w(x,0) = 0$, $w_{yy}(x,0) = 0$, $w(x,b) = 0$, $w_{yy}(x,b) = 0$ p. 439

4. Circle $\Omega_c = \{(r,\phi): 0 < r < a, \ 0 \le \phi < 2\pi\}$:

4a. $w(a,\phi) = 0$, $w_r(a,\phi) = 0$ pp. 266, 273

4b. $w(a,\phi) = 0$, $aw_{rr}(a,\phi) + \sigma w_r(a,\phi) = 0$ pp. 274, 276

V Plane Problem of Theory of Elasticity:

1. Semi-strip $\Omega_{ss} = \{(x,y): 0 < x < \infty, \ 0 < y < b\}$:

1a. $u_y(x,0) = 0$, $v(x,0) = 0$, $u_y(x,b) = 0$, $v(x,b) = 0$
$u(0,y) = 0$, $v(0,y) = 0$, $|u(\infty,y)| < \infty$, $|v(\infty,y)| < \infty$ pp. 439-440

1b. $u_y(x,0) = 0$, $v(x,0) = 0$, $u_y(x,b) = 0$, $v(x,b) = 0$
$u(0,y) = 0$, $v_x(0,y) = 0$, $|u(\infty,y)| < \infty$, $|v(\infty,y)| < \infty$ p. 299

2. Rectangle $\Omega_r = \{(x,y): 0 < x < a, \ 0 < y < b\}$:

2a. $u_y(x,0) = 0$, $v(x,0) = 0$, $u_y(x,b) = 0$, $v(x,b) = 0$
$u(0,y) = 0$, $v(0,y) = 0$, $u(a,y) = 0$, $v(a,y) = 0$ pp. 440-442

2b. $u_y(x,0) = 0$, $v(x,0) = 0$, $u_y(x,b) = 0$, $v(x,b) = 0$
$u(0,y) = 0$, $v_x(0,y) = 0$, $u(a,y) = 0$, $v(a,y) = 0$ pp. 295-297

APPENDIX A: CATALOGUE OF GREEN'S FUNCTIONS   411

3. Circle $\Omega_c = \{(r,\phi): 0 < r < a,\ 0 \le \phi < 2\pi\}$:

$u(a,\phi) = 0,\ \ v(a,\phi) = 0$ pp. 300-301

4. Orthotropic Semi-strip $\Omega_{ss} = \{(x,y): 0 < x < \infty,\ 0 < y < b\}$:

$u_y(x,0) = 0,\ \ v(x,0) = 0,\ \ u_y(x,b) = 0,\ \ v(x,b) = 0$
$u(0,y) = 0,\ \ v_x(0,y) = 0,\ \ u(a,y) = 0,\ \ v(a,y) = 0$ pp. 305-306

VI Fields of Potential in Compound Regions:

1. Compound infinite strip $\Omega_s = \{(x,y): -\infty < x < \infty,\ 0 < y < b\}$ consisting of two semi-strips $\Omega_1 = \{(x,y): -\infty < x < 0,\ 0 < y < b\}$ and $\Omega_2 = \{(x,y): 0 < x < \infty,\ 0 < y < b\}$, composed of different materials whose conductivities are λ_1 and λ_2, respectively, with boundary conditions imposed as:

1a. $u_1(x,0) = u_2(x,0) = u_1(x,b) = u_2(x,b) = 0$ pp. 324-325

1b. $u_1(x,0) = u_2(x,0) = (u_1)_y(x,b) = (u_2)_y(x,b) = 0$ p. 442

2. Compound rectangle $\Omega = \{(x,y): -a < x < a,\ 0 < y < b\}$ that consists of two other rectangles: $\Omega_1 = \{(x,y): -a < x < 0,\ 0 < y < b\}$ and $\Omega_2 = \{(x,y): 0 < x < a,\ 0 < y < b\}$ occupied with materials whose conductivities are specified with constants λ_1 and λ_2, respectively, and boundary conditions are imposed as:

2a. $u_1(-a,y) = u_1(x,0) = u_1(x,b) = 0$
       $u_2(a,y) = u_2(x,0) = u_2(x,b) = 0$ p. 443

2b. $u_1(-a,y) = u_1(x,0) = u_1(x,b) = 0,$
       $(u_2)_x(a,y) = u_2(x,0) = u_2(x,b) = 0$ pp. 444-445

3. Compound semi-circle $\Omega_{sc} = \{(r,\phi): 0 < r < b,\ 0 < \phi < \pi\}$ consisting of the semi-circle $\Omega_1 = \{(r,\phi): 0 < r < a,\ 0 < \phi < \pi\}$ and half-ring $\Omega_2 = \{(r,\phi): a < r < b,\ 0 < \phi < \pi\}$ as shown in Figure 7.3. Assume that Ω_1 and Ω_2 are occupied with materials whose conductivities are specified with constants λ_1 and λ_2, respectively, and boundary conditions are imposed as:

3a. $u_2(b,\phi) = 0,\ \ u_1(r,0) = u_2(r,0) = u_1(r,\pi) = u_2(r,\pi) = 0$ pp. 327-328

3b. $(u_2)_r(b,\phi) = 0,\ \ u_1(r,0) = u_2(r,0) = u_1(r,\pi) = u_2(r,\pi) = 0$ pp. 445-446

412   APPENDIX A: CATALOGUE OF GREEN'S FUNCTIONS

4. Compound thin-walled structure $\Omega = \Omega_1 \bigcup \Omega_2$ consisting of the hemisphere $\Omega_1 = \{(\phi, \theta): 0 < \phi < \pi/2, 0 \le \theta < 2\pi\}$ and the cylinder $\Omega_2 = \{(x, y): 0 < x < a, 0 \le y < 2\pi\}$, both of unit radius (as shown in Figure 7.4). Fragments Ω_1 and Ω_2 are occupied with materials with conductivities λ_1 and λ_2, respectively. The free edge $x = a$ of the cylinder is subjected to the boundary condition:

4a. $\alpha(u_2)_x(a, y) + \beta u_2(a, y) = 0$ $\boxed{\text{pp. 336-338}}$

4b. $u_2(a, y) = 0$, while $\lambda_1 = \lambda_2$ $\boxed{\text{p. 339}}$

5. Compound thin-walled hemisphere $\Omega = \{(\phi, \theta): 0 < \phi < \pi, 0 < \theta < \pi\}$ of unit radius, consisting of two congruent fragments: $\Omega_1 = \{(\phi, \theta): 0 < \phi < \pi/2, 0 < \theta < \pi\}$ and $\Omega_2 = \{(\phi, \theta): \pi/2 < \phi < \pi, 0 < \theta < \pi\}$ occupied with different materials whose conductivities are λ_1 and λ_2, respectively (see Figure 7.5). The free edge of the structure is subjected to the boundary condition:

$$u_1(\phi, 0) = u_1(\phi, \pi) = 0, \quad u_2(\phi, 0) = u_2(\phi, \pi) = 0 \quad \boxed{\text{pp. 342-343}}$$

6. Compound thin-walled structure $\Omega = \Omega_1 \bigcup \Omega_2$ consisting of a cylindrical shell (Ω_1) of a unit radius and of length a closed at one edge with a circular plate (Ω_2) as shown in Figure 7.6. The conductivities of the materials of which the shell and the plate are composed are λ_1 and λ_2, respectively. The boundary condition on the cylinder's edge $x =$ is imposed as:

$$u_1(a, y) = 0 \quad \boxed{\text{pp. 344-345}}$$

7. Compound thin-walled structure $\Omega = \Omega_1 \bigcup \Omega_2$ consisting of a circular plate of radius R that occupies the region $\Omega_1 = \{(r, \theta): 0 < r < R, 0 \le \theta < 2\pi\}$ joined to a toroidal shell, with a circular meridional cross section, occupying the region $\Omega_2 = \{(\phi, \theta): 0 < \phi < \phi_0, 0 \le \theta < 2\pi\}$, where $0 < \phi_0 < \pi$. The meridional cross section of the shell represents a unit circle and the plate's radius $R > 1$ (see Figure 7.7). Both fragments of Ω are composed of the same homogeneous material. The boundary condition on the free edge of the structure is imposed as:

$$u_2(\phi_0, \theta) = 0 \quad \boxed{\text{pp. 349-350}}$$

8. Compound thin-walled structure $\Omega = \Omega_1 \bigcup \Omega_2$ consisting of a hemispheric shell $\Omega_1 = \{(\varphi, \vartheta): 0 < \varphi < \pi/2, 0 \le \vartheta < 2\pi\}$ of radius R_1, joined to a spherical belt $\Omega_2 = \{(\varphi, \vartheta): \phi_0 < \varphi < \pi/2, 0 \le \vartheta < 2\pi\}$ of radius R_2 that is greater than R_1. The conductivities of the materials of which the fragments are composed are λ_1 and λ_2, respectively. The boundary condition on the free edge of the structure is imposed as:

$$u_2(\pi/2, \vartheta) = 0 \quad \boxed{\text{p. 446}}$$

9. Compound thin-walled structure $\Omega = \Omega_1 \bigcup \Omega_2 \bigcup \Omega_3$ consisting of a thin hemispheric shell $\Omega_1 = \{(\varphi, \vartheta): 0 < \varphi < \pi/2, 0 \le \vartheta < 2\pi\}$ of a unit radius, joined to a thin annular plate $\Omega_2 = \{(r, \vartheta): 1 < r < b, 0 \le \vartheta < 2\pi\}$, and to another thin hemispheric shell $\Omega_3 = \{(\varphi, \vartheta): \pi/2 < \varphi < \pi, 0 \le \vartheta < 2\pi\}$ to form a 'Saturn' type construction (see Figure 7.8). Each of the fragments of Ω is composed of

the same homogeneous conductive material. The free edge $r=b$ of Ω_2 is imposed to the boundary condition:
$$u_2(b, \vartheta) = 0 \qquad \boxed{\text{pp. 354-355}}$$

VII Heat Equation:

1. **Thin infinite rod** $\boxed{\text{p. 380}}$

2. **Thin semi-infinite $(0 < x < \infty)$ rod:**

2a. $u(0, t) = 0$ $\boxed{\text{p. 382}}$

2b. $u_x(0, t) = 0$ $\boxed{\text{p. 384}}$

2c. $u_x(0, t) - \beta u(0, t) = 0$ $\boxed{\text{p. 384}}$

3. **Thin finite $(0 < x < a)$ rod:**

3a. $u(0, t) = 0, \ u(a, t) = 0$ $\boxed{\text{pp. 386, 388}}$

3b. $u(0, t) = 0, \ u_x(a, t) = 0$ $\boxed{\text{p. 446}}$

4. **Thin half plane-shaped plate** $\Omega_{hp} = \{(r, \phi) : 0 < r < \infty, \ 0 < \phi < \pi\}$:

$u(r, 0, t) = 0, \ u(r, \pi, t) = 0$ $\boxed{\text{p. 393}}$

5. **Thin infinite strip-shaped plate** $\Omega_s = \{(x, y) : -\infty < x < \infty, \ 0 < y < b\}$:

$u(x, 0, t) = 0, \ u(x, b, t) = 0$ $\boxed{\text{p. 447}}$

6. **Thin semistrip-shaped plate** $\Omega_{ss} = \{(x, y) : 0 < x < \infty, \ 0 < y < b\}$:

6a. $u(0, y, t) = 0, \ u(x, 0, t) = 0, \ u(x, b, t) = 0$ $\boxed{\text{pp. 391-392}}$

6b. $u_x(0, y, t) = 0, \ u(x, 0, t) = 0, \ u(x, b, t) = 0$ $\boxed{\text{p. 447}}$

7. **Assembly of three semi-infinite rods:** $\boxed{\text{pp. 398-399}}$

8. **Compound infinite strip-shaped plate:** $\boxed{\text{p. 447}}$

Appendix B
Answers and Comments

CHAPTER 1: Green's Function for ODE

1.1: (a) Yes; (b) Yes; (c) Yes; (d) Yes; (e) Yes; (f) Yes; (g) No, because the homogeneous boundary value problem in the statement has infinitely many solutions (any function taken from the set $y = C$, where C is an arbitrary constant, represents its solution); (h) No, because any linear function $y = mx + b$ is a solution of the given boundary value problem.

1.2:

$$\text{(a)} \quad g(x,s) = \begin{cases} -x, & \text{for } x \leq s \\ -s, & \text{for } x \geq s \end{cases}$$

$$\text{(b)} \quad g(x,s) = \frac{1}{1+ha} \begin{cases} x[h(s-a)-1], & \text{for } x \leq s \\ s[h(x-a)-1], & \text{for } x \geq s \end{cases}$$

If $h = 0$, then the boundary conditions imposed in the statement of this problem reduce to the ones in the preceding problem, and, consequently, the above Green's function reduces to that of exercise 1.2(a).

$$\text{(c)} \quad g(x,s) = \frac{1}{H} \begin{cases} (1+h_1 x)[h_2(s-a)-1], & \text{for } x \leq s \\ (1+h_1 s)[h_2(x-a)-1], & \text{for } x \geq s \end{cases}$$

where $H = h_2 + h_1(1 + h_2 a)$. If $h_1 = 0$, then the present Green's function reduces to that of *EXAMPLE 3* of Section 1.1.

$$\text{(d)} \quad g(x,s) = \frac{1}{M} \begin{cases} \ln[(ms+p)/(ma+p)] \ln[(mx+p)/p], & x \leq s \\ \ln[(mx+p)/(ma+p)] \ln[(ms+p)/p], & x \geq s \end{cases}$$

where $M = m \ln[(ma+p)/p]$.

$$\text{(e)} \quad g(x,s) = \frac{1}{\beta(e^{-\beta a}-1)} \begin{cases} (e^{-\beta x}-1)(e^{-\beta a}-e^{-\beta s}), & \text{for } x \leq s \\ (e^{-\beta s}-1)(e^{-\beta a}-e^{-\beta x}), & \text{for } x \geq s \end{cases}$$

(f) $g(x,s) = \dfrac{1}{\beta} \begin{cases} e^{-\beta x}-1, & \text{for } x \leq s \\ e^{-\beta s}-1, & \text{for } x \geq s \end{cases}$

(g) $g(x,s) = \dfrac{1}{k\sin(ka)} \begin{cases} \sin(k(a-s))\sin(kx), & \text{for } x \leq s \\ \sin(k(a-x))\sin(ks), & \text{for } x \geq s \end{cases}$

(h) $g(x,s) = \dfrac{1}{6} \begin{cases} x^2(x-3s), & \text{for } x \leq s \\ s^2(s-3x), & \text{for } x \geq s \end{cases}$

1.3: (a) Yes; (b) Yes; (c) No, because the condition of self-adjointness in eqn (1.54) is not met in this case; (d) No, because the condition in eqn (1.54) is not met in this case; (e) Yes.

1.4: (a) The integrating factor e^{-2x} reduces the given equation to the following self-adjoint form

$$e^{-2x}y''(x) - 2e^{-2x}y'(x) + 4e^{-2x}y = 0;$$

(b) The integrating factor $e^{x^2/2}$ reduces the given equation to the following self-adjoint form

$$e^{x^2/2}y''(x) + xe^{x^2/2}y'(x) - x^2 e^{x^2/2}y(x) = 0;$$

(c) The integrating factor x^{-3} reduces the given equation to the following self-adjoint form

$$x^{-1}y''(x) - x^{-2}y'(x) + x^{-3}y(x) = 0;$$

(d) the integrating factor x^{-1} reduces the given equation to the following self-adjoint form

$$xy''(x) + y'(x) - x^{-1}y(x) = 0.$$

1.5: (a) Yes; (b) Yes; (c) Yes; (d) Yes; (e) Yes; (f) No.

1.6: By applying the integrating factor e^{3x}, the original nonself-adjoint boundary value problem, whose Green's function

$$g(x,s) = \dfrac{1}{7} \begin{cases} e^{-(5x+2s)} - e^{2(x-s)}, & \text{for } x \leq s \\ e^{-(5x+2s)} - e^{-5(x-s)}, & \text{for } x \geq s \end{cases}$$

appears to be nonsymmetric, reduces to the following self-adjoint form

$$e^{3x}y''(x) + 3e^{3x}y'(x) - 10e^{3x}y(x) = 0, \quad y(0)=0, \quad y(\infty)<\infty,$$

whose Green's function

$$g(x,s) = \dfrac{1}{7} \begin{cases} e^{-2(x+s)} - e^{5x-2s}, & \text{for } x \leq s \\ e^{-2(x+s)} - e^{5s-2x}, & \text{for } x \geq s \end{cases}$$

is symmetric.

APPENDIX B: ANSWERS AND COMMENTS 417

1.7: The Green's function is obtained in the form

$$g(x,s) = \frac{1}{k\cos(k)} \begin{cases} \sin(kx)\cos(k(s-1)), & \text{for } x \leq s \\ \sin(ks)\cos(k(x-1)), & \text{for } x \geq s \end{cases}$$

The standard construction procedure based on Theorem 1.1, which uses the *standard* fundamental solution set: $y_1 = \sin(kx)$ and $y_2 = \cos(kx)$, is more cumbersome compared to that based on Theorem 1.2, because the latter does not imply a direct satisfaction of the boundary conditions (these are taken care of in advance when a special fundamental solution set: $y_1(x) = \sin(kx)$ and $y_2(x) = \cos(k(x-1))$ is obtained).

1.8: (a) Yes; (b) Yes; (c) Yes; (d) Yes.

1.9:

(a) $g(x,s) = \begin{cases} -\ln(s/a), & \text{for } x \leq s \\ -\ln(x/a), & \text{for } x \geq s \end{cases}$

(b) $g(x,s) = \dfrac{1}{k(e^{ka}+\lambda e^{-ka})} \begin{cases} (e^{kx}+\lambda e^{-kx})\sinh(k(s-a)), & \text{for } x \leq s \\ (e^{ks}+\lambda e^{-ks})\sinh(k(x-a)), & \text{for } x \geq s \end{cases}$

where $\lambda = (k-h)/(k+h)$.

(c) $g(x,s) = \dfrac{1}{6} \begin{cases} x^2(s-1)^2[s(2x-3)+x], & \text{for } x \leq s \\ s^2(x-1)^2[x(2s-3)+s], & \text{for } x \geq s \end{cases}$

(d) $g(x,s) = \dfrac{1}{6} \begin{cases} x(1-s)[(1-s)^2+(x^2-1)], & \text{for } x \leq s \\ s(1-x)[(1-x)^2+(s^2-1)], & \text{for } x \geq s \end{cases}$

1.10:

(a) $y(x) = \dfrac{e^{2x}}{5} - \dfrac{1}{5}[h\sin(b-a)+\cos(b-a)]^{-1}$
$\times \left\{ e^{2a}[h\sin(b-x)+\cos(b-x)]+(h+2)e^{2b}\sin(x-a) \right\}$;

(b) $y(x) = \dfrac{1}{(h+1)e^{x+2b}+(h-1)e^{x+2a}} \left\{ (hb+1)\left(e^{2x+b}-e^{2a+b}\right) \right.$
$\left. - (h+1)\left(xe^{x+2b}-xe^{x+2a}-ae^{a+2b}\right)-(h-1)ae^{2x+a} \right\}$;

(c) $y(x) = \dfrac{1}{2(b-a)}\left[(b-1-x)e^{a-x}\sin(a)-(a-1-x)e^{b-x}\sin(b)\right] - \dfrac{1}{2}\cos(x)$;

(d) $y(x) = \dfrac{1}{360}x^2(x-1)^2(x^2+2x+3)$;

(e) $y(x) = \dfrac{1}{2}(x-3)e^{x-1}+\dfrac{1}{2}(5x+3)e^{-(x+1)}-2(x+1)e^{-x}+2.$

CHAPTER 2: Influence Functions for Beams

A word of caution is appropriate with regard to exercises 2.1 through 2.3. The answers for these exercises are not unique because a set of any n linearly independent particular solutions of a homogeneous equation can be considered as its fundamental solution set. Consequently, the reader's answers must not necessarily be in the same form as those below.

2.1:

(a) $w_1(x) = \sin(k(x-a))$, $w_2(x) = \sin(k(x-b))$;

(b) $w_1(x) = \sin(k(x-a))$, $w_2(x) = \cos(k(x-b))$;

(c) $w_1(x) = \cos(k(x-a))$, $w_2(x) = k\cos(k(x-b)) + h\sin(k(x-b))$;

(d) $w_1(x) = k\cos(k(x-a)) - h\sin(k(x-a))$, $w_2(x) = \sin(k(x-b))$.

2.2:

(a) $w_1(x) = \sinh(k(x-a))$, $w_2(x) = \sinh(k(x-b))$;

(b) $w_1(x) = \sinh(k(x-a))$, $w_2(x) = \cosh(k(x-b))$;

(c) $w_1(x) = \cosh(k(x-a))$, $w_2(x) = (k-h)e^{k(x-2b)} + (k+h)e^{-kx}$;

(d) $w_1(x) = (k-h)e^{k(x-2a)} + (k+h)e^{-kx}$, $w_2(x) = \sinh(k(x-b))$.

2.3:

(a) $w_1(x) = \dfrac{1}{4k^3}[\cosh(kx)\sin(kx) - \sinh(kx)\cos(kx)]$;

$w_2(x) = \dfrac{1}{2k^2}\sinh(kx)\sin(kx)$;

$w_3(x) = \dfrac{1}{4k^3}[\cosh(k(x-a))\sin(k(x-a)) - \sinh(k(x-a))\cos(k(x-a))]$;

$w_4(x) = \dfrac{1}{2k}[\cosh(k(x-a))\sin(k(x-a)) + \sinh(k(x-a))\cos(k(x-a))]$.

(b) $w_1(x) = \dfrac{1}{4k^3}[\cosh(kx)\sin(kx) - \sinh(kx)\cos(kx)]$;

$w_2(x) = \dfrac{1}{2k}[\sinh(kx)\cos(kx) + \cosh(kx)\sin(kx)]$;

$w_3(x) = \dfrac{1}{4k^3}[\cosh(k(x-a))\sin(k(x-a)) - \sinh(k(x-a))\cos(k(x-a))]$;

$w_4(x) = \dfrac{1}{2k}[\cosh(k(x-a))\sin(k(x-a)) + \sinh(k(x-a))\cos(k(x-a))]$.

(c) $w_1(x) = \dfrac{1}{2k}[\sinh(kx)\cos(kx) + \cosh(kx)\sin(kx)]$;

$w_2(x) = \cosh(kx)\cos(kx)$;

$w_3(x) = \dfrac{1}{4k^3}[\cosh(k(x-a))\sin(k(x-a)) - \sinh(k(x-a))\cos(k(x-a))]$;

$w_4(x) = \dfrac{1}{2k^2}\sinh(k(x-a))\sin(k(x-a))$.

APPENDIX B: ANSWERS AND COMMENTS 419

2.4:
$$g(x,s) = \frac{1}{6a}\begin{cases} x(a-s)[(a-s)^2 + (x^2-a^2)], & x \leq s \\ s(a-x)[(a-x)^2 + (s^2-a^2)], & s \leq x \end{cases}$$

2.5: The routine based on the modification of the classical method is notably less time-consuming because it operates with a fundamental solution set, each component of which satisfies in advance the corresponding boundary conditions in the statement.

2.6:
$$g(x,s) = \frac{1}{12a^3}\begin{cases} x(a-s)^2[s(a^2-x^2) - 2a(x^2-as)], & x \leq s \\ s(a-x)^2[x(a^2-s^2) - 2a(s^2-ax)], & s \leq x \end{cases}$$

2.7:
$$g(x,s) = \frac{1}{6a^2 k_0 k_a}\begin{cases} k_a(a-s)\{(x-a) + k_0 ax[(x^2-a^2) + (s-a)^2]\} - k_0 xs, & x \leq s \\ k_a(a-x)\{(s-a) + k_0 as[(s^2-a^2) + (x-a)^2]\} - k_0 xs, & x \geq s \end{cases}$$

where $k_0 = k_0^*/6EI$ and $k_a = k_a^*/6EI$.

2.8: (a) Clearly, the boundary conditions of elastic support reduce to the free edge conditions, if the coefficient of support goes to zero. Thus, the limit of the influence function derived in exercise 2.7, as either k_0^* or k_a^* (or both) approach zero, is undefined. This reflects an obvious physical interpretation of the influence function. Indeed, in such cases (as it follows from the relations in eqn (2.21)) the corresponding boundary conditions reduce to the free edge conditions. And the influence function, in the sense it is introduced here, is undefined for a beam, one edge of which is elastically supported, while the other one is free (or both edges are free).

(b) The boundary condition of elastic support reduces to the condition of simple support if the coefficient of support tends to infinity. Thus, if both the coefficients k_0^* and k_a^* approach infinity, then the influence function derived in exercise 2.7 reduces to that derived in exercise 2.4 for a simply supported beam.

(c) If k_0^* approaches infinity while k_a^* remains finite, then the influence function derived in exercise 2.7 reduces to that derived below in exercise 2.9 for a beam whose left edge is simply supported, while the right edge is elastically supported.

2.9:
$$g(x,s) = \frac{1}{6a^2}\begin{cases} x\{a(a-s)[(x^2-a^2) + (s-a)^2] - s/k_a\}, & x \leq s \\ s\{a(a-x)[(s^2-a^2) + (x-a)^2] - x/k_a\}, & x \geq s \end{cases}$$

where $k_a = k_a^*/6EI$.

The above influence function can be constructed by either the method of variation of parameters or by the modification of the classical method, proposed in the text. It is worth noting that it can also be obtained as a particular case of that influence function derived in exercise 2.7 for the beam, both edges of which are elastically supported. Indeed, the boundary conditions of elastic support, presented in eqn (2.21), can be reduced to the conditions of simple support

$$w(0)=0, \quad \frac{d^2w(0)}{dx^2}=0$$

as the coefficient k_0 approaches infinity. Hence, to derive the influence function in the present exercise, one takes the limit of the influence function in exercise 2.7, as k_0 approaches infinity. The influence function for a simply supported beam

$$g(x,s)=\frac{1}{6a}\begin{cases} x(a-s)[(x^2-a^2)+(s-a)^2], & x \leq s \\ s(a-x)[(s^2-a^2)+(x-a)^2], & x \geq s \end{cases}$$

constructed earlier in exercise 2.4, can, in turn, be obtained from the influence function in the present exercise by taking the limit of it as the coefficient k_a^* approaches infinity. As k_a^* approaches zero, the influence function in the present exercise is undefined.

2.10:
$$g(x,s)=\frac{1}{6}\begin{cases} x(3s^2+x^2-6as), & x \leq s \\ s(3x^2+s^2-6ax), & s \leq x \end{cases}$$

2.11:
$$g(x,s)=\frac{1}{12a}\begin{cases} x^2(3s^2+2ax-6as), & x \leq s \\ s^2(3x^2+2as-6ax), & s \leq x \end{cases}$$

2.12:

(a) $w(x)=\dfrac{q_0}{92160EI}xa^3(57a^2-103x^2)$, $M(x)=-\dfrac{103}{15360}q_0a^3 x$,

$Q(x)=-\dfrac{103}{15360}q_0a^3$, for $0 \leq x \leq a/4$;

$w(x)=\dfrac{q_0}{1474560EI}(12288x^5a-11520x^4a^2+3472x^3a^3-4096x^6-1200x^2a^4$

$-7a^6+1056xa^5)$, $M(x)=-\dfrac{q_0}{15360}[1280x^2(x-a)^2+a^2(160x^2-217ax+25a^2)]$,

$Q(x)=-\dfrac{q_0}{15360}(5120x^3-7680ax^2+2880a^2x-217a^3)$, for $a/4 \leq x \leq 3a/4$;

$w(x)=\dfrac{q_0}{92160EI}a^3(217x-46a)(x-a)^2$, $M(x)=\dfrac{217q_0a^3}{46080}(3x-2a)$,

$Q(x)=\dfrac{217}{15360}q_0a^3$, for $3a/4 \leq x \leq a$.

(b) $w(x)=-\dfrac{x}{162EI}[P_1(6a^2-14x^2)+P_2(5a^2-4x^2)]$,

$M(x)=\dfrac{2}{27}x(7P_1+2P_2)$, $Q(x)=\dfrac{2}{27}(7P_1+2P_2)$, for $0 \leq x \leq a/3$;

APPENDIX B: ANSWERS AND COMMENTS

$$w(x) = -\frac{1}{162EI}[P_1(a-x)^2(13x-a) + P_2 x(5a^2 - 4x^2)],$$

$$M(x) = \frac{1}{27}[P_1(9a-13x) + 4P_2 x], \quad Q(x) = -\frac{1}{27}(13P_1 - 4P_2), \quad \text{for } a/3 \le x \le 2a/3;$$

$$w(x) = -\frac{(a-x)^2}{162EI}[P_1(13x-a) + P_2(23x-8a)], \quad M(x) = \frac{1}{27}[P_1(9a-13x)$$

$$+ P_2(18a - 23x)], \quad Q(x) = -\frac{1}{27}(13P_1 + 23P_2), \quad \text{for } 2a/3 \le x \le a.$$

(c) $\displaystyle w(x) = -\frac{x}{36aEI}[8x^2 M_1 + (5x^2 - 3a^2)M_2], \quad M(x) = -\frac{x}{6a}(8M_1 + 5M_2),$

$$Q(x) = -\frac{1}{6a}(8M_1 + 5M_2), \quad \text{for } 0 \le x \le a/3;$$

$$w(x) = -\frac{1}{36aEI}[2(6a^2 x - 9x^2 a - a^3 + 4x^3)M_1 + x(5x^2 - 3a^2)M_2],$$

$$M(x) = -\frac{1}{6a}[5M_2 x + 2M_1(4x - 3a)],$$

$$Q(x) = -\frac{1}{6a}(8M_1 + 5M_2), \quad \text{for } a/3 \le x \le 2a/3;$$

$$w(x) = -\frac{(a-x)^2}{36aEI}[2(4x-a)M_1 + (5x-8a)M_2],$$

$$M(x) = -\frac{1}{6a}[2M_1(4x-3a) + M_2(5x-6a)],$$

$$Q(x) = -\frac{1}{6a}(8M_1 + 5M_2), \quad \text{for } 2a/3 \le x \le a.$$

2.13: In this exercise, the parameter k_a is introduced as $k_a = k_a^*/6EI$.

(a) $\displaystyle w(x) = \frac{q_0 x}{360 EI}(x^5 - 3ax^4 + 5a^3 x^2 - 3a^5 - \frac{5a^2}{k_a}),$

$$M(x) = \frac{q_0}{12} x(x-a)(x^2 - ax - a^2), \quad Q(x) = \frac{q_0}{12}(2x-a)(2x^2 - 2ax - a^2).$$

(b) $\displaystyle w(x) = \frac{x}{96a^2 EI}\left\{M_1[a(16x^2 + 11a^2) + \frac{16}{k_a}] + 4M_2[a(4x^2 - a^2) + \frac{4}{k_a}]\right\},$

$$M(x) = \frac{x}{a}(M_1 + M_2), \quad Q(x) = \frac{1}{a}(M_1 + M_2), \quad \text{for } 0 \le x \le a/4;$$

$$w(x) = \frac{1}{96a^2 EI}\left\{M_1[a(a-x)(13a^2 - 16(a-x)^2) + \frac{16x}{k_a}]\right.$$

$$\left. + 4M_2 x[a(4x^2 - a^2) + \frac{4}{k_a}]\right\}, \quad M(x) = \frac{1}{a}[M_1(x-a) + M_2 x],$$

$$Q(x) = \frac{1}{a}(M_1 + M_2), \quad \text{for } a/4 \le x \le a/2;$$

$$w(x) = \frac{1}{96a^2 EI}\left\{M_1[a(a-x)(13a^2 - 16(a-x)^2) + \frac{16x}{k_a}] + 4M_2[a(a-x)\right.$$

$$\times (a^2-4(a-x)^2)+\frac{4x}{k_a}]\}, \quad M(x)=\frac{x-a}{a}(M_1+M_2),$$

$$Q(x)=\frac{1}{a}(M_1+M_2), \quad \text{for } a/2 \leq x \leq a.$$

(c) $w(x)=\dfrac{x}{162aEI}\left\{P_1[2a(5a^2-9x^2)+\dfrac{9}{k_a}]+P_2[a(8a^2-9x^2)+\dfrac{18}{k_a}]\right\},$

$$M(x)=-\frac{x}{3}(2P_1+P_2), \quad Q(x)=-\frac{1}{3}(2P_1+P_2), \quad \text{for } 0\leq x\leq a/3;$$

$$w(x)=\frac{1}{162aEI}\left\{P_1[a(x-a)(9(x-a)^2-8a^2)+\frac{9x}{k_a}]+P_2x[a(8a^2-9x^2)+\frac{18}{k_a}]\right\},$$

$$M(x)=\frac{1}{3}[(x-a)P_1+xP_2], \quad Q(x)=\frac{1}{3}(P_1-P_2), \quad \text{for } a/3\leq x\leq 2a/3;$$

$$w(x)=\frac{1}{162aEI}\left\{P_1[a(x-a)(9(x-a)^2-8a^2)+\frac{9x}{k_a}]\right.$$

$$\left.+2P_2[a(x-a)(9(x-a)^2-5a^2)+\frac{9x}{k_a}]\right\}, \quad M(x)=\frac{x-a}{3}(P_1+2P_2),$$

$$Q(x)=\frac{1}{3}(P_1+2P_2), \quad \text{for } 2a/3\leq x \leq a.$$

2.14: In this exercise, we denote $k_0=k_0^*/6EI$, $k_a=k_a^*/6EI$, and $K=6a^2k_0k_aEI$.

(a) $w(x)=\dfrac{q_0}{360EI}\left\{x^6-3ax(x^4+a^4)+\dfrac{5a^2}{k_0k_a}[x(k_a-k_0)-ak_a(1-k_0x^3)]\right\};$

(b) $w(x)=\dfrac{M_0}{K}\left\{[k_a(x-a)+axk_0k_a(x^2-a^2)+k_0x]+\dfrac{3}{4}a^3xk_0k_a\right\}$

$+\dfrac{q_0}{24EI}\left\{x(x-a)(x^2-ax-a^2)+\dfrac{2}{k_0k_a}[ak_a+x(k_0-k_a)]\right\}, \quad \text{for } x\leq a/2;$

$w(x)=\dfrac{M_0}{K}\left\{[k_a(x-a)+(1+axk_0(x-2a))+k_0x]+\dfrac{3}{4}a^3(x-a)k_0k_a\right\}$

$+\dfrac{q_0}{24EI}\left\{x(x-a)(x^2-ax-a^2)+\dfrac{2}{k_0k_a}[ak_a+x(k_0-k_a)]\right\}, \quad \text{for } x\geq a/2.$

(c) $w(x)=\dfrac{M_0}{K}\left\{[k_a(x-a)+axk_0k_a(x^2-a^2)+k_0x]+\dfrac{3}{4}a^3xk_0k_a\right\}$

$-\dfrac{P_0a}{64K}\left\{3k_a[16(x-a)+axk_0(16(x^2-a^2)+9a^2)]-16xk_0\right\}, \quad \text{for } x\leq a/4;$

$w(x)=\dfrac{M_0}{K}\left\{[k_a(x-a)+axk_0k_a(x^2-a^2)+k_0x]+\dfrac{3}{4}a^3xk_0k_a\right\}$

$+\dfrac{P_0a}{64K}\left\{k_a(x-a)[k_0a(16(x-a)^2-15a^2)-48]+16xk_0\right\}, \quad \text{for } a/4\leq x\leq a/2;$

$w(x)=\dfrac{M_0}{K}\left\{[k_a(x-a)+(1+axk_0(x-2a))+k_0x]+\dfrac{3}{4}a^3(x-a)k_0k_a\right\}$

APPENDIX B: ANSWERS AND COMMENTS

$+\dfrac{P_0 a}{64K}\left\{k_a(x-a)[k_0 a(16(x-a)^2-15a^2)-48]+16xk_0\right\}$, for $a/2 \le x \le a$.

(d) $w(x)=\dfrac{q_0}{24ak_0}\left\{\sqrt{\pi}\left[axk_0(2x^2+3)\left(\mathrm{erf}(x)-\mathrm{erf}(a)\right)-2\left(x\mathrm{erf}(x)-a\mathrm{erf}(a)\right)\right]\right.$

$+\dfrac{2}{ak_a}\left[ak_a-(k_0+k_a)x-k_0k_a ax(x^2+1)\right]\left(e^{-a^2}-1\right)$

$\left. + 2ak_0(x^2+1)e^{-x^2}-2ak_0(3x^2-2ax+1)\right\}$

where $\mathrm{erf}(x)$ represents a special function which is referred to as the *error function* [3, 17, 23, 29, 31, 68]. It is defined as follows

$$\mathrm{erf}(x) = \dfrac{2}{\sqrt{\pi}}\int_0^x e^{-t^2}\,dt$$

The error function is well tabulated and can be readily computed in an environment of every contemporary software created for either mathematical physics or statistics. Therefore, this special function is as convenient in practical computation as elementary functions.

2.15: In exercises 2.15 through 2.18, the parameter k is introduced as $k = (k_0/4EI)^{1/4}$.

(a) $w(x)=\dfrac{q_0}{8k^6 EI}\left\{e^{k(x-a)}[k(k+a(1+ka))\cos(k(x-a))\right.$

$+(1+ka)\sin(k(x-a))]-[k(k+b(1+kb))\cos(k(x-b))$

$\left. +(1+kb)\sin(k(x-b))]e^{k(x-b)}\right\}$, for $x \le a$;

$w(x)=-\dfrac{q_0}{8k^6 EI}\left\{e^{k(a-x)}[k(k-a(1-ka))\cos(k(x-a))\right.$

$+(1-ka)\sin(k(x-a))]+[k(k+b(1+kb))\cos(k(x-b))$

$\left. +(1+kb)\sin(k(x-b))]e^{k(x-b)}-2k^2(x^2+1)\right\}$, for $a \le x \le b$;

$w(x)=\dfrac{q_0}{8k^6 EI}\left\{[k(k+b(kb-1))\cos(k(x-b))+(kb-1)\sin(k(x-b))]e^{-k(x-b)}\right.$

$\left. -[k(k+b(ka-1))\cos(k(x-a))+(ka-1)\sin(k(x-a))]e^{-k(x-a)}\right\}$, $x \ge b$.

(b) $w(x)=\dfrac{P_1}{8k^3 EI}e^{k(x-a)}[\cos(k(x-a))-\sin(k(x-a))]$

$+\dfrac{P_2}{8k^3 EI}e^{k(x-b)}[\cos(k(x-b))-\sin(k(x-b))]$, for $x \le a$;

$w(x)=\dfrac{P_1}{8k^3 EI}e^{k(a-x)}[\cos(k(x-a))+\sin(k(x-a))]$

$+\dfrac{P_2}{8k^3 EI}e^{k(x-b)}[\cos(k(x-b))-\sin(k(x-b))]$, for $a \le x \le b$;

$$w(x) = \frac{P_1}{8k^3 EI} e^{k(a-x)} [\cos(k(x-a)) + \sin(k(x-a))]$$
$$+ \frac{P_2}{8k^3 EI} e^{k(b-x)} [\cos(k(x-b)) + \sin(k(x-b))], \text{ for } x \geq b.$$

(c) $w(x) = \dfrac{q_0}{8k^4 EI} \left[e^{k(x-a_1)} \cos(k(x-a_1)) - e^{k(x-a_2)} \cos(k(x-a_2)) \right]$

$$+ \frac{P_0}{8k^3 EI} e^{k(x-b)} [\cos(k(x-b)) - \sin(k(x-b))], \text{ for } x \leq a_1;$$

$$w(x) = \frac{q_0}{8k^4 EI} \left[2 - e^{-k(x-a_1)} \cos(k(x-a_1)) - e^{k(x-a_2)} \cos(k(x-a_2)) \right]$$

$$+ \frac{P_0}{8k^3 EI} e^{k(x-b)} [\cos(k(x-b)) - \sin(k(x-b))], \text{ for } a_1 \leq x \leq a_2;$$

$$w(x) = \frac{q_0}{8k^4 EI} \left[e^{-k(x-a_2)} \cos(k(x-a_2)) - e^{-k(x-a_1)} \cos(k(x-a_1)) \right]$$

$$+ \frac{P_0}{8k^3 EI} e^{k(x-b)} [\cos(k(x-b)) - \sin(k(x-b))], \text{ for } a_2 \leq x \leq b;$$

$$w(x) = \frac{q_0}{8k^4 EI} \left[e^{-k(x-a_2)} \cos(k(x-a_2)) - e^{-k(x-a_1)} \cos(k(x-a_1)) \right]$$

$$+ \frac{P_0}{8k^3 EI} e^{-k(x-b)} [\cos(k(x-b)) + \sin(k(x-b))], \text{ for } x \geq b.$$

(d) $w(x) = \dfrac{1}{4k^2 EI} \left[M_1 e^{k(x-a)} \sin(k(x-a)) + M_2 e^{k(x-b)} \sin(k(x-b)) \right], \ x \leq a;$

$w(x) = \dfrac{1}{4k^2 EI} \left[M_1 e^{k(a-x)} \sin(k(x-a)) + M_2 e^{k(x-b)} \sin(k(x-b)) \right], \ a \leq x \leq b;$

$w(x) = \dfrac{1}{4k^2 EI} \left[M_1 e^{k(a-x)} \sin(k(x-a)) + M_2 e^{k(b-x)} \sin(k(x-b)) \right], \ x \geq b.$

2.16:

(a) $w(x) = \dfrac{\cosh(kx)}{2k^2 EI} \left[M_1 e^{-ka} \sin(k(x-a)) + M_2 e^{-kb} \sin(k(x-b)) \right], \ x \leq a;$

$$w(x) = \frac{M_1}{2k^2 EI} e^{-kx} \sin(k(x-a)) \cosh(ka)$$

$$+ \frac{M_2}{2k^2 EI} e^{-kb} \sin(k(x-b)) \cosh(kx), \text{ for } a \leq x \leq b;$$

$w(x) = \dfrac{e^{-kx}}{2k^2 EI} [M_1 \sin(k(x-a)) \cosh(ka) + M_2 \sin(k(x-b)) \cosh(kb)], \ x \leq b.$

(b) $w(x) = \dfrac{P_0}{8k^3 EI} \{ e^{k(x-a)} [\cos(k(x-a)) - \sin(k(x-a))]$

$$- e^{-k(x+a)} [\sin(k(x+a)) - 2\cos(kx)\cos(ka) - \cos(k(x-a))] \}$$

$$+ \frac{M_0}{2k^2 EI} e^{-kb} \sin(k(x-b)) \cosh(kx), \text{ for } x \leq a;$$

$$w(x) = \frac{P_0}{8k^3 EI} \left\{ e^{k(a-x)} [\cos(k(x-a)) - \sin(k(a-x))] \right.$$
$$\left. - e^{-k(x+a)} [\sin(k(x+a)) - 2\cos(kx)\cos(ka) - \cos(k(x-a))] \right\}$$
$$+ \frac{M_0}{2k^2 EI} e^{-kb} \sin(k(x-b)) \cosh(kx), \quad \text{for } a \le x \le b;$$

$$w(x) = \frac{P_0}{8k^3 EI} \left\{ e^{k(a-x)} [\cos(k(x-a)) - \sin(k(a-x))] \right.$$
$$\left. - e^{-k(x+a)} [\sin(k(x+a)) - 2\cos(kx)\cos(ka) - \cos(k(x-a))] \right\}$$
$$+ \frac{M_0}{2k^2 EI} e^{-kx} \sin(k(x-b)) \cosh(kb), \quad \text{for } x \ge a.$$

2.17:

$$g(x,s) = \frac{1}{8k^3} \begin{cases} e^{k(x-s)}[\sin(k(x-s)) - \cos(k(x-s))] \\ \quad + e^{-k(x+s)}[\sin(k(x+s)) + \cos(k(x+s))], & \text{for } x \le s \\[4pt] e^{k(s-x)}[\sin(k(s-x)) - \cos(k(s-x))] \\ \quad + e^{-k(x+s)}[\sin(k(x+s)) + \cos(k(x+s))], & \text{for } s \le x \end{cases}$$

where $k = (k_0/4EI)^{1/4}$.

2.18:

(a) $$w(x) = \frac{M_0}{2k^2 EI} e^{-ka} \sin(k(x-a)) \cosh(kx) + \frac{P_0}{8k^3 EI} \left\{ e^{k(x-b)} [\cos(k(x-b)) \right.$$
$$\left. - \sin(k(x-b))] - e^{-k(x+b)}[\sin(k(x+b)) + \cos(k(x+b))] \right\}, \quad \text{for } x \le a;$$

$$w(x) = \frac{M_0}{2k^2 EI} e^{-kx} \sin(k(x-a)) \cosh(ka) + \frac{P_0}{8k^3 EI} \left\{ e^{k(x-b)} [\cos(k(x-b)) \right.$$
$$\left. - \sin(k(x-b))] - e^{-k(x+b)}[\sin(k(x+b)) + \cos(k(x+b))] \right\}, \quad \text{for } a \le x \le b;$$

$$w(x) = \frac{M_0}{2k^2 EI} e^{-kx} \sin(k(x-a)) \cosh(ka) + \frac{P_0}{8k^3 EI} \left\{ e^{k(b-x)} [\cos(k(x-b)) \right.$$
$$\left. + \sin(k(x-b))] - e^{-k(x+b)}[\sin(k(x+b)) + \cos(k(x+b))] \right\}, \quad \text{for } x \ge b.$$

(b) $$w(x) = \frac{\cosh(kx)}{2k^3 EI} \left[M_1 e^{-ka} \sin(k(x-a)) + M_2 e^{-kb} \sin(k(x-b)) \right], \quad x \le a;$$

$$w(x) = \frac{1}{2k^3 EI} \left[M_1 e^{-kx} \cosh(ka) \sin(k(x-a)) \right.$$
$$\left. + M_2 e^{-kb} \cosh(kx) \sin(k(x-b)) \right], \quad \text{for } a \le x \le b;$$

$$w(x) = \frac{e^{-kx}}{2k^3 EI} \left[M_1 \cosh(ka) \sin(k(x-a)) + M_2 \cosh(kb) \sin(k(x-b)) \right], \quad x \ge b.$$

2.19:

(a) $w(x) = \dfrac{q_0}{12m^4} \Big\{ mx \big[6(ma+b)^2 + mx(3(ma+b) + m(3a-x)) \big]$

$\qquad 6(ma+b)^2(mx+b)\ln(\dfrac{b}{mx+b}) \Big\}.$

(b) $w(x) = \dfrac{P_1}{2m^3} \Big\{ 2(ma_1+b)(mx+b)\ln(\dfrac{b}{mx+b}) + mx[mx + 2(ma_1+b)] \Big\}$

$+ \dfrac{P_2}{2m^3} \Big\{ 2(ma_2+b)(mx+b)\ln(\dfrac{b}{mx+b}) + mx[mx + 2(ma_2+b)] \Big\}, \quad \text{for } x \leq a_1;$

$w(x) = \dfrac{P_1}{2m^3} \Big\{ 2(ma_1+b)(mx+b)\ln(\dfrac{b}{ma_1+b}) + ma_1[ma_1 + 2(mx+b)] \Big\}$

$+ \dfrac{P_2}{2m^3} \Big\{ 2(ma_2+b)(mx+b)\ln(\dfrac{b}{mx+b}) + mx[mx + 2(ma_2+b)] \Big\}, \quad a_1 \leq x \leq a_2;$

$w(x) = \dfrac{P_1}{2m^3} \Big\{ 2(ma_1+b)(mx+b)\ln(\dfrac{b}{ma_1+b}) + ma_1[ma_1 + 2(mx+b)] \Big\}$

$+ \dfrac{P_2}{2m^3} \Big\{ 2(ma_2+b)(mx+b)\ln(\dfrac{b}{ma_2+b}) + ma_2[ma_2 + 2(mx+b)] \Big\}, \quad x \geq a_2.$

(c) $w(x) = \dfrac{M_1}{m^2} \Big[mx + (mx+b)\ln(\dfrac{b}{mx+b}) \Big]$

$+ \dfrac{M_2}{m^2} \Big[mx + (mx+b)\ln(\dfrac{b}{mx+b}) \Big], \quad \text{for } x \leq a_1;$

$w(x) = \dfrac{M_1}{m^2} \Big[ma_1 + (mx+b)\ln(\dfrac{b}{ma_1+b}) \Big]$

$+ \dfrac{M_2}{m^2} \Big[mx + (mx+b)\ln(\dfrac{b}{mx+b}) \Big], \quad \text{for } a_1 \leq x \leq a_2;$

$w(x) = \dfrac{M_1}{m^2} \Big[ma_1 + (mx+b)\ln(\dfrac{b}{ma_1+b}) \Big]$

$+ \dfrac{M_2}{m^2} \Big[ma_2 + (mx+b)\ln(\dfrac{b}{ma_2+b}) \Big], \quad \text{for } x \geq a_2.$

2.20:

$g(x,s) = \dfrac{1}{2m} \begin{cases} 2xs[1 + \ln(a/s)] - 2a(x+s) + (a^2+s^2), & x \leq s \\ 2xs[1 + \ln(a/x)] - 2a(x+s) + (a^2+x^2), & s \leq x \end{cases}$

2.21:

$w(x) = \dfrac{P_0}{2m} \Big[2a_1 x(1 + \ln\big(\dfrac{a}{a_1}\big)) - 2a(x+a_1) + (a^2+a_1^2) \Big]$

$+ \dfrac{M_0}{m} \Big[x\ln\big(\dfrac{a}{a_2}\big) + (a_2-a) \Big], \quad \text{for } x \leq a_1;$

$w(x) = \dfrac{P_0}{2m} \Big[2a_1 x(1 + \ln\big(\dfrac{a}{x}\big)) - 2a(x+a_1) + (a^2+x^2) \Big]$

APPENDIX B: ANSWERS AND COMMENTS

$$+ \frac{M_0}{m}\left[x\ln\left(\frac{a}{a_2}\right)+(a_2-a)\right], \text{ for } a_1 \leq x \leq a_2;$$

$$w(x) = \frac{P_0}{2m}\left[2a_1x(1+\ln\left(\frac{a}{x}\right))-2a(x+a_1)+(a^2+x^2)\right]$$

$$+\frac{M_0}{m}\left[x\ln\left(\frac{a}{x}\right)+(x-a)\right], \text{ for } x \geq a_2.$$

2.22:
$$g(x,s) = \frac{1}{mb^3}\begin{cases}[b(x+s)-2]-b^2xs+e^{-bx}[b(x-s)+2], & x \leq s \\ [b(x+s)-2]-b^2xs+e^{-bs}[b(s-x)+2], & s \leq x\end{cases}$$

2.23:
$$w(x) = \frac{P_0}{mb^3}\left\{\left[b(x+\frac{a}{4})-2\right]-\frac{1}{4}ab^2x+e^{-bx}\left[b(x-\frac{a}{4})+2\right]\right\}$$

$$-\frac{q_0 a}{8mb^3}\left\{4+(bx-1)(3ab-4)+[3ab-4(bx+2)]e^{-bx}\right\}, \text{ for } x \leq a/4;$$

$$w(x) = \frac{P_0}{mb^3}\left\{\left[b(x+\frac{a}{4})-2\right]-\frac{1}{4}ab^2x+e^{-ab/4}\left[b(\frac{a}{4}-x)+2\right]\right\}$$

$$-\frac{q_0 a}{8mb^3}\left\{4+(bx-1)(3ab-4)+[3ab-4(bx+2)]e^{-bx}\right\}, \text{ for } a/4 \leq x \leq a/2;$$

$$w(x) = \frac{P_0}{mb^3}\left\{\left[b(x+\frac{a}{4})-2\right]-\frac{1}{4}ab^2x+e^{-ab/4}\left[b(\frac{a}{4}-x)+2\right]\right\}$$

$$-\frac{q_0}{8mb^4}\left\{4\left[b^2(x-a)^2+2(3+2b(x-a))\right]e^{-bx}\right.$$

$$\left.+4[2(bx-2)-(ab+2)]e^{-ab/2}+ab[3ab(bx-1)-4(bx-2)]\right\}, \text{ for } x \geq a/2.$$

CHAPTER 4: Assemblies of Elements

4.1: (a) Yes; (b) Yes; (c) Yes.

4.2: (a) The entries $g_{ij}(x,s)$, $(i,j=1,2)$ of the matrix of Green's type are defined in the form:

$$g_{11}(x,s) = \frac{1}{H}\begin{cases}(1-\lambda ks)(x+a), & \text{for } -a \leq x \leq s \leq 0 \\ (1-\lambda kx)(s+a), & \text{for } -a \leq s \leq x \leq 0\end{cases}$$

$$g_{12}(x,s) = \frac{\lambda k(x+a)}{H}e^{-ks}, \text{ for } -a \leq x \leq 0, \ 0 \leq s < \infty$$

$$g_{21}(x,s) = \frac{\lambda k(s+a)}{H}e^{-kx}, \text{ for } -a \leq s \leq 0, \ 0 \leq x < \infty$$

$$g_{22}(x,s) = \frac{1}{2H}\begin{cases}He^{k(x-s)}-(1-\lambda ka)e^{-k(x+s)}, & 0 \leq x \leq s < \infty \\ He^{k(s-x)}-(1-\lambda ka)e^{-k(s+x)}, & 0 \leq s \leq x < \infty\end{cases}$$

where $H = 1+\lambda ka$.

(b) The entries $g_{ij}(x,s)$, $(i,j=1,2)$ of the matrix of Green's type are defined in the form:

$$g_{11}(x,s) = \frac{1}{k\Delta^*} \begin{cases} \sinh(k(x+a))(\cosh(ks)-\lambda\sinh(ks)), & \text{for } -a\leq x\leq s\leq 0 \\ \sinh(k(s+a))(\cosh(kx)-\lambda\sinh(kx)), & \text{for } -a\leq s\leq x\leq 0 \end{cases}$$

$$g_{12}(x,s) = \frac{\lambda}{k\Delta^*} e^{-ks} \sinh(k(x+a)), \quad \text{for } -a\leq x\leq 0\leq s<\infty$$

$$g_{21}(x,s) = \frac{1}{k\Delta^*} e^{-kx} \sinh(k(s+a)), \quad \text{for } -a\leq s\leq 0\leq x<\infty$$

$$g_{22}(x,s) = \frac{1}{k\Delta^*} \begin{cases} e^{-ks}(\lambda\sinh(ka)\cosh(kx)+\cosh(ka)\sinh(kx)), & 0\leq x\leq s<\infty \\ e^{-kx}(\lambda\sinh(ka)\cosh(ks)+\cosh(ka)\sinh(ks)), & 0\leq s\leq x<\infty \end{cases}$$

where $\Delta^* = \lambda\sinh(ka)+\cosh(ka)$.

(c) The entries $g_{ij}(x,s)$, $(i,j=1,2)$ of the matrix of Green's type are defined in the form:

$$g_{11}(x,s) = \frac{1}{\Delta^*} \begin{cases} \sin(k(x+a))(\lambda\sin(ks)\cos(ka)-\sin(ka)\cos(ks)), & x\leq s \\ \sin(k(s+a))(\lambda\sin(kx)\cos(ka)-\sin(ka)\cos(kx)), & s\leq x \end{cases}$$

$$g_{12}(x,s) = \frac{\lambda}{\Delta^*} \sin(k(x+a))\sin(k(s-a)), \quad -a\leq x\leq 0\leq s\leq a$$

$$g_{21}(x,s) = \frac{1}{\Delta^*} \sin(k(x-a))\sin(k(s+a)), \quad -a\leq s\leq 0\leq x\leq a$$

$$g_{22}(x,s) = \frac{1}{\Delta^*} \begin{cases} \sin(k(s-a))(\lambda\sin(ka)\cos(kx)+\sin(kx)\cos(ka)), & x\leq s \\ \sin(k(x-a))(\lambda\sin(ka)\cos(ks)+\sin(ks)\cos(ka)), & s\leq x \end{cases}$$

where $\Delta^* = k(1+\lambda)\sin(ka)\cos(ka)$.

(d) The diagonal entries $g_{ii}(x,s)$ of the matrix of Green's type are defined as follows:

$$g_{ii}(x,s) = \frac{1}{H_0} \begin{cases} e^{-ks}[h_1\cosh(kx)+h_2\sinh(kx)+h_3\sinh(kx)], & x\leq s \\ e^{-kx}[h_1\cosh(ks)+h_2\sinh(ks)+h_3\sinh(ks)], & s\leq x \end{cases}$$

where $i=1,2,3$ and $H_0 = k(h_1+h_2+h_3)$. The peripheral entries $g_{ij}(x,s)$, $(i\neq j)$ of this matrix are defined in the form:

$$g_{ij}(x,s) = H_0 h_j e^{-k(x+s)}, \quad i\neq j, \quad 0\leq x,s<\infty.$$

4.3:

$$y_1(x) = \frac{1}{\pi} H h_1(x-1) - \frac{1}{\pi^2}\sin(\pi x); \quad y_2(x) = \frac{1}{\pi} H h_1(x-1);$$

$$y_3(x) = \frac{1}{\pi} H h_1(x-1) \quad \text{where} \quad H = (h_1+h_2+h_3)^{-1}.$$

APPENDIX B: ANSWERS AND COMMENTS

4.4:

(a) $w_1(x) = \dfrac{q_0 x^2(x-a)}{120aEI}[(x-2a)^2+10a^2],$

$M_1(x) = \dfrac{q_0}{30aEI}[5(x-a)^3+2a^2(6x-a)],$

$Q_1(x) = \dfrac{q_0}{10aEI}[5(x-a)^2+4a^2], \quad \text{for } 0 \le x \le a;$

$w_2(x) = \dfrac{q_0(x-a)}{120aEI}[(x-a)^4-10a^2(x-2a)^2+21a^4],$

$M_2(x) = \dfrac{q_0(x+a)}{6aEI}(x-2a)^2, \quad Q_2(x) = \dfrac{q_0 x}{2aEI}(x-2a), \quad \text{for } a \le x \le 2a.$

(b) $w_1(x) = \dfrac{1}{48EI}(x^2(x-a)[q_1(2x-3a)+6q_2a],$

$M_1(x) = \dfrac{1}{8EI}[q_1(4x^2-5ax+a^2)+2q_2 a(3x-a)],$

$Q_1(x) = \dfrac{1}{8EI}[q_1(8x-5a)+6q_2 a], \quad \text{for } 0 \le x \le a;$

$w_2(x) = \dfrac{(a-x)}{48EI}[q_1 a^3 - q_2(2x^3-14ax^2+34a^2 x-16a^3)],$

$M_2(x) = \dfrac{q_2}{2EI}(x-2a)^2, \quad Q_2(x) = \dfrac{q_2}{EI}(x-2a), \quad \text{for } a \le x \le 2a.$

(c) $w_1(x) = \dfrac{q_0 x^2(a-x)}{720 a^2 EI}(2x^3-10ax^2-10a^2 x-15a^3),$

$M_1(x) = -\dfrac{q_0}{24 a^2 EI}(2x^4-8ax^3-a^3 x+a^4),$

$Q_1(x) = -\dfrac{q_0}{24 a^2 EI}(8x^3-24ax^2-a^3), \quad \text{for } 0 \le x \le a;$

$w_2(x) = \dfrac{q_0(a-x)}{720 a^2 EI}(2x^5-10ax^4-10a^2 x^3+150 a^3 x^2 - 330 a^4 x+165 a^5),$

$M_2(x) = \dfrac{q_0(x+2a)}{12 a^2 EI}(2a-x)^3, \quad Q_2(x) = -\dfrac{q_0(x+a)}{3a^2 EI}(x-2a)^2, \quad \text{for } a \le x \le 2a.$

4.5: In exercises 4.5 through 4.7, we denote $R = EI_1 + EI_2$, $R_1 = EI_1 R$, and $R_2 = EI_2 R$.

(a) $w_1(x) = \dfrac{x(x+a)}{720 R_1} \{ R[q_0(2x^4-2ax^3+2a^2 x^2+12a^3 x)+q_2(30x^2+45ax)]$

$\qquad + q_1[EI_1(6x^3-6ax^2-7a^2 x+28a^3)+EI_2(6x^3-6ax^2-21a^2 x)]\},$

$M_1(x) = \dfrac{(x+a)}{120 R_1} \{ R[q_0(10x^3-10ax^2+10a^2 x+4a^3)+q_2(60x+15a)]$

$\qquad + q_1[EI_1(20x^2-20ax+7a^2)+EI_2(20x^2-20ax-7a^2)]\},$

$$Q_1(x) = \frac{1}{120R_1}\left\{R[q_0(40x^3+14a^3)+q_2(120x+75a)]\right.$$
$$\left. + q_1[EI_1(60x^2-13a^2)+EI_2(60x^2-27a^2)]\right\}, \quad \text{for } -a \leq x \leq 0;$$

$$w_2(x) = \frac{x(x-a)}{720R_2}\left\{R[q_0(2x^4+2ax^3+2a^2x^2-12a^3x)+q_2(30x^2-45ax)]\right.$$
$$\left. + q_1[EI_1(6x^3+6ax^2-21a^2x)+EI_2(6x^3+6ax^2-7a^2x-28a^3)]\right\},$$

$$M_2(x) = \frac{(x-a)}{120R_2}\left\{R[q_0(10x^3+10ax^2+10a^2x-4a^3)+q_2(60x-15a)]\right.$$
$$\left. + q_1[EI_1(20x^2+20ax-7a^2)+EI_2(20x^2+20ax+7a^2)]\right\},$$

$$Q_2(x) = \frac{1}{120R_2}\left\{R[q_0(40x^3-14a^3)+q_2(120x-75a)]\right.$$
$$\left. + q_1[EI_1(60x^2-27a^2)+EI_2(60x^2-13a^2)]\right\}, \quad \text{for } 0 \leq x \leq a.$$

4.6: The entries $g_{ij}(x,s)$, $(i,j=1,2)$ of the influence matrix are defined in the form:

$$g_{11}(x,s) = \frac{1}{12a^3 R_1}\begin{cases} s(a+x)^2[EI_1(3a^2x+2as^2-xs^2) \\ \qquad +2EI_2s(as-3ax-2xs)], \quad \text{for } -a \leq x \leq s \leq 0 \\ x(a+s)^2[EI_1(3a^2s+2ax^2-sx^2) \\ \qquad +2EI_2x(ax-3as-2xs)], \quad \text{for } -a \leq s \leq x \leq 0 \end{cases}$$

$$g_{12}(x,s) = \frac{1}{6a^3 R}xs(a-s)^2(a+x)^2, \quad \text{for } -a \leq x \leq 0 \leq s \leq a$$

$$g_{21}(x,s) = \frac{1}{6a^3 R}xs(a-x)^2(a+s)^2, \quad \text{for } -a \leq s \leq 0 \leq x \leq a$$

$$g_{22}(x,s) = \frac{1}{12a^3 R_2}\begin{cases} x(a-s)^2[2EI_1x(3as-ax-2xs) \\ \qquad +EI_2(3a^2s-2ax^2-x^2s)], \quad \text{for } 0 \leq x \leq s \leq a \\ s(a-x)^2[2EI_1s(3ax-as-2xs) \\ \qquad +EI_2(3a^2x-2as^2-xs^2)], \quad \text{for } 0 \leq s \leq x \leq a \end{cases}$$

4.7:
$$\text{(a) } w_1(x) = \frac{1}{240R}ax((x+a)^2(2aq_0+5q_1),$$

$$M_1(x) = \frac{a}{120R}(3x+2a)(2aq_0+5q_1), \quad Q_1(x) = \frac{a}{40R}(2aq_0+5q_1), \quad \text{for } -a \leq x \leq 0;$$

$$w_2(x) = \frac{x(x-a)^2}{240R_2}\{2q_0[Rx(x+2a)+EI_2a^2]+5q_1(2Rx+EI_a)\},$$

$$M_2(x) = \frac{1}{120R_2}\{2q_0[EI_1(10x^3-9xa^2+2a^3)+2EI_2x(5x^2-3a^2)]$$

$$+5q_1[2EI_1(6x^2-6ax+a^2)+3EI_2x(4x-3a)]\},$$

$$Q_2(x)=\frac{1}{40R_2}\{2q_0[EI_1(10x^2-3a^2)+2EI_2(5x^2-a^2)]$$

$$+5q_1[4EI_1(2x-a)+EI_2(8x-3a)]\}, \text{ for } 0\leq x\leq a.$$

(b) $w_1(x)=\dfrac{x(x+a)^2}{120R_1}\{q_0[Rx(x-2a)+2EI_1a^2]+5q_1Rx\},$

$$M_1(x)=\frac{1}{60R_1}\{q_0[EI_1(10x^3-3a^2x+2a^3)+EI_2(10x^3-9a^2x-2a^3)]$$

$$+5q_1R(6x^2+6ax+a^2)\}, \quad Q_1(x)=\frac{1}{20R_1}\{10q_1R(2x+a)$$

$$+q_0[EI_1(10x^2-a^2)+EI_2(10x^2-3a^2)]\}, \text{ for } -a\leq x\leq 0;$$

$$w_2(x)=\frac{x(x-a)^2}{120R_2}\{q_0[Rx(x+2a)+2EI_1a^2]+5q_1Rx\},$$

$$M_2(x)=\frac{1}{60R_2}\{q_0[EI_1(10x^3-9a^2x+2a^3)+EI_2(10x^3-3a^2x-2a^3)]$$

$$+5q_1R(6x^2-6ax+a^2)\}, \quad Q_2(x)=\frac{1}{20R_2}\{10q_1R(2x-a)$$

$$+q_0[EI_1(10x^2-3a^2)+EI_2(10x^2-a^2)]\}, \text{ for } 0\leq x\leq a.$$

In exercises 4.8 and 4.9, we denote $R_0=4EI_1+3EI_2$, $R_1=EI_1R_0$, and $R_2=EI_2R_0$.

4.8: The entries $g_{ij}(x,s)$, $(i,j=1,2)$ of the influence matrix are defined in this case in the form:

$$g_{11}(x,s)=-\frac{1}{6a^3R_1}\begin{cases}s(a+x)^2[2EI_1(s^2x-3a^2x-2as^2)\\ \quad+3EI_2s(2xs+3ax-as)], \text{ for } -a\leq x\leq s\leq 0\\ x(a+s)^2[2EI_1(x^2s-3a^2s-2ax^2)\\ \quad+3EI_2x(2sx+3as-ax)], \text{ for } -a\leq s\leq x\leq 0\end{cases}$$

$$g_{12}(x,s)=\frac{1}{2a^3R_0}xs(2a-s)(a-s)(a+x)^2, \text{ for } -a\leq x\leq 0\leq s\leq a$$

$$g_{21}(x,s)=\frac{1}{2a^3R_0}xs(x-a)(x-2a)(a+s)^2, \text{ for } -a\leq s\leq 0\leq x\leq a$$

$$g_{22}(x,s)=\frac{1}{6a^3R_2}\begin{cases}x(s-a)[2EI_1x(2a^2x+2axs-xs^2-6a^2s+3as^2)\\ \quad+3EI_2a^2(x^2-2as+s^2)], \text{ for } 0\leq x\leq s\leq a\\ s(x-a)[2EI_1s(2a^2s+2axs-x^2s-6a^2x+3ax^2)\\ \quad+3EI_2a^2(s^2-2ax+x^2)], \text{ for } 0\leq s\leq x\leq a\end{cases}$$

4.9:

(a) $w_1(x) = \dfrac{x(x+a)^2}{24R_1}\{q_1[2EI_1(2x-a)+3xEI_2]+q_2aEI_1\},$

$M_1(x) = \dfrac{1}{4R_1}\{q_1[2EI_1x(4x+3a)+EI_1(6x^2+6ax+a^2)]+q_2EI_1a(2a+3x)\},$

$Q_1(x) = \dfrac{1}{4R_1}\{2q_1[EI_1(8x+3a)+3EI_2(2x+a)]+3q_2aEI_1\}, \quad -a \leq x \leq 0;$

$w_2(x) = \dfrac{x(x-a)^2}{24R_2}$
$\times \{q_1EI_2a(2a-x)+q_2[2EI_1x(2x-3a)+3EI_2(x^2-ax-a^2)]\},$

$M_2(x) = \dfrac{1}{4R_2}(a-x)\{q_1aEI_2-2q_2[EI_1(4x-a)+3x]\},$

$Q_2(x) = \dfrac{1}{4R_2}\{-q_1aEI_2+2q_2[EI_1(8x-5a)+3EI_2(2x-a)]\}, \quad 0 \leq x \leq a;$

(b) $w_1(x) = \dfrac{q_0 a}{15\pi^4 R_1}\{EI_1[\pi(1+60\pi)x(x+a)^2+240(a^2-x^2))$

$\quad - 1920a^3\sin^2(\dfrac{\pi x}{4a})] + 480x(x^2-3a^2)$

$\quad + 360EI_2[x^2((4-\pi)(x+a)-a)-4\sin^2(\dfrac{\pi x}{4a})]\},$

$M_1(x) = \dfrac{2q_0 a}{15\pi^4 R_1}\{EI_1[180(\pi^2-4\pi+8)x+\pi^4(3x+2a)+240a\pi^2\sin^2(\dfrac{\pi x}{4a})]$

$\quad + 90EI_2[12(4-\pi)x-4(\pi-6)a-a\pi^2\cos(\dfrac{\pi x}{2a})]\},$

$Q_1(x) = \dfrac{2q_0 a}{5\pi^4 R_1}\{EI_1[60(\pi^2-4\pi+8)+\pi^4+20\pi^3\sin(\dfrac{\pi x}{2a})]$

$\quad + 15EI_2[24(4-\pi)+\pi^3\sin(\dfrac{\pi x}{2a})]\}, \quad \text{for } -a \leq x \leq 0;$

$w_2(x) = \dfrac{q_0}{120a\pi^4 R_2}x(x-a)(x-2a)\{4EI_1\pi^4x(2a-x)+EI_2[240a^2(\pi^2+4\pi+24)$

$\quad + \pi^4(4a^2+6ax-3x^2)]\}, \quad M_2(x) = \dfrac{q_0(a-x)}{30a\pi^4 R_2}\{4EI_1\pi^4(5x^2-10ax+2a^2)$

$\quad + 15EI_2[a^2(576-24\pi^2-96\pi)+\pi^4 x(x-2a)]\};$

$Q_2(x) = \dfrac{q_0}{10a\pi^4 R_2}\{4EI_1\pi^4(5x^2-10ax+4a^2)$

$\quad + 5EI_2[120a^2(24-4\pi-\pi^2)+\pi^4(3x^2-6ax+2a^2)]\}, \quad \text{for } 0 \leq x \leq a;$

APPENDIX B: ANSWERS AND COMMENTS

4.10: The entries $g_{ij}(x,s)$, $(i,j=1,2,3)$ of the influence matrix are defined in the form:

$$g_{11}(x,s) = \frac{1}{12a^3 EI_1} \begin{cases} x^2(a-s)[s(x-3a)(s-2a)-2a^2x], & \text{for } 0 \le x \le s \le a \\ s^2(a-x)[x(s-3a)(x-2a)-2a^2s], & \text{for } 0 \le s \le x \le a \end{cases}$$

$$g_{12}(x,s) = \frac{1}{4aEI_1} x^2(a-s)(a-x), \text{ for } 0 \le x \le a \le s \le 2a;$$

$$g_{13}(x,s) = \frac{1}{4aEI_1} x^2(a-s)(a-x), \text{ for } 0 \le x \le a, \ 2a \le s \le 3a$$

$$g_{21}(x,s) = \frac{1}{4aEI_1} s^2(a-s)(a-x), \text{ for } 0 \le s \le a \le x \le 2a$$

$$g_{22}(x,s) = \frac{1}{12EI_1} \begin{cases} (a-x)[(2x-a)(a-3s)+2x^2], & \text{for } a \le x \le s \le 2a \\ (a-s)[(2s-a)(a-3x)+2s^2], & \text{for } a \le s \le x \le 2a \end{cases}$$

$$g_{23}(x,s) = \frac{1}{12EI_1}(a-x)[(2x-a)(a-3s)+2x^2], \text{ for } a \le x \le 2a \le s \le 3a;$$

$$g_{31}(x,s) = \frac{1}{4aEI_1} s^2(a-s)(a-x), \text{ for } 2a \le x \le 3a, \ 0 \le s \le a$$

$$g_{32}(x,s) = \frac{1}{12EI_1}(s-a)[(a-3x)(a-2s)-2s^2], \text{ for } a \le s \le 2a \le x \le 3a$$

$$g_{33}(x,s) = -\frac{1}{12EI_1 EI_2} \begin{cases} 2EI_1[x^2(x-3s)+12(xs-ax-as)+16a^3] \\ +3aEI_2[7a(x+s)-5xs], \text{ for } 2a \le x \le s \le 3a \\ 2EI_1[s^2(s-3x)+12(xs-as-ax)+16a^3] \\ +3aEI_2[7a(x+s)-5xs], \text{ for } 2a \le s \le x \le 3a \end{cases}$$

4.11: The entries $g_{ij}(x,s)$, $(i,j=1,2,3)$ of the influence matrix are defined in the form:

$$g_{11}(x,s) = -\frac{1}{6EI} \begin{cases} (a-s)(s^2+as-4a^2+5ax-3xs), & \text{for } 0 \le x \le s \le a \\ (a-x)(x^2+ax-4a^2+5as-3xs), & \text{for } 0 \le s \le x \le a \end{cases}$$

$$g_{12}(x,s) = -\frac{1}{6aEI}(a-s)(2a-s)(3a-s)(x-a), \text{ for } 0 \le x \le a \le s \le 2a;$$

$$g_{13}(x,s) = \frac{a}{6EI}(2a-s)(x-a), \text{ for } 0 \le x \le a, \ 2a \le s \le 3a$$

$$g_{21}(x,s) = -\frac{1}{6aEI}(a-s)(x-a)(x-2a)(x-3a), \text{ for } 0 \le s \le a \le x \le 2a$$

$$g_{22}(x,s) = -\frac{1}{6aEI} \begin{cases} (x-a)(2a-s)[(x+s-2a)^2+2x(a-s)], & a \le x \le s \le 2a \\ (s-a)(2a-x)[(s+x-2a)^2+2s(a-x)], & a \le s \le x \le 2a \end{cases}$$

$$g_{23}(x,s) = -\frac{x}{6aEI}(x-a)(x-2a)(2a-s), \text{ for } a \le x \le 2a \le s \le 3a;$$

$$g_{31}(x,s) = \frac{a}{6EI}(a-s)(x-2a), \quad \text{for } 2a \leq x \leq 3a, \ 0 \leq s \leq a$$

$$g_{32}(x,s) = \frac{s}{6aEI}(a-s)(2a-s)(x-2a), \quad \text{for } a \leq s \leq 2a \leq x \leq 3a$$

$$g_{33}(x,s) = -\frac{1}{6EI}\begin{cases}(x-2a)(x^2+2ax-3xs+4as-4a^2), & 2a \leq x \leq s \leq 3a \\ (s-2a)(s^2+2as-3xs+4ax-4a^2), & 2a \leq s \leq x \leq 3a\end{cases}$$

CHAPTER 5: Potential and Related Fields

In exercises 5.1 through 5.3, we use the following notations:

$$E(p) = \left|1-\exp\left(\frac{\pi p}{b}\right)\right|, \quad E_1(p) = \left|1+\exp\left(\frac{\pi p}{2b}\right)\right|, \quad E_2(p) = \left|1-\exp\left(\frac{\pi p}{2b}\right)\right|,$$

and $z = x+iy$, $\zeta = s+it$, with the bar on ζ denoting its conjugate $\overline{\zeta} = s-it$.

5.1:

(a) $G(x,y;s,t) = \dfrac{1}{2\pi}\ln\dfrac{E(z-\overline{\zeta})}{E(z-\zeta)};$

(b) $G(x,y;s,t) = \dfrac{1}{2\pi}\ln\dfrac{E_1(z-\zeta)E_2(z-\overline{\zeta})}{E_2(z-\zeta)E_1(z-\overline{\zeta})}.$

5.2:

(a) $G(x,y;s,t) = \dfrac{1}{4\pi}\ln\dfrac{E_1(z-\zeta)E_1(z+\zeta)E_2(z-\overline{\zeta})E_2(z+\overline{\zeta})}{E_2(z-\zeta)E_2(z+\zeta)E_1(z-\overline{\zeta})E_1(z+\overline{\zeta})};$

(b) $G(x,y;s,t) = \dfrac{1}{4\pi}\ln\dfrac{E_1(z-\zeta)E_2(z+\zeta)E_2(z-\overline{\zeta})E_1(z+\overline{\zeta})}{E_2(z-\zeta)E_1(z+\zeta)E_1(z-\overline{\zeta})E_2(z+\overline{\zeta})};$

(c) $G(x,y;s,t) = M(x,s) + \dfrac{1}{2\pi}\ln\dfrac{E(z+\zeta)E(z+\overline{\zeta})}{E(z-\zeta)E(z-\overline{\zeta})},$

where

$$M(x,s) = -\frac{1}{b}\begin{cases}x, & x \leq s \\ s, & x \geq s,\end{cases}$$

(d) $G(x,y;s,t) = -\dfrac{1}{2\pi}\ln\left(E(z-\overline{\zeta})E(z-\zeta)E(z+\overline{\zeta})E(z+\zeta)\right)$

$\qquad + M(x,s) - \dfrac{2\beta}{b}\sum_{n=1}^{\infty}\dfrac{e^{-\nu(x+s)}}{\nu(\beta+\nu)}\cos(\nu y)\cos(\nu t), \quad \nu = \dfrac{n\pi}{b},$

where

$$M(x,s) = \frac{1}{b}\begin{cases}\beta^{-1}+2x+s, & x \leq s \\ \beta^{-1}+2s+x, & x \geq s.\end{cases}$$

5.3:

(a) $$G(x,y;s,t) = \frac{1}{2\pi} \ln \frac{E(z-\bar{\zeta})E(z+\bar{\zeta})}{E(z-\zeta)E(z+\zeta)}$$
$$+ \frac{2}{b} \sum_{n=1}^{\infty} \frac{\sinh(\nu x)\sinh(\nu s)}{\nu e^{\nu a} \cosh(\nu a)} \sin(\nu y) \sin(\nu t), \quad \nu = \frac{n\pi}{b};$$

(b) $$G(x,y;s,t) = \frac{1}{2\pi} \ln \frac{E(z-\bar{\zeta})E(z+\bar{\zeta})}{E(z-\zeta)E(z+\zeta)}$$
$$- \frac{2}{b} \sum_{n=1}^{\infty} \frac{(\beta-\nu)\sinh(\nu x)\sinh(\nu s)}{\nu e^{\nu a}(\beta \sinh(\nu a) + \nu \cosh(\nu a))} \sin(\nu y) \sin(\nu t), \quad \nu = \frac{n\pi}{b};$$

(c) $$G(x,y;s,t) = \frac{1}{2\pi} \ln \frac{E_2(z-\bar{\zeta})E_2(z+\bar{\zeta})E_1(z-\zeta)E_1(z+\zeta)}{E_2(z-\zeta)E_2(z+\zeta)E_1(z-\bar{\zeta})E_1(z+\bar{\zeta})}$$
$$- \frac{2}{b} \sum_{n=1}^{\infty} \frac{(\beta-\nu)\sinh(\nu x)\sinh(\nu s)}{\nu e^{\nu a}(\beta \sinh(\nu a) + \nu \cosh(\nu a))} \sin(\nu y) \sin(\nu t), \quad \nu = \frac{(2n-1)\pi}{2b};$$

(d) $$G(x,y;s,t) = \frac{1}{2\pi} \ln \frac{E(z-\bar{\zeta})E(z+\zeta)}{E(z-\zeta)E(z+\bar{\zeta})}$$
$$- \frac{2}{b} \sum_{n=1}^{\infty} \frac{(\beta-\nu)\cosh(\nu x)\cosh(\nu s)}{\nu e^{\nu a}(\beta \cosh(\nu a) + \nu \sinh(\nu a))} \sin(\nu y) \sin(\nu t), \quad \nu = \frac{n\pi}{b}.$$

In exercises 5.4 and 5.5, we use the following notations:

$$z = r(\cos(\phi) + i\sin(\phi)), \quad \zeta = s(\cos(t) + i\sin(t))$$

with the bar on ζ denoting its conjugate $\bar{\zeta} = s(\cos(t) - i\sin(t))$ and with z^p denoting the principal value of a function.

5.4:

(a) $$G(r,\phi;s,t) = \frac{1}{2\pi} \ln \frac{|z^p - (\bar{\zeta})^p|}{|z^p - \zeta^p|}, \quad p = \frac{\pi}{\alpha}.$$

From this, the Green's function

$$G(r,\phi;s,t) = \frac{1}{2\pi} \ln \frac{|z - \bar{\zeta}|}{|z - \zeta|}$$

of the Dirichlet problem for a half-plane follows as a particular case, when $\alpha = \pi$.

(b) $$G(r,\phi;s,t) = \frac{1}{2\pi} \ln \frac{|z^p - (\bar{\zeta})^p||z^p + \zeta^p|}{|z^p + (\bar{\zeta})^p||z^p - \zeta^p|}, \quad p = \frac{\pi}{2\alpha}.$$

From this, the Green's function

$$G(r,\phi;s,t) = \frac{1}{2\pi} \ln \frac{|z^{1/2} - (\bar{\zeta})^{1/2}||z^{1/2} + \zeta^{1/2}|}{|z^{1/2} + (\bar{\zeta})^{1/2}||z^{1/2} - \zeta^{1/2}|}$$

of the mixed (Dirichlet–Newmann) problem for a half-plane follows as a particular case, when $\alpha = \pi$.

5.5:

(a) $$G(r,\phi;s,t) = \frac{1}{2\pi} \ln \frac{|z^p - (\bar{\zeta})^p||(R^2z)^p - (r^2\bar{\zeta})^p|}{|z^p + (\bar{\zeta})^p||(R^2z)^p - (r^2\zeta)^p|}$$
$$- \frac{2R\beta}{\alpha} \sum_{n=1}^{\infty} \frac{1}{\nu(R\beta+\nu)} \left(\frac{rs}{R^2}\right)^\nu \sin(\nu\phi)\sin(\nu t), \quad p = \frac{\pi}{\alpha}, \quad \nu = np;$$

(b) $G(r,\phi;s,t)$
$$= \frac{1}{2\pi} \ln \frac{|z^p - (\bar{\zeta})^p||(R^2z)^p + (r^2\bar{\zeta})^p||z^p + \zeta^p||(R^2z)^p - (r^2\zeta)^p|}{|z^p + (\bar{\zeta})^p||(R^2z)^p - (r^2\bar{\zeta})^p||z^p - \zeta^p||(R^2z)^p + (r^2\zeta)^p|};$$

(c) $$G(r,\phi;s,t) = M(x,s) + \frac{1}{2\pi} \ln \frac{|(R^2z)^p - (r^2\bar{\zeta})^p||(R^2z)^p - (r^2\zeta)^p|}{(r^2s^2)^p|z^p - (\bar{\zeta})^p||z^p - \zeta^p|},$$

where
$$M(x,s) = \frac{1}{\alpha} \begin{cases} \ln(s/R), & r \leq s \\ \ln(r/R), & r \geq s. \end{cases}$$

In exercises 5.6(a) and 5.6(b), the parameters ν, γ and the functions $H_n(x,s)$ and $F_n(x,s)$ are defined as:

$$\nu = \frac{n\pi}{b}, \quad \gamma = \sqrt{\nu^2 + k^2}, \quad H_n(x,s) = \frac{1}{\nu\gamma}\left[\gamma e^{\nu(x-s)} - \nu e^{\gamma(x-s)}\right],$$

and
$$F_n(x,s) = \frac{1}{\nu\gamma}\left[\nu e^{-\gamma(x+s)} - \gamma e^{-\nu(x+s)}\right].$$

5.6:

(a) $$G(x,y;s,t) = \frac{1}{2\pi} \ln \frac{E(z+\zeta)E(z-\bar{\zeta})}{E(z-\zeta)E(z+\bar{\zeta})}$$
$$+ \frac{1}{b}\sum_{n=1}^{\infty}[H_n(x,s) + F_n(x,s)]\sin(\nu y)\sin(\nu t);$$

(b) $$G(x,y;s,t) = \frac{1}{2\pi} \ln \frac{E(z+\zeta)E(z-\bar{\zeta})}{E(z-\zeta)E(z+\bar{\zeta})}$$
$$+ \frac{1}{b}\sum_{n=1}^{\infty}[H_n(x,s) + F_n(x,s)]\sin(\nu y)\sin(\nu t) - \frac{2\beta}{b}\sum_{n=1}^{\infty}\frac{e^{-\gamma(x+s)}}{\gamma(\gamma+\beta)}.$$

In exercises 5.7 through 5.9, I_n and K_n represent the modified Bessel functions of the first and second kind, respectively, of order n; z and ζ introduced as:

$$z = r(\cos(\phi) + i\sin(\phi)), \quad \zeta = s(\cos(t) + i\sin(t))$$

denote the observation and source points, respectively.

5.7:

$$G(r,\phi;s,t) = \frac{2}{\alpha} \sum_{n=1}^{\infty} g_n(r,s) \sin(\nu\phi) \sin(\nu t)$$

where

$$g_n(r,s) = \begin{cases} I_\nu(kr)K_\nu(ks), & r \leq s \\ I_\nu(ks)K_\nu(kr), & r \geq s \end{cases} \qquad \nu = \frac{(2n-1)\pi}{\alpha}$$

As a particular case of the above representation (as $\alpha = \pi$), one obtains the Green's function

$$G(r,\phi;s,t) = \frac{1}{2\pi}\left[K_0(k|z-\zeta|) - K_0(k|z-\bar{\zeta}|)\right]$$

$$- \frac{2}{\pi} \sum_{n=1}^{\infty} g_n^*(r,s) \sin(2n\phi) \sin(2nt)$$

of the mixed (Dirichlet–Newmann) problem for a half-plane, where

$$g_n^*(r,s) = \begin{cases} I_{2n}(kr)K_{2n}(ks), & r \leq s \\ I_{2n}(ks)K_{2n}(kr), & r \geq s \end{cases}$$

5.8:

(a) $G(r,\phi;s,t) = \dfrac{1}{2\pi}\left[K_0(k|z-\zeta|) - K_0(k|z-\bar{\zeta}|)\right]$

$$- \frac{2}{\pi}\sum_{n=1}^{\infty}[h_n^*(r,s)\sin((2n-1)\phi)\sin((2n-1)t) + g_n^*(r,s)\sin(2n\phi)\sin(2nt)],$$

where

$$h_n^*(r,s) = \frac{I_{2n-1}(kr)I_{2n-1}(ks)K_{2n-1}(kR)}{I_{2n-1}(kR)}$$

and

$$g_n^*(r,s) = \begin{cases} I_{2n}(kr)K_{2n}(ks), & r \leq s \\ I_{2n}(ks)K_{2n}(kr), & r \geq s \end{cases}$$

(b) $G(r,\phi;s,t) = \dfrac{1}{2\pi}\left[K_0(k|z-\zeta|) - K_0(k|z-\bar{\zeta}|)\right]$

$$- \frac{2}{\pi}\sum_{n=1}^{\infty}[h_n^*(r,s)\sin((2n-1)\phi)\sin((2n-1)t) + g_n^*(r,s)\sin(2n\phi)\sin(2nt)],$$

where

$$h_n^*(r,s) = \frac{I_{2n-1}(kr)I_{2n-1}(ks)K'_{2n-1}(kR)}{I'_{2n-1}(kR)}$$

and

$$g_n^*(r,s) = \begin{cases} I_{2n}(kr)K_{2n}(ks), & r \leq s \\ I_{2n}(ks)K_{2n}(kr), & r \geq s \end{cases}$$

(c) $G(r,\phi;s,t) = \dfrac{1}{2\pi}\left[K_0(k|z-\zeta|) - K_0(k|z-\overline{\zeta}|)\right]$

$$-\dfrac{2}{\pi}\sum_{n=1}^{\infty}[h_n^*(r,s)\sin((2n-1)\phi)\sin((2n-1)t) + g_n^*(r,s)\sin(2n\phi)\sin(2nt)],$$

where

$$h_n^*(r,s) = \dfrac{I_{2n-1}(kr)I_{2n-1}(ks)[K_{2n-1}'(kR) + \beta K_{2n-1}(kR)]}{I_{2n-1}'(kR) + \beta I_{2n-1}(kR)}$$

and

$$g_n^*(r,s) = \begin{cases} I_{2n}(kr)K_{2n}(ks), & r \le s \\ I_{2n}(ks)K_{2n}(kr), & r \ge s \end{cases}$$

In exercises 5.9(a) and 5.9(b), we use the following notations:

$$g_0^*(r,s) = \begin{cases} I_0(kr)K_0(ks), & r \le s \\ I_0(ks)K_0(kr), & r \ge s, \end{cases} \qquad \varepsilon_n = \begin{cases} 1, & n=0 \\ 2, & n>0 \end{cases}$$

5.9:

(a) $G(r,\phi;s,t) = \dfrac{1}{2\pi}\Bigg[K_0(k|z-\zeta|) + g_0^*(r,s)$

$$-\sum_{n=0}^{\infty}\varepsilon_n \dfrac{I_n(kr)I_n(ks)K_n(kR)}{I_n(kR)}\cos(n(\phi-t))\Bigg],$$

(b) $G(r,\phi;s,t) = \dfrac{1}{2\pi}\Bigg[K_0(k|z-\zeta|) + g_0^*(r,s)$

$$-\sum_{n=0}^{\infty}\dfrac{\varepsilon_n I_n(kr)I_n(ks)}{I_n'(kR) + \beta I_n(kR)}[K_n'(kR) + \beta K_n(kR)]\cos(n(\phi-t))\Bigg]$$

CHAPTER 6: Problems of Solid Mechanics

6.1:
$$G(x,y;s,t) = \dfrac{2}{\pi}\sum_{n=1}^{\infty}\dfrac{1+\nu|x-s|}{4\nu^3}e^{-\nu|x-s|}\sin(\nu y)\sin(\nu t), \qquad \nu = \dfrac{n\pi}{b}$$

6.2:

(a) $G(x,y;s,t) = \dfrac{1}{b}\sum_{n=1}^{\infty}\dfrac{g_n(x,s)}{\nu^3 \sinh(2\nu a)}\sin(\nu y)\sin(\nu t), \qquad \nu = \dfrac{n\pi}{b},$

where, for $x \le s$, the coefficient $g_n(x,s)$ is defined as

$$g_n(x,s) = \sinh(\nu a)\{\sinh(\nu x)\cosh(\nu(s-a)) - \sinh(\nu(s-a))$$
$$\times[\nu s \sinh(\nu x) + \nu x \cosh(\nu x)]\} + \nu a \sinh(\nu x)\sinh(\nu s),$$

while the expression for $g_n(x,s)$ that is valid for $x \ge s$ can be obtained from that above by interchanging x with s;

APPENDIX B: ANSWERS AND COMMENTS 439

(c) $G(x,y;s,t) = \dfrac{1}{4b}\sum_{n=1}^{\infty}\dfrac{g_n(x,s)}{\nu^3[1+2\nu^2a^2-\cosh(2\nu a)]}\sin(\nu y)\sin(\nu t),$

where $\nu = n\pi/b$ and the coefficient $g_n(x,s)$ is defined, for $x\le s$, as

$$g_n(x,s) = 2\sinh(\nu x)\left\{e^{\nu s}\left[(\nu s-1)(1+2\nu a-e^{-2\nu a})-2\nu^2 a^2\right]\right.$$

$$-e^{-\nu s}\left[(\nu s+1)(1-2\nu a-e^{2\nu a})+2\nu^2 a^2\right]\right\}$$

$$+\nu x e^{-\nu x}\left\{e^{\nu s}\left[1+2\nu a+4\nu^2 a(a-s)\right]\right.$$

$$+e^{-\nu s}\left[(1+2\nu s)(1-e^{2\nu a})-2\nu a\right]\right\}$$

$$-\nu x e^{\nu x}\left\{e^{\nu s}\left[(2\nu s-1)(1-e^{-2\nu a})-2\nu a\right]\right.$$

$$-e^{-\nu s}\left[1-2\nu a+4\nu^2 a(a-s)-e^{2\nu a}\right]\right\},$$

while the expression for $g_n(x,s)$ that is valid for $x\ge s$ can be obtained from that above by interchanging x with s;

(d) $G(x,y;s,t) = \dfrac{1}{b}\sum_{n=1}^{\infty}\dfrac{g_n(x,s)}{\nu^3[\sinh(2\nu a)-2\nu a]}\sin(\nu y)\sin(\nu t),\quad \nu = \dfrac{n\pi}{b},$

where, for $x\le s$, the coefficient $g_n(x,s)$ is defined as

$$g_n(x,s) = \nu x\cosh(\nu x)\left[2\nu(s-a)\cosh(\nu s)-\sinh(\nu s)-\sinh(\nu(s-2a))\right]$$

$$-\sinh(\nu x)\left[\nu s\cosh(\nu(s-2a))-\sinh(\nu(s-2a))\right.$$

$$+\nu(s-2a)\cosh(\nu s)+(2\nu^2 a(s-a)-1)\sinh(\nu s)\right]$$

while the expression for $g_n(x,s)$ that is valid for $x\ge s$ can be obtained from that above by interchanging x with s.

6.6: The entries $G_{ij}(x,y;,s,t)$, $(i,j=1,2)$ of the influence matrix are found in this case in the form:

$$G_{11}(x,y;s,t) = \dfrac{2}{b}\left[\dfrac{1}{2}g_{11}^0(x,s)+\dfrac{1}{\lambda+2\mu}\sum_{n=1}^{\infty}g_{11}^n(x,s)\cos(\nu y)\cos(\nu t)\right],$$

$$G_{12}(x,y;s,t) = \dfrac{2}{b\mu}\sum_{n=1}^{\infty}g_{12}^n(x,s)\cos(\nu y)\sin(\nu t),$$

$$G_{21}(x,y;s,t) = \dfrac{2}{b(\lambda+2\mu)}\sum_{n=1}^{\infty}g_{21}^n(x,s)\sin(\nu y)\cos(\nu t),$$

and
$$G_{22}(x,y;s,t) = \frac{2}{b\mu} \sum_{n=1}^{\infty} g_{22}^n(x,s) \sin(\nu y) \sin(\nu t),$$

where λ and μ represent Lame coefficients of the material of which the semi-strip $\Omega_{ss} = \{(x,y): 0 < x < \infty, 0 < y < b\}$ is composed and the coefficients $g_{ij}^n(x,s)$ are defined, for $x \le s$, as

$$g_{11}^n(x,s) = \frac{\lambda+\mu}{2\mu} e^{-\nu s} \left[x \left(\cosh(\nu x) + \frac{\lambda+\mu}{\lambda+3\mu} \nu s e^{-\nu x} \right) \right.$$
$$\left. - \left(s + \frac{\lambda+3\mu}{\nu(\lambda+\mu)} \right) \sinh(\nu x) \right],$$

$$g_{12}^n(x,s) = \frac{\lambda+\mu}{2(\lambda+2\mu)} e^{-\nu s} \left[\frac{\lambda+\mu}{\lambda+3\mu} \nu x s e^{-\nu x} - (s-x)\sinh(\nu x) \right],$$

$$g_{21}^n(x,s) = \frac{\lambda+\mu}{2\mu} e^{-\nu s} \left[(s-x)\sinh(\nu x) + \frac{\lambda+\mu}{\lambda+3\mu} \nu x s e^{-\nu x} \right],$$

and

$$g_{22}^n(x,s) = \frac{\lambda+\mu}{2(\lambda+2\mu)} e^{-\nu s} \left[(s-x)\cosh(\nu x) - \frac{\lambda+3\mu}{\nu(\lambda+\mu)} \sinh(\nu x) \right.$$
$$\left. + s \left(\frac{\lambda+\mu}{\lambda+3\mu} \nu x - 1 \right) e^{-\nu x} \right].$$

The component $g_{11}^0(x,s)$ of $G_{11}(x,y;s,t)$ is found as

$$g_{11}^0(x,s) = -\frac{1}{\lambda+2\mu} \begin{cases} x, & x \le s \\ s, & x \ge s. \end{cases}$$

Note that the series in the above expressions for $G_{ij}(x,y;s,t)$ are summable and their sums can readily be obtained with the aid of the technique widely used in Chapter 5.

6.7: The entries $G_{ij}(x,y;,s,t)$, $(i,j=1,2)$ of the influence matrix are found in this case in the form:

$$G_{11}(x,y;s,t) = \frac{2}{b} \left[\frac{1}{2} g_{11}^0(x,s) - \frac{a}{\lambda+2\mu} \sum_{n=1}^{\infty} \frac{g_{11}^n(x,s)}{\Delta^*} \cos(\nu y)\cos(\nu t) \right],$$

$$G_{12}(x,y;s,t) = -\frac{2a}{b\mu} \sum_{n=1}^{\infty} \frac{g_{12}^n(x,s)}{\Delta^*} \cos(\nu y)\sin(\nu t),$$

$$G_{21}(x,y;s,t) = -\frac{2a}{b(\lambda+2\mu)} \sum_{n=1}^{\infty} \frac{g_{21}^n(x,s)}{\Delta^*} \sin(\nu y)\cos(\nu t),$$

and

$$G_{22}(x,y;s,t) = -\frac{2a}{b\mu} \sum_{n=1}^{\infty} \frac{g_{22}^n(x,s)}{\Delta^*} \sin(\nu y)\sin(\nu t),$$

where λ and μ represent Lame coefficients of the material of which the rectangle $\Omega_r = \{(x, y): 0 < x < a, 0 < y < b\}$ is composed, the coefficient Δ^* is given as

$$\Delta^* = \left[1 - \left(\frac{\lambda+3\mu}{\nu a(\lambda+\mu)}\right)^2 \sinh(\nu a)\right],$$

and the functions $g_{ij}^n(x, s)$ are defined, for $x \leq s$, as

$$g_{11}^n(x, s) = S_{11}^n(a, s)\Phi_1^n(x, a) + S_{21}^n(a, s)\Phi_2^n(x, a),$$

$$g_{12}^n(x, s) = S_{12}^n(a, s)\Phi_1^n(x, a) + S_{22}^n(a, s)\Phi_2^n(x, a),$$

$$g_{21}^n(x, s) = S_{11}^n(a, s)\Phi_3^n(x, a) + S_{21}^n(a, s)\Phi_4^n(x, a),$$

and

$$g_{22}^n(x, s) = S_{12}^n(a, s)\Phi_3^n(x, a) + S_{22}^n(a, s)\Phi_4^n(x, a),$$

while for $x \geq s$, we have

$$g_{11}^n(x, s) = S_{11}^n(a, x)\Phi_1^n(s, a) + S_{21}^n(a, x)\Phi_2^n(s, a),$$

$$g_{12}^n(x, s) = S_{12}^n(a, x)\Phi_1^n(s, a) + S_{22}^n(a, x)\Phi_2^n(s, a),$$

$$g_{21}^n(x, s) = S_{11}^n(a, x)\Phi_3^n(s, a) + S_{21}^n(a, x)\Phi_4^n(s, a),$$

and

$$g_{22}^n(x, s) = S_{12}^n(a, x)\Phi_3^n(s, a) + S_{22}^n(a, x)\Phi_4^n(s, a),$$

The functions $S_{ij}^n(a, s)$ are found in the form

$$S_{11}^n(a, s) = \frac{1}{2a\mu}\left[\frac{\lambda+3\mu}{\nu}\sinh(\nu(a-s)) - (\lambda+\mu)(a-s)\cosh(\nu(a-s))\right],$$

$$S_{12}^n(a, s) = \frac{\lambda+\mu}{2a(\lambda+2\mu)}(s-a)\sinh(\nu(a-a)),$$

$$S_{21}^n(a, s) = \frac{\lambda+\mu}{2a\mu}(a-s)\sinh(\nu(a-a)),$$

and

$$S_{22}^n(a, s) = \frac{1}{2a(\lambda+2\mu)}\left[\frac{\lambda+3\mu}{\nu}\sinh(\nu(a-s)) + (\lambda+\mu)(a-s)\cosh(\nu(a-s))\right],$$

while for the functions $\Phi_i^n(x, a)$, we have

$$\Phi_1^n(x, a) = \frac{x}{a}[\cosh(\nu(x-a)) + \omega\sinh(\nu a)\cosh(\nu x)]$$

$$- \omega\sinh(\nu x)[\cosh(\nu a) + \omega\sinh(\nu a)],$$

$$\Phi_2^n(x, a) = \frac{1}{a}[\omega(x-a)\sinh(\nu a)\sinh(\nu x) - x\sinh(\nu(x-a))],$$

$$\Phi_3^n(x,a) = -\frac{x}{a}[\sinh(\nu(x-a)) + \omega \sinh(\nu a)\sinh(\nu x)]$$
$$- \omega[\cosh(\nu a)\cosh(\nu x) - \cosh(\nu(x-a))],$$

and

$$\Phi_4^n(x,a) = \frac{x}{a}[\cosh(\nu(x-a)) - \omega \sinh(\nu a)\cosh(\nu x)]$$
$$+ \omega \sinh(\nu(x-a)) + [\cosh(\nu x) - \omega \sinh(\nu x)]\sinh(\nu a),$$

where

$$\omega = \frac{\lambda + 3\mu}{\nu a(\lambda + \mu)}.$$

The component $g_{11}^0(x,s)$ of $G_{11}(x,y;s,t)$ is found as

$$g_{11}^0(x,s) = -\frac{1}{a(\lambda+2\mu)}\begin{cases} x(s-a), & x \le s \\ s(x-a), & x \ge s. \end{cases}$$

CHAPTER 7: Compound Media

7.1: The entries $G_{ij}(x,y;,s,t)$, $(i,j=1,2)$ of the influence matrix have individual domains. That is, they are defined for $(x,y) \in \Omega_i$ and $(s,t) \in \Omega_j$ and are found in this case in a closed form as:

$$G_{11}(x,y;s,t) = \frac{1}{2\pi}\left[\ln\frac{E_1(z-\zeta)E_2(z-\overline{\zeta})}{E_2(z-\zeta)E_1(z-\overline{\zeta})}\right.$$
$$\left. + \frac{\lambda-1}{\lambda+1}\ln\frac{E_1(z+\zeta)E_2(z+\overline{\zeta})}{E_2(z+\zeta)E_1(z+\overline{\zeta})}\right],$$

$$G_{12}(x,y;s,t) = \frac{\lambda}{\pi(\lambda+1)}\ln\frac{E_1(z-\zeta)E_2(z-\overline{\zeta})}{E_2(z-\zeta)E_1(z-\overline{\zeta})},$$

$$G_{21}(x,y;s,t) = \frac{1}{\pi(\lambda+1)}\ln\frac{E_1(z-\zeta)E_2(z-\overline{\zeta})}{E_2(z-\zeta)E_1(z-\overline{\zeta})},$$

and

$$G_{22}(x,y;s,t) = \frac{1}{2\pi}\left[\ln\frac{E_1(z-\zeta)E_2(z-\overline{\zeta})}{E_2(z-\zeta)E_1(z-\overline{\zeta})}\right.$$
$$\left. - \frac{\lambda-1}{\lambda+1}\ln\frac{E_1(z+\zeta)E_2(z+\overline{\zeta})}{E_2(z+\zeta)E_1(z+\overline{\zeta})}\right],$$

where $z = x+iy$ and $\zeta = s+it$ represent the observation and the source point, respectively, λ represents the ratio λ_2/λ_1 of the heat conductivities of the materials of which the fragments Ω_2 and Ω_1 are composed, and E_1 and E_2 are introduced as

$$E_1(p) = \left|1 + \exp\left(\frac{\pi p}{2b}\right)\right|, \quad E_2(p) = \left|1 - \exp\left(\frac{\pi p}{2b}\right)\right|.$$

7.2: **(a)** The entries $G_{ij}(x,y;s,t)$, $(i,j=1,2)$ of the influence matrix are defined for $(x,y)\in\Omega_i$ and $(s,t)\in\Omega_j$. After splitting off the singularities available in the diagonal entries, $G_{ij}(x,y;s,t)$ are found in the form:

$$G_{11}(x,y;s,t)=\frac{1}{2\pi}\ln\frac{E(z-\overline{\zeta})[E(z+\zeta)]^k}{E(z-\zeta)[E(z+\overline{\zeta})]^k}$$

$$-\frac{4}{\pi}\sum_{n=1}^{\infty}g_{11}^n(x,s)\sin(\nu y)\sin(\nu t),$$

$$G_{12}(x,y;s,t)=\frac{4}{\pi}\sum_{n=1}^{\infty}g_{12}^n(x,s)\sin(\nu y)\sin(\nu t),$$

$$G_{21}(x,y;s,t)=\frac{4}{\pi}\sum_{n=1}^{\infty}g_{21}^n(x,s)\sin(\nu y)\sin(\nu t),$$

and

$$G_{22}(x,y;s,t)=\frac{1}{2\pi}\ln\frac{E(z-\overline{\zeta})[E(z+\overline{\zeta})]^k}{E(z-\zeta)[E(z+\zeta)]^k}$$

$$-\frac{4}{\pi}\sum_{n=1}^{\infty}g_{22}^n(x,s)\sin(\nu y)\sin(\nu t),$$

where

$$E(p)=\left|1-\exp\left(\frac{\pi p}{b}\right)\right|,\quad \lambda=\frac{\lambda_2}{\lambda_1},\quad k=\frac{1-\lambda}{1+\lambda},\quad \nu=\frac{n\pi}{b}.$$

The coefficients $g_{ij}^n(x,s)$ of the above expansions are defined over individual domains. $g_{11}^n(x,s)$ is given, for $-a\leq x\leq s<0$, as

$$g_{11}^n(x,s)=\frac{\sinh(\nu a)\cosh(\nu x)\cosh(\nu s)+\lambda\cosh(\nu a)\sinh(\nu x)\sinh(\nu s)}{(1+\lambda)ne^{\nu a}\sinh(2\nu a)},$$

while for $-a\leq s\leq x<0$, it can be obtained from that above by interchanging x with s. The coefficient $g_{12}^n(x,s)$ is defined for $-a\leq x\leq 0<s<a$ as

$$g_{12}^n(x,s)=\lambda\frac{\sinh(\nu(x+a))\sinh(\nu(a-s))}{(1+\lambda)n\sinh(2\nu a)}.$$

The coefficient $g_{21}^n(x,s)$ is defined for $-a\leq s\leq 0<x<a$ as

$$g_{21}^n(x,s)=\frac{\sinh(\nu(s+a))\sinh(\nu(a-x))}{(1+\lambda)n\sinh(2\nu a)}.$$

The coefficient $g_{22}^n(x,s)$ is given, for $0\leq x\leq s<a$, as

$$g_{22}^n(x,s)=\frac{\sinh(\nu x)\cosh(\nu a)\sinh(\nu s)+\lambda\cosh(\nu x)\sinh(\nu a)\cosh(\nu s)}{(1+\lambda)ne^{\nu a}\sinh(2\nu a)},$$

while for $0\leq s\leq x<a$, it can be obtained from that above by interchanging x with s.

(b) The entries $G_{ij}(x,y;s,t)$, $(i,j = 1,2)$ of the influence matrix are defined for $(x,y) \in \Omega_i$ and $(s,t) \in \Omega_j$. After splitting off the singularities available in the diagonal entries, $G_{ij}(x,y;s,t)$ are found in the form:

$$G_{11}(x,y;s,t) = \frac{1}{2\pi} \ln \frac{E(z+\bar{\zeta})[E(z-\bar{\zeta})]^k}{E(z+\zeta)[E(z-\zeta)]^k}$$

$$+ \frac{2}{\pi} \sum_{n=1}^{\infty} g_{11}^n(x,s) \sin(\nu y) \sin(\nu t),$$

$$G_{12}(x,y;s,t) = \frac{2}{\pi} \sum_{n=1}^{\infty} g_{12}^n(x,s) \sin(\nu y) \sin(\nu t),$$

$$G_{21}(x,y;s,t) = \frac{2}{\pi} \sum_{n=1}^{\infty} g_{21}^n(x,s) \sin(\nu y) \sin(\nu t),$$

and

$$G_{22}(x,y;s,t) = \frac{1}{2\pi} \ln \frac{E(z+\zeta)[E(z-\bar{\zeta})]^k}{E(z+\bar{\zeta})[E(z-\zeta)]^k}$$

$$- \frac{2}{\pi} \sum_{n=1}^{\infty} g_{22}^n(x,s) \sin(\nu y) \sin(\nu t),$$

where

$$E(p) = \left| 1 - \exp\left(\frac{\pi p}{b}\right) \right|, \quad \lambda = \frac{\lambda_2}{\lambda_1}, \quad k = \frac{1-\lambda}{1+\lambda}, \quad \nu = \frac{n\pi}{b}.$$

The coefficient $g_{11}^n(x,s)$ is given, for $-a \le x \le s < 0$, as

$$g_{11}^n(x,s) = \frac{\lambda e^{\nu(x+s)} \cosh^2(\nu a)}{n(\sinh^2(\nu a) + \cosh^2(\nu a))}$$

$$- \frac{\sinh(\nu a) \sinh(\nu x) \sinh(\nu s) + \lambda \cosh(\nu a) \cosh(\nu x) \cosh(\nu s)}{n e^{\nu a}(\sinh^2(\nu a) + \cosh^2(\nu a))},$$

while for $-a \le s \le x < 0$, it can be obtained from that above by interchanging x with s. The coefficient $g_{12}^n(x,s)$ is defined for $-a \le x \le 0 < s < a$ as

$$g_{12}^n(x,s) = \lambda \frac{\sinh(\nu(x+a)) \cosh(\nu(a-s))}{n(\sinh^2(\nu a) + \cosh^2(\nu a))}.$$

The coefficient $g_{21}^n(x,s)$ is defined for $-a \le s \le 0 < x < a$ as

$$g_{21}^n(x,s) = \frac{\lambda \cosh(\nu(a-x)) \sinh(\nu(a+s))}{n(\sinh^2(\nu a) + \cosh^2(\nu a))}.$$

The coefficient $g_{22}^n(x,s)$ is given, for $0 \le x \le s < a$, as

$$g_{22}^n(x,s) = \frac{\lambda e^{\nu(a-x-s)} \cosh(\nu a)}{2n(\sinh^2(\nu a) + \cosh^2(\nu a))}$$

APPENDIX B: ANSWERS AND COMMENTS 445

$$-\frac{\sinh(\nu a)\cosh(\nu x)\cosh(\nu s)+\lambda\cosh(\nu a)\sinh(\nu x)\sinh(\nu s)}{n e^{\nu a}(\sinh^2(\nu a)+\cosh^2(\nu a))},$$

while for $0 \leq s \leq x < a$, it can be obtained from that above by interchanging x with s.

7.4: The entries $G_{ij}(r,\phi;,s,t)$, $(i,j=1,2)$ of the influence matrix are defined for $(r,\phi) \in \Omega_i$ and $(s,t) \in \Omega_j$. After splitting off the singularities in the diagonal entries, $G_{ij}(r,\phi;s,t)$ are found in the form:

$$G_{11}(r,\phi;s,t) = \frac{1}{2\pi}\ln\frac{|z-\overline{\zeta}|}{|z-\zeta|} + \frac{2}{\pi}\sum_{n=1}^{\infty} g_{11}^n(r,s)\sin(n\phi)\sin(nt),$$

$$G_{12}(r,\phi;s,t) = \frac{2}{\pi}\sum_{n=1}^{\infty} g_{12}^n(r,s)\sin(n\phi)\sin(nt),$$

$$G_{21}(r,\phi;s,t) = \frac{2}{\pi}\sum_{n=1}^{\infty} g_{21}^n(r,s)\sin(n\phi)\sin(nt),$$

and

$$G_{22}(r,\phi;s,t) = \frac{1}{2\pi}\ln\frac{|z-\overline{\zeta}|}{|z-\zeta|} + \frac{2}{\pi}\sum_{n=1}^{\infty} g_{22}^n(r,s)\sin(n\phi)\sin(nt),$$

where $z=r(\cos(\phi)+i\sin(\phi))$ and $\zeta=s(\cos(t)+i\sin(t))$ represent the observation and the source point, respectively, and λ denotes the ratio of the heat conductivities of the materials of which Ω_2 and Ω_1 are composed, $\lambda=\lambda_2/\lambda_1$. The coefficient $g_{11}^n(r,s)$ is defined, for $0 \leq r,s \leq a$, as

$$g_{11}^n(r,s) = -\frac{(1+\lambda)a^{2n}+(1-\lambda)b^{2n}}{2n\Delta^* a^{2n} b^{2n}} r^n s^n.$$

The coefficient $g_{12}^n(r,s)$ is defined, for $0 \leq r \leq a \leq s \leq b$, as

$$g_{12}^n(r,s) = \frac{\lambda}{n\Delta^*}\left[\left(\frac{r}{s}\right)^n + \left(\frac{rs}{b^2}\right)^n\right].$$

The coefficient $g_{21}^n(r,s)$ is defined, for $0 \leq s \leq a \leq r \leq b$, as

$$g_{21}^n(r,s) = \frac{1}{n\Delta^*}\left[\left(\frac{s}{r}\right)^n + \left(\frac{rs}{b^2}\right)^n\right].$$

The coefficient $g_{22}^n(r,s)$ is defined, for $a \leq r \leq s \leq b$, as

$$g_{22}^n(r,s) = -\frac{1}{2n\Delta^*}\left\{(1-\lambda)\left(\frac{a}{b}\right)^{2n}\left[\left(\frac{r}{s}\right)^n+\left(\frac{s}{r}\right)^n\right]\right.$$

$$\left.+\left[(1+\lambda)\left(\frac{rs}{b^2}\right)^n-(1-\lambda)\left(\frac{a^2}{rs}\right)^n\right]\right\},$$

while for $a \leq s \leq r \leq a$, it can be obtained from that above by interchanging r with s. The parameter Δ^* is defined as

$$\Delta^* = \left[(1-\lambda)(a/b)^{2n}+(1+\lambda)\right].$$

7.6: The entries $G_{ij}(\phi,\theta;\psi,\tau)$, $(i,j=1,2)$ of the influence matrix are defined for $(\phi,\theta) \in \Omega_i$ and $(\psi,\tau) \in \Omega_j$. After the series expansions that represent $G_{ij}(\phi,\theta;\psi,\tau)$ are summed up, the latter are found in a closed form as:

$$G_{11}(\phi,\theta;\psi,\tau) = \frac{1}{4\pi} \ln \frac{\Phi_0^{-2} - 2\Phi\Psi\Theta + (\Phi\Psi\Phi_0)^2}{\Phi^2 - 2\Phi\Psi\Theta + \Psi^2},$$

$$G_{12}(\phi,\theta;\psi,\tau) = \frac{1}{4\pi} \ln \frac{1 - 2\Phi\Psi\Phi_0\Theta + (\Phi\Psi\Phi_0)^2}{\Phi^2 - 2\Phi\Psi\Phi_0\Theta + (\Psi\Phi_0)^2},$$

$$G_{21}(\phi,\theta;\psi,\tau) = \frac{1}{4\pi} \ln \frac{1 - 2\Phi\Psi\Phi_0\Theta + (\Phi\Psi\Phi_0)^2}{\Psi^2 - 2\Phi\Psi\Phi_0\Theta + (\Phi\Phi_0)^2},$$

and

$$G_{22}(\phi,\theta;\psi,\tau) = \frac{1}{4\pi} \ln \frac{1 - 2\Phi\Psi\Theta + (\Phi\Psi)^2}{\Phi^2 - 2\Phi\Psi\Theta + \Psi^2},$$

where, for the sake of compactness, we use the following notations

$$\Phi = \tan\left(\frac{\phi}{2}\right), \quad \Psi = \tan\left(\frac{\psi}{2}\right),$$

$$\Phi_0 = \tan\left(\frac{\phi_0}{2}\right), \quad \Theta = \cos(\theta - \tau).$$

CHAPTER 8: Heat Equation

8.1:
$$G(x,t;s) = \frac{1}{2\sqrt{\pi\kappa t}}\left[\exp\left(-\frac{(x-s)^2}{4\kappa t}\right) + \exp\left(-\frac{(x+s)^2}{4\kappa t}\right)\right].$$

8.2:
$$G(x,t;s) = \frac{1}{2\sqrt{\pi\kappa t}}\sum_{n=-\infty}^{\infty}(-1)^n\left[\exp\left(-\frac{(x-s+2na)^2}{4\kappa t}\right) + \exp\left(-\frac{(x+s+2na)^2}{4\kappa t}\right)\right].$$

In exercises 8.3 through 8.5, $\theta_3(u,v)$ denotes the Jacobi Theta function of the third kind whose series expansion is given [1, 29, 68] as

$$\theta_3(u,v) = 1 + 2\sum_{n=1}^{\infty}\exp(-n^2\pi^2 v)\cos(2n\pi u).$$

APPENDIX B: ANSWERS AND COMMENTS

8.3:
$$G(x,y,t;\xi,\eta) = \frac{1}{4b\sqrt{\pi\kappa t}} \exp\left(-\frac{(x-s)^2}{4\kappa t}\right)$$
$$\times \left[\vartheta_3\left(\frac{y-\eta}{2b}, \frac{\kappa t}{b^2}\right) - \vartheta_3\left(\frac{y+\eta}{2b}, \frac{\kappa t}{b^2}\right)\right].$$

8.4:
$$G(x,y,t;\xi,\eta) = \frac{1}{4b\sqrt{\pi\kappa t}}\left[\exp\left(-\frac{(x-s)^2}{4\kappa t}\right) + \exp\left(-\frac{(x+s)^2}{4\kappa t}\right)\right]$$
$$\times \left[\vartheta_3\left(\frac{y-\eta}{2b}, \frac{\kappa t}{b^2}\right) - \vartheta_3\left(\frac{y+\eta}{2b}, \frac{\kappa t}{b^2}\right)\right].$$

8.5: The entries $G_{ij}(x,y,t;\xi,\eta)$, $(i,j=1,2)$ of the influence matrix are defined for $(x,y) \in \Omega_i$ and $(\xi,\eta) \in \Omega_j$ and are found in a closed form as:

$$G_{11}(x,y,t;\xi,\eta) = \frac{1}{4(1+\lambda)b\sqrt{\pi\kappa t}}\left[(1-\lambda)\exp\left(\frac{(x+s)^2}{4\kappa t}\right)\right.$$
$$\left.+(1+\lambda)\exp\left(-\frac{(x-s)^2}{4\kappa t}\right)\right]\left[\vartheta_3\left(\frac{y-\eta}{2b}, \frac{\kappa t}{b^2}\right) - \vartheta_3\left(\frac{y+\eta}{2b}, \frac{\kappa t}{b^2}\right)\right],$$

$$G_{12}(x,y,t;\xi,\eta) = \frac{\lambda}{4(1+\lambda)b\sqrt{\pi\kappa t}}\exp\left(-\frac{(x-s)^2}{4\kappa t}\right)$$
$$\times \left[\vartheta_3\left(\frac{y-\eta}{2b}, \frac{\kappa t}{b^2}\right) - \vartheta_3\left(\frac{y+\eta}{2b}, \frac{\kappa t}{b^2}\right)\right],$$

$$G_{21}(x,y,t;\xi,\eta) = \frac{1}{4(1+\lambda)b\sqrt{\pi\kappa t}}\exp\left(-\frac{(x-s)^2}{4\kappa t}\right)$$
$$\times \left[\vartheta_3\left(\frac{y-\eta}{2b}, \frac{\kappa t}{b^2}\right) - \vartheta_3\left(\frac{y+\eta}{2b}, \frac{\kappa t}{b^2}\right)\right],$$

and

$$G_{22}(x,y,t;\xi,\eta) = \frac{1}{4(1+\lambda)b\sqrt{\pi\kappa t}}\left[(1+\lambda)\exp\left(-\frac{(x-s)^2}{4\kappa t}\right)\right.$$
$$\left.-(1-\lambda)\exp\left(-\frac{(x+s)^2}{4\kappa t}\right)\right]\left[\vartheta_3\left(\frac{y-\eta}{2b}, \frac{\kappa t}{b^2}\right) - \vartheta_3\left(\frac{y+\eta}{2b}, \frac{\kappa t}{b^2}\right)\right],$$

where λ represents a ratio $\lambda = h_2/h_1$ of the heat conductivities of the materials of which the fragments Ω_2 and Ω_1 are composed.

References

1. Abramovitz, M. and Stegun, I. (eds) (1972) *Handbook of Mathematical Functions*, 10th ed., Washington, D.C., National Bureau of Standards.

2. Achenbach, J.D. and Wang, C.-Y. (1994) 2-D time domain BEM for scattering of elastic waves in anisotropic solids, *Proc. of the Int. Conference BETECH-94*, Orlando, University of Central Florida, 157-163.

3. Andrews, L.C. (1985) *Special Functions for Engineers and Applied Mathematicians*, New York, Macmillan.

4. Atkinson, K. (1989) *An Introduction to Numerical Analysis*, 2nd ed., New York, John Wiley.

5. Banerjee, P.K. and Butterfield, R. (1981) *Boundary Element Method in Engineering Science*, London, McGraw-Hill.

6. Barton, G. (1989) *Elements of Green's Functions and Propagation*, Oxford, Clarendon Press.

7. Beck, J.V., [et al]. (1992) *Heat Conduction Using Green's Functions*, London-Washington, DC-Philadelphia, Hemisphere Publishing Corp.

8. Berger, J.R. (1994) Boundary element analysis of anisotropic bimaterials with special Green's functions, *Engineering Analysis with Boundary Elements*, **14**, 123-131.

9. Bergman, S. and Schiffer, M. (1953) *Kernel Functions and Elliptic Differential Equations in Mathematical Physics*, New York, Academic Press.

10. Biggs, J.M. (1986) *Introduction to Structural Engineering Analysis and Design*, N.J., Prentice–Hall, Englewood Cliffs.

11. Blevins, R.D. (1979) *Formulas for Natural Frequency and Mode Shape*, New York, Van Nostrand Reinhold.

12. Boley, B.A. (1956) A method for the construction of Green's functions, *Quarterly of Applied Mathematics*, **14**, 249-257.

13. Brebbia, C.A. (1978) *The Boundary Element Method for Engineers*, London, Pentech Press; New York, Halstead Press.

14. Butkovsky, A.G. (1982) *Green's Functions and Transfer Functions Handbook*, Translation from Russian, New York, Halstead Press.

15. Chen-To, Tai (1994) *Dyadic Green Functions in Electromagnetic Theory*, 2nd ed., New York, IEEE Press.

16. Collatz, L. (1960) *Numerical Treatment of Differential Equations*, 3rd ed., Berlin, Springer.

17. Courant, R. and Hilbert, D. (1953) *Methods of Mathematical Physics*, Vols. 1 and 2, New York, Interscience.

18. Crandall, S.H., Dahl, N.C., and Lardner, T.J. (1972) *An Introduction to the Mechanics of Solids*, 2nd ed., New York, McGraw-Hill.

19. Cruse, T.A. (1977) Mathematical formulation of the boundary integral equations method in solid mechanics, *Report No. AFOSR-TR-77-1002*, Pratt and Whitney Aircraft Group.

20. Cruse, T.A. (1988) *Boundary Element Analysis in Computational Fracture Mechanics*, Dordreicht, Kluwer Academic Publisher.

21. Davydov, I.A., Melnikov, Yu.A., and Nikulyn, V.A. (1978) Green's functions of a steady state heat conduction operator for some shells of revolution, *J. of Engineering Physics*, **34**, 723-728 (in Russian).

22. Denda, M. (1989) Two-dimensional singularity solutions in the presence of an elliptic hole, *Journal of Applied Mechanics*, **56**, 231-233.

23. Dennery, Ph. and Krzywicki, A. (1967) *Mathematics for Physicists*, New York, Harper and Row.

24. Dolgova, I.M. and Melnikov, Yu.A. (1978) Construction of Green's functions and matrices for equations and systems of the elliptic type, *Translation from Russian PMM (J. Appl. Math. Mech.)*, **42**, 740-746.

25. Dolgova, I.M. and Melnikov, Yu.A. (1989) Green's matrix for the plane problem in theory of elasticity for an orthotropic strip, *Translation from Russian PMM (J. of Appl. Math. Mech.)*, **53**, 102-106.

26. Economou, E.N. (1983) *Green's Functions in Quantum Physics* (2nd ed.), Berlin, Springer-Verlag.

27. Elias, Z.M. (1986) *Theory and Methods of Structural Analysis*, New York, John Wiley.

REFERENCES

28. Gavelya, S.P. (1969) On one method of construction of Green's matrices for joint shells, *Reports of the Ukrainian Academy of Sciences*, Ser. A, 12, 1107-1111 (in Russian).

29. Gradstein, I.S. and Ryzhik, I.M. (1980) *Tables of Integrals, Series, and Products*, New York, Academic Press.

30. Greenberg, M.D. (1971) *Application of Green's Functions in Science and Engineering*, N. J., Prentice-Hall, Englewood Cliffs.

31. Haberman, R. (1987) *Elementary Applied Partial Differential Equations*, N. J., Prentice-Hall, Englewood Cliffs.

32. Hasebe, N., Jun Quin, and Yizhou Chen (1996) Fundamental solutions for half plane with an oblique edge crack, *Engineering Analysis with Boundary Elements*, **17**, 263-267.

33. Hwu, C. and Yen, W. (1991) Green's functions of two-dimensional anisotropic plate containing an elliptic hole, *Int. Journal of Solids and Structures*, **27**, 1705-1719.

34. Irschik, H. and Ziegler, F. (1980) Application of the Green's function method to thin elastic polygonal plates, *Acta Mechanica*, **39**.

35. Kamke, E. (1961) *Differentialgleichungen, Losungmethoden und Losungen*, 7th ed., Leipzig, Akademishe Verlagsgesellschaft.

36. Kantorovich, L.V. and Krylov, V.I. (1964) *Approximate Methods of Higher Analysis*, New York, Interscience.

37. Kupradze, V.D. (1965) *Potential Method in the Theory of Elasticity*, New York, Davey.

38. Lebedev, N.N., Skal'skaya, I.P., and Uflyand, Ya.S. (1966) *Problems of Mathematical Physics*, New York, Pergamon Press.

39. Lekhnitskii, S.G. (1963) *Theory of Elasticity for an Anisotropic Body*, New York, Holden Day.

40. Mahan, G.D. (1990) *Many-Particle Physics*, 2nd ed., New York-London, Plenum Press.

41. Melnikov, Yu.A. (1970) Computation of heat potentials, *Proc. of the V Ukrainian Conference for Graduate Students in Maths*, Kiev, Ukrainian Academy of Sciences Publishers, 221-222 (in Russian).

42. Melnikov, Yu.A. (1977) Some applications of the Green's function method in mechanics, *Int. J. Solids and Structures*, **13**, 1045-1058.

43. Melnikov, Yu.A. (1982) A basis for computation of thermo-mechanical fields in elements of constructions of complex configuration, *Thesis Dr. Tech. Sciences*, Moscow Institute of Civil Engineering (in Russian).

44. Melnikov, Yu.A. (1991) *Green's Functions and Matrices for Elliptic Equations and Systems*, Dniepropetrovsk, Dniepropetrovsk State University Publishers (in Russian).

45. Melnikov, Yu.A. (1995) *Green's Functions in Applied Mechanics*, Boston-Southampton, Computational Mechanics Publications.

46. Melnikov, Yu.A. (1996) Influence functions for 2-D compound regions of complex configuration, *Computational Mechanics*, **17**, 297-305.

47. Melnikov, Yu.A. (1998) Green's function formalism extended to systems of mechanical differential equations posed on graphs, *Journal of Engineering Mathematics*, **34**, 3.

48. Melnikov, Yu.A., Bobylyov, A.A., and Shubenko, V.V. (1995) Green's function approach to the nonlinear bending of closely-spaced plates, *Int. J. Solids and Structures*, **32**, 1771-1791.

49. Melnikov, Yu.A. and Bobylov, Ye.A. (1996) Green's function method solution of the Reissner's plate problem, *Engineering Analysis with Boundary Elements*, **17**, 255-262.

50. Melnikov, Yu.A. and Hall, M.T. (1997) Influence matrices for systems of equations posed on graphs, *Proc. of the Int. Conference BETECH-97*, Knoxville, University of Tennessee, 289-298.

51. Melnikov, Yu.A., Hughes, S., and McDaniel, S. (1996) Boundary element approach based on Green's function, *Proc. of the Int. Conference BETECH-96*, University of Hawaii, 166-174.

52. Melnikov, Yu.A. and Koshnarjova, V.A. (1994) Green's matrices and 2-D elasto-potentials for external boundary value problems, *Applied Mathematical Modelling*, **18**, 161-167.

53. Melnikov, Yu.A. and Krasnikova, R.D. (1981) *Construction of Green's Functions for Problems of Mathematical Physics*, Dniepropetrovsk State University Publishers (in Russian).

54. Melnikov, Yu.A., Loboda, V.V., and Govorukha, V.B. (1997) Field of potential in a compound rectangle containing a linear inclusion, *Quarterly of Applied Mathematics*, **55**, 299-311.

55. Melnikov, Yu.A. and Shirley, K.L. (1994) Matrices of Green's type for the potential equation on a cylindrical surface joined to a hemisphere, *Applied Mathematics and Computations*, **65**, 241-252.

56. Melnikov, Yu.A. and Powell, J.O. (1997) A Green's function method for detection of a cavity from a boundary measurement, *Proc. of the Int. Conference BETECH-97*, Knoxville, University of Tennessee, 121-130.

57. Melnikov, Yu.A., Shirley, K.L., and Worsey, A.J. (1994) Computing the potential field from a point source on joined surfaces of revolution, *Proc. of the Int. Conference BETECH-94*, Orlando, University of Central Florida, 279-286.

58. Melnikov, Yu.A. and Titarenko, S.A. (1993) On a new approach to 2-D eigenvalue shape design, *Int. J. for Numerical Methods in Engineering*, **36**, 2017-2030.

59. Melnikov, Yu.A. and Titarenko, S.A. (1995) Green's function BEM for 2-D optimal shape design, *Engineering Analysis with Boundary Elements*, **15**, 1-10.

60. Melnikov, Yu.A. and Tsadikova, E.Ts. (1978) Computation of an elastic equilibrium of a cylindrical shell with nonsmall holes by the method of integral representations, *Reports of the Ukrainian Academy of Sciences*, Ser. A, 12, 1107-1112 (in Russian).

61. Mikhlin, S.G. and Smolitskiy, K. (1967) *Approximate Methods for Solutions of Differential and Integral Equations*, New York, Elsevier.

62. Morjaria, M. and Mukherjee, S. (1981) Numerical analysis of plane time dependent inelastic deformation of plates with cracks by boundary element method, *Int. Journal of Solids and Structures*, **17**, 127-143.

63. Morse, P.M. and Feshbach, H. (1953) *Methods of Theoretical Physics*, Vols. 1 and 2, New York-Toronto-London, McGraw-Hill.

64. Olsen, G.A. (1982) *Elements of Mechanics of Materials*, 4th ed., Prentice-Hall, Englewood Cliffs.

65. Reissner, E. (1945) The effect of transverse shear deformation on the bending of elastic plates, *ASME J. of Appl. Mechanics*, **12**, 69-77.

66. Roach, G.F. (1982) *Green's Functions*, 2nd ed., New York, Cambridge University Press.

67. Shames, I.H. (1964) *Mechanics of Deformable Solids*, N. J., Prentice-Hall, Englewood Cliffs.

68. Smirnov, V.I. (1964) *A Course of Higher Mathematics*, Vols. 2–4, Oxford-New York, Pergamon Press.

69. Snyder, M.D. and Cruse, T.A. (1975) Boundary-integral equation analysis of cracked anisotropic plates, *Int. J. Fract. Mechs.*, **2**, 315-328.

70. Stakgold, I. (1980) *Green's Functions and Boundary Value Problems*, New York, John Wiley.

71. Tewary, V.K. (1991) Elastic Green's function for a bimaterial composite solid containing a free surface normal to the interface, *Journal of Materials Research*, **6**, 2592-2608.

72. Timoshenko, S. and Woinowsky-Krieger, S. (1959) *Theory of Plates and Shells*, 2nd ed., New York, McGraw-Hill.

73. Timoshenko, S., Young, D., and Weaver, W. (1974) *Vibration Problems in Engineering*, New York, John Wiley.

74. Timoshenko, S.P. and Goodier, J.N. (1970) *Theory of Elasticity*, 3rd ed., New York, McGraw-Hill.

Index

A

Abramovitz, M., 449
absolute
   convergence, 285
   integrability, 288
absolutely rigid plane, 315
Achenbach, J.D., 449
acoustics, v
additive component, 202
adjoint
   equation, 23
   operator, 23
adjustments, 81
allowable
   function, 27, 194, 238
   vector-function, 256, 318, 357
alternative
   approach, vi, 385
   form, 388
   formulation, 116
   method, 35, 203, 225, 266
   representation, 227, 371
analytic
   differentiation, 69, 90
   integration, 64, 85, 100, 175
   solution, 9, 100, 103, 114, 130, 309
analogy, 1
Andrews, L.C., 449
angular variable, 346
annular plate, 6, 351, 370
answers, 6, 415
applied
   mathematics, vi
   mechanics, v, vi, 3, 14, 44, 49, 149, 193, 371, 401
approximate
   differentiation, 70
   integration, 133, 175
   solution, 113, 115, 119
a priori estimation, 79

arbitrary constants, 12, 102, 293, 311, 332, 363, 381, 395
area of science, 1
Aspinwall, L, vi
assembly of
   elements, 149, 157, 175, 177, 184, 187, 427
   thin rods, 393, 398, 413
   three thin-walled fragments, 351
Atkinson, K., 449
axial
   cross-section, 328, 339
   direction, 378
   plane, 308
axis of revolution, 235

B

backward substitution, 62, 162
Banerjee, P.K., 449
Barton, G., 449
beam
   of a unit length, 20, 120
   problems, 8, 49, 111, 139
   resting on elastic foundation, 4, 49, 81, 84, 87, 90, 108, 137
   theory, 20, 49, 64, 121
Beck, J.V., 449
bending
   load, 64, 69
   moment, 4, 49, 50, 68, 69, 70, 72, 73, 75, 79, 81, 84, 90, 94, 97, 100, 107, 162, 167, 173, 188, 261, 264
   of a beam, 20, 49, 50, 61, 64, 112, 160
   of a thin plate, 3, 6, 255
Berger, J.R., 449
Bergmann, S., 449
Betti's reciprocal work theorem, 285
Biggs, J.M., 449
biharmonic
   equation, 2, 6, 193, 255, 257, 266, 401
   function, 258
bilinear combination, 25, 28
Blevins, R.D., 449
Bobylov, Ye.A., 452

Bobylyov, A.A., 452
body
   force vector, 290
   load, 302
Boley, B.A., 449
boundary, 203, 276
boundary conditions, 1, 8, 20, 26, 50, 51, 53, 55, 57, 81, 82, 87, 96, 101, 112, 115, 122, 129, 136, 141, 146, 150, 152, 162, 170, 175, 184, 194, 203, 220, 230, 259, 274, 277, 281, 297, 303, 309, 315, 319, 329, 334, 353, 377, 381
   physically feasible, 2, 112, 129, 276, 294
boundary element method, v, vi
boundary value problem, vi, 3, 20, 34, 45, 50, 52, 54, 60, 69, 75, 77, 79, 91, 98, 111, 113, 127, 129, 135, 149, 172, 176, 187, 194, 218, 238, 251, 255, 260, 278, 313, 318, 325, 330, 355, 384, 392
   mixed, 203, 218, 227
bounded function, 26, 52, 198, 258, 288, 297, 302, 392
boundedness, 20, 39, 40, 42, 82, 89, 220, 267, 322, 332, 352, 395
Brebbia, C.A., 450
buckling, 111, 122, 128, 137
   failure shape, 129, 132, 138
Butkovsky, A.G., 450
Butterfield, R., 449

C

calculus, vi, 255
cantilever beam, 61, 63, 71, 94, 97, 101, 108, 146, 191
   compound, 160, 169
Cartesian
   coordinates, 195, 230, 258
   Laplacian, 290, 329, 356
Cauchy problem, 310
Cauchy-Euler equation, 154
central angle, 227, 230
characteristic equation, 41, 99
Chen-To, Tai, 450

circle, 219, 300, 407, 409–411
circular
   natural frequency, 115
   plate, 6
   sector, 225, 230, 252, 300, 407
clamped
   beam, 55, 57, 76, 79, 89, 105, 108, 136, 140, 144, 189
   edge, 8, 20, 54, 57, 60, 72, 87, 94, 106, 118, 141, 162, 258, 263, 267, 273, 276, 310, 314
   end-point, 2, 50
closed
   form, 77, 201, 234, 237, 241, 266, 273, 297, 325, 328, 386
   interval, 8
coefficient matrix, 10, 20, 21, 29, 41, 54, 62, 89, 113, 132, 133, 135, 161, 179, 186, 280, 292, 297, 311, 334, 362
   five-diagonal, 81
Collatz, L., 450
combination of loads, 4, 74, 97, 107, 108, 164, 167, 171, 174, 175
compact form, 24, 30, 82, 142, 182, 196, 214, 221, 227, 244, 312, 359
complementary solution, 332, 340, 378, 381, 395
complete summation, 5, 193, 216, 229, 297, 342, 350
complex
   conjugate, 197, 325
   plane, 228, 373
   variables, 196, 202, 214, 224, 241, 244, 261, 299, 325, 373
      imaginary part, 202, 224
      method, 258
      modulus, 202, 224
      real part, 202, 224, 265
compound
   bar, 151
   beam, 160, 165, 167, 171, 189
   hemispherical shell, 339, 412
   infinite strip, 324, 411, 413
   media, 6, 317, 357, 442
   plate, 399

INDEX

plate-shell structures, 231, 412
rectangle, 357, 369, 411
region, 401
semi-circle, 325, 369, 411
compressive axial force, 121, 125, 128, 135, 137
computational
algorithm, 3, 115, 139, 285
efficiency, 76
error, 200
experiment, 78, 114, 117, 130, 134
mechanics, v, 5, 12, 49, 147, 193
potential, 81, 119, 132
procedure, 114, 119, 121, 127, 131
computer
algebra, 103
software, 210, 387
concentrated bending moment, 68, 71, 107, 165
conductive
material, 157, 176, 393
properties, 393
conductivity, 176, 184, 317, 320, 326, 339, 352, 369, 393
congruent fragments, 339
configuration, 119, 121, 193, 247, 300
conformal manner, 203, 209
conservation of energy, 177
consistent system, 10
constant
multiple, 28
of elastic support, 51, 72
constraints, 24
construction
of Green's function, 7, 9, 30, 32, 35, 52, 57, 234, 237, 247
procedure, 14, 16, 22, 55, 57, 87, 169, 194
contact
conditions, 149, 150, 152, 162, 170, 175, 177, 184, 319, 329, 334, 353, 364, 368, 394
frictionless, 358
points, 149
problem, 4, 139, 144

zone, 139, 144, 146
continuity, 10, 31, 60, 83, 359, 395
continuous
derivative, 9, 178
function, 8, 30, 36, 53, 56, 141, 150, 176, 317, 359, 370, 393
continuously
differentiable, 25, 27, 176
distributed load, 66, 69, 71, 84, 188
continuum mechanics, v, 7, 176, 193, 255, 317
convectional (Newtonean) cooling, 382
conventional notations, 8
convergence, 78, 103, 205, 211, 214, 242, 261
non-uniform, 209, 218
convergent series, 201, 388
Convolution theorem, 374
coordinate systems, 5, 193
Courant, R., 450
CPU time, 104
Crandall, S.H., 450
critical
force, 134, 137
value, 129, 132, 135, 138
cross-section, 50, 69, 70, 92, 167, 175
variable, 120
cross-sectional area, 112, 115, 120
Cruse, T.A., 450, 453
cubature formula, 286
cubic function, 120

D

Dahl, N.C., 450
Davydov, I.A., 450
definite integral, 67, 142, 313
containing a parameter, 69, 142
deflection function, 2, 50, 52, 57, 63, 65, 66, 68, 71, 73, 75, 78, 81, 84, 90, 94, 97, 103, 107, 133, 135, 142, 162, 172, 188, 257, 263, 274, 277, 288
degree
of freedom, 33
of a vertex, 177

Denda, M., 450
Dennery, Ph., 450
derivative, 16, 94
  of a vector function, 280, 292, 362
detachment, 291, 296, 302, 315, 358
determinant, 10, 29, 41, 54, 89, 136, 179, 186, 292, 334, 362
diagonal entries, 178, 181
differentiable function, 23, 33
differential
  form, 231
  operator, 7, 183, 233
differentiation order, 10
Dippery, R., vii
Dirac delta function, 51, 65
  properties, 51, 66
direct method, 31
Dirichlet
  conditions, 198, 208, 228, 338
  problem, 195, 200, 203, 209, 214, 219, 223, 226, 234, 242, 244, 247, 325
discontinuity, 60, 83, 176
discontinuous function, 9, 53, 56, 178, 184
discontinuously variable, 317, 393, 399
disk, 223, 227, 253
displacement, 1, 50, 307, 358, 370
  formulation, 8, 64, 193, 255, 289, 302, 356
  vector, 290, 296, 300, 302, 307, 315, 356
    normal component, 359
    tangential component, 359
distributed forces, 65, 67, 76, 168, 356
Dolgova, I.M., 450
domain integral, 285, 289
double integral, 288
double-spanned beam, 167, 189

E

Economou, E.N., 450
edge, 208, 261, 275, 296, 308
  conditions, 4, 50, 52, 54, 59, 69, 71, 73, 82, 87, 90, 112, 118, 121, 136, 310, 314
  incident to a vertex, 177
  of a graph, 5, 175, 186
eigenfunction, 113, 114, 122
eigenvalue problem, 112, 114, 117, 122, 128, 130, 132, 138
  of linear algebra, 114, 117, 125, 129, 131, 132, 134
eigenvalue spectrum, 114, 116, 118, 129, 132, 135
elastic
  beam, 52, 140
  constant of foundation, 81, 87, 108, 137
  equilibrium, 6, 256, 307, 357
  material, 50, 277, 290, 370
  medium, 289, 315, 358
  plate, 3
  shell of revolution, 314
  support, 169, 170
elastically supported edge, 8, 50, 57, 58, 60, 72, 105, 276, 314
elasticity modulus, 50, 257, 277, 301, 370
electric potential, 177
electromagnetism, v
electro-statics, 230
element of area, 223
elementary
  functions, 17, 77, 100, 297, 338, 373
  load, 70, 74
  rectangle, 286, 289
  segment, 133
Ellias, Z.M., 450
elliptical PDEs, vi, 5, 193, 256, 371, 384
end-point, 11, 19, 23, 50, 149, 353, 363, 368, 380, 385, 399
  of a beam, 2, 129
  of a graph, 5, 175
energy, 256
  flow, 230
engineering practice. 69, 175
environment, 103
equation of potential, 328
equality, 26, 32

equidimensional, 280
equilibrium state, 54, 82, 98, 277, 302, 308
error
  estimation, 78, 117, 287
  function, 423
    complementary, 376
estimate, 206, 208, 214, 242, 246, 276, 287
estimation of a remainder, 194, 214, 248
Euclidean space, 194, 238, 256, 317, 355, 376
Euler
  elastic buckling force, 122, 129, 132
  formulation of buckling problems, 4, 128, 132, 135
Euler-Bernoulli equation, 50, 92, 141
exact solution, 17, 64, 75, 77, 81, 103, 115, 122, 138, 288
existence
  of a Green's function, 7, 9, 36, 41, 54
  of a matrix of Green's type, 5, 155, 158, 175, 178
exponential
  form, 39
  function, 15, 85, 98, 175, 206, 213, 233, 322, 373
    term, 204

F

face surface, 230
feasible combination, 69
Feshbach, H., 453
field point, 57, 207, 247, 261, 275, 318, 325
finite
  difference scheme, 78, 81, 117
  difference method, 70, 78, 85, 114, 117
  element method, 70
  interval, 19
  length, 2
  sum, 70
  weighted graph, 5, 149, 175, 185

fixed edge, 296
flexural rigidity, 2, 3, 49, 50, 52, 64, 81, 94, 97, 101, 108, 120, 160, 164, 189, 257, 266, 277
  variable, 4, 49, 92, 94, 103, 112, 119, 127, 130, 134, 137, 140
fluid flow, v
four-point posed problem, 188, 352
Fourier expansion (series), 5, 193, 198, 203, 209, 219, 228, 232, 240, 259, 266, 278, 291, 320, 324, 330, 346, 360, 386, 389, 391
  coefficients, 193, 199, 204, 219, 232, 241, 259, 267, 275, 291, 321, 330, 347, 360, 387, 389
    fundamental rule, 222, 295, 305, 324, 335, 387
Fourier-Euler formula, 199, 260, 295, 313, 335, 387, 390
free edge, 8, 50, 58, 61, 90, 108, 162, 261, 276, 370
Fresnal integrals, 77
friction, 291, 296, 302, 315
function
  defined in pieces, 31, 38, 64, 141, 181, 282, 295, 318, 356, 379
  of exponential order, 373
  of a limited variation, 52, 66
functional series, 207
fundamental
  natural frequency, 116
  solution, 250, 257, 357, 377, 380, 393
  solution set, 9, 10, 12, 15, 17, 18, 20, 22, 30, 34, 38, 41, 52, 54, 57, 82, 95, 96, 99, 102, 104, 124, 155, 178, 183, 220, 232, 267, 279, 291, 303, 309, 321, 331, 340, 348, 353, 361

G

Gavelya, S.P., 451
general solution, 12, 20, 33, 36, 38, 41, 59, 62, 87, 95, 152, 158, 183, 220, 279, 304, 310, 362
generality, 28

generalized functions, 52
geographic coordinates, 231, 235
geometric
  constraint, 51
  linearity, 307
  nonlinearity, 94
  progression, 385
Goodier, J.N., 454
governing differential equation, 9, 17, 28, 57, 140, 155, 176, 182, 226, 256, 317
Govorukha, V.B., 452
Gradstein, I.S., 451
graph, 176, 180, 352
  theory, 5, 175
gravitation, 230
Greenberg, M.D., 451
Green's
  formula, 26, 27
  function, v, vi, 12, 13, 21, 29, 35, 40, 43, 45, 47, 51, 58, 60, 64, 82, 87, 95, 101, 111, 113, 115, 120, 123, 128, 135, 145, 160, 193, 199, 203, 210, 215, 221, 226, 230, 233, 238, 244, 250, 267, 295, 325, 379, 383, 389
    catalogue, 6, 401
    defining properties, 3, 5, 7, 21, 53, 61, 83, 89, 178, 193, 221, 268
    formalism, v, 14, 44, 160, 175
    method, v, vi, 44, 49, 103
    for ordinary differential equation, 1, 7, 415
    for partial differential equation, 1, 44, 193,
Green's matrix, 193, 256, 295, 305, 310
grid-points, 117, 132, 135
grouping, 32, 37, 40, 56, 63, 162

H

Haberman, R., 451
half-disk, 227
half-open interval, 8
half-plane, 242, 250, 392, 408
  absolutely rigid, 291, 296, 302, 358
Hall, M.T., 452

harmonic
  function, 195, 258, 319
  series, 206
    generalized, 206
Hasebe, N., 451
heat
  conduction, 376, 388, 392
  conductivity, 2, 152, 157, 393
  equation, 6, 371, 378, 386, 393, 401, 413, 446
  flow, 378, 388, 394
  flux, 393, 395
  transfer, v
    coefficient, 382, 384
hemisphere, 328, 339, 351
higher order derivatives, 32
highest eigenvalue, 127
Hilbert, D., 450
homogeneity, 113, 175
homogeneous
  boundary value problem, 3, 8, 9, 20, 36, 38, 54, 58, 62, 93, 98, 100, 150, 157, 161, 165, 177, 187, 194, 199, 203, 228, 240, 256, 291, 295, 305, 318, 331, 348, 356, 368, 379, 383, 389
  boundary conditions, 8, 20, 30, 36, 38
  equation, 1, 8, 30, 34, 36, 51, 54, 82, 94, 99, 101, 113, 183, 232, 333, 340
  material, 8, 50, 119, 184, 266, 277, 290, 317, 325, 328, 346, 356, 376, 378, 385, 388, 393
  media, 149, 160
  system of linear equations, 12, 20
Hooke's law, 68
horizontal partitioning, 280, 292, 362
Hughes, S., 452
Hwu, C., 451
hyperbolic
  PDEs, 371
  form, 40
hyperbolic-exponential form, 40

I

ideal contact, 157, 317, 340, 393

identity, 32
  matrix, 134
illustrative example, 54, 70, 139, 169, 393, 399
improper integral, 259, 285, 372
  convergent, 372
improvement of convergence, 194, 200, 204, 216
inconsisnency, 117
indirect applications, 4, 112
inequality, 214, 242
infinite
  beam, 81, 85, 108
  interval, 14
  media, 14
  product, 210
  strip, 218, 240, 251, 300, 314, 319, 325, 369, 406, 408, 409
influence function, v, 1, 2, 6, 44, 49, 55, 57, 58, 60, 62, 64, 66, 68, 70, 74, 76, 82, 86, 91, 101, 112, 118, 120, 123, 126, 132, 134, 141, 145, 160, 177, 193, 230, 235, 255, 260, 263, 266, 273, 314, 372, 376, 380, 382, 386, 390, 401
  analytic expressions, 4, 276
  costruction, v, 3, 54, 55, 61, 82, 90, 93, 96, 101, 191, 230, 276, 307, 371, 377, 384
  formalism, 4, 6, 49, 84, 111, 119, 355, 371
    for beams, 50, 418
    of the second order, 68, 70, 73, 75, 94
  method, v, 3, 49, 64, 67, 69, 76, 78, 80, 81, 86, 92, 94, 103, 111, 115, 121, 123, 130, 132, 139, 149, 165, 169, 175, 276, 289, 307, 371, 401
influence lines, 49
influence matrix, 4, 44, 149, 161, 165, 168, 171, 175, 177, 184, 189, 237, 255, 278, 283, 287, 291, 295, 299, 306, 314, 318, 327, 330, 339, 345, 355, 369, 372, 393, 398, 399, 401
inhomogeneity, 176
initial
  approximation, 142, 146
  conditions, 59, 88, 101, 377
  equilibrium state, 4
  estimation, 140
  temperature, 376, 378, 382, 385, 388, 392
  value problem, 59, 87, 102, 103, 387
initial-boundary value problems, 6, 371, 377, 380, 384, 386, 388, 392
instant point source, 377, 390, 398
integrable, 256, 258, 260, 318, 357
integral containing a parameter, 31, 35, 181, 293, 364
integral equation, vi, 113, 115, 118, 121, 123, 129, 131, 132, 134, 138, 142
  Fredholm of the second kind, 113, 128, 142
  nonlinear, 142
integral-differential equation, 131, 132
integral
  representation, 35, 38, 65, 77, 82, 127, 130, 157, 159, 182, 187, 260, 283, 286, 294, 304, 311, 318, 323, 357, 367, 377
  transforms, 372
integrand, 36, 131, 142, 182
integrating factor, 25, 46, 220, 232, 267
integration, 335, 373, 387, 390
  by parts, 131
interaction, 140
interface, 318, 340, 356, 359, 370
interior problem, 300
intermediate support, 160, 169, 189
interpretation, 2, 157
interval of integration, 69
invariant, 30, 245, 248
inverse
  Laplace transform, 373, 380, 383, 385, 390, 393, 397
  matrix, vi, 280, 292, 311, 362, 366
Irschik, H., 451
isotropic
  material, 50, 119, 266, 277, 290, 307, 356, 376, 388
  media, 6, 255, 289
iterations, 136, 142, 146

J

Jacobi Theta function, 388, 391, 446
jump of discontinuity, 73, 151
Jun Quin, 451

K

Kamke, E., 451
Kantorovich, L.V., 451
kernel
   function, 35, 38, 43, 44, 63, 86, 124, 132, 138, 153, 183, 194, 200, 221, 223, 226, 239, 260, 323, 333, 370, 379, 380, 387
   matrix, 150, 159, 184, 256, 280, 283, 295, 298, 304, 313, 318, 357, 367, 396,
Kirchhoff's beam theory, 2, 49, 64, 71, 81, 92, 111, 128, 139, 147, 160
Klein-Gordon equation, 5, 193, 238, 244, 247, 250, 252, 255, 376, 389, 392, 401, 408
   static, 238
Koshnarjova, V.A., 452
Krasnikova, R.D., 452
Krylov, V.I., 451
Krzywicki, A., 450
Kupradze, V.D., 451

L

Lagrange's method, 7, 30, 32, 36, 39, 47, 62, 152, 161, 220, 279, 292, 310, 319, 332, 361
Lame's
   coefficients, 290, 356, 370
   operator, 356
   system, 5, 193, 290, 300, 357, 401
Laplace's
   equation, 5, 193, 200, 203, 215, 219, 223, 226, 234, 239, 242, 249, 255, 325
   operator, 194, 231, 238, 257, 318, 329, 339, 377, 392
Laplace transform, 6, 372, 382, 389, 392, 394
   existence, 372
      sufficient conditions, 373
   method, 372, 385
   properties, 373, 378, 389
Lardner, T.J., 450
lateral
   deflection, 257
   displacement, 1
   surface, 378, 388, 392, 394
latitude, 231, 235
latitudinal coordinate, 307
leading
   coefficients, 9, 23, 26, 150, 314
   principles, 372
Lebedev, N.N., 451
Lekhnitskii, S.G., 451
length of an edge, 176
level of accuracy, 78, 85, 86, 100, 104, 114, 116, 125, 137, 142, 147, 207, 218, 229, 243, 247, 289
Levy's method, 258
like integrals, 42
limitation, 17, 82, 230, 234, 256
line integral, 227
linear
   algebra, vi
   boundary value problem, 7
   combination, 10, 53, 56, 87, 94, 102, 362, 373
linearity, 84, 373
linearly independent
   boundary conditions, 277, 308
   forms, 8, 30, 150
   functions, 9, 27, 29, 32, 52, 54, 55, 57, 58, 60, 87, 89, 96, 101, 120, 178
Lipschitz condition, 287
loaded interval, 65, 75, 85, 90, 93, 100
loading
   function, 76, 82, 85, 100, 103, 142, 164, 175, 258, 278
   vector, 286, 307
Loboda, V.V., vi, 452
local
   coordinates, 181
   maxima, 52
   minima, 52
logarithmic

INDEX

function, 196
   singularity, 195, 209, 216, 223, 239, 249, 264, 319, 327, 338
'long' shell, 314
longitude, 231, 235
longitudinal coordinate, 120, 176, 307
loss of stability, 4, 137, 204
lowest natural frequency, 116, 127

M

Macdonald function, 239, 376
magneto-statics, 230
Mahan, G.D., 451
Maple V, 103
mass density, 1, 112, 115
mathematical
   model, 1, 195
   physics, 1, 2, 175, 238, 371
   prototype, 7
matrix
   rectangular, 279, 364
   square, 280
matrix form of a system, 21
matrix of Green's type, 4, 149, 158, 161, 165, 175, 179, 183, 186, 318, 324, 327, 331, 342, 348, 352, 368, 396
   construction procedure, 180
matrix-operator, 256, 290, 297, 300, 302, 307, 356, 361, 364
Maxwell's reciprocity, 57
McDaniel, S., vi, 452
mean value theorem of integration, 133, 286
mechanical engineering, vi, 112
mechanics of materials, 49
Melnikov, Yu.A., 450–453
meridional
   coordinate, 307
   cross-section, 235, 346
method of
   eigenfunction expansion, vi, 5, 193, 196, 258, 386, 388
   undetermined coefficients, 77

variation of parameters, 5, 30, 50, 61, 77, 95, 106, 152, 158, 161, 172, 183, 185, 193, 221, 268, 304, 353, 378, 381, 384, 394
methodology, 49, 219, 238
middle
   plane, 257, 273, 277
   surface, 307, 339
midpoint, 78, 133, 144, 189
Mikhlin, S.G., 453
Miller, D., vi
mixed
   boundary conditions, 208, 228
   boundary value problem, 215, 218, 231, 234, 243, 250, 252
modification, 27, 52, 57, 105
modified Bessel function
   of the first kind of order n, 249, 436
   of the second kind of order n, 249, 250, 436
   of the second kind of order zero, 239, 376, 393, 437
modulus, 214, 242
moment resultant, 288
moment of inertia, 50
Morjaria, M., 453
Morse, P.M., 453
Mukherjee, S., 453
multi-point posed problem, 4, 44, 149, 157, 160, 172, 176, 185, 187, 394
multiplicity, 99
multi-spanned beam, 4, 149, 160, 171

N

natural vibrations, 112, 127, 132
   frequencies, 4, 111, 112, 114, 122
   mode shapes, 4, 112, 114
Navier's method, 258
Neumann conditions, 208
Nikulyn, V.A., 450
nonhomogeneous
   boundary value problem, 183, 199
   equation, 30, 36, 38, 101
   material, 2

nonintegral term, 31, 37, 132, 182
nonlinear
  eigenvalue problem, 135, 138
  function, 135
  problem, 139
nonlinearity, vi, 68, 94, 111, 139, 142
nonsingular matrix, 180
nonsingularity, 179
nonsymmetrical form, 22
nontrivial solution, 102, 112, 114
non-uniformly convergent series, 218, 241, 264, 338
nonzero constant, 14
normal
  direction, 194, 230, 238, 318, 377
  form, 309
  stress, 359
null-matrix, 180
numerical
  differentiation, 79, 288
  integration, 79, 85, 100
  methods, vi, 70, 77, 113, 114, 129, 134, 142, 309, 373
    procedures, 9, 78
    solution, 3, 76, 101, 103, 124, 134, 139, 175, 276, 314

O

observation point, 57, 65, 75, 90, 94, 103, 133, 178, 194, 197, 208, 211, 216, 224, 239, 249, 263, 273, 285, 299, 318, 352, 377, 393
Oden, J.T., vii
Olsen, G.A., 453
one-to-one relationship, 2
open interval, 8
operative coordinate system, 119
operator, 24, 50, 150, 256, 267, 281, 291, 310
optimal shape design, vi
order of
  accuracy, 78, 117
  convergence, 217
  operations, 200, 260

system, 277, 309
  terms, 33
ordinary differential equations, v, vi, 7, 17, 44, 111, 113, 193, 199, 219, 228, 232, 255, 267, 279, 291, 303, 309, 326, 330, 352, 384, 401
  of higher order, 22, 50, 161, 177, 255
  of separable type, 95
  with constant coefficients, 9, 309
  with variable coefficients, 309
original, 372
orthotropic
  material, 301, 306
  media, 6, 255, 301
  semi-strip, 411
orthotropy, 307
  principal directions, 301
outer contour, 317, 356

P

parabolic PDEs, 6, 371
parameter, 31, 115, 129, 135, 242, 267, 274, 279, 289, 297, 304, 320, 350, 370
parametric equations, 231
parametrization, 235
partial
  contact, 140
  derivative, 10, 68, 83, 263, 285, 359
  differential equations, v, vi, 44, 113, 193, 255, 307, 401
    of higher order, 277
  summation, 5, 193, 211, 218, 270, 274, 355
particular solution, 9, 34, 52, 55, 57, 58, 59, 87, 95, 101, 348
partitioned diagonal form, 179
partitioning, 364
  parameter, 78, 81, 103, 134, 136, 139, 287
  points, 78, 114, 116
    equally spaced, 114
  step, 117
perfect contact, 359, 370
perfectly insulated, 238

periodic functions, 330, 352
periodicity, 266, 330, 346
   conditions, 8, 16, 219, 236
peripheral entries, 179, 181, 327
phenomenon in mechanics, 1
physical
   interpretation, 51, 54, 113, 115, 131
   nature, 230
   nonlinearity, 68, 94
   problem, 69
   properties, 302, 317, 319
piecewise
   continuous, 6, 372
   format, 44, 75, 91, 97, 166, 168
   homogeneous, 4, 149, 160, 175, 176, 319
   smooth, 194, 238, 256, 276, 376
   uniform, 188, 191
plane problem, 5, 6, 8, 193, 255, 289, 300, 306, 355, 401, 410
plate, 6, 238, 328
   circular, 266, 268, 273, 276, 314, 343
      clamped, 266, 273
      simply supported, 273, 275
   half-plane-shaped, 392, 413
   infinite strip-shaped, 399, 413
   rectangular, 277, 283, 314
   semi-circular, 273
   semi-strip-shaped, 258, 261, 388, 399, 413
   square simply supported, 288
   of various shapes, 6
plate problems, 257, 276
plate-shell structures, 230
point source, 159, 193, 369
Poisson's equation, 197, 219, 317, 329, 339, 346
Poisson ratio, 257, 274, 277, 301, 370
Poisson-Kirchhoff's plate theory, 2, 6, 255, 257, 263, 314
polar
   coordinates, 219, 230, 249, 266, 300, 325
   Laplacian, 300, 326, 392
pole, 352

Polyakov, N.V., vi
polynomial, 55, 85, 175
potential
   field, 5, 193, 230, 237, 255, 317, 329, 339, 346, 369, 411, 434
   on surfaces, 230
   problems, 230, 234, 258
Powell, J.O., vi, 453
practical solvability, 4, 94, 111
principal
   diagonal, 178, 327
   value of a function, 227, 230
problem class, v, 81, 92, 111, 128, 144, 218, 230, 255, 289, 301, 307, 401
problems posed on graphs, 175, 180
product rule, 24, 25, 33, 41, 99
properties of materials, 8

Q

quadratic equation, 146
quadrature
   coefficients, 114
   formulae, 77, 100, 114, 124, 128, 131, 142, 175
   technique, 114
qualitative
   analysis, 79
   theory, v, 3

R

radial bending moment, 274
radicand, 123
rate of convergence, 134, 142, 146, 204, 209, 215, 260, 275, 338, 387
real-valued function, 197, 214, 241, 244, 261, 265, 299, 325
reciprocal value, 134
rectangular
   cross-section, 119
   region, 209, 214, 218, 247, 252, 261, 290, 297, 300, 315, 407, 409, 410
reflection method, 203, 225, 250, 258, 393
region configuration, 5

regular component of influence function, 5, 194, 211, 216, 227, 230, 239, 249, 257, 338, 355,
Reissner plate theory, 6, 255, 276, 283, 315, 453
relative
   accuracy, 121
   error, 289
remainder, 205, 209, 214, 242, 245, 276
   estimation, 206, 209, 229, 276
repeated roots, 41
response, 1, 57, 66, 68, 84, 89, 94, 108, 112, 171, 177, 187, 377
restrictions, 28
resultant
   deflection, 68, 75, 97, 166, 168
   equation, 25
   stress, 277, 287
review exercises, 6, 45, 102, 187, 251, 314, 359, 399
Riemann mapping theorem, 196, 203, 209
right-hand term, 8, 11, 26, 36, 41, 44, 51, 65, 77, 82, 103, 113, 123, 130, 132, 131, 150, 180, 198, 203, 214, 219, 234, 258, 267, 283, 292, 297, 310, 360, 396
Roach, G.F., 453
rod, 2, 157, 184, 187, 378, 399, 413
rotation, 50, 277, 288
Roubides, L., vii
routine algebra, 57, 96, 168
Runge-Kutta procedure, 103
Ryzhik, I.M., 451

S

sandwich type
   media, 4, 355
      assembly, 4, 175
'Saturn' type construction, 351, 355
scalar multiple, 65, 87, 113, 173
Schiffer, M., 449
second order derivative, 131
self-adjoint

boundary value problem, 27, 28, 46, 210, 225, 228, 235, 240, 259, 268, 378, 381, 385, 389
   equation, 23, 45
   operator, 23, 267
self-adjointness, 7, 24, 27, 43, 57, 60, 136, 211, 219, 222, 226, 233, 259, 262, 297, 385
semi-circle, 250, 253, 325, 328, 409
semi-infinite
   beam, 87, 89, 108
   rod, 380, 413
semi-strip, 195, 200, 203, 208, 211, 218, 244, 251, 258, 297, 300, 302, 306, 315, 319, 369, 388, 406, 408–410
separation of variables, 198, 353
series of segments, 157
Shames, I.H., 453
shape complexity, vi
shear
   force, 51, 69, 70, 72, 74, 75, 79, 84, 90, 94, 98, 100, 107, 162, 167, 173, 188, 261, 277, 359
   modulus, 301
   resultant, 288
   stress, 359
shell problems, 8
shell, 6, 256, 307, 328
   conical, 309
   cylindtrical, 6, 309, 328, 338, 343
   hemispherical, 353, 369
   spherical, 6, 230
   toroidal, 346
Shirley, K.L., 453
Shubenko, V.V., 452
simply connected region, 3, 194, 238, 256, 376
simply supported
   edge, 8, 20, 50, 58, 61, 74, 98, 101, 106, 118, 191, 258, 263, 273, 276, 283, 310, 314, 370
   beam, 74, 98, 105, 114, 120, 121, 123, 130, 132, 136, 139, 146, 167,
   plate, 262
simplifying assumption, 33, 81

INDEX

single-span beam, 50, 63, 114, 117, 121, 140, 146
singular
   case, 26
   component of influence function, 5, 194, 200, 216, 230, 250, 257, 276, 338, 355,
      matrix, 12, 29, 41
      point, 19, 155
singularity, 203, 211, 216, 249, 352
   removable, 266
   method, 49
Skal'skaya, I.P., 451
sliding edge, 50, 106, 119
Slowey, E., vii
slowly convergent series, 204, 275
Smirnov, V.I., 453
Smolitskiy, K., 453
smooth contour, 3
Snyder, M.D., 453
solid mechanics, 6, 111, 255, 307, 438
solution vector, 256, 278, 283, 286, 288, 292, 304, 397
Somigliana tensor, 357
source point, 57, 178, 179, 187, 194, 197, 207, 211, 216, 224, 239, 247, 257, 261, 273, 275, 286, 299, 318, 325, 352, 393
spacial variables, 6, 184, 317, 372, 386, 393, 399
special functions, 373, 387
spherical
   belt, 370
   cap, 328, 338
   coordinates, 231, 329, 340
   segment, 234, 408
   triangle, 231, 408
spring constant, 169
Stakgold, I., 454
standard
   approach, 59, 76, 82, 93, 99, 203
   form, 24
   summation formula, 207, 208, 218, 229, 234, 237, 243, 246, 261, 265, 275, 298, 306, 324

static equilibrium, 92, 129, 137, 160, 301, 307, 314
statics, 50, 64
steady-state
   formulation, 371
   heat conduction, 2, 151, 157, 184, 230
   mass conduction, 230
Stegun, I., 449
stiffness, 314
straightforward
   algorithm, 9, 84
   differentiation, 182, 285
   integration, 34, 43, 95, 152, 220, 285, 293, 321, 348, 362, 373
   substitution, 199
   truncation, 261
strategy, 198
strength of materials, 49
stress components, 287, 358, 370
stress-strain
   relationship, 69
   state, 4, 49, 79, 81, 85, 92, 97, 112, 140, 144, 147, 165, 174, 189, 258
stress tensor, 290
   normal component, 276
   shear component, 296
string, 1
structural analysis, 49
subinterval, 66, 78, 116, 132, 149
submatrix, 179, 280, 364, 365
subvector, 280, 362, 365
successive integration, 18, 77
sum of a series, 214, 216, 223
summable series, 226, 233, 237, 250, 275, 298, 306, 345
summand, 338
summation, 211, 213, 223, 226, 233, 241, 261, 299, 306, 336, 345, 369, 386, 390
superiority, 79, 289
superposition, 68
surfaces of revolution, 230, 307
   conical, 231
   cylindrical, 231
   hyperboloidal, 231

joined, 237
paraboloidal, 231
spherical, 230, 234
toroidal, 230, 235
surrounding medium, 290, 382
symmetric, 30, 57, 60, 267, 302, 386
symmetry, 78, 144, 296, 308
symmetry of Green's function (matrix), 3, 7, 22, 23, 24, 27, 46, 52, 259, 297
system of linear algebraic equations, 10, 17, 29, 36, 41, 54, 60, 62, 78, 84, 88, 96, 113, 117, 133, 152, 156, 161, 179, 183, 279, 292, 311, 362, 366
homogeneous, 113, 134

T

tangential variable, 274
temperature, 2, 177, 377, 380, 395
tensile axial force, 121, 125
term-by-term
differentiation, 263, 285
integration, 246, 285
terminology, 3
test problem, 77, 195, 288
Tewary, V.K., 454
theory of
elasticity, v, 5, 8, 193, 255, 289, 300, 306, 319, 355, 401, 410
potential, 6, 193, 401, 406
thermal diffusivity, 376, 378, 388, 392, 393
thickness, 257, 370
uniform, 258, 266, 277, 288
thin
plate problems, 6, 276
shells, 6, 307
thin-walled structures, 6, 328
joined, 328, 346, 355, 370
three-point posed problem, 169, 320, 347, 361
time-consuming computation, 85
time-independent problem, 371
Timoshenko, S.P., 454
Titarenko, S.A., 453

toroidal
sector, 236
shell, 6
transcendental equation, 136
transformation matrix, 308, 360
Translation theorem, 374
transverse
distributed load, 1, 2, 49, 50, 64, 71, 81, 85, 87, 90, 93, 98, 107, 140, 161, 168, 257, 266, 276, 288
forces, 4, 49, 57, 61, 69, 74, 75, 87, 97, 165, 171
normal stress, 276
shear deformation, 276
unit concentrated force, 1, 52, 54, 57, 60, 62, 64, 68, 112, 118, 129, 139, 165, 189, 258, 263, 283, 314, 370
vibrations, 111, 112, 119
trapezoid rule, 78, 79, 100, 114, 116, 120, 124
triangle inequality, 242
triangular matrix, 21, 62, 161
trigonometric
factor, 242
function, 85, 175, 206, 207, 288
series, 211, 223, 233, 236, 242, 257, 266, 308, 328, 342, 352
triple-spanned beam, 191
trivial solution, 9, 13, 18, 19, 22, 27, 30, 36, 45, 46, 52, 54, 150, 158, 178, 187, 194, 238, 256, 318, 356
truncation of series, 200, 204, 211, 216, 228, 241, 243, 247, 263, 276, 289, 328
truncating parameter, 208, 243
Tsadikova, E.Ts., 453
twisting moment, 277

U

Uflyand, Ya.S., 451
unbounded function, 15, 155
underdetermined system, 10
undetermined
coefficients, 179, 184, 221
functions, 279

INDEX

uniform
   flexural rigidity, 50, 87, 92, 101, 114, 129, 132, 136, 144, 146, 160, 191
   partitioning, 78, 116, 120, 124, 288
uniformly
   convergent series, 5, 200, 204, 228, 230, 243, 248, 263, 275, 285, 289, 327, 338, 355
   distributed load, 74
unique
   solution, 10, 29, 30, 32, 50, 89, 178, 194, 238, 256, 280, 318, 356, 377
   solvability, 59, 258
uniqueness, 364
   of Green's function, 7, 9, 18, 36, 41, 54
   of matrix of Green's type, 5, 155, 158, 175, 178
unit
   circle, 203, 209, 346
   concentrated body force, 306, 315, 370
   force, 74
   source, 2, 177, 187, 324
   sphere, 329
universal approach, 49, 57, 111

V

validation example, 114, 132, 138, 219, 266, 276, 288
variable
   coefficients, 7, 17
   of integration, 65, 72
   limits, 31, 35
variation of parameters method, 3, 7, 32, 362
vector-function, 150, 180, 256, 280, 291, 318, 356
Ventsel, E., vii
vertices of a graph, 5, 175, 353

W

Wang, C.-Y., 449
Weaver, W., 454
wedge, 249, 252, 407, 409
Weierstrass functions, 209
well-posed
   matrix, 13, 152
   system, 54, 56, 60, 84, 88, 96, 117, 153, 156, 163, 179, 183, 292, 321, 334, 396
well-posedness, 179, 364
Winkler foundation, 4, 137, 140, 147
Woinowsky-Krieger, S., 454
Worsey, A.J., 453
Wronskian, 10, 29, 41, 54, 55, 89, 179, 183, 280, 292, 362

X, Y, Z

Yen, W., 451
Yizhou Chen, 451
Young, D., 454
zero
   matrix, 364
   solution, 9
   vector, 280
Ziegler, F., 451
Zijlstra, J., vi